International Conference on Rural Development and Entrepreneurship (ICORE 2019)

SHS Web of Conferences Volume 86 (2020)

Yogyakarta, Indonesia
20 – 21 August 2019

Editors:

W.R. Adawiyah
Suliyanto
N.C. Afif
N.A. Prayogo
T. Harisam

ISBN: 978-1-7138-2108-3

Printed from e-media with permission by:

Curran Associates, Inc.
57 Morehouse Lane
Red Hook, NY 12571

Some format issues inherent in the e-media version may also appear in this print version.

For additional information, please contact EDP Sciences – Web of Conferences at the address below.

EDP Sciences – Web of Conferences
17, Avenue du Hoggar
Parc d'Activité de Courtabœuf
BP 112
F-91944 Les Ulis Cedex A
France

Phone: +33 (0) 1 69 18 75 75
Fax: +33 (0) 1 69 28 84 91

contact-edps@webofconferences.org

Additional copies of this publication are available from:

Curran Associates, Inc.
57 Morehouse Lane
Red Hook, NY 12571 USA
Phone: 845-758-0400
Fax: 845-758-2633
Email: curran@proceedings.com
Web: www.proceedings.com

TABLE OF CONTENTS

Statement of Peer review

In submitting conference proceedings to *Web of Conferences*, the editors of the proceedings certify to the Publisher that

1. They adhere to its **Policy on Publishing Integrity** in order to safeguard good scientific practice in publishing.
2. All articles have been subjected to peer review administered by the proceedings editors.
3. Reviews have been conducted by expert referees, who have been requested to provide unbiased and constructive comments aimed, whenever possible, at improving the work.
4. Proceedings editors have taken all reasonable steps to ensure the quality of the materials they publish and their decision to accept or reject a paper for publication has been based only on the merits of the work and the relevance to the journal.

Title, date and place of the conference

ICORE 2019 - The International Conference on Rural Development and Entrepreneurship
(August 20 - 21, 2019 - Yogyakarta, Indonesia)

Proceedings editor(s):

Taufan Harisam

Date and editor's signature

Date : November 2, 2020

Communication challenges with older adults patients during clinical learning: A qualitative study among students nurse in Indonesia

R Setiyani[1*1], M Sumarwati[1], A Iskandar[1] and I Rismawati[1]

[1] Faculty of Health Sciences, Jenderal Soedirman University, Indonesia

Abstract. One of consequences of population aging is the increase of health care utilization by elderly population. Student nurse who will work as a professional nurse in the future are expected to be able to communicate effectively with elderly patients, in order to delivering high quality care. This research aimed at exploring communication challenges with older adult patients experienced by students nurse during clinical practice in Indonesia. In-depth interview conducted among 6 final year students nurse, who selected using purposive sampling method. Data were analyzed using Charmaz's constructivist grounded theory. There are four categories of participant's experience communicating with elderly patients were emerged from the data analysis: 1) *"less functioning"; 2) 'kromo' language as a symbol of respect;* 3) *"bridging the gap"* and 4) *"intimate connection".* The present findings revealed that communication challenges could be either from patient or nurse's students. Improving communication skills by learning the basics of elderly first language (Jawa "Kromo") are necessary to students nurse for an effective communication with elderly. Developing constructive view to elderly is also pivotal for successful intergenerational communication.

1 Introduction

The growing of elderly population worldwide has brought profound implications. One of the consequences of population aging is the increase in health care service utilization by older people, mostly due to aging related chronic diseases [1]. Long-term care service is also projected to be highly demanded [2]. Those situations lead to the increasing demand of health care workforce who will work closely to older people [1]. With no exception, most nurses will also be involved in caring for older people.

Communication is one of important aspects of patient care. Communication is involved throughout nursing process, starting from assessment to interventions, such as health promotion and education, prevention, treatment and rehabilitation [3].

*Corresponding author: rahmi.setiyani@unsoed.ac.id

Nurses use communication as a means to establish helping relationship with patients and families [4].

Communication is also essential for greater patient outcome. There has been growing studies showing the benefits of effective communication between nurses and patients. A literature review concludes that patient-centered communication style has positive impacts to patient satisfaction to care services, patient health status and patient adherence to care plan [5].

Nurses are expected to develop good communication skills to all age groups considering its enormous benefits. However, communication to older people could be challenging. Communication becomes the major problem nurses have to face when caring for elderly with cognitive deficit such as dementia. There are lack of understanding of the meaning of message sent by the patient and of language agreement between nurse and patient [6]. Patient related factors are such as hearing deficit, being tired, mistrust, and memory problem and nurse related factors such as lack of intimacy, formal and insincere attitude has been identified to be barriers for effective communication between nurses and elderly patients [7].

Although many researches has been done to identify communication barrier to older adult, study that consider cultural related factor in communication is still rare. Indonesia is an example of multicultural society, which consists of many ethnic groups. Many of them share different value, belief and spoken languages.

Students nurse who will work as a professional nurse in the future are expected to be able to communicate effectively with elderly patients, in order to delivering high quality care. Thus, this research aims at exploring communication challenges with older adult patients during clinical practice of students nurse of a public university in Central Java, Indonesia.

2 Methods

2.1. Study design

In this qualitative study, students nurse were interviewed in depth about their experiences regarding communicating with older adult patients during clinical practice. Data were collected after students finished all clinical rounds to make sure that they have had experience caring for elderly patients. Then, data were analyzed using Charmaz' Constructivist Grounded Theory [8].

2.2.Participants

Participants are 6 final year nursing students of a public university in Central Java, Indonesia. This location was chosen since there are many students from various ethnic and cultural backgrounds attend this university, creating a culturally diverse society. Final year students were chosen as they have already has experience caring for patients from different age groups, including older people. They also have passed clinical practicums in various health care settings, including hospital, long term facilities, and communities. Participants were recruited using purposive sampling technique.

2.3. Ethical consideration

Ethical approval for the research was gained from Ethics Committee of Faculty of Medicine of X University (Reg no. 0816060538). Participants were informed about aim and nature of the study, and the right to refuse and withdrawn from the participation at any time before signing the written consent. Confidentialy was secured by limiting access to the data only to reseachers. Initial was used to labeling the data to preserve anonymity.

2.4. Data collection

Interviews were conducted face to face in a mutually agreed upon setting after signing the informed consent. The interviews were conducted in Bahasa Indonesia and audio-recorded, each lasting 45 minutes to 60 minutes.

2.5. Data analysis

Constructivist grounded theory was used to guide data analysis. Analysis began through process of coding. Coding was conducted in 3 steps: initial coding, focused coding, and theoretical coding [8]. Information from 6 participants was considered adequate to generate categories. Interviewing more participants would not give more information nor generate new category.

3 Results

There are four categories that emerged from the data analysis: 1) *"less comprehend and less able to hear"*; 2) *'Kromo' language as a symbol of respect;* 3) *"bridging the gap"* and 4) *"intimate connection"*. Those categories depict perceived barriers, strategy to deal with challenges, and perceived nature of interaction:

3.1. Category 1: Less functioning

This category captures communication barriers as result of decline in physical functions, particularly in cognitive and sensory function. Elderly showed lower understanding to complex sentences and instruction compared to other age group as a result of cognitive decline. Hearing loss experienced by elderly was found to be another factor that makes communication become less effective.

"It is quite easy communicating with adult patients. When we talk about something, they usually understand what we mean. But, talking to elderly patients is different. Even when we have already used simple sentences, they still don't understand what we really mean (R.1)."

"Yes, they (elderly patients) sometimes couldn't really catch what we mean (R.2)."

"In my opinion, the biggest barrier (to interact with elderly patients) is in communication because their hearing function has already declined (R.2)."

"Hmm...mostly elderly patients suffer from hearing decline. That is the problem. Most often, they couldn't get what we really mean (R.6)."

3.2. Category 2: 'Kromo' language as a symbol of respect

Javanesse is the spoken language speaks by most of elderly in this study. Javanesse is known to have different speech styles for different social context. In Java culture, people are expected to speech in "Kromo" style as symbol of respect when they communicate to people with "higher" social status, including to elderly. However, lack in "Kromo" language mastering by students nurse means lack in medium to deliver the symbol. Students in this study truly understand that the use of other languages which understood well by both parties such as Bahasa or Jawa Ngoko style would never be able to give the similar "sense of respect" as 'Kromo' language does.

"When we have to speak Kromo language, we find it is difficult to do. As opposite, when we speak Bahasa, they (the elderly patients) find difficulties to understand. If we use Ngoko style, we don't think it is appropriate. We are afraid if it (the Ngoko language) will hurt them (the elderly patients). That is the problem. We know we should respect the elderly, but we couldn't speak Kromo language well (R.1)."

"We should use 'appropriate' language, which is Kromo language since there are social norms to be considered. We are afraid of becoming impolite if we speak in Ngoko language (R.2)."

"We usually adjust the way we talk when we interacting with elderly patient. If they (elderly patients) speak in "Ngoko" style, that is really okay for us. But when they use Kromo language, we are really in trouble. We will be shamed if we can't adjust our self to use the similar style (R.5)."

3.3. Category 3: Bridging the gap

This category highlights strategies used by students to overcome communication barrier with elderly patients. Communication gap, primarily lack in "Kromo" language proficiency is perceived to cause ineffective communication between students and elderly patients. Thus, students use "bridge" to close the gap. They mobilized family and friends to "bridge" the communication.

"Most often, we asked family member (of the patient) for help. We explained what we want or what what we are going to do and then they (family members) will re-explain it to the patient. It was like that. They are just like a medium (R.1)."

"Sometimes, the family members have deeper understanding of what patient said compared to us.

Thus we use them as a medium to communicate with the patient (R.2)."

"My friend is from West Java. She was really in trouble when she has to see elderly patient who couldn't speak Bahasa. The patient couldn't understand what she said. She really needs a translator. She usually asks another student (who could speak Java) for companion (R.3)."

"Sometimes, we ask our friend (who could speak Java) to accompany us when we have to see patient. If they are not available, we will talk to the family members instead (R.4)."

Other techniques used by students to achieve successful communication with elderly were using simple sentences, using non-verbal language (gesture and facial expression) and reducing physical distant to patient.

"So, when we had to talk to elderly patient, we moved closer or showed our mimic (expression). It was like that (R.2)."

"If we talked to adult patient, we like to keep distant at certain level. But when we talked to elderly patient, we come closer just to show that we care. Sometimes we touch their hands to express our caring when we care for female elderly patients (R.1)."

"If we want to explain something to patient, we have to make sure that it (the language) should be easy to understand. We shouldn't use difficult terms (R.5)."

3.4. Category 4: Intimate connection

This last category depicts positive view of students toward communication with elderly amid the challenges. One participant of this study found kind of "sense of intimacy" when interacting with elderly patients. This sense couldn't be found when interacting with client from other age groups.

For a matter of communication, I admit that caring for adult patient is far easier (compared to elderly patient). They have very good comprehension, their body function is still good and they could also understand many languages (not only Jawa language). However, at some point, I experience different feeling when I care for elderly patient. I couldn't really explain what it is. But I feel "close". It couldn't be found when you care for adult patient (R.6).

Hmm…I don't really know how to say my feeling. I just feel at ease, feel very comfortable (when communicate to elderly patient). I really have very good time with them (elderly patients). Most of the time, I experience all those feelings when I care for elderly patient (R.6).

4 Discussion

This study aimed to gain a better understanding of challenges faced by students nurse when communicating with elderly clients during clinical practice. The present study found that communication with elderly patients was apparently more challenging than communication with individual from other age groups. This finding supported the tendency as suggested by previous studies that inter-generational (different age group) communication is apparently more challenging than intra-generational (the same age group) communication. A quantitative study suggested that conversation with non-family elders was perceived as least satisfying, followed by family elder and same-age peers [9]. Age difference was found to be one of influential factors affecting communication between nurses and patients [10].

In this study barriers for an effective communication with elderly patients were perceived to be from either the patients or students nurses themselves. Patient related factors are including declining in hearing and cognitive function, while students' related factor is lack in "Kromo" language skill.

In communication with elderly patients, participants in this study found that declining the hearing function and thinking ability affects patients' speech perception and daily communication. Hearing impairment is one of the main problems in aging. The majority of hearing loss in older adult is sensory neural [11]. Few characteristics of age-related hearing loss are high frequency hearing loss, reduced hearing sensitivity and speech understanding in noisy environment. Hearing loss could have a great impact on elderly's daily activities, cognition, relationships, psychological health, and quality of life [12].

Effective communication also relies on an intact cognitive function. With normal aging, certain types of memory decline and cognitive processing slows. The length and complexity of sentences is also reduced [12].

Communication is heavily influenced by culture. Cultural groups have distinct linguistic and communication patterns. These patterns reflect core cultural values of a society [4]. Nurses are expected to have general knowledge of other cultures, especially those commonly encountered in their work area and its related skills to ensure the delivery of culturally congruent care.

Language commonality was seen as an important factor for effective communication [10]. However, in certain cultures where hierarchical language is applied, like in Javanese culture, effective communication is goes beyond "sharing the similar language". In Javanese culture, communicating with elderly using Kromo language is an expression of respect. Elderly in Javanese social structure is a group that earns an individual "*hormat*" (respect) from others. Hildred Geertz defines "*hormat*" as a Javanese term that means the recognition of superior rank by means of the appropriate forms of etiquette [13]. In Javanese culture, age is one of standard to define the rank. Nurses are typically younger than older adult patients. As a consequence, in this intergenerational-type communication it is crucial for them to show the respect by choosing the language level spoken.

There are numerous factors causing communication problem with older adult patients. Thus, it is important to provide mechanisms to help with communication issues such as adapting the environment, using assistive listening device and having communication partner adjust how they communicate with the individual [12]. Participants in this study have modified the way they communicate with elderly patient with hearing problem by positioning themselves close to the patients to reduce the distance and using non-verbal language such as gesture, facial expression and touch. Touch is one of the most effective forms of communication. It delivers many messages such as encouragement, affection, emotional support, and attention [4]. Meanwhile, when communicate with elderly patients with cognitive decline, it is necessary to communicate directly to topic with simple statement and provide adequate time to perceive, process, and respond to information [12].

Furthermore, in order to facilitating effective communication, companions during interaction may give some benefits. A previous study found that companions made a significant contribution to the communication between patient and health care personnel [14]. In this study, companions are family and friends who are able to speak Kromo to overcome language barrier.

Communication with elderly, from the participant's experience, may create the feeling of intimacy and comfort. This is an interesting finding though majority of students of this study didn't have the similar experience. To feel close and comfortable with our communication partner, these would be definitely helpful for an effective communication. Previous study had identified that lack of intimacy and official behavior were the most important nurse-related factors affecting communication [7]. Another study suggested that nurse' reluctant to communicate with the patients was one of important barriers, particularly from the viewpoint of patients [10]. It is expected that the presence of intimate or closeness feeling when interacting with elderly patient could give students positive experience which is important in building positive view and counter negative stereotype of elderly people.

5 Conclusion

This study explored the experience of nursing students' communication with elderly patient. The present results indicated that communication barriers could be either from patient or nurse's students. Decline in sensory and cognitive functions of elderly patients and lack of local language proficiency by students nurse were perceived to be barriers for effective communication. Thus, to overcome these barriers, students applied certain techniques and used help from patient's family members or friends. Finally, amidst the challenges; however, students viewed that interaction with elderly patients was pleasing and comforting.

The findings of this study provided larger perspective of intergenerational communication in the context of nursing practice in multicultural society. With the growing number of elderly patients, nurse should become skilled at communication technique and basic local language spoken by older adult patient. Cultural awareness should be developed considering that Indonesia is a multicultural country. Developing constructive view of elderly people is also essential for successful communication.

References

[1] T. Dall, P. Gallo, R. Chakrabart, T. West, A. Semilla and M. Storm, Health Aff. **32,** 2013–2020 (2013)

[2] J. Ansah, R. Eberlein, S. Love, M. Bautista, J. Thompson and Malhotra, *Health Policy (New. York).* **116,** 105–13 (2014)

[3] L. Kourkouta and I. Papathanasiou,. *Socio Medica.* **26,** 65–7

[4] P. Potter, A. Perry, P. Stockert and A. Hall *Fundamental of Nursing* (St. Louis, Missouri: Elsevier Mosby, 2013)

[5] C. Charlton, K. Dearin, J. Berry and M. Jhonson, *J.* Am. Assoc. Nurse Pract. **20,** 382–8 (2008)

[6] J. Wang, P. Hsieh and C. Wang, *Korean.* Soc. Nurs. Sci. **7,** 99–103 (2013)

[7] M. Agha and L. Hasani, J. Hormozgan Univ. **14,** 312–8 (2011)

[8] K. Charmaz, *Constructing Grounded Theory: A Practical Guide through Qualitative Analysis* (London: SAGE Publication, 2006)

[9] H. Giles, K. Noels, A. William, H. Ota, T. Lim and S. Ng, J. Cross Cult. Gerontol. **18**, (2003)

[10] Norouzinia R, Aghabarari M, Shiri M, Karimi M and Samami E 2016 Communication barriers perceived by nurses and patients *Glob. J. Health Sci.* **8** 65–74

[11] S. Grossman and C. Porth, *Porth's Pathophysiology: Concept of Altered Health States* (Philadelphia: Lippincott Williams & Wilkins, 2014)

[12] K. Mauk, *Gerontological Nursing Competencies for Care* (USA: Jones and Bartlett Publisher, 2010)

[13] D. Irawanto, P. Ramsey and J. Ryan, Challenge of leading in Javanese culture *Asian Ethn.* **12**, 125–39 (2011)

[14] H. Ishikawa, D. Roter, Y. Yamazaki and T. Takayama, *Soc. Sci. Med.* **60**, 2307–20 (2005)

The Effect of *Sikkomodo* (Combination of Music, Humor, and Prayer) Formulation Toward Blood Pressure of Hypertension Patient on The Elderly Group in Banyumas Regency

S Awaludin[1], ASumeru[1], G N Alivian[1], D Novitasari[2]

[1] Health Science Faculty. Jenderal Soedirman University

[2]Faculty of Health Sciences, Harapan Bangsa University

Abstract. Uncontrolled hypertension can lead to heart disease, kidney disease, and stroke. Hence, it needs therapy. There are complementary therapies that can reduce high blood pressure. Music provides a relaxing effect so that blood vessel dilates and blood pressure drops. Humor therapy can provide a relaxation response and decrease blood pressure. Prayer therapy has a positive effect on body physiology that can lower blood pressure. The purpose of this study was to identify the effect of combination of music, humor, and prayertherapy on blood pressure. This study used true experiment pre and post control group design. The number of samples in this study were 131 respondents with primary hypertension on the elderly in Banyumas Regency. The sampling technique used was simple random sampling. Wilcoxon was used to test the difference of blood pressure before and after the intervention in one group, while the difference test among groups employed Mann Whitney test. The result indicated that respondent's hypertension is categorized into moderate. There is difference on the mean of blood pressure before and after SIKKOMODO therapy and combination of music, humor, and humor therapy. There is a significant difference in term of systolicblood pressure betweentreatment and control group, and there is no difference on the systolic blood pressure between the two groups.

1 Introduction

An individual's health is a physical, mental, and social condition that is not only free from problems of illness and weakness, but also free from all problems[1]. A health problem that mostly occurs in the society and is not recognized by the patient is hypertension or high blood pressure[2]. Theoccurrence of hypertension in Indonesia is quite high. The result of Basic Health Research (Riskesdas) in 2013 showed that most of hypertension cases in the community had not been diagnosed. The prevalence of hypertension in Indonesia is 26.5%, in which only 9.5% of the population realized having hypertension and only 0.1% of them who took hypertension medication at the age of 18 and over. This incidence increases as thepeople grow older. The result of the study from Awaludin,et.al (2016) indicated that the incidence of hypertension on the elderly in RW 01 Mersivillage Purwokerto was more than 90%.

Hypertension is a cardiovascular disorder characterized by blood pressure of more than 140/90 mmHg[3]. High blood pressure is one of the most influential risk factors for the incidence of heart disease and blood vessel disease. Hypertension often shows no symptoms, so it is just realized if it has caused organ impairment such as impaired heart function or its capillary. The majority of people with hypertension find that they have hypertension accidentally such as at the moment of their medical check- up orwhen it comes with other complaints. A person with hypertension in the blood vessels experiences a narrowing or atherosclerosis. With this narrowing, it takes a great pressure to supply blood throughout the body. In addition, it also requires a fast heartbeat. If it lasts a long time, the patient will experience headache, weakness, and discomfort, and can even lead to a decrease in consciousness and cause death[1].

A few of the complementary therapies for lowering blood pressure are music therapy, humor therapy, and prayer therapy[4]. Geer (2009) defines music therapy as the use of music to get relaxed, accelerate healing, improve mental function, and create a sense of welfare. Music can affect several physiological functions such as respiration, heart rate, and blood pressure. It stimulates the release of endorphins. Humor therapy is an action to stimulate someone to laugh. It can promote the release of endogenous opiates called endorphins. The benefit of endorphins is to make relaxation which has an impact on the dilation of blood vessels and as a result it can decrease blood pressure. With the relaxed situation, it will also make the heart rate normal. Humor therapy is done using several ways, for example watching funny movies, listening to comedians, looking at funny cartoons, comics and caricatures, and reading a collection of funny storie[5]. While prayer therapy which is preceded by *dhikr* is a therapy carried out by persuading, conveying the requests utterances, and asking for His protection sincerely[6].Prayer can influence a positive emotional response that will flow to the body and is accepted by the brainstem. After being formatted by the brain language, prayer is transmitted to one part of the cerebrums, thalamus, then the thalamus transmits the hippocampal impulse (a vital center of memory which coordinates everything that is absorbed by the senses) to secrete Gama Amino Batiric Acid (GABA). GABA serves as a controller of emotionalresponse and blockasetylcholine, serotonin, and other neurotransmitters that produce cortisol secretion which leads tohomeostasis process and cause relaxation. As a result, blood pressure can decrease[7].

The three therapies are combined into one therapy formula called SIKKOMODO (*musik kombinasi humor dan doa*), a combination of music, humor, and prayer therapy. This therapy formulawill work together to speed up blood pressure reduction. This formula is expected to be an invention to reduce high blood pressure non-pharmacologically. The purpose of this study was to find out the effect of SIKKOMODO therapy toward blood pressure to hypertension patients on the elderly group in Banyumas Regency.

2 Research Methods

This study used true experiment pre and post control group design. It observed two times, before and after the treatment. The researcher divided the respondents into two groups, one treated using SIKKOMODO therapy as the experimental group and another one given only humor- music combination therapy as the control group. The population were all the inhabitants of Mersi, Ledug, Arcawinangun, and Sumbang villages who had hypertension. The sample of this study was 120 respondents consisting of 60 patients for experimental group (A) and 60 patients for control group (B). The sampling technique used was simple random sampling. The sample in this study was the subjects who met the inclusion criteria. The inclusion criteria were the respondents with the following conditions: willing to be observed, suffering from primary hypertension, having age range 45 - 75 years old, having composmentis. Exclusion criterion in this study was patient having mental disorders. Blood pressure measurement was carried out by using OMRON digital sphygmomanometer that was calibrated. The measurement result was recorded in the observation sheet. Blood pressure measurement in the experimental and control group was assessed twice, before the therapy and right after the therapy. The respondents in group A were given SIKKOMODO therapy by having them listen to instrumental music for 5 minutes, watch a humor therapy video for 10 minutes, and finally have prayer therapy for 5 minutes. Group B respondents only listened to instrumental music for 5 minutes, watched a humor therapy video for 10 minutes, and had no prayer therapy.

Kolmogorov Smirnov test result showed that p value of systolic pressure from pre group A was 0.059 and the one from pre group B was 0.110. The p value of systolic pressure from post group B was 0.310 and p value of diastolic pressure from post group B was 0.378; those data had normal distribution. Apart from them, the data was not normally distributed. The difference test between two means of equal groups, diastolic pressure pre and post treatment, in the control group employed parametric test (dependent t test).

The others used a difference test between two means of equivalent groups with non-parametric test (Wilcoxon). Difference test of different group for systolic blood pressure used parametric test (independent t test). Others used difference test between two means of different group with non- parametric test (Mann Withney).

3 Result

1. Blood Pressure Description of Pre- SIKKOMO Therapy

Table 1. Mean Description of Respondent's Blood Pressure of Pre SIKKOMODO Therapy

Variable	Treatment Group Mean ± SD	Control Group Mean± SD
Systolic blood pressure	170.84 ± 15,09	173.89 ± 17.61

(mmHg)		
Diastolic blood pressure (mmHg)	98.34 ± 8,37	96.39 ± 5.81

Table 1 indicates that the mean of blood pressureis categorized into moderate hypertension (systolic blood pressure is 160-179 mmHg)in the two groups.

2. Blood Pressure Description of Post-SIKKOMODO Therapy

Table 2. Mean Description of Respondent's Blood Pressure of Post SIKKOMODO Therapy

Variable	Treatment Group Mean ± SD	Control Group Mean ± SD
Systolic blood pressure (mmHg)	153.01 ± 16.41	169.40 ± 17.55
Diastolic blood pressure (mmHg)	96.93 ± 15.38	96.52 ± 7.16

Table 2 shows that in the treatment group the mean of blood pressure is classified into mild hypertension (systolic blood pressure is 140-159 mmHg)and in the control group the mean of blood pressure belongs to moderate hypertension category (systolic blood pressure is 160-179 mmHg).

3. The Difference Test Result of Blood Pressure between Pre- and Post in the Treatment and Control Group

Table 3. The Difference Test Result of Blood Pressure Between Pre- and Post Treatment

Variable	*p value*Treatment	*p value* Control
Systolic blood pressure	0.000 [a] (Z: -6.96)	0,000 [b]
Diastolic blood pressure	0.000 [a] (Z: -4.77)	0,016 [a] (Z: -2.41)

a = non- parametric test, b = parametric test

Table 3 shows that the result of test difference of blood pressure in the treatment and control group is significantly different ($p<0,05$). The blood pressure of systolic and diastolic in the control group indicates a significant difference, but Z value is negative. It means blood pressure in the control group significantly.

4. The Effect of SIKKOMODO Therapy on the Respondent's Blood Pressure between Groups

Table 4. The Effect of SIKKOMODO Therapy on the Respondent's Blood Pressure between **Groups**

Variable	Homogeneity Test	*p value*
Systolic blood pressure	0.285*	0.000
Diastolic blood pressure	0.220*	0.486*

*= not significantly different

Table 4 shows that the data of blood pressure pre- and post in the treatment and control group is not significantly different or homogenous, hence the difference test from the data post between groups can be done. The systolic blood pressure in the treatment and control group is significantly different ($p < 0.05$), while diastolic pressure on the two groups is not significantly different ($p > 0.05$).

4. Discussion

Table 1 shows that blood pressure in the treatment and control groups is within moderate hypertension category (systolic blood pressure 160-179 mmHg). Table 2 shows that the result of difference test on systolic and diastolic blood pressure before and after treatment in the treatment group indicates significant difference ($p = 0.000$), while in the control group it also shows significant difference but the Z value is negative, which means that blood pressure in the control group increases significantly. As a matter of fact, what is expected is a decline.

Music therapy can influence several physiological functions such as respiration, heart rate, and blood pressure[8]. Slendro gamelan music can decrease blood pressure significantly on hypertension patients. Instrumental music can reduce stress levels on patients with ventilators[9]. Decreased stress is followed by vagal outflow and decreased activity of the sympathetic nervous system[10]. Decreased sympathetic activity to the smooth muscle of blood vessels will cause relaxation and vasodilation so that blood pressure will drop[11]. Music provides a positive stimulus to the limbic system in the brain[12]. The limbic system is associated with emotional behavior (anger and fear), thereby it reduces the activity of the autonomic nervous system, creating a relaxed feeling. A relaxed feeling will stimulate the production of Nitric Oxide (NO) hormone[13]. NO hormone is synthesized by vascular endothelial cells from the arginine amino acid which is frequently found in plasma. The effect of NO on the cardiovascular system is as a vasodilator agent, so it can reduce blood pressure[14]. The therapy of music and humor combination (SIKKOMO) is able to synergize and strengthen blood pressure reduction.

Table 3 shows that there is significant difference on the systolic blood pressure of experimental group and control group ($p <0.05$), while the diastolic blood pressure on the experimental group and control group is not significantly different ($p> 0.05$). An individual who grows older and enters the elderly period (> 60 years) will experience peripheral resistance occur due to narrowing of the arteries and arterioles, decreased capillary blood vessel density, and decreased peripheral elasticity. This increases venus return and preload which in turn causes systolic and diastolic dysfunction[15]. This thing allows no significant differences between the two groups. Other possibilities include basic data from the two groups of respondents showing varied blood pressure and all respondents aged> 60 years (elderly).groups was not significantly different ($p> 0.05$). A person who experiences aging and goes into the elderly (> 60 years) peripheral resistance will occur due to narrowing of the arteries and arterioles, decreased capillary blood vessel density, and also a decrease in peripheral elasticity. This increases venus return and preload which in turn causes systolic and diastolic dysfunction[15]. This allows no significant differences between the two groups. Other possibilities include basic data from the two groups of respondents which shows various blood pressure and similar age range of all respondents that is over 60 years old.

An individual who prays fervently will have a positive effect on the physiology of the body that is received to the brainstem. After formatted with the brain language, pray is transmitted to one part of the cerebrum, the thalamus. Then the thalamus transmits the hippocampal impulse (a vital memory center coordinating everything absorbed by the senses) to secrete Gama Amino Batiric Acid (GABA) which acts as a controller of emotional responses, and block acetylcholine, serotonin and other neurotransmitters that produce cortisol secretion that leads to the occurrence of homeostasis and able to reduce blood pressure[7]. Table 3 shows that the group receiving SIKKOMODO therapy and the one which only receives a combination of two therapies, music and humor, have a significant difference in their systolic blood

pressure. The combination of three types of appropriate therapy can have a strengthening effect to lower the blood pressure rather than just combining two types of complementary therapies in reducing blood pressure.

Music can influence several physiological functions such as respiration, heart rate and blood pressure[8]. It can also reduce myocardial oxygen demand and improve sleep quality. A relaxed condition will stabilize the metabolic rate and the body's oxygen demand, so that vital signs such as pulse, respiratory rate, and blood pressure are within normal limits[16]. In addition, it can decrease the adrenaline andnon-adrenaline secretion. The effect of adrenaline decrease on the cardiovascular system is a decrease on the heart rate and blood pressure, while non-adrenaline decrease causes a decrease in blood pressure[17]. Music stimulates the central nervous system directly and will affect the subject's cognitive and physiological situation. Besides, it affects the expression of opiate receptors, morphine-6 glucuronate, and interleukin-6 levels on healthy volunteers. The effect can be moderated by nitric oxide[18]. Music can improve emotional responses, aggressive or receptive conditions, and stimulate the limbic system. The harmony structure, interval quality, and tone color will be recognized by the non-dominant hemisphere (mostly in the right hemisphere) temporarily and permanently. Short-term melodies in music, such as volume variations, tone counts, and lyrics are recognized by the dominant brain hemisphere[19].

When someone laughs, the release of endorphins hormone occurs. This hormone is believed to help fixing the condition of blood vessels. The release of this hormone in large quantities also occurs after a person exercises. In addition, laughing will stimulate the release of nitric oxide, which is a gas that can relax the endothelium (one part of the blood vessels). In one study, it is found that laughter can improve the immune system up to 40%. Therefore, health professionals should consider humor as a therapy[5].

People who laugh a lot will feel more free to breathe, because laughter speeds up the release of saturated air from the body which is immediately replaced by fresh air. This alteration will enrich the blood with oxygen and clean the respiratory parts or organs. Strong breathing will also train the heart muscle, improve blood circulation, accelerate the flow of oxygen, and nutrient contraction of the heart muscle will be better trained in term of rhythmic rhythms of the brain. As a result, the bloodstream is getting better. Blood within the vessels will transport oxygen and nutrients faster to meet their needs throughout the body and improve the nutritional function of the body circulation. During laughter, antibodies and white blood cells actively prevent infection, while hormones can improve alertness and memory function[5].

People who often laughs or gets relaxed will experience the release of endorphins and enkephalin that function as endogenous opiates. Endorphin provides relaxation. A relaxed condition will stabilize the metabolic rate and the body's oxygen demand, so that vital signs such as heart rate, respiratory frequency, and blood pressure are within normal limits[16]. A relaxed feeling can reduce the secretion of adrenaline and non-adrenaline. The effects of adrenaline on the cardiovascular system include an increase on heart rate and blood pressure, while non-adrenaline causes an increase in blood pressure[17].

Praying is a form of physical, psychological, and religious compassion so that all elements of the body become more relaxed. Someone who prays and does dhikr by mentioning the name of Allah, he actually invigorates the qualities and names of Allah that have extraordinary power inside himself. As a consequence, he has the spiritual power that makes him feel peaceful and balanced. A peaceful soul can normalize the organs of the body, arouse the balance and positive thoughts, and affect healing. The prayer can be done through the servitude process by uttering the name of Allah repeatedly. Then it can be continued by confessing that all power belong to Allah only by saying "La haullawalaquwata ill billah" along with appreciation. Dhikr can also normalize the work of the sympathetic nerves and the heart rate[20]. While in diastolic blood pressure, it does not show a significant difference because elderly has experienced vascular dysfunction which results in diastolic dysfunction[15].

5. Conclusion

The average of respondent's blood pressure is categorized as moderate hypertension (systolic blood pressure is 160-179 mmHg).There is a difference in blood pressure before and after SIKKOMODO therapy.There is a difference in blood pressure before and after music and humor combination therapy.There is a significant difference in systolic blood pressure between the treatment and the control groups, however there is no significant difference in diastolic blood pressure between the two groups.

6. Suggestion

SIKKOMODO therapy can be used as part of nursing intervention in attempt to reduce blood pressure and keep the heart rate in normal conditions in nursing services.It is used as a recommendation for curriculum development for complementary therapy subject, especially SIKKOMODO therapy for nursing education institutions.For further research, it is necessary to have a combination with other therapies lower blood pressure to the normal limits.

Acknowledgments

We greatly acknowledge to Rector, Head of LPPM JenderalSoedirman University, all patient and all parties who have helped in this research.

References

[1] Smeltzer, C. Suzanne, *Buku Ajar MedikalBedah Brunner &Suddarth* / editor, Suzanne C. Smeltzer, Brenda G. Bare ; alihbahasa, AgungWaluyo....[et al] ; editor edisibahasa Indonesia, Monica Ester. – **Ed. 8**. Vol 2 – Jakarta: EGC (2008)

[2] S. Awaludin, I. Purnawan,, G.N. Alivian, Pengaruh terapi humor untuk menurunkan nyeri kepala akibat hipertensi di Kelurahan Mersi Purwokerto.. Kumpulan Abstrak Semnas . LPPM Universitas Jenderal Soedirman (2016)
[3] WHO, *Report of hypertensions*. WHO/CDS/RBM/2001.35. Geneva 28-30 March 2009
[4] B. Kozier, *Fundamental of nursing : Concept, process and practice*, 7thedition. New Jersey: Pearson Education Inc. (2014)
[5] L. Keegan, Delmar Thomson Learning (2001)
[6] A.F. Elkaysi, *Pengobatan doa-zikir dan ruqyah*. Jakarta :MutiaraMedia (2012)
[7] M. Sholeh, *Agama sebagai terapi telaah menuju kedokter anholistik*. Yogyakarta :PustakaPelajar (2005)
[8] E. Geer, A.C. Vink, J.M. Schols, J.P. Slaets, International Phychogeriatrics. **21**, 86-93 (2009)
[9] S. Ismail, C. Kritpracha, P. Thaniwattananan, *The effects of music intervension on aniety reduction in patients with ventilator support*. Makalah dipresentasikan pada *Java International Nursing Conference* tanggal 2-3 Oktober, Semarang (2010).
[10] J. M. White, Critical Care Nurse Clinical North America. **12** (2), 219-225 (2000)
[11] S. Oparil, A.M. Zaman, D.A. Calhoun, Ann Intern Med. 139, 761-776 (2003)
[12] Hidayat. (2006). *Terapi musik teori dan aplikasi.* Cetakan pertama. Yogyakarta: Galang Press (2006)
[13] Champbell, *Efek Mozart: Memanfaatkan kekuatan musik untuk mempertajam pikiran,meningkatkan kreativitas dan menyehatkan tubuh.* Jakarta: Gramedia Pustaka Utama (2001)
[14] L.P. Hooi, The effect of music intervention on anxiety and pain in female surgical patients.Unpublished Dissertation, Prince of Songkla University, Hatyai (2007)
[15] E.L. Schiffrin, *Hypertension*19 (2 Suppl): II1–9. PMID 1735561 (2000)
[16] A.G. Perry, and P.A. Potter, *Konsep, proses danpraktik.(***4**,vol 2. St.Louis: Mosby, 2005)
[17] N. Ulrica, Associations of operating room nurses (AORN) Journals. **83** (4); 780 (2008)
[18] Kemper, J. Kathi, Danhauer, C. Suzzane, C. Journal of Clinical Nursing, **6** (5), 341 – 346 (2005)
[19] E.T. Chang, H.L. Lai, P.W. Chen, Y.M. Hsieh, L.H. Lee, International Journal of Nursing Studies. 928 (2018)
[20] N. Yanti, *Perbandingan efektifitas terapi zikir dengan relaksasi Benson terhadap kadar glukosa darah pasien Diabtes Melitus di Sumatera Barat*. Universitas Indonesia (2012)

Socio Demographic Characteristics Of Pregnant Women Who Are Experiencing Nausea Vomiting In Rural Areas Of Banyumas Regency

Lutfatul Latifah[11*], *Nina* Setiawati[1], *Aprilia* Kartikasari[1], *Eli* Kusmiati[1]

[1]Department of Nursing, Faculty of Health Sciences, Universitas Jenderal Soedirman

Abstract. Nausea vomiting in pregnancy or commonly referred to as morning sickness is a common complaint in the first trimester, although it can also occur in the second trimester or all trimester. The purpose of this study was to determine the characteristic of pregnant women who experience morning sickness in rural areas. Respondents in this study were pregnant women who experienced morning sickness during July-September 2018 in rural areas of Banyumas District, Central Java Province, Indonesia. This study have used quantitative descriptive design.The results of univariate analysis of 61 pregnant women involved, 77% age of mothers were at low risk, 39,3% were junior high school education, 82% were not working, 60% were pregnant in the first trimester, 61,7% were multigravida, 55.7% did not have a history of nausea and vomiting, and 73,8% in the category of mild nausea and vomiting. It can be concluded that pregnant women who experience nausea and vomiting in rural areas are the majority of those who are of low risk, do not work, are pregnant for the first time, have early pregnancy, and with mild nausea and vomiting.

1 Introduction

The Pregnancy begins with the conception or meeting of eggs and sperm. During pregnancy physical and psychological changes occur in pregnant women as a form of pregnancy adaptation. Changes that occur during pregnancy cause various discomforts felt by pregnant women, one of which is nausea and or vomiting, known as morning

*Corresponding outhor: latifah.lutfatul@gmail.com/lutfatul.latifah@unsoed.ac.id

sickness. Although referred to as morning sickness, symptoms of nausea and vomiting can occur throughout the day and often are early symptoms of pregnancy (1). The time and duration of nausea and vomiting are different in some pregnant women, some experience non-stop nausea with or without vomiting and some experience day or night and often last throughout the day. Morning sickness can begin to be felt at the beginning of the second week of pregnancy or at 8 or 12 weeks' gestation and disappear when gestational age reaches the twentieth week, but 20-30% of pregnant women continue to experience symptoms for more than 20 weeks of gestation until delivery (2,3).

The cause of morning sickness is still unclear but its high frequency shows that nausea and vomiting are normal events in early pregnancy (4). Possible causes are a result of increased levels of human chorionic gonadotropin (hCG) and estrogen and with thyroxine, prostaglandin E2, and prolactin as additional supporting factors (2). Apart from that emotional condition where mothers in the first trimester experience mood swings, ambivalence or rejection of pregnancy also contributes (5,6).

In several studies, the incidence of nausea and vomiting in pregnant women was associated with increased pregnancy hormones, older maternal age, type of work, level of education, smoking behavior, infant sex, and stress levels (2). In addition, reproductive history such as increased gravidity, parity, and a history of abortion are also said to increase the risk of nausea and vomiting. Chan et al. (2) also conveyed the duration of symptoms of long-term nausea and vomiting, which is longer than 4 months and is usually more common in younger women, women with multiple pregnancies, and in multigravidas. The results of several studies have considered demographic characteristics, maternal and psychosocial factors that can cause nausea and vomiting in general. Therefore, the aim of our study was to determine the prevalence and characteristic features of pregnant women who experience nausea and vomiting specifically in rural area populations.

2 Research methods

This study has received ethic approval from the Ethics Commission of the Faculty of Medicine Unsoed. This study uses descriptive analysis method with a quantitative approach. Data for this study were collected from July to October 2018 in rural populations in 3 sub-districts with 6 working areas of health centers in Banyumas Regency. During this period there were 61 pregnant women with nausea and vomiting who examined the condition of their pregnancy to Baturraden I and II Health Centers, Sumbang I and II Health Centers, and Kembaran I and II Health Centers stated they were willing to become respondents.

Demographic variables (such as maternal age, gestational age, gravida, parity, history of nausea vomiting, abortion history, education, employment and economic status) were reported to be associated with the incidence of nausea and vomiting in pregnant women (2,7,8) taken using a questionnaire filled out by the respondents themselves. While the variable severity of vomiting nausea was measured using Unique Quantification of Emesis and Nausea (PUQE) -24 scoring system developed by Ebrahimi, Maltepe, Bournissen, & Koren (9) and has been translated into Indonesian and used in previous research (10). Descriptive statistics were used to analyze the description of each socio-demographic variable and the severity of nausea and vomiting.

3 Result

The results of descriptive statistical analysis of several demographic variables and severity of nausea and vomiting can be seen in Table 2. below.

Table 1. Table of Frequency Distribution of Characteristics of Pregnant Women With Vomiting Nausea in Rural Areas in July-October 2018 (n = 61)

Variable	f	Percentage (%)
Age of mothers		
a. High risk (<20 years and > 35 years	14	33
b. Low risk (20-35 years) mothers	47	77
Level ofeducation in		
a. elementary	18	29,5
b. junior high school	24	39,3
c. senior high school	16	26,2
d. higher education	3	4,9
Job Status		
a. Not working	50	82
b. Working	11	18
Income		
a. Low income		55,7
b. High income		44,3
Age pregnancy		
a. First trimester	36	60.0
c. b. Second trimester	21	35.0
d. Thirdtrimester	3	5.0
Gravida		
a. Primigravida	23	38.3
b. Multigravida	38	61.7
History of nausea and vomiting		
a. Never	34	55,7
b. Ever	27	44,3
Level of Nausea Vomiting		
a. Mild	45	7.,8
b. Moderate	16	26.2
c. Weight	0	0.0

Based on Table 1. it can be seen that the majority of pregnant women who experience nausea vomiting in rural areas are those aged between 20-35 years or in the category low risk as many as 47 respondents (77 %) with 39.3% of respondents educated last junior high school or as many as 24 respondents. The majority of mothers who experience nausea and vomiting also have a non-working status that is as many as 50 respondents (82%) and followed by more than 50% of respondents have low income (34 respondents). Mothers who experience nausea vomiting on average with gestational age in the first trimester (60%) and second trimester (35%) and the majority are multigravida (61.7%). More than half of respondents (55.7 %)

have no history of previous nausea and vomiting and 73.8% or as many as 45 respondents experienced mild nausea and vomiting.

4 Discussion

Age is one of the risk factors for NVP, where NVP occurs more at a younger age (11), while increasing age is associated with a decrease in nausea and vomiting in pregnancy (12). The age of respondents in this study was mostly at low risk gestational age. The results of this study are not different from previous studies where the average age of pregnant women who experience NVP is in the age range of 20-35 years. Kugahara and Ohashi (13) reported that the average age of pregnant women experiencing nausea and vomiting was 25.2 years while the average age of pregnant women suffering from NVP reported by Matok et al. (14) is 29.7 years. In the results of the Suwarni study (15), 80% of mothers were aged 21-35 years.

In this study, most pregnant women who experience nausea and vomiting were at 9-12 weeks' gestation. Lacroix's (16) study found that the majority of pregnant women, about 90%, experience nausea vomiting at 8 weeks' gestation and peak at 11-13 weeks. At 14 weeks' gestation, 50% of pregnant women have started to reduce vomiting, but 90% continue to experience nausea and vomiting until the age of 22 weeks. Some studies have found a link between the production of hCG (human chorionic gonadotropin) and the incidence of nausea and vomiting.

The Kugahara & Ohashi study (13) divides the gestational age of mothers who experience morning sickness to 4-7 weeks, 8-11 weeks, 12-15 weeks and 16-19 weeks. He said that at the age of 16-19 weeks the mother had reduced her vomiting. In this study there were also third trimester pregnant women who still experience nausea vomiting, which is as much as 5%. The incidence of nausea and vomiting reported in the third trimester is significantly lower than in the first or second trimester of pregnancy and along with increasing gestational age the symptoms of severity of nausea and vomiting diminish. This is in accordance with that expressed by Gadsby et al. (12) where the symptoms of nausea and vomiting decrease after 20 weeks' gestation. However, on the contrary, Chou et al. (12) who have conducted a prospective longitudinal study of 91 pregnant women measured using Index of Nausea, Vomiting, and Retching (INVR) reported that NVP in the second and third trimesters was significantly lower than the first trimester, so the average value of INVR in the second trimester is not much different from the third trimester.

The highest education level of respondents is junior high school. According to data from Central The Bureau of Statistics province of Central Java (12), the average length of education of the population of Central Java is 7.27 and the population of Banyumas Regency is 7.4. The results of this study indicate that respondents had higher education than the average population of Central Java and especially Banyumas Regency, but were lower than the average length of education of the Indonesian population at 8.32 years (17). The results of this study are in line with the research (10) where 40% of respondents were educated in junior high school. It is different from the research of Suwarni (15) and Heitmann, Nordeng, Havnen, Solheimsnes, & Holst (18) where the highest education level of respondents is in the category of higher education. This difference is possible because of the research area of Suwarni (15) and Heitmann et al. (18) located in urban areas while this research is located in rural areas.

NVP is associated with low income levels and mothers do not work (16). In line with this study, it was found that the majority of respondents did not work and more than half had income below the regional minimum wage (UMR). UMR of Banyumas Regency amounted to Rp 1,589,000, and the family income of respondents was a majority below that figure. The researcher analyzes that more than 50% of the respondents are below the poverty line and are between 17.52% of the poor population of Banyumas Regency and 14.88% of the population in Central Java with an income level of Rp 322,489 per month (17,19). However, the results of this study are different from findings from the Heitmann et al. (18) who reported that 80% of mothers who experienced NVP were working mothers and only 7.4% did not work. Piwko's, Koren, Babashov, Vicente, & Einarson (20) reported that NVP caused an increase in economic burden. The results of this study found that mothers with majority NVP did not work and family income was below the UMR. This can make the picture that mothers who experience nausea vomiting in rural areas will have an increasingly heavier economic burden.

Pregnant women who have a history of nausea and vomiting in previous pregnancies will be at risk of developing nausea and vomiting in subsequent pregnancies compared to those who do not have a history of nausea and vomiting. This statement is in line with Trogstad, Stoltenberg, Magnus, Skjærven, & Irgens's research (21) which states that 15.2% of mothers with a history of nausea and vomiting in previous pregnancies are at risk of experiencing nausea and vomiting in subsequent pregnancies. However, this statement is not in line with what was found in this study, where among 61 mothers who experienced nausea and vomiting, more than 50% of mothers had never experienced nausea and vomiting before. Meanwhile Fejzo, Macgibbon, Romero, Goodwin, & Mullin (22) found 57 multigravida mothers, 81% of whom reported that they experienced severe nausea and vomiting in their second pregnancy. Likewise with the results of this study where of 61 mothers who experienced nausea vomiting, 61.7% of them were multigravida mothers. But the results of a recent study conducted by Nurmi, Rautava, Gissler, Vahlberg, and Polo-Kantola (23), concluded that in the majority of pregnant women with a history of nausea and vomiting, there is no recurrence in subsequent pregnancies. Even so the risk of recurrence cannot be predicted for each individual (pregnant woman).

The severity of NVP in pregnant women varies. The results of this study the majority of respondents experienced mild NVP and none experienced severe NVP. In contrast to the research of Heitmann et al. (18) who reported that of a total of 712 respondents, the majority of 61.66% of respondents experienced moderate NVP, as many as 29.5% of respondents experienced severe NVP and only 0.09% of respondents experienced mild NVP. In line with Heitmann's study, the average NVP in the Matok et al. (14) of 9 which means that the average respondent experiences moderate nausea and vomiting. The study both used the PUQE-24 score in categorizing NVP.

5 Conclusion

The conclusion that can be drawn from this study is that mothers who experience nausea vomiting in rural areas are mostly low educated followed by low economic status. This makes the burden of life more severe so that appropriate treatment is needed to overcome nausea and vomiting in pregnant women in rural areas.

6 Acknowledgement

The authors acknowledge the Director of Research and Community Engagement of Universitas Jenderal Soedirman for providing funding. And also the authors express gratitude to the participants and all other parties involved in this study.

References

[1] S. L. Ward and S. M. Hisley, *Maternal-child nursing care [electronic book]: Optimizing outcomes for mothers, children, and families*. (Philadelphia: F.A. Davis Co., 2009)

[2] R. L. Chan *et al.*, "Maternal influences on nausea and vomiting in early pregnancy,"*Matern. Child Health J.*, vol. 15, no. 1, pp. 122–127, 2011.

[3] N. M. Lee, "Nausea and Vomiting of Preg nancy Nausea Vomiting Pregnancy Hyperemesis gravidarum," *Gastroenterol. Clin. NA*, vol. 40, no. 2, pp. 309– 334, 2011.

[4] J. Kramer, A. Bowen, N. Stewart, and N. Muhajarine, "NAUSEA and VOMITING of PREGNANCY : Prevalence , Severity and Relation," pp. 21–27, 2013.

[5] McKinney, E.S., S. R. James, S. S. Murray, and J. . Ashwil, *Maternal-Child Nursing*, 4th ed. St. Louis, Missouri: Saunders Elsevier, 2013.

[6] S. E. P. Lowdermilk, Deitra Leonard. and K. Cashion, *Maternity Nursing*, 8th ed. Maryland Heights, MO: Mosby/Elsevier, 2016.

[7] C. A. Reichner, "Insomnia and sleep deficiency in pregnancy," *Obstet. Med.*, vol. 8, no. 4, pp. 168–171, 2015.

[8] N. Taskiran, "Pregnancy and Sleep Quality," *J. Turkish Soc. Obstet. Gynecol.*, vol. 8, no. 3, pp. 181–187, 2011.

[9] N. Ebrahimi, C. Maltepe, F. G. Bournissen, and G. Koren, "Nausea and Vomiting of Pregnancy: Using the 24-hour Pregnancy-Unique Quantification of Emesis (PUQE-24) Scale," *J. Obstet. Gynaecol. Canada*, vol. 31, no. 9, pp. 803–807, 2009.

[10] L. Latifah, N. Setiawati, and E. H. Dwi, "Efektifitas Self Management Module dalam Mengatasi Morning Sickness Effectiveness Self Management Module in Overcoming Morning Sickness," *Jkp*, vol. 5, no. 1, pp. 10–18, 2017.

[11] Klebanoff MA, K. PA, K. R, and R. GG., "Epidemiology of vomiting in early pregnancy," *Obs. Gynecol*, vol. 66, no. 5, pp. 612–6, 1985.

[12] M. Weigel and R. Weigel, "The association of reproductive history, demographic factors, and alcohol and tobacco consumption with the risk of developing failure and communication in early pregnancy," *Am. J. Epidemiol.*, vol. 127, no. 3, pp. 562–570, 1988.

[13] T. Kugahara and K. Ohashi, "Characteristics of nausea and vomiting in pregnant Japanese women," *Nurs. Heal. Sci.*, vol. 8, no. 3, pp. 179–184, 2006.

[14] I. Matok *et al.*, "Characteristics of women with nausea and vomiting of pregnancy who chose to continue compassionate use of placebo after a randomised trial," *J. Obstet. dna Gynaecol.*, vol. 33, no. August, pp. 557–560, 2013.

[15] Suwarni, "HUBUNGAN KARAKTERISTIK MORNING SICKNESS DI POLIKLINIK KEBIDANAN DAN PENYAKIT KANDUNGAN BADANPELAYANAN KESEHATAN RSU DR ZAINOEL ABIDIN BANDA ACEH TAHUN 2007," pp. 50–59, 2010.

[16] R. Lacroix, E. Eason, and R. Melzack, "Nausea and vomiting during pregnancy: A prospective study of its frequency, intensity, and patterns of change," *Am. J. Obstet. Gynecol.*, vol. 182, no. 4, pp. 931–937, 2000.

[17] Ministry of Health of the Republic of Indonesia, *Health Profile of Indonesia 2015*. Ministry of Health of the Republic of Indonesia, 2015.

[18] K. Heitmann, H. Nordeng, G. C. Havnen, A. Solheimsnes, and L. Holst, "The burden of nausea and vomiting during pregnancy: severe impacts on quality of life, daily life functioning and willingness to become pregnant again - results from a cross-sectional study," *BMC Pregnancy Childbirth*, vol. 17, no. 1, p. 75, 2017.

[19] Central Bureau of Statistics of Central Java, "Human development index of Central Java Province." 2018.

[20] C. Piwko, G. Koren, V. Babashov, C. Vicente, and T. R. Einarson, "Economic burden of nausea and vomiting of pregnancy in the USA," *J. Popul. Ther. Clin. Pharmacol.*, vol. 20, no. 2, pp. 149–160, 2013.

[21] L. I. S. Trogstad, C. Stoltenberg, P. Magnus, R. Skjærven, and L. M. Irgens, "Recurrence risk in hyperemesis gravidarum," *BJOG An Int. J. Obstet. Gynaecol.*, vol. 112, no. 12, pp. 1641–1645, 2005.

[22] M. S. Fejzo, K. W. Macgibbon, R. Romero, T. M. Goodwin, and P. M. Mullin, "Recurrence Risk of Hyperemesis Gravidarum," *J. Midwifery Women's Heal.*, vol. 56, no. 2, pp. 132–136, 2011.

[23] M. Nurmi, P. Rautava, M. Gissler, T. Vahlberg, and P. Polo-Kantola, "Recurrence patterns of hyperemesis gravidarum.," *Am. J. Obstet. Gynecol.*, vol. 219, no. 5, p. 469.e1-469.e10, 2018.

Maternal Serum Matrix Metalloproteinase 14 (MMP14) in Early Onset Preeclampsia and Normal Pregnancy

H Sumawan[1*], Sutrisno[1]

[1]Medical Faculty, Jenderal Soedirman University, Purwokerto, Indonesia

Abstract. Matrix Metalloproteinases 14 presumably play for cytotrophoblast migration and invasion of the uterine wall and in the remodeling of the spiral arteries in pregnancy. Inadequate trophoblastic invasion leads to an inappropriate vascular remodeling, which generates conditions of hypoxia and increased oxidative stress in the placenta early onset preeclampsia. Therefore, it is particularly important to investigating whether MMP14 altered and can be used as biomarker of preeclampsia. There have been no studies done to measure MMP14 in serum maternal between early onset preeclampsia and normal pregnancy The purpose of this study was to analyse whether maternal serum matrix metalloproteinases 14 levels differ in early onset preeclampsia and uncomplicated pregnancies. This crosssectional study was carried out in 20 subjects with early onset preeclampsia and 20 subjects of normotensive pregnant women range 24 up to 34 weeks of gestation. The study was conducted in Margono Hospital Purwokerto, Indonesia. Level of MMP 14 was measured in maternal serum using an enzyme-linked immunosorbent assay (ELISA). The mean difference was statically analysed by independent samples T-test and ROC curve to determine sensitivity and specificity of MMP 14.Women age, gestational age, parity and body mass index showed a non significant difference between both groups. In this study level MMP 14 in serum was higher in pregnant women with preeclampsia compared to the normotensive (266.41 vs 46.80 pg/dl ; $p<0.00$). Moreover, the area under curve of serum MMP 14 was 0.936, standard error 0.043, $p<0.00$. The optimal cut-off value of serum MMP at 110.73 pg/dl showed a high diagnostic value in preeclampsia with a sensitivity of 90 % and a specificity of 90%. Maternal serum MMP 14 was higher in preeclampsia and the important finding is the MMP 14 probably become a marker to predict early onset preeclampsia.

*Corresponding author: h_sumawan@yahoo.com

1 Introduction

Preeclampsia is a major complication in pregnancy and a leading cause of death and also maternal and perinatal morbidity. Globally, 10-15% of maternal deaths are directly related to preeclampsia and eclampsia. The incidence of preeclampsia is different in each country, it ranges between 2-8% of all pregnancy (Duley, 2009). Maternal mortality as reported in East Java as a cause of death by preeclampsia reached 31% and in Central Java 34% (Dinas Kesehatan Jawa Tengah, 2016; Dinas Kesehatan Jawa Timur, 2016). Based on the Indonesian Demographic and Health Survey (IDHS) the tendency of maternal mortality increase from 228 per 100,000 live births (2007) to 359 per 100,000 live births in 2012. The Population Survey (SUPAS) reported maternal mortality rate that declined slightly to 305 per 100,000 live births in 2015 (Kementrian Kesehatan, 2017).

Preeclampsia is still described as a "disease of theories" because of the etiology and pathogenesis have not been fully proven. Poor placentation, with shallow remodelling of uteroplacental spiral arteries during the first half of pregnancy had known source of early onset preeclampsia. The ensuing uteroplacental malperfusion, placental oxidative and endoplasmic reticulum stress, then causes placental hypersecretion of pro-inflammatory and antiangiogenic factors into the maternal circulation (Redman C.W.G and Staff A.C., 2018). One of antiangiogenic factors is soluble endoglin (sEng) which has been identified as a central factor to induce endothelium dysfunction of preeclampsia, while its specific mechanism is unclear. Previous studies from laboratory research in trophoblastic BeWo cells found that MMP-14 is the likely cleavage protease of endoglin in the setting of preeclampsia (T. J. Kaitu'u-Lino et al., 2013; T. u. J. Kaitu'u-Lino et al., 2012) MMP-14 has been evidenced in syncitiotrophoblasts, and are known to be downregulated by endothelin-1 during first trimester inhibiting trophoblastic migration and invasion (Majali-Martinez A., 2017).

To our knowledge, there is no data on the role of MMP 14 as biomarker for early onset preeclampsia in serum maternal. We aimed in our study to compare MMP 14 levels in women with early onset preeclampsia to corresponding levels in a healthy pregnant control group.

2 Materials and Methods

This cross-sectional study was performed between July 2018 and September 2018 after ethical approval by Ethic Committee Medical Faculty Jenderal Soedirman University.

2.1 Patients.

The study was undertaken with 20 patients with early onset preeclampsia and 20 normal pregnant women with gestational age between 24 – 34 weeks who visit or admitted in Margono Hospital Purwokerto. Written informed consent was obtained from all participants. Preeclampsia was defined as the development of hypertension (systolic blood pressure (BP) $\geq$ 140 mmHg and/or diastolic BP $\geq$ 90 mmHg on two occasions at least 4 hours apart) and proteinuria ($\geq$0.3 grams in a 24-hour specimen or urine protein/creatinine ratio (uPCR) $\geq$ 0.3) after the 20th week of gestation in a previously normotensive patient. Severe preeclampsia was defined as preeclampsia complicated by either a systolic BP ≥ 160 mm and/or a diastolic BP ≥ 110 mmHg (on 2 occasions at least 4 hours apart while the patient was on bed rest) and/or pulmonary edema and/or renal abnormality (progressive renal insufficiency; serum creatinine > 1.1 mg/dL) and/or cerebral/visual symptoms (persistent headaches, neurological symptoms, and visual disturbances) and/or hepatic abnormality (severe epigastric or right upper quadrant pain and/or liver transaminases at least twice the normal concentration) and/or platelet count < 100,000/microliter. According to the new preeclampsia criteria , patients with new onset hypertension (BP ≥140/90) without proteinuria were accepted to have severe preeclampsia if they had one of the above criteria.(Cunningham FG. et al., 2014) Exclusion criteria included premature rupture of membrane, twin or multiple pregnancies or any evidence of previous medical disease.

2.2 Laboratory Measurements

Venous blood samples from patients were obtained to measure serum maternal MMP 14. Participants were asked to collect a urine sample for estimating the level of proteinuria. Blood samples for MMP 14 measurements were collected in serum gel separator tubes and centrifuged for 15 minutes at a speed of 1000 g. Supernatant plasma was removed into polypropylene plastic tubes. Samples were taken and stored at −80oC until the time of analysis. The concentration of MMP14 in serum was calculated with an enzyme linked immunosorbent assay (ELISA) kit, according to the protocols provided by the manufacturers (Human MMP 14, EIAab Science, Wuhan, China). A Microplate photometer was used for readings at 450 nm. The results were expressed as pg/mL. This test was conducted at Biomedical laboratory Medical Faculty, Jenderal Soedirman University.

2.3 *Statistical Analysis.*

All variables were expressed as mean ± SD. $p < 0.05$ was considered to be statistically significant. Comparisons between two groups were assessed with Student's unpaired t-test and Mann- Whitney test , as appropriate. ROC curve was used to assess predictors of MMP 14 levels.

3 Result and discussion

The subject characteristics are presented in Table 1. The age, parity, gestational age and body mass index, of the subjects were higher in women with early preeclampsia than in women with normal pregnancy. As shown in Table 1, the mean concentration of serum MMP 14 was higher in early onset preeclampsia than normal pregnancy (2.02±0.99 ng/mL vs. 3.24±2.67 ng/mL; p<0.05). A ROC analysis lead to AUC of 0.936 for MMP 14 standard error 0.043, p<0.00. The optimal cut-off value of serum MMP at 110.73 pg/dl showed a high diagnostic value in preeclampsia with a sensitivity of 90 % and a specificity of 90%.

Table 1. The subject characteristics of the study.

	Preeclampsia	Normal Pregnancy	P value
	Mean±SD	**Mean ± SD**	
Age	32,4 (8,0)	29,1 (6,0)	0,145
Gestasional Age	30,9 (2,8)	30,4 (3,0)	0,595
BMI	24,8 (5,2)	27,6 (4,7)	0,089
Parity	1,25 (1,06)	0,85 (0,87)	0,201
MMP 14	266,4(135,4)	46,8 (39,9)	0,000

*p values are obtained by independent t-test.

Large cohort of singleton deliveries in Washington State, USA, concluded that African-American descent, chronic hypertension and congenital anomalies were more strongly associated with very- early-onset pre-eclampsia (delivery before GW 34), while younger maternal age, nulli parity and diabetes mellitus were more associated with late-onset disease (Lisonkova & Joseph, 2013). That phenomena are the same as the results of the characteristics of the sample obtained. Early onset occurs at a young age and has a history of having children. The pathology of early onset comprises lesions of uteroplacental malperfusion. These are associated with maladaptation of the uteroplacental spiral arteries in early pregnancy (8–18 weeks, poor placentation) such that they are too small and too contractile to sustain the non-pulsatile, high volume, low pressure flow needed by the third trimester placenta. The result is oxidative stress and even infarction that damage placental tissue.(Redman

C.W.G and Staff A.C., 2018) . MMPs and their inhibitors play a major role in trophoblast invasion into the uterine wall. The profound changes in uterine microarchitecture required to transform the spiral vessels and create an optimum environment for embryonic development involve a grounding transformation in which MMPs are essential. This MMPs Probably contribute to the transformation of the 120–140 spiral arteries that are necessary to supply the placenta. (Pollheimer, Fock, & Knöfler, 2014).

Matrix metalloproteinases (MMPs) comprise a family of 23 zinc and calcium-dependent proteases that degrade different components of the extracellular matrix. These enzymes are classified into collagenases, stromelysins, matrilysins, membrane-anchored MMPs, and others according to their specific substrate. (Nagase H., 2006). Many previous studies focusing in MMP 2 and MMP 9 and showed altered in women with preeclampsia. It has been reported that MMP-2 plasma levels are elevated in preeclampsia and that this event is mediated by the vascular endothelial growth factor (VEGF), which controls vascular permeability. MMP-2 elevation can be detected from the second trimester onwards in the plasma of women who subsequently develop preeclampsia. MMP-9 increases along both, normal and preeclampsia-complicated pregnancy, while its inhibitor TIMP-1

increases in preeclamptic vs. normal pregnant women. A significant increase of MMP-9 concentrations in serum from women who subsequently developed preeclampsia can be detected during the first and third trimester of gestation. MMP-2/MMP-9 ratio provided better compromise between specificity and sensitivity in distinguishing PE from normal pregnancies, than either of the two MMPs alone (Feng et al., 2017; Palei et al., 2012) .

MMP 14, also known as MT1 MMP, is localized in invadopodia, a structure at the cellular edge of ECM degradation and invasion progression. This MMP expression in cytotrophoblasts is maintained through the first trimester of pregnancy and suggests that, together with its role as pro- MMP2 activator, MMP 14 may be involved in trophoblast invasion. In the present study, we established the normal reference value of the serum concentration of MMP 14 from the second half of pregnancy. We also made two important observations. First, we found that there were women with persistently low MMP 14 levels during the second half of pregnancy would not develop preeclampsia. All sample from normal pregnancy that we measure MMP 14 didn't show both early onset and late onset preeclampsia events. Second, high level of MMP 14, women with early-onset showed almost 5.7 fold than normal pregnancy. AUC value showed 0,93 indicate that this marker probably can be used to predict early onset preeclampsia, which sensitivity and specifity 90 percent. This result suggest that MMP 14 is potential marker to identify or predict the preeclampsia beside MMP 2 and MMP 9.

The previous study demonstrated that MMP-14 is a cleavage protease of endoglin. Alteration endoglin release is associated with the onset of preeclampsia, while the regulatory mechanism is unknown.(Baumann et al., 2008; Hirashima et al., 2008). Endoglin is co-localized with MMP-14 in syncytiotrophoblast cells and is significantly upregulated in placental tissue of patients with preeclampsia (T. J. Kaitu'u-Lino et al., 2013; T. u. J. Kaitu'u-Lino et al., 2012; Zhang XH, 2018). They proposed that MMP-14 cleavage on Endoglin may occur in the maternal-fetal cross-section of the placenta and, then, release its extracellular domain into the maternal circulation to form sEng.(Zhang XH, 2018) This result also showed that MMP 14 not only has high expression in placenta but also release in maternal circulation like soluble endoglin.

Beside some potential result from this research, limited data and sample become shortcoming of our study, and other group like late onset preeclampsia probably can give another comprehensive information related with prediction of preeclampsia.

4 Conclusion

Our study demonstrates that the MMP 14 concentration was higher in pregnant women with early onset preeclampsia compared to normal pregnant women. The MMP 14 may play a role as a marker in preeclampsia. There is a need for further studies to clarify the potential role of MMP 14 in early onset and or late onset preeclampsia which has different pathogenesis of preeclampsia and to determine whether it can be regarded as a predictor of the development of preeclampsia.

Conflict of Interests

The authors report no conflict of interests. The authors alone are responsible for the content and writing of the paper.

Reference

[1] Baumann, M. U., Bersinger, N. A., Mohaupt, M. G., Raio, L., Gerber, S., & Surbek, D. V. (2008). First-trimester serum levels of soluble endoglin and

soluble fms-like tyrosine kinase-1 as first-trimester markers for late-onset preeclampsia. *American Journal of Obstetrics & Gynecology, 199*(3), 266.e261-266.e266. doi:10.1016/j.ajog.2008.06.069

[2] Cunningham FG., Leveno KJ., Bloom S., Hauth JC., Rouse D., & Spong C. (2014). Pregnancy Hypertension *Williams Obstetrics* (Vol. 24th ed, pp. 728). United states of America: Mc Graw Hill Medical Companies.

[3] Dinas Kesehatan Jawa Tengah. (2016). *Profil Kesehatan Jawa Tengah, 2015*. Retrieved from http://www.depkes.go.id/resources/download/profil/PROFIL_KES_PROVINSI_2016/13_Jate ng_2016.pdf.

[4] Dinas Kesehatan Jawa Timur. (2016). *Profil Kesehatan Jawa Timur, 2015*. Retrieved from http://www.depkes.go.id/resources/download/profil/PROFIL_KES_PROVINSI_2015/15_Jati m_2015.pdf.

[5] Duley, L. (2009). The Global Impact of Pre-eclampsia and Eclampsia. *Seminars in Perinatology, 33*(3), 130-137. doi:https://doi.org/10.1053/j.semperi.2009.02.010

[6] Feng, H., Wang, L., Zhang, M., Zhang, Z., Guo, W., & Wang, X. (2017). Ratio of matrix metalloproteinase-2 to -9 is a more accurate predictive biomarker in women with suspected pre-eclampsia. *Bioscience reports, 37*(2), BSR20160508. doi:10.1042/BSR20160508

[7] Hirashima, C., Ohkuchi, A., Matsubara, S., Suzuki, H., Takahashi, K., Usui, R., & Suzuki, M. (2008). Alteration of Serum Soluble Endoglin Levels after the Onset of Preeclampsia Is More Pronounced in Women with Early-Onset. *Hypertension Research, 31*, 1541. doi:10.1291/hypres.31.1541

[8] Kaitu'u-Lino, T. J., Tuohey, L., Ye, L., Palmer, K., Skubisz, M., & Tong, S. (2013). MT- MMPs in pre-eclamptic placenta: Relationship to soluble endoglin production. *Placenta, 34*(2), 168-173. doi:10.1016/j.placenta.2012.11.034

[9] Kaitu'u-Lino, T. u. J., Palmer, K. R., Whitehead, C. L., Williams, E., Lappas, M., & Tong, S. (2012). MMP-14 Is Expressed in Preeclamptic Placentas and Mediates Release of Soluble Endoglin. *The American Journal of Pathology, 180*(3), 888-894.
doi:10.1016/j.ajpath.2011.11.014

[10] Kementrian Kesehatan. (2017). *Profil Kesehatan Indonesia 2016,*. Jakarta Retrieved from http://www.depkes.go.id/resources/download/pusdatin/profil-kesehatan-indonesia/Profil- Kesehatan-Indonesia-2016.pdf.

[11] Lisonkova, S., & Joseph, K. S. (2013). Incidence of preeclampsia: risk factors and outcomes associated with early- versus late-onset disease. *American Journal of Obstetrics & Gynecology, 209*(6), 544.e541-544.e512. doi:10.1016/j.ajog.2013.08.019

[12] Majali-Martinez A., V. P., Pollheimer J., Knofler M., Yung H.W., Burton G.J., Tabrizi-Wizsy N.G., Lang U., Hiden U., Desoye G., Dieber-Rotheneder M. . (2017). Endothelin-1 down- regulates matrix metalloproteinase 14 and 15 expression in human first trimester trophoblasts via endothelin receptor type B. *Hum. Reprod, 32*, 46-54. doi:10.1093/humrep/dew295

[13] Nagase H., V. R., Murphy G. . (2006). Structure and function of matrix metalloproteinases and TIMPs. *Cardiovasc. Res., 69*, 562-573. doi:10.1016/j.cardiores.2005.12.002

[14] Palei, A. C. T., Sandrim, V. C., Amaral, L. M., Machado, J. S. R., Cavalli, R. C.,

Duarte, G., & Tanus-Santos, J. E. (2012). Association between matrix metalloproteinase (MMP)-2 polymorphisms and MMP-2 levels in hypertensive disorders of pregnancy. *Experimental and Molecular Pathology, 92*(2), 217-221. doi:https://doi.org/10.1016/j.yexmp.2012.01.008

[15] Pollheimer, J., Fock, V., & Knöfler, M. (2014). Review: The ADAM metalloproteinases – Novel regulators of trophoblast invasion? *Placenta, 35*, S57-S63. doi:10.1016/j.placenta.2013.10.012

[16] Redman C.W.G and Staff A.C. (2018). The Differences Between Early- and Late-Onset Pre- eclampsia. In S. Saito (Ed.), *Preeclampsia Basic, Genomic, and Clinical* (pp. 157-112): Springer, Singapore.

[17] Zhang XH, Z. H., Lu S, Jiang LL, Wu J, Yang YL, Zhang SA. (2018). MMP-14 aggravates onset of severe preeclampsia by mediating soluble endoglin release. *Eur Rev Med Pharmacol Sci, 22*(5), 1209-1215. doi:10.26355/eurrev_201803_14460

Commercialization of Work Relation Between Land Owner and Landless Peasant in Central Java

Imam Santosa [1*], *Muslihudin* Muslihudin [2], *Wiwiek* R. Adawiyah[3] and *Dinda* Dewi Aisyah [4]

[1] Faculty of Social and Political Sciences, Jenderal Soedirman University, Purwokerto, Indonesia
[2] Faculty of Social and Political Sciences, Jenderal Soedirman University, Purwokerto, Indonesia
[3] Department of Management, Faculty of Economic and Business, Jenderal Soedirman University, Purwokerto, Indonesia
[4] Department of Agricultural Economic, Faculty of Agriculture, Gajahmada University, Yogyakarta, Indonesia

Abstract. This research aims to explore the shifting of work relations between land owners and landless peasants in Central Java. Besides, it also proposed to find the trend of the impact of shifting work relations between them. The research is designed using the semi-grounded method and phenology based on qualitative approach. This research is intentionally conducted in Purbalingga and Banyumas Regency, Central Java Province. Based on the research results, farmers in rural areas that are relatively far from the city have working relationships that tend to be exploited and are asymmetrical. On the other hand, the relationship of farmers in rural areas near cities tends to be more rational and commercial and symmetrical. The suggestion that can be given is the working relationship between the land owner and the cultivator who is beneficial to both parties needs to be maintained, but for the exploitative nature of work relationship there needs to be continuously empowered so that it does not bring damage to each of the party

1 Introduction

Globalization in every aspect of life has led to the uprooting of local wisdom values and the widening of social inequality and ecological degradation [1]. One variant of economic globalization is commercialization. Any goods or services tend to be profit oriented. This phenomenon is becoming increasingly prevalent not only in urban areas, but also in the hinterland area of the countryside. As a result, the patron client relationship has suffered a setback. Initially, a patron with all of his political power in economy able to received relatively cheap labor services, but he had an obligation to protect his clients when facing various life crises, such as: famine (crisis period), paying school fees for children, death of a husband / wife and or children and so on, but such values are now increasingly slackened by the rationality of the owners of farmers who are increasingly become rational. The logic

* Corresponding author: scokronegoro@yahoo.com

that [2] said about farmers' subsistence is still relevant to explain that they will put safety first. This applies not only to farmers who own land, especially poor farmers (land tenants) and so on, but such values are now increasingly loosened because of the rationality of the farmers and the increasingly rational mindset.

In the sociological dimension, [3] at the beginning of the emergence of peasant social relations, the right theory of reciprocity is characterized by reciprocal symmetrical social exchange between interacting parties. According to [3] social exchanges between individuals or groups can be formed due to the motive of interdependence. In this context, exchange becomes balanced and equitable. Each party places itself in the exchange of goods and services. In ideal reciprocity a balanced and just social exchange emerges. [4] thinking about reciprocity is different from [3] , because social exchange takes place unevenly when it enters power. The reciprocity of the farming community varies in giving rewards such as plant seeds, agricultural tools and farming advice [5]. The results of this study have not explained the balance of reciprocity. In certain farming communities, reciprocity is based on a mix of logical lead contexts from both culture and markets used in agricultural and environmental business governance [6]. Farmers' behavior in managing agricultural business is inseparable from a network of collaboration that utilizes reciprocity or reciprocity [7].

Vulnerable reciprocal relations used by various parties can put the damage on farmers. The issue of unbalanced reciprocity is vulnerable to farm laborers. These problems are currently can be found in farming communities in rural Indonesia. The existence of reciprocity proposed by [3] and [4] is irrelevant to the conditions of rural farmers due to the existence of polarization which is difficult to control. The burden of farm laborers is heavy with the maro and mertelu systems. All of the production activities burdened as a responsibility of farm laborers. Landowners only have to bear with the economic burden for some production facilities in the form of land. The economic burden of farm workers is getting heavier when crop failure happened to their farm. There is no guarantee of substitute insurance for a certain amount of income. The pattern of reciprocity relations which put more of the burden of the responsibility on the farm worker make them trapped in poverty that similar to a cycle that seems unbroken. These problems motivate the theme of this research. The purpose of this study is to explore changes in work relations between the farmer owners and farm workers. In addition, it also reflects the trends and changes in relations in the present era. The author suspects that the relationship between farmer groups will be increasingly rational and increasingly commercial. As a result, the roots of farmer community relations are grow uprooted.

2 Research Method

The research location is purposively determined at in the rural-urban and in the hinterland area both Purbalingga and Banyumas Regency, Central Java Province. The research location is selected based on consideration that the marginal area with higher population density and in hinterland area are less population density. Both of them have located in the land and cultivation as the main earning for living patterns as farmers. Based on the research's results, [8] figure out that the majority (> 30 percent) of farmers are classified into landowner and landless peasant with the status as farming labors. The types of data collected consist of primary and secondary data. The data are collected through in-depth interview, observation, and secondary data analysis. The research population twoe group around the research location. The respondent data sources are taken with a purposive technique fulfilling the criteria as landowner and lessland farmers who first implement in agricultural world. The qualitative data are analyzed by utilizing simple classification, categorization and analysis for getting trend of work relation.

3 Result and Discussion

There have been many references [9], that the main drivers of social change include changes in population and technological change. Changes in population cause technological changes and conversely technological change causes changes in population. There is a reciprocal relationship between the two. For Southeast Asian countries, including Indonesia, there is a population change trend which stimulates an increasing number of people who have needs that must be fulfilled. The limited land area stimulates spatial experts to improve technology. In the case of farmers in the rural areas of Banyumas and Purbalingga, it seems likely that they face the problem of land scarcity [10]. Changes in population affect the use of increasingly intensive agricultural technology. One of them is shown by the change in land plowing tools that used rakes and plows, but now use hand tractors [11]. Another indication that also shows changes in production technology is the emergence of various types of fertilizers, both organic and non-organic fertilizers. Likewise, the emergence of various pest control drugs that are increasingly varied. Intensive technology requires expensive costs that burden farmers. Changes in farming techniques were also followed by changes in work relations.

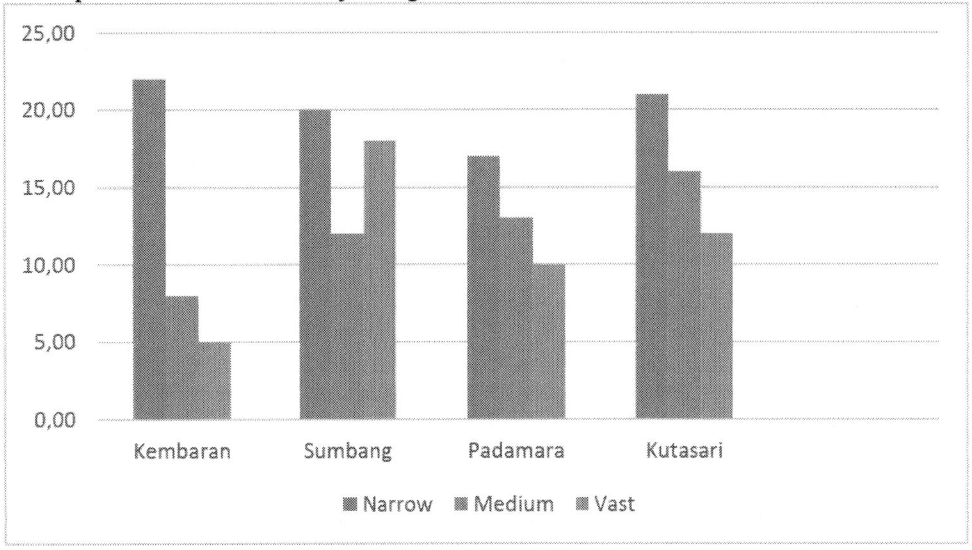

Fig 1. Average Land Ownership of Farmers in Four Sub-districts

The narrowing of agricultural land experienced by farmers in the villages of Kembaran District and Sumbang Subdistrict, both of which are in the Banyumas Regency area also occurred in the villages of Padamara Subdistrict and Kutasari District, Purbalingga Regency. Informants in four sub-districts had similar problems in the ownership status of agricultural land. The most severe pressure from land scarcity is faced by farmers who live in rural areas that are relatively close to the city area. There is a tendency for more and more cities in the area, the more farmers who have increasingly narrow land. The problem of land scarcity for farmers can cause serious conflicts [12]. This tendency can suppress the limit of the average land ownership of farmers in the four sub-districts can be seen in Figure 1.

The bar chart shown in figure 1 provides information about the percentages of the average of agricultural land owned by the peasant in Kembaran, Sumbang, Padamara, and Kutasari. Kembaran is the closest region to the city. Kembaran has the narrowest agricultural land compare to Sumbang, Padamara and Kutarasari. On the other hand, the extensive land owned by the peasant has the smallest percentage compared to the other three region. Meanwhile, Sumbang is the third region which closest to the city, and the following chart shows that the peasants in Sumbang has the vastest agricultural land, amounts to 18 % and the narrow agricultural land has 20%. Padamara is the second closest region to the city rather than Sumbang and Kutasari. The rates indicates that the agricultural land owned by each of the peasant in Padamara have shorter gap of the breadth. Narrow land area is the last ranked compared to Kembaran, Sumbang, and Kutasari. Notwithstanding, Kutasari the most remote area to the city of Purbalingga. The chart shows that the rates of narrow agricultural land owned by the peasant in Kutasari reach almost similar with Kembaran percentage (around 21%). The medium agricultural land owned by peasant in Kutasari occupy the highest position compared to Kembaran, Sumbang, and Padamara which amount 16%. All in all, the distance between regions to cities does indeed indicate the influence of the location of the area on agricultural land owned by the farmers, but if it is seen that remote area have bigger percentage of vast agricultural compared to the one which close to the city, it can be denied because in fact, from the bar chart above, it shows that the vast agricultural land in Kutasari is lower than in Kembaran, the region which is closer to the city.

The unequal distribution of land in the four research locations is due to the effect of agricultural commercialization that is unable to keep up with the pace of industrial development, the price of agricultural products is volatile and does not favor the farmers. The market for agricultural products is not guaranteed, so as to avoid the risk of rot, farmers choose to sell their crops at a minimum price. This condition is certainly detrimental to farmers, making it difficult to get profits. This issue includes the source of the causes of farmers choosing to sell land to village elites and investors from outside the village. Amount of money received from agricultural crops can only be used by farmers for a certain period to fulfilled the needs of their family. Increasing transactions in agricultural land is indeed difficult to avoid. The effects caused tend to harm farmers because of uncertainty in prices and unclear contracts such as those that occur in Rural Bulgaria [13]. The seriousness of the problem of land ownership status also takes place in research informants, as shown in Figure 1.

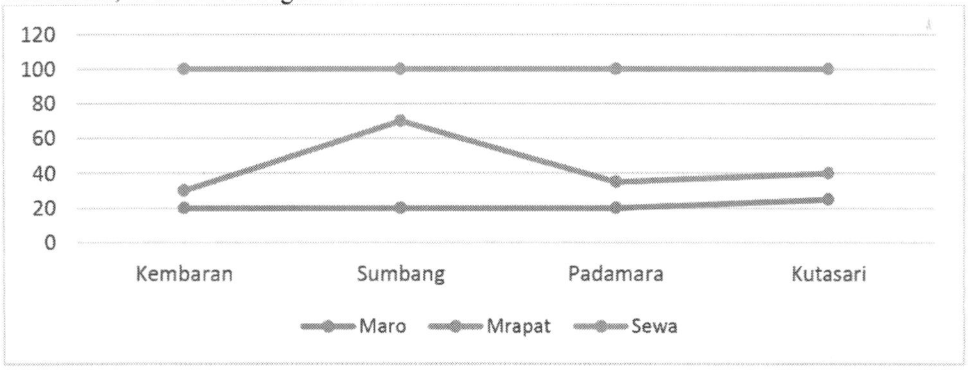

Fig 2. The Pattern of Work Relations Between Land Owner and Laborer

The shifting of technology and agricultural land ownership lead to changes in employment relations. The relationship between the owner of the farm (land owner) and the cultivator (landless) initially is often called a patron client relationship. In this context, historically, a patron initially protected its clients. Patron does expect rewards in the form of work given to his client. Of course, with the inclusion of technology and increasing necessities of life there have been many changes in agricultural employment relations in the four regions studied. This is reflected in the area that is not densely populated and its urban touch is still not strong. We can see it in Figure 2 below.

Figure 2 give the information about the pattern of work relationship between landlords and cultivators of the field. The system of rice fields cultivation divided into three types, maro, mrapat and sewa. Pic 2. above illustrates that in all relations trend in each regions, the relation of landlords and cultivators have to do with leasing land. As for the second choice, which means that the harvest result is divided into four with a notes that ¼ portion for cultivators and ¾ portion is for the owners, but all production costs are borne by the landowner. This pattern is most favored by the cultivators in Sumbang sub-district Banyumas Regency.

The maro pattern or the rice field system with the results are divided in two with the notes that the production costs borne by the cultivators are still found in Sumbang sub-district (Banyumas) and Kutasari sub-district (Purbalingga), but the numbers are very rare to be found. There is a tendency of the more a region leads into a city, the work relation will become more rational. Each of the party did not want to get loss. The peasants who does not have the land also become more rational. The group of clients are incresingly become more autonomous and independent. When there is more patront that can give more benefit, they do not get hesitate to move to another landowners. Some social relations between farmer owners and tenants have long been underway and are considered conventional networks [14]. Some work relationships found in the four research locations are the result of the effects of commercialization in the countryside [15]. The research team also explained the condition of the commercialization of working relations of farmers in West Java characterized by the use of land in rational ways. Characteristic of the rational way in the process of managing agricultural businesses, such as: the spread of social capital, land considered as the basis of the management of farmers, but the size of land is not the only requirement for the success of agricultural management. Limited land actually encourages progressive farmers to consolidate land through family relations and close neighbors. An increasingly commercial and rational relationship pattern is shown by the change in the bawon system to the daily and bulk wage systems [16]. For each region the level of commercialization is different [15]. Of the four villages per sub-district studied the commercialization system has been evenly distributed and almost the same, but specifically for Sumbang Subdistrict they have found the most suitable form and not detrimental to the cultivator, namely the mrapat system (the results are divided into four) where the production input are burdened to the land owner. It shows that the bargaining position of cultivator in rural areas of Sumbang District is better, because it has autonomy in management with smaller risk. The motives underlying social relations in four locations are different [16]. It appears that commercial motives are more prominent. In addition, there are motives that foster social harmony. There are other motives related to the role as a sharecropper because of compulsion and because they have not found alternative employment.

4 Conclusion

Work relations between farmer groups in suburban villages tend to be more rational, commercial and symmetrical (mutual benefit), while social relations in rural hinterland areas tend to be more asymmetrical and more profitable for large-scale farmers. Suggestions can be given, among others, that the working relationship between the owner farmer and the cultivating farmer that is beneficial to both parties still needs to be maintained, but for the exploitative nature of the work relationship there needs to be continuous assistance and empowerment.

4.1 Acknowledgements

Authors are very grateful to Directorate of Research and Social Service, Ministry of Research and Technology and Higher Education for necessary fund provided. Without this, The research can't be implemented. Appreciation goes to peasant's community who help us for collecting and completing data honestly.

References

[1] E. Goldsmith and J. Mander, *The Case Against the Global Economy: And for a Turn toward the Local* (Sierra Club Books, 1996)
[2] J. C. Scott, *Moral Ekonomi Petani. Pergolakan Dan Subsistensi Petani Di Asia Tenggara* (LP3ES, Jakarta, 1989)
[3] G. C. Homans, *Social Behavior: Its Elementary Forms* (Harcourt Brace, New York, 1974)
[4] P. Blau, Int. J. Bus. Soc. Sci. (1964)
[5] R. Jana, S. Bandyopadhyay, and A. K. Choudhuri, J. Hum. Ecol. **41**, 45 (2013)
[6] F. Bétrisey and G. Mouline, J. Mt. Res. Dev. **34**, 369 (2014)
[7] D. Baldassarri, Am. J. Sociol. **121**, 355-395 (2017)
[8] I. Santosa and A. Suyanto, J. Arts Humanit. **7**, (2018)
[9] E. Boserup, *Population Push and Technological Change* (The University of Chicago Press, Chicago, 1981)
[10] B. D. Dumasari, W.Darmawan, D. D. Aisyah, I. Santosa, J. Art Humanit. (2019)
[11] A. Gaitán-Barrera and G. K. Azeez, *Autonomy in the Caribbean Coast: Neoliberalim, Landless Peasant and Ethnic Conflict.* (Lexington Books, 2018)
[12] M. Georgiev, Agrar. Nauk. **3**, 53 (2011)
[13] I. Santosa, *Dampak Perubahan Organisasi Ekonomi Petani Terhadap Perubahan Kelembagaan Petani* (Sekolah Pacasarjana IPB, Bogor, 1994)
[14] H. Tarigan, R.N. Suheti, R.S. Rivai, S. Suhartini, V. Darwis, *Strengthening Smallholders Institution Toward Commercial Farmers* (Indonesian Center for Agricultural Socio Economic and Policy Stuides, 2018)
[15] S. H. Susilowati, in *Forum Penelit. Agro Ekon.* 48 (2005)
[16] W. D. Dumasari, A. Iqbal, B. Dharmawan, I. Santosa, Int. J. Adv. Sci. Eng. Inf. Technol.**9**,717 (2019)

Religiosity and Purchase Intention of Purwokerto Halal Mart

Pramono Hari Adi[1*], *Feronica* Via Christiany[2]

[1]Lecturer of Faculty of Economics and Bussines, Universitas Jenderal soedirman
[2]Student of Faculty of Economics and Bussines, Universitas Jenderal soedirman

Abstract. The research on which entitle Religiosity and Purchase Intension of Purwokerto Halal Mart was aim to analyze the relationship between religiosity and trust in the purchase intention of halal mart retail in Purwokerto. This research is a survey of Taklim Assembly members of the Great Mosque of Baitussalam, Jenderal Soedirman Mosque, Wijayakusuma Mosque, Darussalam Mosque, Bhayangkara Mosque and Fatimatuzahra Mosque. The number of sample were 225 with simple random sampling technique. The statistical test performed is using SEM data analysis. The results show that several dimensions of religiosity, namely experiential, ritualistic, intellectual have positive influence on trust. While the ideological and consequential dimensions do not have a positive effect on trust. For the influence of the dimensions of religiosity on purchase intention namely the experiential, ritualistic, ideological and consequential dimensions have a positive effect on purchase intention, but the intellectual dimension does not have a positive effect on purchase intention. And trust also has a positive effect on purchase intention. This result could inform halal mart and consumers that religiosity empirically influences consumers' trust and purchase intentions, and also consumers' trust influence purchase intention in the halal mart.

1 Background

Many consumers are starting to take notice of products labeled as halal today. As Muslims majority, not only food products that got attention but also all products that are used in fulfilling everyday needs, such as cosmetics, bath soap, laundry soap, toothpaste, medicine and many others. This phenomenon is interesting among Muslim businessmen to develop Halal Mart which has an advantage of all of the products are guaranteed halal. These halal-guaranteed products are expected to be in demand by Muslim consumers. As the servants of Allah, Muslim have an obligation to use halal goods, they also believe that halal goods have positive values and trusted safety to be used. Although it does not rule out the possibility of non-Muslim consumers who are also keen on using them, because everything which is halal is good for health.

However, the development of Halal Mart is still less known compared to other retailers that sell through supermarkets or hypermarkets, and the fact that many Muslims do not

* Corresponding author: hariadipramono@yahoo.com

know the existence of Halal Mart's retail and its products. Therefore, this study needs to be conducted in order to determine what are the factors that can affect people's purchase intention on the products of Halal Mart.

Purchase intention is essentially derived from a process of learning and thought processes, and or an experience that forms a perception. This purchase intention creates a motivation which continues to be recorded in the minds of consumers and becomes a very strong desire that in the end when a consumer must meet his needs then he will actualize what is in his mind (Deighton, 1994). To understand and attract customers, the company must have a strong belief in the minds of consumers. Trust is a foundation in a business process. A transaction between two or more parties will occur if both sides trust each other. Belief in a business cannot come instantly but must be built from the beginning of the business' establishment (Hendrata et al, 2013).

Public interest in the retail of halal mart should be great, that one of the indicators is the amount of purchase intention of the society that can be seen from the growth of sales in the halal retail which increased from the previous year, but the fact stated otherwise. The data showed the growth of halal retail in 2017 decreased from the previous year. Even though the majority of the population in Indonesia, including in Banyumas regency, are Muslims, amounting to 1.76095 million of a total population of 1,801,477 (Data of BPS of Central Java in August 2016). Logically, the purchase intention of society in retail that sells halal products is also big, especially since Muslims are instructed to consume halal products. Therefore, religiosity factor can influence the purchase intention in halal retail which all of its products are guaranteed halal like halal mart. Therefore, the factor of religiosity is the next factor that can affect people's buying interest in the retail of halal mart. The research of Adi and Adawiyah (2018), Glock (1972) also Ateeq-ur-Rehman and M. Shahbaz Shabir (2010) suggest that religiosity is divided into six dimensions, ideological, ritualistic, intellectual, consequential, experiential. However, the researchers will only discuss religiosity through three dimensions wglockhich are considered the most important and essential, that are ideological, ritualistic, intellectual.

The gap phenomenon above is the reason for the researchers to conduct research on the purchase intention of Muslim society on halal mart retail. This study uses the research of Arshia Mukhtar and Muhammad Mohsin Butt (2012) as its main journal, which conducted research on the influence of subjective norms, attitudes, and religiosity toward purchase intention. The results show that subjective norms, attitudes, and religiosity strongly influence the purchase intention of halal products. On the other hand, a research done by Kusumawardhani et al (2016) who studied the effect of religiosity, product quality, islamic branding and value consciousness toward purchase intention. The results showed that only value consciousness that significantly influences purchase intention, while religiosity, product quality, branding Islam do not influence the purchase intention of halal products. Further research conducted by Hashim (2017) underlying the use of trust variable in this study, in which the result is that trust can also mediate the effect of the product benefits and corporate image toward purchase intention. These results indicate that variable trust can be used as a mediator.

Therefore, the researchers are interested in conducting research on the effect of religiosity toward purchase intention in Halal mart which is mediated by trust. The research novelty is the purchase intention of consumers who use Multi-Level Marketing (MLM).

2 Formulation of the research problem

1. Does religiosity influence the consumers' trust in Halal Mart?
   a. Does ideological dimension, ritualistic dimension and intellectual dimension influence trust?

2. Does religiosity influence the purchase intention in Halal Mart?
    a. Does ideological dimension, ritualistic dimension and intellectual dimension influence on purchase intention?
3. Does trust influence on the purchase intention in Halal Mart?
4. Does trust mediate the effect of religiosity on purchase intention in Halal Mart?

3 Objective

The objective of this study is to investigate and analyze:

1. The effect of religiosity on consumers' trust on Halal Mart.
    a. The effect of ideological dimension, ritualistic dimension and intellectual dimension on trust.
2. The effect of religiosity on buying intention in Halal Mart.
    a. The effect of ideological dimension, ritualistic dimension and intellectual dimension on purchase intention.
3. The effect of trust on purchase intention in Halal Mart.
4. The mediation of trust in the effect of religiosity on purchase intention in Halal Mart.

4 Research model

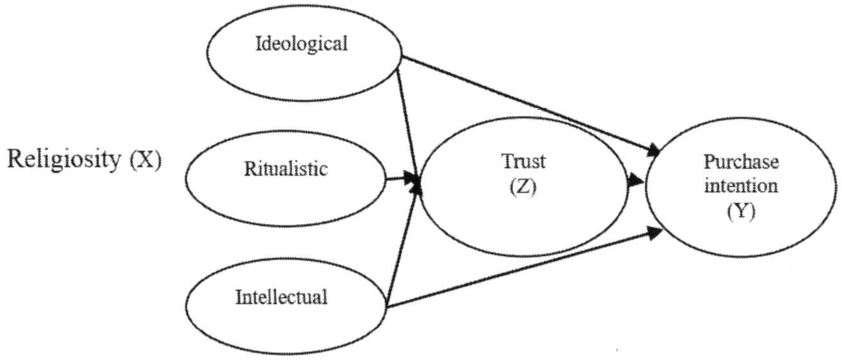

Fig. 1. Research Model

5 Research hypothesis

H1: Religiosity influences consumers' trust.
H1.a. (H1): ideological dimension influences trust.
H1.b. (H2): ritualistic dimension influences trust.
H1.c. (H3): intellectual dimension influences trust.
H2: Religiosity influences the purchase intention in the halal mart.
H2.a. (H4): ideological dimension influences purchase intention.
H2.b. (H5): ritualistic dimension influences purchase intention.
H2.c. (H6): intellectual dimension influences purchase intention.
H7: Trust influences the purchase intention in halal mart.
H8: The trust in mediating the effect of religiosity on the purchase intention in the halal mart.

6 Research method

6.1 Objects of research

This research was conducted in the area of Purwokerto city, precisely in Baitussalam Great Mosque, General Sudirman Mosque, Wijayakusuma Mosque, Bhayangkara Mosque, Darussalam Mosque, and Fatimatuzzahra Purwokerto Mosque.

6.2 Population and research sample

The population in this study are all talim assembly members of the Baitussalam Great Mosque, General Sudirman Mosque, Wijayakusuma Mosque, Bhayangkara Mosque, Darussalam Mosque, and Fatimatuzzahra Purwokerto Mosque.

According to Hair, et al. (2010) who suggested that by using Equational Structural Model (SEM), the minimum sample size is 100. It is suggested that the sample size is 5-10 times the parameter.

Sampling in mosque using cluster random sampling method. The cluster used was mosques in the city of Purwokerto. Based on this method, the mosques which were selected as the samples are Baitussalam Great Mosque, General Sudirman Mosque, Wijayakusuma Mosque, Bhayangkara Mosque, Darussalam Mosque, and Fatimatuzzahra Purwokerto Mosque. The next step in selecting the sample of talim assembly member using a simple random sampling method.

6.3 Data types and sources

a. Primary Data
In this study, primary data were directly obtained from respondents through the questionnaire on religiosity, trust, and purchase intention.
b. Secondary Data
Secondary data of this study were obtained by doing reviews of the works of literature; Fundamentals of Marketing, Marketing Management, Journal of Islamic Marketing, Journal of Marketing, Scientific Journal of Economics and Management Student, and other journals besides management journals as supporter.

6.4 Conceptual and operational definitions

6.4.1 Variable of purchase interest (Y)

a. Conceptual Definition: Consumer behavior in response to an object that shows a desire to make a purchase (Kotler and Keller, 2012).
b. Operational Definition: The desire of Muslim consumers to buy in the halal mart.
c. Indicators (Kotler and Keller, 2012): The urge to buy in the halal mart; The notion of halal mart; The knowledge of halal mart products; The confidence in buying in the halal mart
d. Measurement: Consumers are encouraged to buy in the halal mart; Consumers understand about halal mart; Consumers know the products in the halal mart; Consumers are confident in buying in the halal mart

6.4.2 Variable of trust (Z)

a. Conceptual Definition: Consumers trust the reliability and integrity of a company (Kotler and Keller, 2012)
b. Operational Definition: Consumers trust in the halal mart
c. Indicators (Robbins and Judge, 2007): Integrity, correctness of the information and services; Competence, meeting consumer needs; Consistency, maintaining product value
d. Measurement: Truth of the product information; Purchase service is as expected; Meeting the needs of consumers for goods; Only selling halal and quality products

6.4.3 Variable of religiosity (X)

a. Conceptual Definition: Someone's degree in believing religious values and practice the values of his religion (Lung et al, 2008)
b. Operational Definition: Consumer's trust toward his religion including dimensions of (Glock, 1972):
c. Indicators
   1) Ideology Dimension: Believes Allah; The Prophet as a role model; heaven/hell; Believes Judgment day; Allah the Helper
   2) Ritual Dimension: Prayer; Reading basmalah; Reading Quran; Ramadan fast; Halal consumption; Istighfar
   3) Intellectual Dimensions: Remembers Allah; Justifying the teachings of Islam; Knows about sin; Obeys religion; Stay away from the forbidden
d. Measurements
   1) Ideology Dimensions: Believes Allah: Prophet Muhammad as a role model; Believes heaven and hell; Believes with the help of Allah
   2) Ritual Dimensions: Pray 5 times/day; Reading Basmalah before every move; Read Quran; Ramadan fasting; Eats and drinks halal food or drink; Istighfar if guilty
   3) Intellectual Dimension: Always remember Allah; Justifying the teachings of Islam; Knowledge of big and small sins; Obeys to the teachings of Islam in life; Stay away from the forbidden

6.5 Data analysis technique

This study uses two kinds of analysis techniques:

a. Confirmatory factor analysis in the SEM used to confirm the factors that are most dominant in a variable group.
b. Regression Weight in the SEM used to determine how much the studied variables influence each other

7 Analysis results

7.1 Classic assumption test

7.1.1 Normality test

In the SEM test, the data must be normal. Univariate and multivariate normality of the data used in this study were tested using AMOS. SEM analysis requires to meet the assumptions of normality, in which according to Ghozali (2008) normality of the data using the criteria of critical ratio (cr) skewness value equals to + 2.58 at the significance level of 0.01. The data can be concluded to have a normal distribution if the value of the critical ratio (cr) skewness value is below the absolute price of 2.58. In table 2 (appendix), it can be seen that the value of the critical ratio (cr) and skewness value of all the indicators are under the absolute price of 2.58. Based on these values, it can be concluded that the assumptions of normality of data in this study are met and can be used in the subsequent analysis phase.

7.1.2 Outliers evaluation

The detection of multivariate outliers was done by looking at the value of Mahalanobis Distance. The determination criterion is to compare the value of Mahalanobis Distance to the value of Chi-Square. At a significance level of p <0.01. Chi-Square value of this study is to look at the degree of freedom equals to 50.892. Based on the analysis using a computer program of AMOS 18 on the Text Output > Observations farthest from the centroid (Mahalanobis Distance), it can be concluded that this analysis did not find multivariate outliers because the value of Mahalanobis Distance is less than X2 (50.892).

7.1.3 Evaluation of multicollinearity and singularity

In this research, it is known that the highest value of Mahalanobis Distance is 69.330, then the result is 2,039. That result is below the limit of 2.5 then it can be concluded that there is no multicollinearity and singularity so that the data in this study are suitable for use.

7.1.4 Model suitability test

Table 1. The result of model suitability test (Goodness of Fit Test)

Goodness – of – fit index	**The result of research model**	*Cut of Value*	**Information**
Chi – Square	556,511	*< 559,438*	Good
Significant probability	0,060	*≥ 0,05*	Good
GFI ≥	0.873	*≥ 0,90*	Marginal
AGFI ≥	0.850	*≥ 0,90*	Marginal
CFI	0.994	*≥ 0,95*	Good
TLI	0.993	*≥ 0,95*	Good
CMIN/DF	1.100	*≤ 2,0*	Good
RMSEA	0.021	*≤ 0,08*	Good

One of the test equipment to measure the overall fit is the likelihood ratio chi-square statistic. Chi-Square is very sensitive to the size of the sample used. The model that will be tested would be considered good or satisfactory when the value of Chi-Square is low. The smaller the chi-square, the better the model is accepted based on the probability of the cut-off value of p> 0.05 (Ferdinand, 2011). In this study, the chi-square value obtained is equal to 436,511 and p = 0.060, thus it can be concluded that the model in this study includes into a very good fit since the calculated chi-square value is greater than the chi-square table and a greater probability value of 0.05, so that the hypothesized model fits the data.

7.1.5 Construct validity test

Table 2. The result of Construct Validity Test

No.	Variabel (*Construct*)	Reliability value (α)
1.	Ideological	0,994
2.	Ritualistic	0,995
3	Intellectual	0,992
4	Trust	0,995
5	Purchase Intention	0,993

The calculated result of construct reliability as presented in the table above, it appears that the seven variables or latent constructs used in this study have | coefficient> 0.70 or greater than the cut-off value (> 0,700), so we concluded that the latent constructs used in this research are really reliable.

7.1.6 Extract variance test

Table 3. The result of extract variance test

No.	Variabel (*Construct*)	The value of *Variance* (α)
1.	Intellectual	0,962
2.	Ritualistic	0,970
3.	Ideological	0,971
4.	Trust	0,980
5.	Purchase Intention	0,972

The calculated result of variance extracted is listed in Table 4 above which shows that the four latent constructs used in this study have | coefficient> 0.50 or greater than the cut-off value (> 0.500), so we concluded that the latent constructs used in this study are really reliable.

7.1.7 Convergent validity test

According to Ghozali (2008), the indicators of a latent construct must converge or share a high proportion of variants and this is what it is called as convergent validity. In the process of measuring construct validity, it can be seen from its loading factor. In cases where there is high construct validity, then a high loading value on a factor (latent constructs) indicates that they converge at one point. The first prerequisite is that the loading factor must be significant.

Therefore, a significant loading factor may still be of low value, then the value of standardized loading estimate must be> 0.50 and it should ideally reach a value of 0.70. The result of convergent validity test can be seen in the table below.

Table 4. The result of Convergent Validity test

Variable	Indicator	Loading Factor
Ideological	X1	0,876
	X2	0,840
	X3	0,841
	X4	0,863
	X5	0,848
Ritualistic	X6	0,817
	X7	0,838
	X8	0,883
	X9	0,873
	X10	0,890
	X11	0,815
Intellectual	X12	0,847
	X13	0,857
	X14	0,826
	X15	0,845
	X16	0,799
Trust	X27	0,874
	X28	0,878
	X29	0,886
	X30	0,862
Purchase Intention	X31	0,848
	X32	0,867
	X33	0,898
	X34	0,815

https://doi.org/10.1051/shsconf/20208601006

From Table 4 above, it is known that the loading factor value of all indicators are statistically significant because loading is greater than the cut-off > 0.50. Therefore, it can be concluded that all the indicators used are valid and can be used in the analysis at the next stage.

7.1.8 Discriminant validity test

The discriminant validity test used to measure a construct that is different from the other constructs. A way to test the discriminant validity is to compare the value of the square root of the average variance extracted (AVE) with the correlation between the composites (constructs). The result of discriminant validity test of each variable in this study is presented in the table below.

Table 5. The result of Discriminant Validity test

	Intellectual	**Ritualistic**	**Ideological**	**Trust**	**Purchase intention**
Intellectual	**0,981**				
Ritualistic	0,916	0,985			
Ideological	0,916	0,920	**0,985**		
Trust	0,936	0,937	0,908	**0,990**	
Purchase intention	0,950	0,938	0,941	0,958	**0,986**

Based on Table 5 above, it is known that the AVE value of Intellectual construct is 0.983, Ritualistic is 0.985, Ideological is 0.985, Trust is 0.990e and Purchase intention is 0.986. Based on these results, it can be concluded that the entire AVE values of constructs for each variable in this study are greater than the square of the correlation between the constructs which meet the requirements of good discriminant validity.

7.1.9 Full model causality test

The test criterion is to compare the value of P with a significance level used in this study which is 95% (0.05). If the value of P> 0.05, we conclude that there is no influencing relationship between the two variables.

Table 6. The result of Full Model Causality Test: Regression Test

			C.R.	t tabel	P
Trust	<---	Ideological	0,651	1,960	0,515
Trust	<---	Ritualistic	3,388	1,960	0,000
Trust	<---	Intellectual	1,999	1,960	0,046
Purchase intention	<---	Trust	2,233	1,960	0,026
Purchase intention	<---	Intellectual	0,213	1,960	0,832
Purchase intention	<---	Ritualistic	-0,034	1,960	0,973
Purchase intention	<---	Ideological	2,132	1,960	0,033
Trust	<---	Religiusitas	18,546	1,960	0,000
Purchase intention	<---	Religiusitas	4,505	1,960	0,000

The analysis result of mediating trust variable on the influence of religiosity which includes the dimension of Ideological, Ritualistic, Intellectual, toward purchase interest in the halal mart. Based on the results obtained by SEM analysis, results about the influence of religiosity which include the dimension of Ideological, Ritualistic, Intellectual, and trust both directly and indirectly (mediation) as well as the total effect toward the purchase intention of the standardized total, direct and indirect effects which are as follow:

Table 7. The result of Total, Direct and Indirect Effects test

					Koefisien Jalur			Information
					Total	Direct	Indirect	
Intellect	→	Trust			0,319	0,319		
Ritual	→	Trust			0,414	0,414		
Ideologi	→	Trust			0,073	0,073		
Religius	→	Trust			0,969	0,969		
Intellect	→	Trust	→	PI	0,130	0,032	0,097	Not mediate
Ritualist	→	Trust	→	PI	0,122	-0,004	0,126	Not mediate
Ideologi	→	Trust	→	PI	0,237	0,215	0,022	Mediate
Religius	→	Trust	→	PI	0,983	0,892	0,091	Mediate
Trust	→	Purchase intention			0,306	0,306		

8 Hypothesis discusion

1. First Hypothesis (H1)
   Ideological dimension has no effect on consumers' trust in the halal mart. In the future, the halal mart expected to be more concerned with the element of ideological dimension in order to build consumers' trust in the halal mart better. For example, by convincing consumers that the sold products and sales system in the halal mart is in accordance with Islamic law.
2. Second Hypothesis (H2)
   Other dimensions that also strengthen the effect of religiosity toward trust positively influence and significant is ritualistic dimension. Deeds like never missed praying five times, always read Basmalah in any activities, foods, and beverages are halal, foster a desire or push to have a view or heart inclined toward things that are in accordance with the rules of Islam, also have high trust in the halal mart.
3. Third Hypothesis (H3)
   Intellectual dimension included in the dimension that has a positive and significant impact on trust. Remembering Allah, believing the truth of His teaching, knowing the big or little sins, always trying to obey the teachings of Islam in all aspects of life and keep yourself from everything that is forbidden will heighten the consumers' trust on the mart more.
4. Fourth Hypothesis (H4)
   Ideological respondents believe in the existence of Allah, heaven and hell, the Day of Judgment, the Prophet Muhammad as a role model and Allah is the Helper which have led to a sense that life must be lived in accordance with Islamic law, as well as using or consuming halal products also have a high intention for buying in the halal mart.
5. Fifth Hypothesis (H5)
   Unlike the ritualistic dimension, this dimension does not miss a practice such as praying five times, reading Quran every day, full fasting during Ramadan, saying istighfar if making a mistake cannot influence the purchase intention in the halal mart. With this result, the halal mart will further strengthen the elements associated with the ritual dimension in the future in order to increase purchase intention of consumers. For example, by adding an easily accessible outlet so that consumers more easily and frequently come to the halal mart that will make consumers more interested and increased purchase intention.
6. Sixth Hypothesis (H6)
   Similar to the ritualistic dimension, intellectual dimension also does not influence purchase intention. By remembering Allah, believing the truth of the teachings of Islam, knowing the big and little sins, always trying to obey the teachings of Islam in every aspect of life are still not able to provide a positive effect on purchase intention. This is a concern for the party of the halal mart to evoke the intellectual dimension of consumers in order to build a positive relationship with the halal mart which will further increase purchase intention. It can be done by providing evidence that the purchase of products in the halal mart is the right step both in terms of health and faith.
7. Seventh Hypothesis (H7)
   From the research results, the description of consumers' trust in the halal mart is high. This suggests that consumers perceive the halal mart as a reliable seller to meet their needs for products with branded halal. This is in line with the views expressed by Kolsaker and Payne (2002) that trust is the dimension of a business relationship that

determines the degree to which each of the parties feels they can rely on the promise offered by the other party. Therefore, trust is a foundation in a business process.
It is proved empirically that the trust influence on the purchase interest of consumers in the halal mart. This can happen because the trust was awakened by the hope that the halal mart will provide services in accordance with the needs and desires of consumers. When someone has trusted others, they believe that the expectations will be met and there will be no disappointment (Ryan, 2002).

8. Eighth Hypothesis (H8)
   The values on the analysis results of all dimensions show that trust mediates the effect of religiosity on purchase intention. This means that consumers' trust strengthens the effect of religiosity on consumers' purchase intention in the halal mart.
   The analysis results of mediation of trust influence on each dimension of religiosity toward consumers' purchase intention which varies between dimensions. The analysis results showed that trust mediates the effect of the dimensions of experiential, ideological, and consequential toward the consumers' purchase intention in the halal mart. However, trust does not mediate ritualistic and intellectual dimension toward the the consumers' purchase intention in the halal mart.
   This shows that trust is not always able to strengthen a Muslim's religiosity to buy in the mart that clearly sell halal products. However, it can also occur due to the possibility of each individual which has a different subjective norm. The extent to which a person's religiosity affect his lives is dependent on the subjective norm in him.

9 Conclusions and implications

9.1 Conclusions

1. The dimension of ideological influences consumers' trust in the halal mart.
2. The dimension of ritualistic effect influences consumers' trust in the halal mart.
3. The dimension of intellectual influences consumers' confidence in the halal mart.
4. The dimension of ideological influence on the purchase intention of consumers in the halal mart.
5. The dimension of ritualistic does not influence the purchase intention of consumers in the halal mart.
6. The dimension of intellectual does not influence the purchase intention of consumers in the halal mart.
7. Trust influences the purchase intention of consumers in the halal mart.
8. Trust mediates the effect of religiosity on purchase intention.

9.2 Implications

1. Inform the halal mart that religiosity gives a positive effect to trust and the purchase intention of consumers in the halal mart and trust gives a positive effect on the purchase interest of consumers in the halal mart.
2. In addition, to give information to consumers that religiosity empirically influences consumers' trust and their purchase intentions in the halal mart, and their trust influence purchase intention in the halal mart.

References

[1] P. Hari Adi and W. R. Adawiyah, J. Islam. Mark. **9**, 841 (2018)

[2] A. Mukhtar and M. Mohsin Butt, J. Islam. Mark. **3**, 108 (2012)
[3] J. Deighton, C. M. Henderson, and S. A. Neslin, J. Mark. Res. **31**, 28 (1994)
[4] A. Ferdinand, *Metode Penelitian Manajemen Pedoman Penelitian Untuk Penulisan Skripsi, Tesis, Dan Disertasi Ilmu Manajemen. Edisi III* (AGF Books. Fakultas Ekonomika dan Bisnis Universitas Diponegoro, Semarang, 2011)
[5] I. Ghozali, *Structural Equation Modelling, Edisi II* (Universitas Diponegoro, Semarang, 2008)
[6] C. Y. Glock, in *J. E. Faulkner (Ed.). Relig. Influ. Contemp. Soc. Readings Sociol. Relig.* (C. E. Merril Publishing Company, Columbus, Ohio, 1972)
[7] Hair, J.F., Black, W.C., Babin, B.J., and R. E. Anderson, *Multivariate Data Analysis. Seventh Edition* (Prentice Hall, 2010)
[8] Hashim, Int. J. Econ. Commer. Manag. **5**, 329 (2017)
[9] Y. K. Hendrata, T. Purbandari, and Mujilan, J. Ris. Manaj. Dan Akunt. (2013)
[10] Kolsaker and C. Payne, Mark. Intell. Plan. **20**, 206 (2002)
[11] P. Kotler and G. Armstrong, *Principles Of Marketing Kotler 14th Edition Pearson* (2012)
[12] Kusumawardhini, S. Sarah, S. R. H. Hati, and S. Daryanti, in *3rd Int. Conf. Bus. Econ. 21 - 23 Sept.* (2016)
[13] C. Lung, L. Chai, and R. Zulkufly, J. Bus. Syst. Gov. Ethics **3**, 43 (2008)
[14] A. Rehman and M. Shahbaz Shabbir, J. Islam. Mark. **1**, 63 (2010)
[15] S. P. Robbins and T. A. Judge, *Organizational Behaviour* (Pearson Education, Inc., Upper River, New Jersey, 2007)
[16] C. Bicchieri, E. Xiao, and R. Muldoon, Polit. Philos. Econ. **10**, 170 (2011)

The Effects of DRRR Policy on Large versus S&M Sized Banks in China

Guo Xiaohui[1*] and *Kang* Shusheng[2]

[1] School of Economics, Hebei University, Baoding, 071002, China
[2] School of Economics, Hebei University, Baoding, 071002, China

Abstract. This paper employs cointegration test to measure the effects of Differentiated Reserve Requirement Ratio (DRRR) policy on large banks versus Small and Medium (S&M) sized banks from 2009M01 to 2013M10. The results confirm the hypothesis that this DRRR policy promotes the economy growth more in the earthquake-stricken counties and a lower DRRR to some extent alleviate the tightening effects of RRR on the S&M sized banks. Due to the big regional disparity, the PBC can consider expand the use of this DRRR policy to more less developed regions to make a more coordinated regional economic development in China.

1 Introduction

In recent years, along with the deepening of economic reform and opening up, to cope with the complicated domestic and international economic situation, the PBC (the People's Bank of China, central bank of China) has begun to make some new tries to the implementation of monetary policy instruments. The PBC has introduced Differentiated Reserve Requirement Ratio (DRRR) policy since April 25th, 2004. The main content of DRRR policy is that the required reserve ratio applicable to a financial institution will be linked with such indicators as its capital adequacy ratio and asset quality. The lower the capital adequacy ratio of a financial institution and the higher its NPL (Non-Performing Loan) ratio, a higher required reserve ratio will be charged. The purpose of this policy is to limit the loan expansion of those financial institutions with inadequate capital and poor asset quality, reduce financial risk and maintain financial stability.

DRRR policy said that financial institutions whose capital adequacy ratios were below 4% would apply the DRRR 0.5% higher than normal level. Since this policy applied, 38 joint-stock banks and city commercial banks such as Guangdong Development Bank, China Everbright Bank and Shenzhen Development Bank have been applied DRRR 0.5% higher than normal level as their loan growth is too fast and the risk is too high. In July, 2006, the Agricultural Bank of China also was forced to apply DRRR 0.5% higher than normal level.

In practice, the PBC also attempted to implement DRRR based on regions and the structure of credit to continuously extend the application of DRRR policy. Since September 21st, 2003, all financial institutions that took deposits from the public have had to meet a

* Corresponding author: nlmh6@126.com

uniform reserve requirement ratio of seven percent, with the exception of the Urban and Rural Credit Cooperatives that still abide by the six percent standard. To support the development of rural area, the PBC did not raise the reserve requirement ratio of Rural Credit Cooperative until 2006. On May 12th, 2008, a big earthquake hit Sichuan province and made lots of damage. To support the reconstruction of the disaster area, the PBC has applied DRRR a little lower than normal level to the disaster area in Sichuan province that suffered most in the earthquake from June 8th, 2008 to June, 2011.

Global financial crisis happened in 2008 made the domestic and overseas economic condition become much more complicated. Facing this situation, the PBC decided to apply DRRR to different sizes of financial institutions. On September 25th, 2008, the PBC applied DRRR to Large banks and Small & Medium sized banks (S&M sized banks). Large banks applied the DRRR 1% higher than S&M sized banks. Large banks occupied a bigger market shares, applying a higher DRRR was helpful for the PBC to control money and credit supply. Considering most of loans made by S&M sized banks flowed to S&M sized enterprises (China Banking Regulatory Commission Annual Report 2011), the PBC applied a little lower DRRR to S&M sized banks so as to solve the difficulty of financing in S&M sized enterprises. Now large and S&M sized banks apply Different Deposit Reserve Ratio separately (see Figure 1).

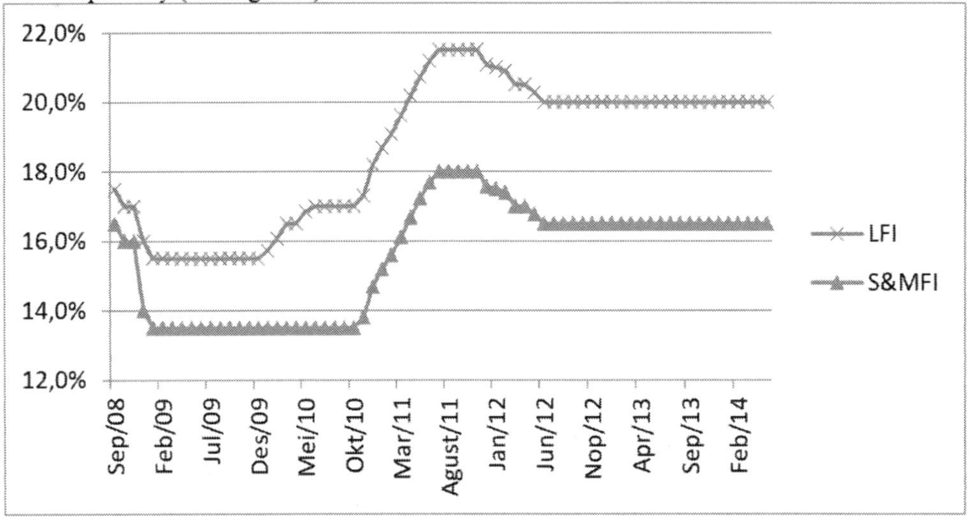

Fig.1. Differentiated Reserve Requirement Ratio in China from 2008M08 to 2014M05, Sources: The People's Bank of China, Notes: LFI: deposit reserve ratio of Large Financial institutions. S&MFI: deposit reserve ratio of Small and Medium sized Financial Institutions.

The initial purpose of DRRR policy is to limit the loan expansion of those financial institutions with inadequate capital and poor asset quality. In its subsequent development, the PBC tries to use this monetary policy instrument to promote the economy growth of certain region (such as the earthquake-stricken area) and adjust the credit structure (DRRR policy to Large and S&M sized bank). If the practice can prove that this policy plays an important role in these areas as expected, then the implementation of DRRR policy has a significant meaning to the PBC: first, the shortcoming of traditional monetary policy is that "one size fit all" practice doesn't consider the unbalance of regional economies. But to some extent this DRRR policy can solve this problem. Second, the PBC can use this policy to adjust the structure of credit. For example, they can apply a higher DRRR to large banks to curb inflation while at the same time implementing a lower DRRR to S&M sized banks to solve the difficulty of financing in S&M sized enterprises.

Therefore, in this study, we will use the earthquake-stricken area and large and S&M sized bank to do case studies in order to examine the effects of DRRR policy. The organization of this study is as follows: the next section discusses briefly literature review. Section three devotes on methodology and section four highlights the empirical results. Section five is conclusion and limitations.

2 Literature Review

The traditional description of monetary policy generally emphasizes the reserve requirement constraint on banks. In this story, banks are an important link in the transmission of monetary policy because changes in bank reserves influence the quantity of reservable deposits held by banks (Peek and Rosengren, 1995). Raising the reserve requirement constraint can limit monetary growth and squeeze excess reserves if not fully neutralized, potentially limiting banks' capacity to lend over time (Ma, Yan and Liu, 2011).

It is generally accepted that reserve requirements act as a tax on banks and thus hurt their profits, affecting cost of credit and possibly inducing them to widen the loan-deposit rate spread (Borio and Disyatat, 2009; Montoro and Moreno, 2011). Banks are motivated to transfer the cost imposed by reserve requirements via attempts to either pay depositors less or charge borrowers more or both — by widening the loan-deposit rate spread (Montoro and Moreno, 2011). This rests on the implicit assumption that the burden can be passed through.

Hein and Jonathan (2002) point out that while it is generally accepted that reserve requirements act as a tax on banks, there is widespread disagreement that as to who bears the tax and who benefits when the tax is reduced. Black (1975) and Fabozzi and Thurston (1986) argue that the reserve requirement tax is passed on to depositors of reservable instruments in the form of lower yields. Cargill and Mayer (2005) find that banks do not respond to the changes in reserve requirements essentially by changing their excess reserves, instead, banks meet a substantial part of their increased reserve requirements by reducing their earning assets, including loans. They prove that the increase in reserve requirements reduces the availability of bank credit.

There is much recent evidence suggesting that at least a portion of the tax is not passed on to borrowers or depositors but rather is passed on to, or absorbed by, bank shareholders. Some authors (Osborne and Zaher, 1992; Cosimano and McDonald, 1998; and Stewart and Hein, 2002) provide evidences suggesting that increases (reductions) in reserve requirements lead to lower (higher) bank stock prices.

According to the bank lending channel thesis, monetary policy actions can also alter the supply of bank loans by changing bank reserves, and thus on the real economy (Kashyap and Stein, 1994). By lowering bank reserves, contractionary policy can decrease the core deposit funding of bank loans (due to reserve requirements) (Van Den Heuvel, 2002). If some banks (such as small banks) are unable to raise nonreservable funds to continue lending, policy will decrease their loan supply and affect their bank-dependent borrowers. Decreases in either loan supply or loan demand lead to reduced investment and consumption expenditures (Kishan and Opiela, 2000).

Ma, Yang and Liu (2011) argue that the PBC finds it easier to reach consensus on reserve requirements than interest rate decisions and enjoys greater discretion in applying this tool in recent years. The effects of reserve requirements should be explored in conjunction with other policy actions. Depending on the policy mix, higher reserve requirements tend to signal a tightening bias, to squeeze excess reserves of banks, to push market interest rates higher, and to help widen net interest spreads, thus tightening domestic monetary conditions. It is well known that reserve requirements is seemed as a tax burden on banks. Banks appear to pass through a big part of the burden to their customers, mostly depositors and small and medium-sized enterprises (SMEs). Yet bank shareholders may also bear part of the burden, if wider margins and spreads in part reflect increased risk premia.

In all, most of the literature analyzes the function of reserve requirements and the link between reserve requirements ratio and bank loans. Only Gray (2011) gives a simple description of differentiated reserve requirement ratio. Gray (2011) indicates that most of countries use a single reserve requirement rate for all reservable liabilities, a few countries use differentiated reserve requirements ratio with clear and achievable goals. Historically, some central banks have applied differential reserve requirements ratio to different types of banks (such as the United States set reserve requirements ratio based on geographic distinctions among member banks and the level of deposits during 1966-1972). This would appear to reflect the taxation aspect of unremunerated reserves, and implies a subsidy to those banks which have a lower reserve requirement.

To our knowledge, no literature tests the effects of DRRR policy on different regions and large VS S&M sized banks. We will do these tests to examine the effects of DRRR policy on the disaster region and the loans made by large VS S&M sized banks in China.

3 Methodology

As we have two objectives, the first one is to check the effects of DRRR policy on the GDP growth of earthquake-stricken area, the second one is to examine the impacts of DRRR policy on the loans made by large VS S&M sized banks in China, we also have two models. The first one is panel data model and the second one is cointegration test.

3.1 The First Model

In the first model, we want to check the role of DRRR policy on the outputs of the quake-stricken areas. As the 2008 Wenchuan earthquake mainly happenes in Sichuan province, we select 39 counties which are badly hurt by the earthquake and enjoy the preferential DRRR policy in Sichuan province and the other 39 counties in Sichuan province which do not apply the DRRR policy to make a comparison to check whether or not the DRRR policy can promote the GDP growth more in the quake-stricken 39 counties than normal RRR (Reserve Requirement Ratio) policy implemented in the other 39 counties. The estimation period is the implementation period of the DRRR policy during 2008-2011.

We construct our first model mainly based on the expenditure approach of GDP. As well known, GDP mainly contains four parts: government expenditure, investment, consumption and net exports. As these 78 counties in Sichuan province belong to the less developed western region, far from the coastal areas, there are very little or no exports in these counties, we neglect net exports in our model. We divide investment into two parts: in the first part, the funds of investment are from bank loans, in the second part, the funds are from other sources. According to the bank lending channel, we know that reserve requirement ratio can affect loans and the change of loans can influence investment, and finally will affect the real economy. We assume that the DRRR can affect the investment through loans, thus we introduce an interaction item: DRRR*Loans, into GDP equation. We assume that the interaction item mainly affects the first part of investment and want to measure the influence of this interaction item on the GDP. Therefore, we get our model like this:

$$GDP_{it}= \alpha_i+Govexp_{it}\,\beta_i+DDRR_{it}{*}Loan_{it}\,\gamma_i +Othinvest_{it}\,\delta_i+Consum_{it}\,\sigma_i+e_{it} \quad (1)$$

Where GDPit is the GDP of 39 counties in Sichuan province, t = 2008, 2009, 2010 and 2011. i = 1, 2, ……, 39. The loanit is the total loan of each county. $Govexp_{it}$ is government expenditure measured by the local government intra-budgetary ordinary expenditure of each county. $Oth\ invest_{it}$ is the investment whose funds are from other sources (except bank loans) measured by total investment in fixed assets (minus bank loans) of each county. $Consumit$ is consumption measured by total retail sales of consumer goods in each county. These variables are real variables adjusted by the CPI index (2008=100). DRRRit is Differentiated Reserve Requirement Ratio. As the counties in Sichuan province are less developed areas located in inland China, they do not have or have very little FDI, thus we do not add FDI as an independent variable. eit are residuals.

We will use panel data model to estimate this equation. The data cover 78 counties in Sichuan province from 2008 to 2011. Firstly we will estimate the model for the 39 earthquake badly hurt counties which enjoy the DRRR policy in Sichuan province, and then we estimate the model for the other 39 counties without the DRRR policy in Sichuan province. We will compare the coefficients: γ_i to check the effects of DRRR policy on the GDP growth of the counties with DRRR policy and counties without DRRR policy. In this model, we do not use dummy variable to differentiate the earthquake-stricken counties and the other counties, because what we want to know is whether or not the preferential DRRR can support the economy growth of the disaster counties, not whether the earthquake can influence the economic growth in these counties or not.

3.2 The Second Model

In the second model, we will examine the role of DRRR policy on the loan made by large and S&M sized banks. We adopt the model developed by Zhang, Ji and Cui (2008), but we make a little modification. The loan supply is mainly influenced by the reserve requirement ratio and the spread between the benchmark loan rate and deposit rate which regulated by the PBC. The loan demand is mainly affected by the economic condition (measured by the GDP). Thus we set our third model as follows:

$$Loans_t= \alpha+DDRR_t\,\beta+Spread_t\,\gamma+GDP_t\,\theta+et \quad (2)$$

Where loans is the total loan made by Large banks and S&M sized banks. $DRRR_{it}$ is Differentiated Reserve Requirement Ratio. Spread is the spread between the benchmark loan rate and deposit rate which regulated by the PBC. As we cannot get the monthly data of GDP, we use real industrial value added as a proxy. e are residuals. We test the model using monthly data from 2009M01 until 2013M10.

Cointegration test is an econometrics method that is applied to investigate the existence of any long-run equilibrium between variables. If a group of time series is cointegrated, the variables are assumed to have a long run equilibrium relationship. Long run equilibrium relationship can be defined as a group of time series that tend to move together through time in a long period of time and the difference between them are stable although in the short run they may diverge from each other. As the estimation period is from 2009M01 to 2013M06 using monthly data, we will use VEC (Vector Error Correction) model to run the cointegration test separately for large banks and the S&M sized banks. We will compare the coefficients β to compare the role of DRRR policy on the large banks versus S&M banks.

4 Empirical Results

4.1 Results of the First Model

In panel data analysis, we estimate three models: pooled regression model, fixed effects model and random effects model. We use Pooled Ordinary Least Squares (POLS) to estimate the pooled regression model, use Least Squares Dummy Variable (LSDV) to estimate the fixed effects and Generalized Least Squares (GLS) to estimate random effects model. The results are in table 1.

In table 1, we estimate model 1 for the 39 earthquake badly hurt counties with DRRR policy, model 2 for the other 39 counties without the DRRR policy in Sichuan province. At the first glance, we can see the coefficient of the interaction part DRRR*Loans of model one is bigger than that of model two in all regressions.

Then we run the Likelihood ratio test and Hausman test to see which model is the best one. For model one, first we run the Likelihood ratio test, the cross-section F-statistic is 37.8345, the P-value is 0.0000, so we reject the null hypothesis that the proper model is pooled regression model. Then we run the Hausman test to compare the fixed effects and random effects, if the Hausman test rejects the null hypothesis, the fixed effects model is appropriate, or the random effects model is estimated (Hausman, 1978). The Chi-Sq. Statistic is 9.8087 and P-value is 0.0438, thus we reject the null hypothesis, the final model is fixed effects model. In model two, for Likelihood ratio test, the cross-section F-statistic is 50.9824, the P-value is 0.0000, for the Hausman test, the Chi-Sq. Statistic is 9.2490 and P-value is 0.0552, the null hypothesis cannot be rejected and we choose the random effects model as the best model.

Table 1. Results of the First Model

	Dependent Variable (DV): The GDP of Each County					
	Pooled Regression		**Fixed Effects**		**Random Effects**	
Method	**POLS**		**LSDV**		**GLS**	
IV	**Model 1**	**Model 2**	**Model 1**	**Model 2**	**Model 1**	**Model 2**
Intercept	8313.620 (0.2483)	-19642.29 (-0.4720)	127673.5*** (4.6397)	141629.2*** (4.5538)	97443.01** (2.5172)	110396.3** (2.5199)
Govexp	0.1075 (0.7446)	1.6609*** (4.0155)	0.0187 (0.2645)	1.1864*** (5.5581)	0.0704 (1.0599)	1.1052*** (5.4733)
DRRR*Loans	2.4510*** (8.1831)	2.1256*** (3.3684)	2.1382*** (9.2374)	1.8848*** (4.8838)	2.2096*** (10.4767)	1.6614*** (4.6499)
Othinvest	0.1086** (1.9971)	0.1422 (1.5602)	0.0096 (0.4023)	0.2344*** (4.7386)	0.0081 (0.3522)	0.2128*** (4.5181)
Consum	1.7354*** (16.0563)	1.4405*** (6.3429)	1.6998*** (13.3535)	1.0650*** (5.0503)	1.7545*** (16.4983)	1.2763*** (6.9874)
Adjusted R^2	0.9315	0.8763	0.9933	0.9909	0.9305	0.9342
Observations	156	156	156	156	156	156

Notes: T-statistics in parentheses. *, **, and *** indicates that a null hypothesis of zero is rejected at the 10%, 5%, and 1% levels, respectively

In the fixed effects model, we can see in model one (the earthquake-stricken counties), the coefficients of government expenditure and other investments are positive and non-significant. The coefficients of the interaction part, consumption and the intercept are positive and statistically significant. In the random effect model, all the coefficients of model two are positive and significant. The coefficient of the interaction part is the combined effect of the DRRR and loans on the GDP. This coefficient in model one (fixed effects model) is bigger than that in model 2 (random effects model). Ceteris paribus, considering all these 78 counties are in the same province Sichuan, located in the adjacent places, almost share the same economic development level, thus we can say that it is the DRRR policy which mainly causes the difference. In earthquake-stricken counties, the combined effect of lower DRRR and more loans is about 28.70% bigger than that in the other 39 counties, indicating that the preferential DRRR in the earthquake-stricken counties is effective and do can help the economy of the less developed disaster counties to gain more growth than the normal RRR in other counties.

4.2 Results of the Second Model

In the second model, we use the cointegration test to examine the role of DRRR policy on the loan made by large and S&M sized banks using monthly data from 2009M01 to 2013M10. The variables used in the estimation must be stationary so that standard statistical theory applies. We conduct augmented Dickey-Fuller (ADF) unit root tests applied to the levels and first-difference of the system's variables in EViews 6.0 (see Table 2). The data of Loans and GDP are seasonally adjusted and expressed in logs. From table 2 we can see that all the variables are I(1), then we run the cointegration test separately for large banks and S&M sized banks. The lag length is 8 for large banks and 4 for S&M sized banks. We can see the results from table 3 and 4.

Table 2. Unit root tests for the variables

Variables	Level		First Difference	
	Constant	Constant with trend	Constant	Constant with trend
Lnbloan_sa	-4.8537^{***}	-7.0690^{***}	-1.7897	-5.9714^{***}
Lnsloan_sa	-2.3818	-2.3671	-7.5485^{***}	-7.9350^{***}
LnGDP_sa	-2.9773^{**}	-5.7259^{***}	-7.3610^{***}	-6.4489^{***}
LDRRR	-1.5162	-1.3165	-3.1017^{**}	-3.2273^{*}
SDRRR	-1.5511	-3.0239	-2.8136^{*}	-2.8393
Spread	-0.3260	-1.6557	-6.1777^{***}	-6.2002^{***}

Notes: Asterisk *, ** and *** indicates the rejection of the null hypothesis at level of confidence 10%, 5% and 1% respectively, critical values are from Davidson and Mackinnon (1993). The null and alternative hypothesis are respectively H0 = unit root (series is non-stationary) and H1 > 0 (series is non-stationary). Based on the Akaike Information Criterion (AIC) and Schwarz Information Criterion (SC), considering the degree of freedom, the lag lenth is two.

Table 3. Cointegration Rank Test (Trace) for Large Banks

Hypothesized No. of CE(s)	Eigenvalue	Trace Statistic	0.05 Critical Value	Prob.**
None *	0.54114	65.76838	47.85613	0.0005
At most 1	0.342734	28.37586	29.79707	0.0722
At most 2	0.114773	8.231883	15.49471	0.4409
At most 3	0.048377	2.380138	3.841466	0.1229

Notes: Trace test indicates 1 cointegrating equation at the 0.05 level. * denotes rejection of the hypothesis at the 0.05 level. ** denotes MacKinnon, Haug and Michelis (1999) p-values.

Table 4. Cointegration Rank Test (Trace) for S&M Sized Banks

Hypothesized No. of CE(s)	Eigenvalue	Trace Statistic	0.05 Critical Value	Prob.**
None *	0.395723	55.71469	47.85613	0.0077

At most 1	0.243278	29.52113	29.79707	0.0538
At most 2	0.179143	15.02565	15.49471	0.0587
At most 3 *	0.087483	4.760528	3.841466	0.0291

Notes: Trace test indicates 1 cointegrating equation at the 0.05 level. * denotes rejection of the hypothesis at the 0.05 level. ** denotes MacKinnon, Haug and Michelis (1999) p-values.

According to table 3 and 4, cointegration tests show that one cointegrating equation exists for large banks and S&M sized banks, indicating the existence of long run equilibrium relationship between these variables. We write the cointegrating equations as follows:

For large banks:

$$Dlnbloan_{sa} = -1.0767DLDDRR + 32.6405DSpread + 1.4284DLnGDP \quad (3)$$
$$(0.1940) \qquad (10.3500) \qquad (0.1769)$$

For S&M sized banks:

$$Dlnsloan_{sa} = -3.4271DSDDRR + 83.5923DSpread + 1.4284DLnGDP \quad (4)$$
$$(1.8793) \qquad (114.5560) \qquad (0.9243)$$

The log likelihood of equation (3) and (4) are 1105.053 and 964.6038. For equation (3), the coefficient of LDRRR is -1.0767, which means when the DRRR for large banks raises one percent, the loans of large banks in logs will reduce 1.0767 million CHY. That is to say, the loans of large banks will reduce 2.9350 million CHY. The loan of large banks is positively related to the spread and the GDP, indicating that when the spread increases and the economy grow fast, the loans will increase. The equation (4) is for S&M sized banks. The coefficient of LDRRR is -3.4271, which means when the DRRR for S&M sized banks raises one percent, the loans of S&M sized banks in logs will reduce 3.4271 million CHY and the loans of large banks will reduce 30.7872 million CHY. Compared with equation (3) and (4), we can see that the S&M sized banks are more easily affected by the reserve requirement ratio, the spread and the real conditions of the economy. When the PBC raises the RRR one percent, this action apparently influence the S&M sized banks more. Therefore, it is proper that the PBC applies the DRRR policy to the banking system, the large banks apply a little higher RRR and the S&M sized banks apply a little lower RRR. However, comparing equation (3) and (4), we can see that although the S&M sized banks apply the lower DRRR, the tightening effects on them (coefficient of SDRRR) is still much bigger than that on large banks. Therefore, there are still some room to reduce the DRRR for the S&M sized banks.

5 Conclusion

This study uses panel data analysis to examine the effects of DRRR policy on the earthquake-stricken area during 2008-2011 and employs the cointegration test to measure the effects on large banks versus S&M sized banks from 2009M01 to 2013M10. The results confirm our hypothesis that this DRRR policy promotes the economy growth more in the earthquake-stricken counties and a lower DRRR to some extent alleviate the tightening effects of RRR on the S&M sized banks.

Our results have a profound policy implication. As China is a huge country and the regional disparity is obvious (Lau, 2010; Fan, Kanbur and Zhang, 2011; Pedroni and Yao, 2006), common monetary policy may have regional effects on different provinces and these regional effects may increase the gap among regions in China. This study provides an empirical proof that we can narrow the gap from the perspective of monetary policy.

When the PBC formulates and implements the common monetary policy, they should take the different effects of monetary policy into account. This Differentiated Deposit Requirement Ratio (DRRR) policy is a new try by the PBC. Our finding proves that this DRRR policy does support the economy growth of the disaster area and a lower DRRR to some extent can alleviate the tightening effects of RRR on the S&M sized banks. Then we can suggest that the PBC can consider expanding to apply this DRRR policy to more less developed provinces (Such as some provinces in the Middle and West) or some rural areas in order to promote a more coordinated regional economic development. For different banks, our results show that the PBC apply a lower DRRR to S&M sized banks is reasonable but not enough. Although the S&M sized banks apply a lower DRRR, the tightening effects of DRRR is also much bigger on the S&M sized banks than that on large banks. The PBC can consider applying much lower DRRR to the S&M sized banks.

Our study also has some limitations. For the first model, in fact, the best data used in the model is actually monthly or seasonal data, but due to the data limitation, we can only get the annual data. For panel data analysis, four year data may be too short. For the second model, as we can just get the total loans made by the large banks and S&M sized banks, we cannot get the monthly data for each bank and analyze in detail. These limitations will be solved when the data are available in the future.

References

[1] J. Peek and E. S. Rosengren, Bank Lending and the Transmission of Monetary Policy, in *Conf. Ser. Reserv. Bank Bost.* 47 (1995)
[2] G. Ma, X. Yan, and X. Liu, *China's Evolving Reserve Requirements* , in Discuss. Pap. Bank Finl. Inst. Econ. Transit.1 (2011)
[3] C. Montoro and R. Moreno, BIS Q. Rev. March (2011)
[4] C. Borio and P. Disyatat, *Unconventional Monetary Policies: An Appraisal* (BIS Working Papers, 2009)
[5] S. E. Hein and J. D. Stewart, Econ. Rev. Reserv. Bank Atlanta **87**, 41 (2002)
[6] F. Black, J. Financ. Econ. **2**, 323 (1975)
[7] F. J. Fabozzi and T. B. Thurston, J. Financ. Quant. Anal. **21**, 427 (1986)
[8] D. K. Osborne and T. S. Zaher, J. Bank. Financ. **16**, 799 (1992)
[9] T. F. Cosimano and B. McDonald, J. Monet. Econ. **41**, 57 (1998)
[10] J. D. Stewart and S. E. Hein, J. Financ. Res. **25**, 367 (2002)
[11] A. K. Kashyap and J. C. Stein, in *Natl. Bur. Econ. Res. Work. Pap. Ser.* (1994)
[12] S. Van den Heuvel, Econ. Policy Rev. **8**, 259 (2002)
[13] R. P. Kishan and T. P. Opiela, J. Money, Credit Bank. 121 (2000)
[14] S. Gray, Int. Montery Fund **11**, (2011)
[15] J. A. Hausman, J. Econom. Soc.**46**,1251 (1978)
[16] C. K. M. Lau, China Econ. Rev. **21**, 293 (2010)
[17] R. Davidson and J. G. MacKinnon, *Estimation and Inference in Econometrics* (Oxford University Press, 1993)
[18] S. Fan, R. Kanbur, and X. Zhang, Rev. Dev. Financ. **1**, 47 (2011)
[19] J. G. MacKinnon, A. A. Haug, and L. Michelis, J. Appl. Econom. **14**, 563 (1999)
[20] X. Zhang, Z. Ji, and Y. Cui, J. Econ. Res. (in Chinese) **7**, 65 (2008)
[21] P. Pedroni and J. Y. Yao, J. Asian Econ. **17**, 294 (2006)

The extent to which participants in the indonesian legislative elections obey the rules

Eko Suyono[1*], *Oman* Rusmana[1], and *Agung* Praptapa [1]

[1] Faculty of Economics and Business, Jenderal Soedirman University, Indonesia

Abstract. This study aims to evaluate the extent to which the compliance of legislative candidates as participants of the 2019 Indonesian general election is related to the rules for reporting campaign funds. To achieve this objective, this study uses election rules indicators that relate to the rules of campaign finance reporting with reference to 21 indicators according to Law No. 7 of 2017 concerning General Elections in the Republic of Indonesia along with all derivative rules. Based on the rules that apply in Indonesia, general election participants who do not comply with the rules for reporting campaign funds can be canceled from the process of election as legislative members. In other words, the results of the compliance audit which concludes that a legislative candidate does not comply with the rules for the reporting of campaign funds could be canceled as a legislative member. This study uses the "X" Political Party legislative members from South Kalimantan Province, including legislative candidates for the DPR-RI, Provincial DPRD, and district/city DPRD. The total number of legislative candidates from Party X is 11 for the DPR-RI from 2 electoral areas, 55 for the Provincial DPRD from 7 electoral areas, and 296 for district/city DPRDs from an average of 5 electoral areas for each district/city. By setting a sample of 20% of the population, this study ended-up with 74 legislative candidates as the sample., i.e, 3 DPR RI legislative candidates, 11 legislative candidates for the provincial DPRD, and 60 legislative candidates for district/city DPRD. Descriptive analysis method is used to answer the problems in this study. Evaluation of compliance with the reporting of campaign funds is carried out by using compliance audit procedures through review of existing supporting documents and interviews with legislative candidates who are selected as the sample as confirmation procedure. The review of documents is carried out carefully and thoroughly to evaluate whether the legislative candidates from political parties follow the rules in the mechanism of reporting the campaign funds for the period February 20, 2018, until April 25, 2019. After the entire evaluation process conducted in May 2019 has been completed, this study concluded that in general, the legislative candidates adhere to the rules of campaign funds reporting in accordance with applicable laws and regulations. This study does not find any material deviations from applicable laws and regulations.

* Corresponding author: ekyo75@unsoed.ac.id

1 Introduction

The general election held in Indonesia on April 17, 2019 is the first simultaneous general elections which simultaneously elect the president and his representatives, legislative members and regional representative councils. This study focuses on the legislative general election where of the 27 parties that register, there are only 16 parties that fulfill the national factual administrative and verification requirements and 4 local parties for the Nangroe Aceh Darussalam Province special area. Verification to determine the requirements for parties that can qualify to participate in the general election includes the existence of the main political party officials at the central level, women's representation of at least 30%, and office domicile at the head office level. Then, at the provincial level, there are additional conditions, namely fulfilling membership in 75% of districts/cities in 34 provinces. Moreover, the last requirement is the status of the distribution of administrators of at least 50% of districts in 75% of districts/cities in 34 provinces.

The political parties participating in the 2019 Legislative Election are as follows: (1) National Awakening Party (Partai Kebangkitan Bangsa/PKB), (2) Greater Indonesia Movement Party (Gerakan Indonesia Raya/GERINDRA), (3) Indonesian Democratic Party of Struggle (Partai Demokrasi Indonesia Perjuangan/PDIP), (4) Golongan Karya Party (GOLKAR), (5) National Democratic Party (NASDEM), (6) Indonesian Change Movement Party (GARUDA), (7) Working Party (BERKARYA), (8) Prosperous Justice Party (Partai Keadilan Sejahtera/PKS), (9) Indonesian Unity Party (PERINDO), (10) United Development Party (Partai Persatuan Pembangunan/PPP), (11) Indonesian Solidarity Party (Partai Solidaritas Indonesia/PSI), (12) National Mandate Party (Partai Amanat nasional/PAN), (13) People's Conscience Party (HANURA), (14) Democratic Party (DEMOKRAT), (15) Aceh Party (PA), (16) Sira Party (SIRA)), (17) Aceh Regional Party (PD Aceh), (18) Nagroe Aceh Party (PNA), (19) Bulan Bintang Party (PBB), and (20) Indonesian Justice and Unity Party (PKPI). Furthermore, the 2019 legislative election which was held on Wednesday (04/17/2019) was to fill 575 seats in the Republic of Indonesia Parliament for the 2019-2024 term. The number increased by 15 seats compared to the previous period in 2014 which only fought for 560 seats (KPU, 2019). In addition, this legislative election was also held simultaneously to fight for seats in the Regional Representatives Council (DPRD) in 34 provinces, 416 regencies and 98 cities in Indonesia.

One of the main issues in the legislative election is the compliance of the election participants with the applicable rules, especially those relating to campaign funds reporting, namely Law No. 7 of 2017 concerning General Elections along with all explanatory rules. Itsis because campaign funds reporting according to the applicable rules is very important to be obeyed by the legislative general election participants to ensure the fulfillment of the principle of free and confidential direct elections according to the mandate of the 1945 Constitution. In order to ensure compliance with such regulations, a compliance audit is required to evaluate how obedient the legislative election participants adhere to the campaign finance reporting rules so that a clean and transparent general election is held without matters that violate the rules.

On the basis of the explanation above, the purpose of this study is to evaluate the level of compliance of legislative election participants in terms of campaign funds reporting in accordance with laws and regulations that apply to the Indonesian Parliament, Provincial DPRD, and Regency/City DPRD in Indonesia.

2 Literature review and related regulations

Compliance audit is a work process that determines whether the auditee has followed certain procedures, standards, and rules set by the competent authority (Othman et al.,

2013; Slobodianyk et al., 2018). So the auditor is tasked with ensuring that the work process carried out is in accordance with certain procedures, standards, and rules. In the other side, compliance audit has a purpose to to evaluate the internal performance of the organization and its goal achieving (Gomes, 2001)

The application of compliance audit to the public sector is in line with the new public management (NPM) theory where the theory emerged in the 1980s
encourage various uses of approaches in the public sector, especially in the field
accounting and auditing (Hood, 1995; Reichborn-Kjennerud, 2013). NPM is generally seen as an approach in public administration that applies the knowledge and experience gained in the world of management and other disciplines to improve efficiency, effectiveness of public service performance in modern bureaucracies.

Mintzberg (1980), while describing various types of structural organizations, notes that many public institutions have a simple and healthy organizational structure that he classifies as Machine Bureaucracy. Organizations that have such structures are characterized by hierarchies and simple processes and operations, which repeat themselves, have a standard and predictable form. Therefore, external controls inside such organizations are associated with determining compliance with regulations, standards and procedures. That is why the Machine Bureaucracy is associated with external compliance audit (Mintzberg, 1980). However, more complex organizational structures require different approaches in the field of control to assess not only compliance with regulated standards but also to determine suitability, effectiveness and efficiency.

The compliance audit for the reporting of campaign funds for 2019 legislative elections in Indonesia is based on: (1) Law No. 7 of 2017 concerning General Elections (2) PKPU No. 7 of 2017 concerning Stages, Programs, and Schedule for the Implementation of General Elections in 2019 (3) PKPU No. 5 of 2018 Amendment to PKPU No. 7 of 2017 concerning Stages, Programs, and Schedule for the Implementation of General Elections in 2019 (4) PKPU No. 32 of 2018 Second Amendment to PKPU No. 7 of 2017 concerning Stages, Programs, and Schedule for the Implementation of General Elections in 2019 (5) PKPU No. 24 of 2018 concerning General Election Campaign Funds (6) PKPU No. 29 of 2018 Amendment to PKPU No. 24 of 2018 concerning General Election Campaign Funds (7) PKPU No. 34 of 2018 Second Amendment to PKPU No. 24 of 2018 concerning General Election Campaign Funds (8) SK No. 1126/PL/01.6-Kpt/03/KPU/IX/2018 concerning Campaign Fund Reporting Technical Guidelines (9) SK No. 1781/PL.01.6-Kpt/03/KPU/XI/2018 concerning Audit Implementation Guidelines on General Election Campaign Fund Reports.

Based on the provisions of General Election Commission (Komisi Pemilihan Umum, hereafter KPU) Regulation Number 24 Year 2018 Article 36, the reporting of political party campaign fund consists of: (1) Campaign Funds Preliminary Report (Laporan Awal Dana Kampanye/LADK) is bookkeeping that contains information on Campaign Fund Special Accounts (Rekening Khusus Dana Kampanye/RKDK), source of initial balance or opening balance, details of income and expenditure calculations obtained before RKDK opening, and receipt of donations sourced from Political Parties or Joint Party Prospective DPD Members or other parties (2) The Report on Receipt of Campaign Fund Donations (Laporan Penerimaan Sumbangan Dana Kampanye/LPSDK) is a bookkeeping containing all receipts received by Election Participants after LADK is submitted to the KPU; and (3) The Campaign Fund Receipt and Expenditure Report (Laporan Penerimaan dan Pengeluaran Dana Kampanye/LPPDK) is a bookkeeping that contains all of the Campaign Funds' receipts and expenses.

The criteria used in this evaluation are in accordance with the provisions stipulated in KPU Regulation Number 24 of 2018 concerning the General Election Funds as amended several times, the latest by the General Election Commission Regulation No. 34 of 2018 concerning General Election Campaign Funds. These criteria are also used as references and applied by Political Parties in the preparation of campaign fund reports. The criteria that must be obeyed by political parties in compiling campaign fund reports are:

(1) The political parties participating in the election open a Special Campaign Fund Account (Rekening Khusus Dana Kampanye/RKDK) on behalf of the political parties which are separate from the personal accounts of political parties at the public bank and are opened jointly by representatives of 2 (two) political party managers according to their level.
(2) The political parties participating in the election open the RKDK not exceeding the opening provisions of the RKDK which is no later than 1 (one) day before the start of the campaign period.
(3) The political parties participating in the election manage the RKDK by appointing RKDK manager who is specifically tasked to manage the RKDK, which is accompanied by a statement from the political parties.
(4) The political parties participating in the election comply with the bookkeeping period of the Initial Campaign Fund Report (LADK), which is from the opening of the RKDK and closes 1 (one) day before the campaign period.
(5) The political parties participating in the election compile LADK which contains information as follows: RKDK; initial balance or opening balance; source of acquisition of initial balance or bookkeeping balance; amount of details of calculation of receipts and expenditures that have been made before submission of LADK, if the initial balance is the remainder of the receipt of funds with campaign allotment obtained before the period of accounting for LADK; acceptance of contributions originating from political parties and other parties; balance per closing of LADK bookkeeping; and taxpayer identification number.
(6) The political parties participating in the election comply with the submission of LADK not exceeding the stipulated time, namely 1 (one) day after the closing period of LADK no later than 18.00 local time to the KPU, which is completed with: Form the LADK1-PARPOL Model or the LADK1.HP-PARPOL Model form; Form the LADK2-PARPOL Model or the LADK2.HP-PARPOL Model form; Form the LADK3-PARPOL Model or the LADK3.HP-PARPOL Model form; Form the LADK4-PARPOL Model or the LADK4.HP-PARPOL Model form; Form the LADK5-PARPOL Model or the LADK5.HP-PARPOL Model form; Form the Model LADK6-PARPOL or form Model LADK6.HP-PARPOL; and Form the LADK7-PARPOL Model or the LADK7.HP-PARPOL Model form;
(7) The political parties participating in the election comply with the bookkeeping period of the Report on the Receipt of Campaign Fund Donations (LPSDK) which starts from 1 (one) day after the closing period of LADK and closes 1 (one) day before the report on receipt of donations is submitted to the KPU.
(8) The political parties participating in the election compile the LPSDK which included the full record for the receipt of Campaign Fund contributions received after the making of LADK.
(9) The political parties participating in the election comply with the submission of the LPSDK not exceeding the time set in accordance with the schedule as stipulated in the KPU Regulation concerning Stages, Programs and Schedules submitted to the KPU no later than 18.00 local time, which is completed with: Form of LPSDK1-Political Party Model; Form of LPSDK2-Political Party Model; Form of LPSDK3-Political Party Model; Form of LPSDK4-Political Party Model
(10) The political parties participating in the election comply with the bookkeeping period of Campaign Funds Receipts and Expenditures Reports (LPPDK) starting 3 (three)

days after the Political Parties are designated as Participants in the Election and are closed 8 (eight) days after the voting day.

(11) The political parties participating in the election prepare LPPDK which includes the bookkeeping of all receipts of Campaign Funds received and all expenses of Campaign Funds in the form of money, goods and/or services that are converted to monetary value, including debt and purchase discounts for goods or services that exceed the fairness of transactions buying and selling in general is accompanied by proof of expenditure that can be accounted for.

(12) The political parties participating in the election submit Campaign Fund Reports no later than 15 (fifteen) days after voting no later than 18.00 local time to KPU, accompanied by LADK and LPSDK and completed with: Form Model LPPDK1-PARPOL; Form Model LPPDK2-PARPOL; Form Model LPPDK3-PARPOL; Form Model LPPDK4-PARPOL; Form Model LPPDK5-PARPOL; Form Model LPPDK6-PARPOL; Form Model LPPDK7-PARPOL; Statement of Outside Individual Donors; Statement of Outside Group Donors; Statement of Donors from non-government business entity; Copy of debt (if any); Proof of receipt and expenditure transactions

(13) The political parties participating in the election are responsible that all candidates for DPR/DPRD members report and attach the reports on recording the receipt and expenditure of Campaign Funds for candidates for DPR/DPRD members.

(14) The political parties participating in the election record and report on all donations (including money, goods, and/or services that are converted to monetary unit, including debt and discounts on purchases of goods or services that exceed the fairness of general buying and selling transactions) that are accepted into the List of Recipients

(15) The political parties participating in the election record and report on all donations (including money, goods, and/or services that are converted to monetary unit, including debt and discounts on purchases of goods or services that exceed the fairness of general buying and selling transactions) reported in LADK, LPSDK, and LPPDK does not exceed the amount below: IDR 2,500,000,000 for donations by outside individual parties; and IDR 25,000,000,000 for donations of groups and/or non-government business entities.

(16) The political parties participating in the election book and place all donations in the form of money received into the RKDK before being used for campaign activities.

(17) The political parties participating in the election do not use funds from prohibited donations, if accepting donations is prohibited then they should comply with the following provisions: Not using the donation; Report the prohibited donations; and Depositing the prohibited contribution to the country no later than 14 days after the campaign period ends and submitting the KPU's report evidence.

(18) The political parties participating in the election record all Campaign Fund expenditure transactions to the List of Campaign Fund Expenditure Activities.

(19) The political parties participating in the election use fair market prices to assess goods and services in the Campaign Fund receipt and expenditure activities.

(20) The political parties participating in the election do not use Campaign Funds for prohibited provisions, such as financing witnesses in voting and counting.

(21) The political parties participating in the election finance all expenses of the Campaign using funds placed in the RKDK.

3 Research methodology

This study uses Political Party "X" with electoral area of South Kalimantan Province which includes Barito Kuala, Tapin, Hulu Sungai Selatan, Hulu Sungai Tengah, Hulu Sungai

Utara, Balangan, Tabalang, Banjar, Tanah Laut, Tanah Bambu, Kotabaru, Banjarmasin City and Banjarbaru City. The total number of legislative candidates from this political party is 11 for the DPR-RI from 2 electoral areas, 55 for the Provincial DPRD from 7 electoral areas, and 296 for district/city DPRD from an average of 5 electoral areas for each district/city. By setting a sample of 20% of the population, this study ended-up with 74 legislative candidates as the sample., i.e, 3 DPR RI legislative candidates, 11 legislative candidates for the provincial DPRD, and 60 legislative candidates for district/city DPRD.

Descriptive analysis method is used to answer the problems in this study. Evaluation of compliance with the reporting of campaign funds is carried out by using compliance audit procedures through review of existing supporting documents and interviews with legislative candidates who are selected as the sample in the confirmation procedure. The review of documents is carried out carefully and thoroughly to evaluate whether the legislative candidates from political parties follow the rules in the mechanism of reporting the campaign funds for the period of February 20, 2018, until April 25, 2019. This study was done during May 2019.

4 Findings and Discussion

This study implements procedures and obtains relevant and limited evidence on information contained in the Campaign Fund Report submitted to us by the "X" Political Party. In the implementation of insurance engagement, there are limitations inherent, for example, sampling tests conducted allow failure to detect non-compliance and fraud that occurs. There are limitations to the nature, moment, scope, variety, and characteristics inherent in the source of funds as the origin of the Campaign Fund, so there is a possibility that the Campaign Fund is not covered in full in the Campaign Fund report.

Our examination is also limited to documents and/or information submitted and/or provided by the Party concerned so that this study does not carry out checks on matters not disclosed in the report. We do not provide a legal determination of the compliance of Political Parties with the relevant laws and regulations in the Campaign Fund Report.

Moreover, Political Parties are responsible for the preparation of Campaign Fund Reports in accordance with applicable laws and regulations. The responsibility also includes the application of valuation methods, assumptions and estimates used in the report, as well as the design of the implementation, maintenance of systems and processes that are relevant for the preparation of the Report.

Political parties are also responsible for compliance with legislative regulations related to campaign funds, as stated in the Assertion of Political Party Campaign Fund reports as contained in the Attachment to KPU Regulation Number 5 of 2017 on the LPPDK Model 1.

This study concludes with sufficient confidence in the Campaign Fund Report based on the procedures performed and the evidence obtained. We plan and carry out our work to get all the evidence and information needed to determine our conclusions. In carrying out the Engagement with sufficient confidence, we carried out a series of procedures and evaluated sufficient and appropriate evidence including the acquisition of an understanding of the Reporting of Campaign Funds. The nature, timing and scope of the procedures we carry out are also based on professional considerations including the level of materiality and risk assessment. The overview of the procedures that we perform include:

(1) Conducting interviews with Political Parties and/or those who compile the "X" Party campaign finance report.
(2) Conduct inspections of information and/or supporting documents used in preparing campaign fund reports to assess compliance with statutory provisions.
(3) Confirm the receipt of campaign fund contributions from individual donors, business entities, associations and/or political parties.

(4) Perform recalculation to ensure mathematical accuracy.
(5) Conduct an examination of the completeness and suitability of the report against the coverage period and the deadline for submission of reports.
(6) Conduct checks and tests on recording the receipt of campaign funds against the suitability of the periodic report.
(7) Check the suitability of RKDK ownership of Political Parties against ownership of campaign funds newspaper accounts in Commercial Banks;
(8) Check the suitability of contributors and test the maximum limit of contributions from each type of contributor received by the Political Party "X";
(9) Checking and testing expenses from campaign activities against conformity to Campaign Fund expenditure limits according to regulations;
(10) Check and test, the timeliness of reporting of LADK, LPSDK and donations that are prohibited to the South Kalimantan KPU, as well as verify the list of donations that are prohibited with existing evidence.
(11) Checking the suitability of the report period of receipts and expenditures of Political Parties and conducting tests on receipts and expenditures.
(12) Checking the suitability of the RKDK opening at the Bank that has been determined and the ownership of the RKDK in accordance with the provisions.
(13) Check the suitability of LADK coverage and the completeness of the information included in LADK.
(14) Check the suitability of the LPSDK coverage and the completeness of the information included in the LPSDK.

Our examination reveals that in general the "X" political party and its legislative candidates adhere to the provisions of legislation and report campaign funds in terms of managing the receipt and expenditure of campaign funds in the form of money originating from donations from Legislative Candidates through the RKDK.

5 Conclusion

This study aims to evaluate the level of compliance of legislative election participants in terms of campaign funds reporting in accordance with laws and regulations in Indonesia. By using 74 legislative candidates form Political Party "X" in the South Kalimantan electoral area, this study concludes that except for some points which are not material, the Political Party in the Campaign Fund Report above, in all material matters, are in accordance with the applicable criteria as stipulated in KPU Regulation Number 24 Year 2018 concerning the General Election Fund as several times last amended by General Election Commission Regulation No. 34 of 2018 concerning General Election Campaign Funds.

References

1. M. B. Gomes, *Performance Audit Argument: A Public Management Policy Analysis about Supreme Audit Institutions Role* (Caracas, 2001)
2. C. Hood, Accounting, Organ. Soc. **20**, 93 (1995)
3. H. Mintzberg, Manage. Sci. **26**, 322 (1980)
4. R. Othman and N. Mahzan, NZICA. 31 (2013)
5. K. Reichborn-Kjennerud, Does Performance Auditing Matter? The Government Administration's Response to the Performance Auditing of the Norwegian Office of the Auditor General, University of Bergen, 2013

6. Y. Slobodianyk, S. Shymon, and V. Adam, Balt. J. Econ. Stud. **4**, 320 (2019)

7. Law No. 7 of 2017 concerning General Elections
8. PKPU No. 7 of 2017 concerning Stages, Programs and Schedule for the Implementation of General Elections in 2019
9. PKPU No. 5 of 2018 Amendment to PKPU No. 7 of 2017 concerning Stages, Programs and Schedule for the Implementation of General Elections in 2019
10. PKPU No. 32 of 2018 Second Amendment to PKPU No. 7 of 2017 concerning Stages, Programs and Schedule for the Implementation of General Elections in 2019
11. PKPU No. 24 of 2018 concerning General Election Campaign Funds
12. PKPU No. 29 of 2018 Amendment to PKPU No. 24 of 2018 concerning General Election Campaign Funds
13. PKPU No. 34 of 2018 Second Amendment to PKPU No. 24 of 2018 concerning General Election Campaign Funds
14. SK No. 1126/PL/01.6-Kpt/03/KPU/IX/2018 concerning Campaign Fund Reporting Technical Guidelines
15. SK No. 1781/PL.01.6-Kpt/03/KPU/XI/2018 concerning Audit Implementation Guidelines on General Election Campaign Fund Reports.

Development of Islamic Microfinance Institutions with Social Capital Mechanism: A Case Study on BMT Tumang, Boyolali

Okta Hadi Nurcahyono [1*], and *Ghufronudin*[2]

[1] Sebelas Maret University, Indonesia
[2] Sebelas Maret University, Indonesia

Abstract. The increasing public interest in sharia-based (non-ribawi) transaction and finance encourages the growth and development of Tumang Baitul Maal wat Tamwil (BMT) as an alternative sharia-based microfinance institution. This article is the result of research aiming to analyze the role of social capital in the development of Islamic microfinance institutions at the Tumang BMT, Boyolali. This research is a qualitative research with a case study approach. The unit of analysis of this study is that Tumang BMT stakeholders were selected by purposive sampling. The techniques of collecting data used were observation, in-depth interviews and documentation. Data validation was carried out using source triangulation, and then data analysis was conducted using an interactive model. The results showed that social capital played a significant role in the growth and development of Tumang BMT's existence in the community. The theoretical analysis used in this paper was Robert Putnam's social capital approach. In his book "Making Democracy Work: Civic Traditions in Modern Italy (1993), Putnam defines social capital as the 'features of social organizations, such as networks, norms, and trust that facilitate coordination and co-operation for mutual benefits. The social capital owned by Tumang BMT includes norms based on Islamic principles that are embedded into members and customers. The members and the community's trust grows through the implementation of principles of corporate good governance (GCG) trust, reflected on the professional service to the community. Reciprocity between BMT and the community could be accomplished through providing financial assistance and sponsorship to various events. Networking (networking) is created through "celengan bumbung' system, Angkat Telepon Moro (ATM), Membangun Keluarga Utama (MKU), organizing football competitions and scholarship programs.

1 Introduction

The Islamic finance industry in the world continues to experience growth from year to year. In 2016, the total assets of the Islamic finance industry in the world reached US $ 2,202

* Corresponding author: okta.hadi@staff.uns.ac.id

billion, an increase from the previous year, which amounted to US $ 2.063 billion. The achievement of these assets recorded a growth of 6.74% compared to the previous year, and is projected to increase every year (OJK; 2017). The growth of the Islamic finance industry is projected to continue to increase, this is in line with the increase in the world's Muslim population, the increase in per capita income and wealth owned by Muslim, as well as increasing awareness of Islamic finance.

Islamic microfinance is at the heart of Islamic finance because it addresses one of the fundamental roles of financial intermediaries demanded by Sharia (Mansori, et.al, 2015. Microfinance institutions have an important role to play in reducing poverty in developing countries the majority of people in developing countries are people low and poor income (Karlan & Goldberg, 2011; Arafah & Nugrono, 2016). According to the Economist (Mansori, et.al, 2015), the population is one third of the world's poor communities. Over the past few decades, many initiatives have been carried out to reduce poverty in the country. Muslim countries, one of them through Islamic micro-finance institutions, Islamic microfinance is an industry that has developed in several Muslim countries such as Bangladesh, Nigeria and Malaysia, according to the Islamic Finance Development Report 2017, ICD-Thomson Reuters, the country with the largest Islamic financial assets is Iran Following in figure 1, there are 10 countries with sharia financial assets.

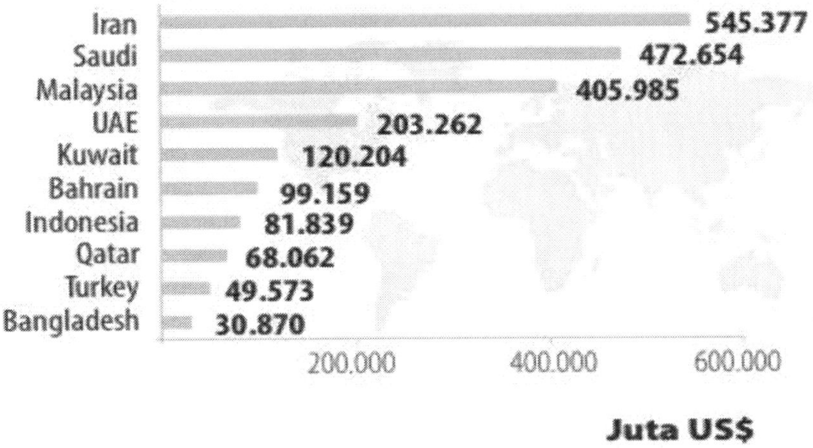

Fig. 1. Countries with the Largest Islamic Financial Assets

Micro Islamic financial institutions have evolved since 1990. In contrast to conventional banking financial institutions that have succeeded in developing and successfully opening their sharia subsidiaries, Islamic microfinance institutions, such as Sharia BPR, have largely failed to prove themselves as efficient microfinance service providers, and dynamic (Seibel, 2008). This is because Islamic microfinance institutions that are not monitored, such as conventional banks, are a direct threat to their customers such as losing savings. For this reason, the challenge for Islamic microfinance institutions is to gain public trust. One of them is the strategy of social capital, so this article is to answer how the development of Islamic micro financial institutions using social capital strategies.

2 Methods

This article is the result of this qualitative research conducted by Tumang BMT, who came from Tumang Village, Cempogo District, Boyolali Regency, Central Java, Indonesia. The approach used in this qualitative research is a case study approach (Yin, 2014). The

Tumang BMT case was interesting because, in the midst of many failures of sharia-based microfinance institutions in Indonesia, Tumang BMT was able to develop rapidly and become The Best Islamic Micro Finance organized by Karim Bussiness Consulting 2014. The rapid development of Tumang BMT as Islamic microfinance is one of them with a social capital strategy this becomes interesting to study.

The research unit of analysis is the Tumang BMT stakeholders which consist of branch managers, marketing staff, customers and the surrounding community. Meanwhile, the sampling technique is done purposively. Data collection techniques with non-participatory observation, in-depth interviews with stakeholders and literature studies. Data validity using source triangulation and data analysis with interactive model analysis (Milles and Huberman, 1984).

3 Result and Discussion

Tumang Baitul Maal wat Tamwil (BMT) now renamed into Sharia-based Financing Save-Loan Cooperatives of Tumang BMT has been initiated vigorously by an initiator of transformation, Adib Zuhairi, to release Tumang people from usury trap. Majority Tumang people work in metal craft sector, such as copper and aluminum. Thus, capital infusion is highly desirable. On the other hand, the condition of people is non-bankable and not familiar with banking industry. It is because Tumang people live in Tumang Village, Cempogo Sub District located in the slope of Mount Merapi; thus their access to bank is difficult. Such condition leads the people to rely highly on usurer service for accessing their business capital.

On October 1, 1998, Tumang BMT was established as new sharia-based financial institution in Boyolali region. The existence of Tumang BMT develops rapidly with a total of 20 subsidiaries distributed in both Central Java and DIY, with total members of 22,531 people in 2016, asset of IDR 98 billions in 2015, increasing to IDR 129 billions in 2016. In 2017, this figure increases to 165 billions. In addition, this sharia-based institution has received many awards from Islamic finance award, among others, for The Best Islamic Micro Finance IDR 25 < Asset < IDR 50 held by Karim Business Consulting on February 2014 (Ricky, 2018).

The existence of Tumang BMT as sharia-based micro financial institution, viewed from sociological perspective can be explained through social capital concept. Putnam in Field (2016: 51) mentions that social capital refers to inter-individual relation, social networking, and reciprocity and reliability norms growing from those relations. From the definition above, it can be concluded that trust, network, norm, and reciprocity is an important aspect in social capital.

3.1 Trust

In its early establishment, the attempt of growing public's trust was, of course, not easy. Many attempts taken to introduce the concept of BMT as a new financial institution in the public needs extra power. The management accomplished it through going to the people's houses one by one to visit them and to invite them all at once to join this institution. However, despite good acceptability, people were still reluctant and unwilling to be the members. Eventually, the management thought of how to grow the people's interest and trust in saving.

An innovative measure taken to attract people's sympathy and enthusiasm with saving is to apply bumbung (tubing) diplomatic system. This option was taken by the management because it was inspired with Tumang people's habit of using the perforated bamboo pillar (cagak) as money box. This traditional habit has deeply rooted within the society; thus it seemed to attract the people's sympathy to save their money easily. Then, management party produces bumbung massively, each of which was equipped with small padlock and sticker for the members' name and number. Technically, the bumbung was distributed for free to the members with the padlock key brought by Tumang BMT management and then Tumang BMT management would take an amount of fund collected in those bumbungs once a week from one house to another.

From this bumbung diplomatic system, the people's trust arises as characterized with many more people interested in saving their money for such reasons as minimum risk and no fee for renting the bumbung. The effect was so great, from the target of 50 bumbungs in the first week increasing to hundred bumbungs in months thereby having implication to the strengthened capital of BMT. From bumbung, the trust grew and new members come continuously due to words of mouth information. This bumbung method is event used continuously until Tumang BMT develops into the one with tens subsidiaries. Not only through bumbung, trust is grown as well through the utilization of strategic opportunities such as establishing business to provide product and service needed by people. The management then thought of establishing a potential business to provide product and service not existing yet there at that time. The choice was to establishing a business of LPG agent, carbide welding agent, providing sewing tools, pulse counter and initiating and catering on phone network establishment.

The greatness of Tumang BMT is also supported by many programs leading to the improved public trust. Tumang BMT serves not only as a financial institution but also as proselytizing (dakwah) and social institution. In these two functions, Tumang BMT functions as baitulmal or the partner of managing tithe (zakat) from dompet dhuafa and collecting money donation (wakaf) from Badan Wakaf Indonesia (BWI) and then Tumang BMT serves as distributor of zakat, infak, and shodaqoh (alms) to the needy people. Tumang BMT keeps improving its system over times to grow trust and to prevent fraud from occurring through implementing Good Governance Corporate (GCG) with five primary principles: transparency, accountability, responsibility, independency, and equality and fairness in both internal personnel management and professional service to its members. The professionalism of Tumang BMT performance then has implication to the growth of public trust as manifested into the achievement of some awards from many institutions. Consequently, in its almost two-decade existence, Tumang BMT has been rewarded with the achievement of A (Very Good) accreditation by Islamic Microfinance Standard (IMS) of Indonesian BMT Association. Therefore, the work space of KSPPS Tumang BMT has fused into national level since October 2016, after achieving the provincial level previously in 2011. Through these attempts, the members' trust in Tumang BMT is getting stronger.

3.2 Norm

Tumang BMT as a sharia-based financial institution plays an important part as role model of sharia value implementation to the public. In running its operational activity, Tumang BMT applies sharia principles to both employees and members as the mutual consensus in achieving the vision of institution. Norm is applied based on Islamic principle inculcated into members and customers, basically, Tumang BMT applies two norm principles: Standard Operating Management (SOM) and Standard Operating Management (SOP). Entire management is required to understand these two principles and to apply them in undertaking its task.

Technical norms include employees' dressing rule according to sharia, behavioral order ad speaking rule with opposite sex, and appeal to give service with smile, greeting, and addressing to customers. Furthermore, the management is required to be disciplined in applying the accounting corresponding to the Standard Financial Accounting Statement standardized into computer application program integrated and connected to the internet. Then, the account officer (AO) on duty serving to receive financing installment or saving should be equipped with mobile printer connected to the server in the office. Thus, this norm can grow professional work culture among the employees in giving service to the customers. It leads to the public's stronger trust in Tumang BMT institution.

3.3 Reciprocity

Reciprocity is accomplished by Tumang BMT, members, and people through many activities and programs, the benefit of which can be felt collectively. Through Tumang BMT serving not only as the financial institution but also as social institution that distributes *zakat, infak,* and *sedekah* to the people, many reciprocal benefits can be felt by both parties. On the one hand, Tumang BMT as an institution benefits from the public's stronger trust in it, and on the other hand, members and people also benefit from the program useful to their social life. The reciprocity between BMT and people can also accomplished through giving financial grant ad sponsorship to many events held by the village, for example, Tumang Fair, *sadranan*, blood donation, 17-an (Indonesia's Independence Day) competition, and other activities. In addition, the presence of membership building program through routine Qur'an study (*pengajian rutin*) for *samawa* family also benefits reciprocally the credibility of Tumang BMT and gives education to the members and people.

3.4 Social Networking

The existence of Tumang BMT is highly supported with the presence of social networking. Social networking involves social relation established between Tumang BMT and its members, employees, and external relation. In relation to social networking, Putnam divides social capital into two basic forms: bridging ad bonding social capital. Putnam (2000) in Slamet (2012: 28) revealed that bonding social capital sees more inward and tends to strengthen exclusive identity and homogeneous group. Meanwhile, bridging social capital can result in broader identity and reciprocity. Wolcock added one more social capital form, linking social capital. This social capital form reaches people in different situation, for example those existing out of community completely, thereby encouraging its members to utilize more resources than the ones available in the community (Wolcock in Field, 2016).

Bonding social capital in Tumang BMT has been implemented to many activities, either internally(employees), and externally (customers). Many internal activities like routine Quran study for samawa family is held monthly in turn from an employee's house to another's, recreation activity including picnic, outbound, family gathering, njagong manten (attending wedding party) and takziah (attending mourning ceremony) are internal activities that can strengthen the employees' solidarity and the institutional identity bond. Meanwhile, bonding with customer networking can be manifested into saving system with 'celengan bumbung (tubing money box)', Angkat Telepon Moro (ATM) constituting the cash withdrawal and depositing service using proactive service during work hour, building Excellent Family (MKU) as a means of building, educating, and habituating the good deed (amal saleh) in order to be Excellent family, facilitation program for remote villages, holding badminton event, giving sponsorship to many village events, holding Tumang BMT Cup football competition with national-level player materials and teams, giving grant and help to destitute and orphanage, giving help to disaster victim, distributing Qurban meat and providing scholarship to the people, from Elementary to Senior High School Levels.

In addition, Tumang BMT also has economic empowerment program named POKUSMA (Community Business Group) as its bonding social capital. This program aims to alleviate poverty through rotating loose fund loan (qardhul hasan) constituting the loan program without any additional return in its payment given to new members to improve their micro business that has run or to establish new business. Additionally, Tumang BMT management also gives both group and personal facilitation to monitor the progress of individual members' and group's business in order to conduct periodical evaluation.

Linking social capital is established through social relation with other BMT. The initiator, in this case the General Director of Tumang BMT, Adib Zuhairi, establishes linking by synergizing with other BMTs. In 1999, he initiated the establishment of BMT Regional Soloraya Association focusing on synergy and joint training. Soon after, Adib along with his two friends initiated the establishment of Central Java BMT Association. The journey still continued until in 2005 an Association of Indonesia BMT (PBMT) was established. The establishment of PBMT itself is the manifestation of commitment to reinforcing and developing BMT movement. The reinforcement is intended to keep improving many internal indicators in order to optimize the achievement of specified vision and mission. Meanwhile, the development is intended to improve the role and benefit of BMT either qualitatively or quantitatively. PBMT then conducts many sustainable activities and consolidation, one of which is to establish an enterprise to maintain and to strengthen the liquidity of BMT as its members. The enterprises established by PBMT include PBMT Ventura (capitalization), PBMT Institute (Human Resource), PBMT Rowasia (Information Technology), PBMT Ta'awun and PBMT Travel.

Bridging social capital of Tumang BMT is manifested into some cooperation established along with external groups or institutions beyond the relation with other BMTs. This presence of social capital can create a new identity broadly and reciprocity from many parties involved within it. Tumang BMT's activities becoming the part of bridging capital are cooperation, socialization, publication and advocacy intended more to external parties such as Ministry of Cooperatives and SMEs, DPR (Legislative Assembly), DPRD (Local Legislative Assembly, proselytizing institution, badan zakat dan wakaf (an institution managing zakat and wakaf) and of course Islamic organization. Adib, as the director, often sees many national figures, attends many events related to BMT, and hold discussion forum. Such bridging social capital will be good for the external relation to support the need for information that cannot be fulfilled by fellow BMTs. This bridging social capital can serve to create a new identity broadly and reciprocity from many parties involved within it. Such relation enables to establish mutually benefiting connection and network with external association or group. Thus, the progress of Tumang BMT can be achieved more easily because idea exchange keeps developing and stimulating the development of BMT institution and of course the stakeholders within it.

Table. 1 Development of Tumang BMT with The Social Capital Mechanism

The Element of Social Capital	Research Data Finding
Trust	1. Offering product by door to door 2. Applying *bumbung* diplomatic system for members to saving 3. Open potential services for goods and services (LPG gas agents, carbide welding agents, providers of sewing tools, telephone services 4. Optimizing roles as financial institutions and also as da'wah and social institutions
Norm	1. Applying sharia principles to both employees and members as a together consensus in achieving the institution's vision 2. Implementing two norm principles, namely the Operational Management Standard (SOM) and the Standard Operating Procedure (SOP)
Reciprocity	1. Optimazing the dual of BMT functions as a financial institution and as a dakwah and social institution makes many reciprocal benefits felt by customers and members 2. Providing financial assistance and sponsorship to various events held by the village 3. The existence of a membership development program through a regular recitation of the samawa family
Social Networking	**Bonding** 1. Implemented in various activities both internal (employees) and external (customers). 2. Various internal activities such as a routine study of the regular Samawa family every month which is held in turn from one house to another employee's house, picnic and outbound, family gatherings, *jagong manten* and *takziah* 3. Whereas bonding with customer networks is manifested in the form of a savings system with a *'Celengan bumbung'* model, *Angkat Telepon Moro (ATM), Membangun Keluarga Utama (MKU)*, assistance program for remote villages, holding badminton events, giving sponsorship funds to various village events, organizing BMT Tumang Cup soccer match for the benefit of the poor and orphaned, giving assistance to victims of disasters, distributing *qurban* meat and providing scholarship

https://doi.org/10.1051/shsconf/20208601009

	programs for people from elementary to high school. **Linking** 1. Manifests through social relationships with other BMT as belonging to the Association of BMT Regional Soloraya, Association Central of BMT (Province) and Association of BMT Indonesia (National). **Bridging** 1. Manifests with diverse forms of cooperation with external groups and institutions outside of a relationship with another BMT 2. Doing cooperation, socialization, publications and advocacy that is more aimed at external parties such as the Ministry of Cooperatives and MSMEs, DPR, DPD, *Dakwah* Institutions, *wakaf* and *zakat* institutions, and Islamic organizations 3. often visit various national figures, attend events that are mostly related to BMT and hold discussion forums

4 Conclusion

Baitul Maal wat Tamwil (BMT) Tumang is one of the microfinance institutions that has developed in Indonesia. This year the Tumang BMT has a total of 20 branches both in Central Java and DIY with 22,531 members in 2016, with assets of Rp 98 billion in 2015, and increasing to Rp 129 billion in 2016, on December 2017 the assets reached Rp 165 billion. Various kinds of awards were won by MICRO Sharia financial institutions, one of which was "The Best Islamic Micro Finance IDR 25 <Asset <IDR 50 held" by Karim Bussiness Consulting. The initial purpose of the existence of the cooperative was to save the Tumang Village people who were in debt by loan sharks. This is because the Tumang community is classified into the structure of rural communities in the slopes of Mount Merapi, so it is not bankable. Besides that, most of them work as metal craftsmen (aluminum and copper), so they need a lot of capital.

The development of Tumang's Baitul Maal wat Tamwil (BMT) is inseparable from the strategy of the social capital it operates. The strategy of social capital according to Robert Putnam in his book "Making Democracy Work: Civic Traditions in Modern Italy (1993), social capital is used as a feature of social organizations, such as networks, norms, and beliefs, which facilitate coordination and cooperation for mutual benefit. As a form of financing that is foreign to the community, Baitul Maal wat Tamwil (BMT) Tumang, builds "trust" in the community by offering products in a doo to door manner. One interesting thing is implementing bumbung diplomacy. This choice was made by management because it was inspired by Tumang community habits who often use bamboo pole or cagak to be used as piggy banks (celengan). As the BMT Islamic micro finance institution Tumang implements "norms or rules" with sharia principles for both employees and members as a joint consensus in both Standard Operating Management (SOM) and Standard Operating Management (SOP). As a micro sharia or Islamic financial institution and as a missionary and social institution (social assistance and sponsorship) make many "mutual relations" benefits arise between Tumang BMT and the community. The "Network" of Tumang BMT with customers is realized in the form of a system of saving money with a model of "celengan bumbung", Raising Moro Phones (ATM), Building a Main Family (MKU), and a village empowerment program.

References

[1] O. Jasa Keuangan, *Laporan Perkembangan Keuangan Syariah 2017* (2017)
[2] S. Mansori, C. S. Kim, and M. Safari, Asian Soc. Sci. **11**,273 (2015)
[3] W. Arafah and L. Nugroho, Int. J. Res. Bus. Stud. Manag. **3**, 7 (2016)
[4] H.D. Seibel, J. Soc. Issues Southeast Asia. **23**, 86 (2008)
[5] R. K. Yin, *Studi Kasus Desain & Metode* (Rajawali Pers, Jakarta, 2014)
[6] M. B. Miles and A. M. Huberman, *Qualitative Data Analysis, 2nd Ed* (sage, 1994)
[7] M. Ricky PD, *Inspirasi Bumbung Bambu (Kisah Juang Adib Zuhairi Membangun BMT Tumang)* (PT. Aksara Solopos, Surakarta, 2018)
[8] J. Field, *Social Capital* (USA: Roudledge, Canada, 2008)
[9] R. D. Putnam, R. Leonardi, and R. Y. Nanetti, *Making Democracy Work: Civic Traditions in Modern Italy* (Princeton university press, 1993)
[10] Y. Slamet, *Modal Sosial Dan Kemiskinan : Tinjauan Teoritik Dan Kajian Di Kalangan Penduduk Miskin Di Perkotaan* (UNS Press, Surakarta, 2012)
[11] S. Usman, *Modal Sosial* (Pustaka Pelajar, Yogyakarta, 2018)

Study of Internal Factors and External Factors of Insurance Companies Towards Company Value (Study on Indonesia Stock Exchange)

Chairul Anam[1*]

[1]Management Study Program, Trunojoyo Madura University, Indonesia

Abstract. The purpose of this study is to determine the influence of the company's external factors with proxies: inflation and interest rates, and the company's internal factors with proxies: Return on Equity, and Debt to Equity Ratio partially and simultaneously to firm value in insurance sector companies listed on the Stock Exchange Indonesia. The research method used in this study is quantitative research methods, with the object of research of insurance companies listed on the Indonesia Stock Exchange, amounting to 14 companies. This study used a purposive sampling technique that produced 8 companies as the research sample. The research data source uses secondary data in the form of documents including data about the company's general description and financial statements of insurance companies on the Indonesia Stock Exchange (IDX) for 5 years. The results of this study indicate inflation, interest rates, Return on Equity, and Debt to Equity Ratio simultaneously have a positive but not significant effect on company value, then partially the other 3 variables, namely inflation, interest rates, and Debt to Equity Ratio have a positive effect but not significant to firm value, while variable Return on Equity has a positive and significant effect on firm value. Based on the coefficient of determination of 0.138 this shows the influence of 4 variables, namely inflation, interest rates, return on equity, debt to equity ratio of 13.8% while the remaining 86.2% is influenced by other factors, for example: the level of competition, policy company, developments in macroeconomic conditions.

1 Introduction

In the era of industrial revolution 4.0 competition between companies was increasingly competitive, including companies in Indonesia competing to enhance high competitiveness and excellence in the business and economic world. Competition between companies is not except with insurance companies. The increasing prevalence of insurance companies in Indonesia makes competition very tight to make its existence in the insurance world. In

* Corresponding author: chairul.anam199@gmail.com

addition, it is undeniable that human beings need protection at any time when they are anywhere and even at any time. Every normal human being must have aspirations to always improve or improve their standard of living in order to pursue physical and spiritual well-being.

The development of the insurance industry in Indonesia is quite prospective, the general insurance industry has the opportunity to grow by double digits in 2019, although this estimate is strongly influenced by the political situation next year. If the election runs well and is conducive, then post-election economic activities are believed to be able to run faster (Julian Noor, 2018).

The development of insurance companies is influenced by external factors and company internal factors. External factors in this study are proxied with inflation and interest rates, while the internal factors of the company are proxied with profits as measured by Return on Equity (ROE) and debt as measured by the Debt Equity Ratio (DER)

In an economic system, inflation is an economic indicator that describes the decline in the value of the rupiah, and this condition is characterized by the increase in the prices of goods needed in the market (Noer Irawan and Hamid, 2013). Increasing inflation can reduce people's purchasing power, because the prices of goods need to increase, while people's income remains. Inflation is often a problem, especially if the government is unsuccessful or unable to control its growth rate.

According to Gusnardi (2017) inflation is an increase in the price of goods continuously. Inflation is low and reflects the stability of good economic growth, the stability and low benefits of inflation for companies will certainly increase profits for companies from rising goods prices, on the contrary high inflation will cause the value of goods and services to rise sharply so that the emergence of speculative actions from investors will detrimental to the company even the economy of a country will be negatively affected, such as: the decline in value and trust in the currency that will decrease, with this in mind the government has one of the tools to control inflation, namely interest rates. Furthermore Demir (2007), states that inflation has a significant negative effect on the profitability of manufacturing companies, as well as Pareira (2010) states that there is a negative relationship between inflation and stock prices.

Hamidah (2015) said that the high level of inflation would reduce wealth in the form of money and would reduce people's purchasing value. The existence of this is caused by several factors, one of which is the stronger the community will be a demand for a particular item, because it increases the cost of producing an item. Inflation that occurs in a country is also very influential on interest rates because every time there is an increase in the inflation rate it will be followed by an increase in BI interest rates. With this, the government policy to reduce the inflation rate by raising interest rates means to reduce the amount of money in circulation and attract public funds so as to reduce inflation. According to Hafiz Syamid Ardhi (2017) Interest rates are a tool for the central bank of a country in maintaining the stability of a country's economic growth, interest rates are also one of the macro variables that need to be considered that can affect the value of the company. An increase in interest rates will reduce corporate profits and cause

In the era of industrial revolution 4.0 competition between companies was increasingly competitive, including companies in Indonesia competing to enhance high competitiveness and excellence in the business and economic world. Competition between companies is not except with insurance companies. The increasing prevalence of insurance companies in Indonesia makes competition very tight to make its existence in the insurance world. In addition, it is undeniable that human beings need protection at any time when they are anywhere and even at any time. Every normal human being must have aspirations to always improve or improve their standard of living in order to pursue physical and spiritual well-being.

The development of the insurance industry in Indonesia is quite prospective, the general insurance industry has the opportunity to grow by double digits in 2019, although this estimate is strongly influenced by the political situation next year. If the election runs well and is conducive, then post-election economic activities are believed to be able to run faster (Julian Noor, 2018).

The development of insurance companies is influenced by external factors and company internal factors. External factors in this study are proxied with inflation and interest rates, while the internal factors of the company are proxied with profits as measured by Return on Equity (ROE) and debt as measured by the Debt Equity Ratio (DER)

In an economic system, inflation is an economic indicator that describes the decline in the value of the rupiah, and this condition is characterized by the increase in the prices of goods needed in the market (Noer Irawan and Hamid, 2013). Increasing inflation can reduce people's purchasing power, because the prices of goods need to increase, while people's income remains. Inflation is often a problem, especially if the government is unsuccessful or unable to control its growth rate.

According to Gusnardi (2017) inflation is an increase in the price of goods continuously. Inflation is low and reflects the stability of good economic growth, the stability and low benefits of inflation for companies will certainly increase profits for companies from rising goods prices, on the contrary high inflation will cause the value of goods and services to rise sharply so that the emergence of speculative actions from investors will detrimental to the company even the economy of a country will be negatively affected, such as: the decline in value and trust in the currency that will decrease, with this in mind the government has one of the tools to control inflation, namely interest rates.

Hamidah (2015) said that the high level of inflation would reduce wealth in the form of money and would reduce people's purchasing value. The existence of this is caused by several factors, one of which is the stronger the community will be a demand for a particular item, because it increases the cost of producing an item. Inflation that occurs in a country is also very influential on interest rates because every time there is an increase in the inflation rate it will be followed by an increase in BI interest rates. With this, the government policy to reduce the inflation rate by raising interest rates means to reduce the amount of money in circulation and attract public funds so as to reduce inflation. According to Hafiz Syamid Ardhi (2017) Interest rates are a tool for the central bank of a country in maintaining the stability of a country's economic growth, interest rates are also one of the macro variables that need to be considered that can affect the value of the company. An increase in interest rates will reduce company profits and cause stock prices fell. According to Sujoko and Ugy (2007) said a high interest rate would reduce investors' interest to invest their funds into the capital market and prefer to invest in deposits so that stock trading activities would decrease and the value of the company would also decline and vice versa, if the value of the company increased then the performance of the company must be improved so that a company can benefit.

Furthermore, William T. Sucuahi et al (2016), states that based on the results of the study show that even though the inflation rate can positively influence the growth of stock prices among diversified companies in the Philippines, this influence is not significant. The inflation rate is not the main factor for the growth of stock prices. Acquiring the company's shares is the best strategy for investors even though inflation fluctuates and is unstable. Thus, the concept that inflation can affect stock prices cannot be applied to companies listed on the Stock Exchange. The results of this study are expected to help investors understand risk and hedge investments in the making and evaluation of monetary policy.

Companies must also have to be able to effectively manage their sources of funds in order to increase net income. As for the type of ratio used to analyze the company's financial performance, one of them is the profitability ratio. This profitability ratio is a ratio to assess and measure the ability of a company's value in obtaining profits or profits by using the capital owned by the company itself. As for the part of the profitability of the profitability ratio is Return On Equity (ROE) which aims to measure the ability of a company to obtain profit after tax by using its own capital owned by the company. The higher Return On Equity (ROE) will increase the demand for company shares. The high stock price will determine the return that will be received by investors. Dewi and Wirajaya (2013: 363) said that the higher the company gets a profit, the greater the return expected by investors so that it can make the company's value better. Prasetyorini (2013) says that profitability has a positive and significant effect on firm value.

Company management must be observant in collaborating with other companies in order to optimize cash in the company. If the company experiences financial difficulties and is threatened with bankruptcy due to excessive use of debt so that it is unable to fulfill its obligations, the loss can be in the form of dividend reduction or bankruptcy, in order to find out how much debt is used to meet company spending. One of the calculation of this ratio is the Debt To Equity Ratio, which is to measure the size of the use of long-term debt when compared with the equity in the company. Noerirawan and Muid (2013) say that Debt to Equity Ratio (DER) has a negative but not significant effect on firm value. Caused by the existence of debt that is too high and has an impact on the lack of public confidence in the company.

Based on the description above, the following hypotheses can be formulated:

H1 : Inflation affects the value of insurance companies

H2 : Interest rates affect the value of insurance companies

H3 : Profitability (Return On Equity) affects the value of the insurance company

H4 : Debt Equity Ratio affects the value of the insurance company

H5 : Inflation,interest rate,return on equity and debt equity ratio affects the value of insurance companies

2 Research methods

2.1 Research data

The type of data used in this study is quantitative data. The dependent variable used in this study is company value which is a value for measuring company performance which is often associated with investor perceptions of the company's stock price. While the independent variables used in this study are the company's external factors which include: Inflation, Interest Rates, and internal company factors which include: profitability as measured by Return On Equity (ROE), and debt as measured by Debt To Equity Ratio (DER) which is associated with an assessment of the value of the company.

2.2 Object of research

This research was conducted at an insurance company with components listed on the Indonesia Stock Exchange. The time of the research conducted in this study is that it starts from 2013 until 2017. According to the data obtained from the Indonesia Stock Exchange which is accessed through the official website of the Indonesia Stock Exchange (www.idx.co.id).

2.3 Population

The population in this study were 14 Insurance Companies Listed on the IDX (Indonesia Stock Exchange) for the period 2013-2017.

Table 1. Population Research

No	Kode Perusahaan	Nama Perusahaan	IPO
1	ABDA	Asuransi Bina Dana Arta Tbk	06 Juli 1989
2	AHAP	Asuransi Harta Aman Pratama Tbk	14 Sep 1990
3	AMAG	Asuransi Multi Artha Guna Tbk	23 Des 2005
4	ASBI	Asuransi Bintang Tbk	29 Nop 1989
5	ASDM	Asuransi Dayin Mitra Tbk	15 Des 1989
6	ASJT	Asuransi Jaya Tania Tbk	23 Des 2003
7	ASMI	Asuransi Mitra Maparya Tbk	16 Januari 2014
8	ASRM	Asuransi Ramayana Tbk	19 Maret 1990
9	JMAS	Asuransi Jiwa Syariah Jasa Mitra Abadi Tbk	18-Desember-2017
10	LPGI	Lippo General Insurance Tbk	06-September-2005
11	MREI	Maskapai Reasuransi Indonesia Tbk	04-September-1989
12	MTWI	Malacca Trust Wuwungan Insurance Tbk	11-Oktober-2017
13	PNIN	Paninvest Tbk	20-September-1983
14	VINS	Victoria Insurance Tbk	28-September-2015

2.4 Sample

By using purposive sampling method that is by limiting the sampling that will be examined based on certain criteria, namely:

a) Insurance companies registered on the IDX from 2013 to 2017.
b) Insurance Companies that issue financial statements during the period 2013-2017.
c) The company's financial statements contain a balance sheet, profit / loss, cash flow that has been audited by a public accountant
d) The company uses the Rupiah unit as the currency in its financial statements.
e) The Company has the end of the financial statement period 31 December.

Table 2. Research Samples

No	Kode Perusahaan	Nama Perusahaan	IPO
1	ABDA	Asuransi Bina Dana Arta Tbk	06 Juli 1989
2	AHAP	Asuransi Harta Aman Pratama Tbk	14 Sep 1990
3	ASBI	Asuransi Bintang Tbk	29 Nop 1989
4	ASJT	Asuransi Jaya Tania Tbk	23 Des 2003
5	ASRM	Asuransi Ramayana Tbk	19 Maret 1990
6	ASDM	Asuransi Dayin Mitra Tbk	15 Des 1989
7	LPGI	Lippo General Insurance Tbk	06 Sept 2015
8	MREI	Maskapai Reasuransi Indonesia Tbk	04 Sept1 989

3 Research results and discussion

3.1 Research result

The research data in the form of secondary data obtained from the Indonesia Stock Exchange was processed and analyzed using multiple regression analysis techniques.

3.1.1 Multiple linear regression analysis

Multiple linear regression analysis is used to measure the effect of independent variables on dependent variables together or partially. This measurement of influence involves four independent variables (X) and a dependent variable (Y),

The regression equation with multiple linear in this study are:

$$Y = a + b1\,X1 + b2\,X2 + b3\,X3 + b4\,X4 + e \qquad (1)$$

Information :
a: Constant
Y: Price Book Value
e: Standard Error Regression X1: Inflation
X2: Interest Rate
X3: ROE
X4: DER
B1-4: Coefficient

3.1.2 Hypothesis testing

3.1.2.1 Test t

The T test is used to determine whether the independent variables partially influence the dependent variable. To test whether each independent variable has a significant effect on the dependent variable partially with $\alpha = 0.05$. Then the way that is done is:

a) If t count <t table then Ho will be accepted, meaning that the independent variable does not have a significant influence on the dependent variable.
b) If t count> t table then Ho will be rejected, meaning that the independent variable has a significant influence on the dependent variable
c) If Ha is accepted if the probability is less than 0.05 (a), it means that there is a significant influence between the independent variables on the dependent variable.

3.1.2.2 Test F

The F test is used to determine whether the independent variables together have a significant effect on the dependent variable. The basis for making test decisions is:

a) If the significant value of the test is F> 0.05 then Ho will be accepted, meaning that the variation of the regression model used does not succeed in explaining the independent variable.
b) If the significant value of the test is F <0.05, Ho will be rejected, meaning that the variation of the regression model used successfully explains the independent variable

The formula used is:

$$F = \frac{R^2/(k-1)}{(1-R)^2/(n-k)} \qquad (2)$$

Information :
F = F count which is then consulted with F table
R2 = Coefficient of determination
n = Number of samples
k = Number of variables

3.1.3 Determination coefficient test (r2)

According to Ghozali (2013: 97) the coefficient of determination (R2) essentially measures how far the model's ability to explain the variation of the dependent variable. The coefficient of determination is between zero and one. The small value of R2 means that the ability of independent variables to explain variations in the dependent variable is very limited. A value close to one means that the independent variables provide almost all the information needed to predict the dependent variable.

3.1.4 Hypothesis testing

Based on the results of multiple regression data analysis techniques using the SPSS statistical program the following results are obtained

Model		Unstandardized Coefficients		Standardized Coefficients	t	Sig.	Correlations			Collinearity Statistics	
		B	Std. Error	Beta			Zero-order	Partial	Part	Tolerance	VIF
1	(Constant)	1,694	2,310		,733	,468					
	Inflasi	,471	,668	,137	,705	,485	,157	,118	,111	,653	1,533
	Suku bunga	-,427	2,237	-,037	-,191	,850	,041	-,032	-,030	,663	1,508
	ROE	1,140	,548	,386	2,078	,045	,330	,331	,326	,714	1,401
	DER	-,341	,443	-,141	-,768	,447	,066	-,129	-,121	,729	1,372
a. Dependent Variable: PBV											

Model Summaryb

Model	R	R Square	Adjusted R Square	Std. Error of the Estimate	Change Statistics					Durbin-Watson
					R Square Change	F Change	df1	df2	Sig. F Change	
1	,372a	,138	,040	,65965	,138	1,402	4	35	,254	,727
a. Predictors: (Constant), inflai, sukubunga, ROE,DER										
b. Dependent Variable: PBV										

ANOVAa

Model		Sum of Squares	df	Mean Square	F	Sig.
1	Regression	2,440	4	,610	1,402	,254b
	Residual	15,230	35	,435		
	Total	17,669	39			
a. Dependent Variable: PBV						
b. Predictors: (Constant), inflasi,sukubunga, ROE,DER						

3.2 Discussion

3.2.1 Test Coefficient of Determination (r2)

The Determination Coefficient Test (R2) shows that the proportion of variation in the dependent variable (Company Value) can be explained by independent variables (Inflation, Interest Rate, Return On Equity, and Debt to Equity Ratio) while the remainder can be explained by other variables not included in research.

The value of R Square is equal to 0.138 which indicates that the independent variable (Company Value) can be explained by the independent variable (Inflation, Interest Rate,

Return On Equity, Debt To Equity Ratio) which is 13.8% and the remaining 86.2% influenced by other factors not explained in the regression model in this study such as: economic conditions, type and quality of products offered, service quality, level of competition in the insurance industry in Indonesia.

3.2.2 Influence of Inflation on Company Values

Based on the results of this study there is an influence of inflation on the value of the company in the insurance sector listed on the Indonesia Stock Exchange. The results of this study indicate that the value of t obtained is 0.705 and a significant level of 0.485 is greater than 0.05 (0.485> 0.05) which means that inflation does not significantly influence the value of companies in insurance sector companies listed on the Indonesia Stock Exchange in the 2013 period until 2017. This shows that if the company is able to increase inflation then the value of the company will decrease but the effect is not significant. non-significant results inform that the increase and decrease in inflation is not in line with the value of the company.

Viewed from the point of view of investors Inflation as a company external factor for that management is more focused and pay attention to how the company produces high profits so that it can produce high returns.

Based on the research data, the inflation value fluctuated from year to year, for example in 2013, which was 4.03 and the Company Value was 428.5644, then in 2014 Inflation increased to 107% to 8.38 and the Company Value was 324,0062, in 2015 moved down 0.2% to 8.36 and the value of the Company decreased as much as 369.3593, until 2016 Inflation decreased by 0.6% to 3.35 and the Company Value was 32.6691 The lowest inflation was in the year 2017, which is a decrease of 9% to 3.02 and a Company Value of 25.4098.

This is in line with the research conducted by Hamidah (2015) where the results show that inflation does not significantly influence the value of the company. Gusptasari (2013) also said that Inflation had no significant effect on Company Value. This is also consistent with Sangkyun's (1997) study, which states that the inflation rate does not have a significant effect on stock returns that have an impact on the stock prices of manufacturing companies.

3.2.3 Effect of Interest Rates on Company Values

Based on the results of this study there is an influence of interest rates on the value of the company in the insurance sector listed on the Indonesia Stock Exchange. The results of this study indicate that the interest rate has no significant effect on the value of the company. With a value of t equal to - 0.191 greater than 0.05, which means that the Interest Rate does not have a significant effect on the Corporate Value of insurance sector companies listed on the Indonesia Stock Exchange from 2013 to 2017.

Based on the results of data analysis can be shown in the interest rate fluctuations in 2013 amounted to 5.77% with a Company Value of 428.5644 then in the period of 2013 the interest rate increased by 7.02% with a Company Value of 324,0062 Interest rates in 2014 also increased by 8.16% with a Company Value of 369.3593, while for 2015 the interest rate decreased by 7.52% with a Corporate Value of 32.6691 up to 2017 Interest Rates also decreased by 6.00% with a Company Value of 25.4098.

This indicates that the BI rate is more likely to control inflation in the rupiah exchange rate decline not to increase or decrease the value of the company. However, capital market players are not affected by the increase in interest rates due to the increase and decrease in the interest rates are only temporary, and capital market players prioritize long-term returns.

The results of this study are in line with Khodir's research (2013), saying that interest rates have a negative and not significant effect on firm value. The results of these studies are also consistent with the research of Sujoko and Ugy (2007); Rakhimsyah and Gunawan (2011) say that the interest rates have a negative effect that is not significant on the value of the company.

3.2.4 Effect of Return On Equity Against Company Values

Based on the results of this study there is an effect of Return On Equity (ROE) on Corporate Values in the insurance sector listed on the Indonesia Stock Exchange. The results of this study indicate that Return On Equity (ROE) does not have a significant effect on Firm Value. with a t value of 2.078 greater than 0.05, which means that Return On Equity (ROE) does not have a significant effect on Company Values in insurance sector companies listed on the Indonesia Stock Exchange period 2013 to 2017

Based on the results of data analysis can be shown in the level of value fluctuations Return On Equity (ROE) with an average in 2013 that is equal to 1.4654 with Corporate Value that is equal to 428.5644, then in 2013 amounted to 1.2989 with Corporate Value that is equal to 324,0062. Then in 2014 amounted to 1.2472 while the Company Value was 369.3593. In 2015 amounted to 1,1954 with a Company Value of 32,6691. Then in 2017, which is equal to 1.0128 with a Company Value of 25.4098

Return On Equity (ROE) has a significant effect on Firm Value, influential Return on Equity (ROE) on Corporate Values is caused because the profits obtained by the company are stable from year to year and tend not to fluctuate so investors do not hesitate and are confident of the results that the company will get in the future future so that it reflects the company's ability to generate high returns for shareholders. The greater the profits obtained, the greater the ability of the company to pay dividends and will have an impact on increasing the value of the Company. high low value of Return On Equity (ROE) because the company experiences fluctuations in the company's net income, each year the Return On Equity (ROE) decreases starting from 2013-2017. a decrease in Return on Equity (ROE) on Company Value.

The results of the hypothesis testing are in line with the research of Siska Adelina (2014), Anthony Wijaya (2015), and Ria Fihendri (2017) who say that Return On Equity (ROE) significantly influences the Firm Value. But the statement is also not in line with the research conducted by Astriana (2014) saying that profitability does not affect the value of the company.

3.2.5 Effect of Debt To Equity Ratio on Company Values

Based on the results of this study there is an influence of the Debt to Equity Ratio (DER) on the Company Value in the insurance sector listed on the Indonesia Stock Exchange. the results of this study indicate that Debt to Equity Ratio (DER) does not significantly influence the value of the company. with a t value of -0.768 greater than 0.05, which means that the Debt to Equity Ratio (DER) does not have a significant effect on the Company Value in insurance sector companies listed on the Indonesia Stock Exchange from 2013 to 2017

Based on the results of data analysis, it can be shown in the value of the fluctuation of Debt to Equity Ratio (DER) with an average in 2013 that is equal to 19.4463 with Company Value that is equal to 428.5644, then in 2013 amounted to 18.9815 with Corporate Value namely amounting to 324,0062. Then in 2014 amounted to 18.8152 while the Company Value was 369.3593. In 2015 it amounted to 17.4232 with a Company Value of 32.6691. Then in 2017, that is equal to 14,8694 with Company Value that is equal to 25,4098 so that it can show that the existence of Debt to Equity Ratio (DER) is not able to increase Company Value.

This shows that the funding decision obtained from debt is not a direct factor that can be considered for investors to invest in the company. Investors need more information on how the management uses these funds as company capital effectively and efficiently to achieve added value for the company.

From year to year the Debt to Equity Ratio (DER) experiences fluctuations that differ only slightly in number so that the value of Debt to Equity Ratio (DER) to Company Value is not significant. the existence of these statements is not in line with the research of Nurchanifia (2011) and Gunawan (2011) saying that Debt to Equity Ratio (DER) has a significant effect on Company Value. But also in line with the research of Pancawati (2009), Sari and Sidiq (2013), Febrianti (2013) said that Debt to Equity Ratio (DER) has a positive effect on Firm Value. Dewi and Wirajaya's research (2013) also found different research results, namely Debt to Equity Ratio (DER) which had a negative and significant effect on Firm Value.

3.2.6 Effect of Inflation, Interest Rates, Return On Equity (ROE), Debt To Equity Ratio (DER) on Company Value.

Based on the results of this study there are influences of Inflation, Interest Rate, Return On Equity (ROE), Debt To Equity Ratio (DER) on Corporate Values in the insurance sector listed on the Indonesia Stock Exchange. The results of this study show that inflation, interest rates, return on equity (ROE), debt to equity ratio (DER) did not significantly influence the value of the company. Based on the analysis using SPSS, obtained F count value of 1.261 and a significant level of F of 0.254> 0.05. So that it can be concluded that simultaneously all the independent variables, namely the Inflation, Interest Rate, Return On Equity (ROE), Debt To Equity Ratio (DER) variables did not significantly influence the Company Value in insurance sector companies listed on the Indonesia Stock Exchange,

The results of the regression analysis show the coefficient of determination (R2) that is equal to 0.138 or 13.8%, changes that occur in Corporate Values can be explained by Inflation, Interest Rate, Return On Equity (ROE), Debt To Equity Ratio (DER) while the remainder is 86.2% is explained by other variables outside the research regression model.

This reflects that it is better that before investing in prospective investors not only pay attention to the internal factors of the company that will be targeted to invest, namely financial performance, but also pay attention to external factors that can also affect the state of the company. Inflation is one indicator to see the economic stability of a region or country that shows an increase in prices of services and goods. Where an economic condition occurs imbalance between demand and demand for products that are higher than the supply so that there is a tendency for price increases. High inflation will cause the purchasing power of money to decline and the level of investor income also decreases. The increase in the inflation value will have an impact on the increase in high operational costs so that the company gets fewer profits and will cause investor interest in investing in the company to decline. Inflation can also be caused by several factors, among others, the increasingly strong desire of the community for a demand for a particular item, due to the increase in bank interest rates, when all prices experience continuous increases in a long time, inflation will occur. In addition, if the interest rate increases, it will affect the industry, especially in credit matters. Thus the increase in BI interest rates will also rise and have an impact on commercial banks so that the lending rates are likely to increase. This proves that whenever there is an increase in the inflation rate, it will be followed by an increase in the BI Rate.

According to Kasmir (2008) said that Return On Equity is a ratio used to measure profit after tax with own capital, high Return On Equity (ROE) reflects the company's ability to generate high profits for shareholders. Then the higher the stock to buy. The existence of this will cause the demand for shares to increase and cause the stock price to continue to rise and the value of the company will also increase. Debt to Equity Ratio (DER) is a value that describes the ability of an issuer to fulfill all of its responsibilities, for example in debt repayment, if there is an increase in debt value in a company it will result in a high Debt To Equity Ratio, which means the higher interest expense companies that later affect the value of net income to be received. The results of the study conducted by Guspitasari (2013) say that simultaneously the inflation rate, interest rate and Debt To Equity Ratio (DER) have a positive effect on Company Value.

4 Conclusion

Based on data analysis and hypothesis testing conclusions can be obtained as follows :

a) Based on the testing of hypotheses using the variable interest rate t test does not have a significant effect on the value of insurance sector companies listed on the Indonesia Stock Exchange.
b) Based on hypothesis testing using t test debt variable measured by Debt Equity Ratio (DER) has no significant effect on the value of insurance sector companies listed on the Indonesia Stock Exchange.
c) Based on hypothesis testing using t test profitability variables measured by Return On Equity (ROE) have a significant effect on the value of insurance sector companies listed on the Indonesia Stock Exchange.
d) Based on testing using the F Test has no significant effect together between the independent variables: Inflation, Interest Rate, Return On Equity Ratio (ROE), and Debt To Equity Ratio (DER) to the value of the insurance sector companies listed on the Indonesia Stock Exchange 2013-2017 period.
e) Based on the coefficient of determination, the results obtained 13.8%, this shows the effect of independent variables on the non-independent variable by 13.8%, the remaining 86.2% is influenced by other factors, such as macroeconomic conditions, the level of competition, type and quality of products etc.

5 Recommendation

a) For investors, it is better to pay more attention to good financial performance that is able to explain the Company Value is greater than the financial performance that is not good and is able to increase its operational costs so that the Company's Value will improve.
b) For companies to manage the insurance sector to be more efficient in operating the company so that an increase in the independent variable can increase the greater value of the Company.
c) For the next researcher, it is expected to add the variables under study in order to obtain comprehensive and integrative results.

References

[1] H. S. Ardhi, H. A. Hasan, Y. Efni, A. Rokhmawati, and Gusnardi, J. Akuntansi, Kewirausahaan, Dan Bisnis (2017)
[2] S. Azizah and M. P. Priyadi, J. Ilmu Dan Ris. Akunt. **5**, (2016)

[3] S. dan Soebiantoro, J. Manaj. Dan Kewirausahaan **9**, 41 (2007)
[4] F. Demir, White Pap. Dep. Econ. Univ. Oklahoma (2007)
[5] A. S. M. Dewi and A. Wirajaya, E-Jurnal Akunt. **4**, 358 (2013)
[6] I. Fahmi, Manajemen Investasi Edisi 2 (Salemba Empat, Jakarta, 2015)
[7] I. Fahmi, Pengantar Teori Portofolio Dan Analisis Investasi (Alvabeta, Bandung, 2015)
[8] I. Ghozali, Aplikasi Analisis Multivariate Dengan Program SPSS (Badan Penerbit Universitas Diponegoro, 2006)
[9] Guspitasari, J. Ekon. Dan Pendidik. **8**, 112 (2013)
[10] H. Hamidah, H. Hartini, and U. Mardiyati, JRMSI-Jurnal Ris. Manaj. Sains Indones. **6**, 395 (2015)
[11] A. Hasymi, Dasar-Dasar Asuransi (Balai Aksara, Jakarta, 1981)
[12] Hery, Analisis Laporan Keuangan (Kompas Gramedia, Jakarta, 2016)
[13] A. Ira Roshita Dewi and L. Sri Artini, E-Jurnal Manaj. Univ. Udayana (2016)
[14] A. Martono and A. Harjito, Manajemen Keuangan. Edisi Pertama (Ekonesia, Yogyakarta, 2008)
[15] M. R. Noerirawan, Pengaruh Faktor Internal dan Eksternal Perusahaan Terhadap Nilai Perusahaan (Studi Empiris Pada Perusah. Manufaktur Yang Terdaftar Di Bursa Efek Indones. Periode 2007-2010) (2012)
[16] S. Nurchanifia, (2012)
[17] S. Park, Financ. Anal. J. **53**, 52 (1997)
[18] D. W. Pereira Garmendia, Available SSRN 1802431 (2010)
[19] T. Pramana, Jawa Timur Sinar Ilmu Publ. (2011)
[20] T. Pramana, Jawa Timur Sinar Ilmu Publ. (2011)
[21] F. PrasetyorinI and B. Fitri, J. Ilmu Manaj. **1**, (2013)
[22] R. Prihantini, (2009)
[23] L. A. Rakhimsyah and B. Gunawan, InFestasi (Jurnal Bisnis Dan Akuntansi) **7**, 31 (2011)
[24] L. A. Rakhimsyah and B. Gunawan, InFestasi (Jurnal Bisnis Dan Akuntansi) **7**, 31 (2011)
[25] H. Suad and E. Pudjiastuti, Dasar-Dasar Manajemen Keuangan Edisi Keenam (UPP STIM YKPN, Yogyakarta, 2012)
[26] H. Suad and E. Pudjiastuti, Penerbit UPP STIM YKPN, Yogyakarta (2012)
[27] W. T. Sucuahi, J. A. E. Alvarez, M. A. M. Gudes, and R. B. B. Parsacala, Int. J. Account. Res. **42**, 1 (2016)
[28] I. M. Sudana, Manajemen Keuangan Perusahaan Teori Dan Praktik (Erlangga, Jakarta, 2011)
[29] P. D. Sugiyono, Metod. Penelit. Kuantitatif, Kualitatif Dan R&D **291**, 292 (2014)
[30] M. Suharli, J. Akunt. Dan Keuang. **7**, 99 (2005)
[31] M. M. Warsono, Manajemen Keuangan Perusahaan Jilid Satu Edisi Tiga (Bayumedia Publishing, Malang, 2003)
[32] M. M. Warsono, Jilid Satu, Ed. Tiga, Cetakan Pertama, Bayumedia Publ. Malang (2003)
[33] R. M. Werner, Analisis Laporan Keuangan Proyeksi Dan Valuasi Saham (Salemba Empat, Jakarta, 2013)

Nugroho Gunawan A. Sarsiti 2015. Analisis Pengaruh Eva dan Mva terhadap Return Saham pada Perusahaan Manufaktur di Bursa Efek Indonesia. Vol 13. No 2.

Ria Fihendri 2016. Pengaruh Profitabilitas Terhadap Nilai Perusahaan Dengan Pengungkapan Corporate Social Responsibility (Csr) Sebagai Variabel Pemoderasi Pada Pt. Semen Indonesia (Tbk). Skripsi, Universitas Trunojoyo Madura : Madura.

Setiawan, Didik. 2015. Analisis Pengaruh Faktor Internal Dan Faktor Eksternal Perusahaan Terhadap Harga Saham Pada Bank Bumn Yang Terdaftar Di BEI. Skripsi Diterbitkan. Universitas Trunojoyo Madura: Jurusan Manajemen.

online
Bisnis.com, Nop 30, 2018

Revitalization Model of Balai Latihan Kerja (Blk) For Improving Knowledge, Education, And Skills Of Migrant Worker Based On Link And Match

RM. Moch. Wispandono [1*], *Eni* Sri Rahayu Ningsih [2], Deni S.B Yuherawan[3], and Netty Dyah[4]

[1]Lecturer in Business and Economics Faculty, University of Trunojoyo Madura, Indonesia
[2]Lecturer in Business and Economics Faculty, University of Trunojoyo Madura, Indonesia
[3]Lecturer in Law Faculty, University of Trunojoyo Madura, Indonesia
[4]Lecturer in Culture and Social Science Faculty, University of Trunojoyo Madura, Indonesia

Abstract. The vocational training center (Balai Latihan Kerja/BLK) has great potential to improve the quality of Indonesian workers. The existence of this BLK is increasingly strategic with the support of the government. This can be seen from the large number of BLKs scattered in each regency city in Indonesia. However, now the conditions of the BLK are very concerning, both in terms of the effectiveness of the use, the quality of the Training Center and the use of the BLK function. On the other hand, the condition and quality of Indonesian Migrant Workers (BMI) is very alarming. Based on research by the Center for Regional Studies of the University of Brawijaya, it was found that the majority of them did not have the skills needed in the destination country. This is compounded by their lack of knowledge about market analysis of needs (skills) in the destination country. From this fact, there is no link and match between migrant workers and the BLK of each region. This study aims to find a model that can minimize the weakness of handling migrant workers and increase their work productivity

1 Introduction

The vocational training center (Balai Latihan Kerja/BLK) has great potential to improve the quality of Indonesian labor. The existence of this BLK is increasingly strategic with the support of the government. The functions of the BLK include (a) preparing plans, programs, budgets, evaluations and reporting; (b) implementation of labor training; (c) conducting training program trials, and labor competency tests; (d) institutional cooperation and preparation of reports. The number of BLKs is spread in each provincial city and district in Indonesia. However, now the condition of the BLK is quite alarming. Based on the standpoint of the Ministry of Manpower and Transmigration, it shows that almost 100 percent of BLKs in Eastern Indonesia are poor, in Central Indonesia region 3.8 percent are

* Corresponding author: wispandono@trunojoyo.ac.id

good, 27.2 percent are moderate and 75 percent are bad and in the West Indonesia 15.7 percent are good, 37 , 3 percent is moderate and 47 percent is bad. Nationally the BLK with good conditions was only 11.1 percent, moderate conditions were 30.2 percent and bad conditions were 58.7 percent.

On the other hand, the condition and quality of Indonesian Migrant Workers (BMI) is very alarming. Some of the problems faced by BMI include culture, language and communication, low performance of BMI, violence, rape, termination of unilateral contracts, substandard wages, overstay, pregnancy and having children and lesbian issues. This happens because the average quality (input) of BMI is very low. Lack of knowledge, education and skills and protection makes them experience violence in the destination country.

This condition of BMI does not really need to happen if these migrant workers are given skills in the BLK of each region. The vocational training center should be the responsible facilitator to take care of the needs of migrant workers who will go abroad. Services provided include administrative services and practical services. But the fact is ironic, the condition of BLK in Indonesia is very alarming starting from the aspect of management, the quality of human resources, facilities and infrastructure, curriculum, and weak networks to the lack of care of the local government. Based on the facts above, this study offers a study of revitalization of vocational training centers to improve the knowledge, education and skills of migrant workers with the concept of link and match.

The revitalization study of the BLK is also very urgent because based on the facts, the condition of the Madurese (migrant workers) workforce is very alarming. Troubled Indonesian Workers from East Java in 2014 were dominated by Madurese with a total of around 3,519 people (metrotvnews.com). Four districts in Madura, according to Disnakertransduk East Java records, enter the five regions with the highest rate of migrant workers in East Java (Data from East Java Disnakertransduk 2015). Throughout 2014, there were 7,493 Indonesian migrant workers (BMI) from East Java who were in trouble, ranging from minor to severe violations. "Of that amount, around 70% percent are from Madura (Edi Purwinarto, 2015). The problematic TKIs included Sampang, Pamekasan and Sumenep. For Sampang there are 15.61 people, Pamekasan is 1129, Sumenep 829 TKI (Head of the East Java Provincial Government's Dinakertransduk, 2015).

The problem of this research is how to build a revitalization model of the Work Training Center (BLK) as a place to increase the knowledge, education and skills of migrant workers who have the concept of link and match.

Based on these problems, the steps taken in this study are: (1) identifying training programs at the Vocational Training Center (2) analyzing the potential and problems of human resources and facilities and infrastructure at the Vocational Training Center (3) identifying, analyzing stakeholders (private, government, academics) who have the opportunity to become BLK partners (4) analyze the training and management models at BLK based on the needs of the labor market and migrant workers, and (5) formulate a link and match based training center institutional strengthening model.

2 Method

The design of the research used is descriptive qualitative with data collection techniques adjusted to the data to be collected and the intended data source. Research locations in Bangkalan, Sampang, Pamekasan and Sumenep. The population in this location is BLK, Disnakertrans and prospective BMI with purposive sampling technique (sample technique). This research is expected to produce a BLK revitalization model as an effort to improve BMI skills and expertise. Data collection in this study was carried out with several steps, namely: In-depth interviews and observations conducted to collect data about the needs and problems of vocational training centers. While data on the level of knowledge, education and skills of BMI were collected through questionnaires, observations and in-depth interviews. To test the validity of the data triangulation method is used, both source triangulation and theory triangulation.

3 Results and discussion

The beginning of the establishment of the BLK came from the idea of establishing the Industrial Training Program Training Center (PLKPPI) in the field of industry in 1953. In 1960, the PLKPPI was directed to become an Employee Job Seekers Training in order to become skilled workers. In 1970, along with the times, there was a change from the Industrial Training Program Training Center (PLKPPI) to become a Training Center under the guidance of the Minister of Manpower of the Republic of Indonesia. And since the regional autonomy of the Vocational Training Center (BLK) has changed to the Regional Work Training Center (BLKD) under the auspices of the Managing Unit (UPT) of the Manpower Service in each region in Indonesia.

The Dinas Technical Implementation Unit (UPTD) of the Social Service, Manpower and Transmigration Office of Bangkalan Regency was established in 2012. The background of the establishment of the BLU UPTD in the service area was due to classic problems that emerged in Bangkalan district which were almost the same as in other regions in Indonesia is a lot of prospective applicants / job seekers who do not have adequate abilities and competencies as requested by users. The encouragement and demands of the workforce / industry (users) so that the government can facilitate the availability of prospective applicants / job seekers who can fulfill the qualifications they need is another reason for the existence of BLK so that the formation of BLK is intended to conduct training and guidance for job seekers to have abilities and competencies as expected by users. This is in accordance with the main task of the UPTD BLK Dinsosnakertrans Bangkalan which has the task of carrying out some technical activities in the field of skills, knowledge and administration training and management of Occupational Training Centers located under and responsible to the Bupati through the Head of Social Services, Manpower and Transmigration.

The BLK in Sampang District was established in 2014. The BLK was originally under the Office of Social, Manpower and Transmigration (Dinsosnakertrans). In this service there is a field of labor. Activities in the workforce are synonymous with training and job creation, including TKI (Indonesian Workers). Because the activities were related to training, it was deemed ineffective if they were in the Social Service Office. Finally, in 2013, they began designing the Vocational Training Center (BLK), then in 2014 the BLK was officially established. (interview with Pak Bisrul Hafil, BLK Sampang). Initially the BLK participated in the Dinsosnakertrans, but in the nomenclature of the 2016-2017 fiscal year, the BLK was sheltered by the Office of Cooperatives, Micro and Manpower Enterprises, to be exact on December 28, 2016. Based on interviews with informants, several data were obtained including strengths, weaknesses, opportunities and threats.

The Job Training Center in Sumenep Regency, which was renamed the Job Training Institute, but later returned to being a Sumenep BLK established since 1981 and is among the oldest on the island of Madura. This BLK was directly under the guidance of the East Java provincial government and so far it has been able to produce 2832 alumni who have been successfully trained in various skill fields, namely: graphic design, automotive, workshop, makeup, sewing, and office tools. Sumenep BLK currently has a total of 18 staff and 6 instructors / trainers (but 4 people have full-time status but are still needed when there are training activities). The BLK built on an area of 2.5 hectares has several buildings including office space (BLK leaders and staff), prayer rooms, and the main room / building for training (workshop space), such as workshop, automotive , make-up, sewing, besides also having classrooms (for the delivery of material in class before practice). Each training is scheduled to be held for one month (holidays on Saturday and Sunday) or 240 hours of effective training for which training materials include theory (20%) and practice (80%) so that training activities are in the classroom (for theory) and at workshops / workshops (for practice). Each type of training has its own classrooms and workshops with a fairly good arrangement of space both in terms of lighting and air circulation so that trainees feel comfortable to attend training

Based on interviews with informants, obtained some data including strengths, weaknesses, opportunities and threats.

3.1 Programs available at the BLK

The training program at the Center for Training is diverse. The training programs in this BLK included embroidery and sewing, there were 3 automotive, there were cars, motorbikes, ships, carbide welding and electric welding and batik. (interview with Pak Bisrul Hafil, BLK Sampang). Some programs that have been held by Job Training Centers (BLK) are manjahit, border, automotive (car, motorcycle, ship), carbide welding, electric welding and batik. The target of training is the general public with criteria: poor people, migrant workers who fail to leave, unemployment, productive age (17-40 years). (interview with Pak Bisrul Hafil, BLK Sampang).

Meanwhile, according to Pak Suhardi (Head of Sumenep BLK) that the routine agenda programmed in the DIPA UPT of Sumenep Work Training is under the auspices of the East Java Provincial Government. This agenda is held annually in various fields of training, where the Sumenep Job Training Unit provides skills in vocational fields: Graphic design, Automotive engineering (R2 and R4), Electric welding techniques, Refrigeration, Beauty (cosmetics), Technical support, Cooking, sewing, electrical engineering, embroidery techniques, building engineering, creative industries, automotive motorbike engineering, screen printing techniques, and business management. If organized based on funds from DIPA, the implementation has been proposed a year before for the field of training needed by the community / world of work where the trainees are job seekers.

The mechanism for disseminating information, namely the BLK, provided socialization in each sub-district attended by the Village Head from each of these sub-districts. So most trainees were recruited from the village head. There are also those who register themselves here, but rarely. The target is after training, the community is able to create jobs with the expertise they get. Because after they finish training they are provided with equipment according to the training program they take. For example, sewing, they get a sewing machine. Motorcycle training, participants get motorcycle and car repair equipment. The equipment is free. Starting from registration, consumption, residence equipment, certificates and transport money.

There are many programs that are in the Job Training Center (BLK). There is training in automotive, electric welding, carbide, sewing, building or trading and computers.

The following is an excerpt of the interview with Mr. Agus Sumarso:

"A lot of training has been given by the BLK to the participants, including training in the automotive sector, which is divided into 3 branches, namely motorbikes, cars, and ships which are often called outboard motors. Then there is electric welding and carbide welding, embroidery and sewing, building or carpentry, batik, also once there was computer training ".

3.2 Stakeholders who have the opportunity to become partners

One of the BLK stakeholders who played a role was PT CTKI. This is because the majority of prospective migrant workers register at PT CTKI. So the one providing training to prospective migrant workers is PT CTKI. This then caused no BLK program from the Office of Cooperatives, Micro Enterprises and Employment specifically for prospective TKI. This is in accordance with the results of interviews conducted with Mr. Bisrul Hafil:

"Until now there hasn't been one. The problem is handled by PT. CTKI. We also collaborate with them. Most prospective migrant workers register directly to PT. CTKI, then from PT. CTKI requested a report to us and also reported prospective TKI who registered at the PT. Some register directly with us, but only a little. Sometimes there are only 15 people registered, but those who get to other countries can be hundreds. Yes it's illegal, bro. Until now, there have been around 90% of TKI illegal in Kab. Sampang itself ".

Based on the data above, most prospective migrant workers register directly with PT CTKI. PT CTKI asks for recommendations to the Office and reports prospective migrant workers who apply to the PT. Some TKIs have registered directly with the Dinas, but very few. Sometimes those who register are only 15 people, but those who get to other countries can be hundreds (illegal). Up to now there are around 90% of illegal TKI from Sampang. The Office does not provide training programs specifically for prospective migrant workers, because PT CTKI has handled the problem of TKI training. The role of Dinas is only to provide information on the Indonesian labor market in the sub-district to the public regarding illegal TKI.

Furthermore, one of the stakeholders who is very instrumental is Trunojoyo Madura University, which has a special handling unit on labor issues (in accordance with one of the fields in UTM's Tehno Park Science namely employment and women) such as the Center for BUSINESS INCUBATORS and Career Development and Entrepreneurship. The Center (CDEC) contributes as an expert in providing training and certification about HR management, training management and experts in legislation related to migrant workers. Experts at CDEC consist of academics and practitioners from Management Sciences, Development Economics, Law, Communication, Engineering from the Faculties in the Trunojoyo Madura University.

3.3 Potential and problems of HR and infrastructure at the job training center

Some of the potential in the Job Training Center (BLK) include:

a. The BLK facilities in the four districts in Madura are adequate and complete. Each room is used for one training (both theory and practice), the room is also adequate, starting in 2016/2017 there is a dormitory for trainees and also a dining room, prayer room, bathroom.
b. The programs have also been quite good and there are additional programs this year, namely batik.
c. The location of the BLK is strategic because it is in the middle of the city and on the edge of the highway so it is easily accessible by the community.
d. According to the observations regarding the services provided to the trainees it was quite good, including the services provided to the guests (when we visited there it was served well).
e. The community is enthusiastic about programs from the BLK. It is proven by each training the participants reach maximum capacity (20 people).
f. There is support from the district and sub-district governments, so the program runs smoothly.

Apart from strengths, there are some weaknesses that are still inherent in the Vocational Training Center (BLK). These weaknesses include:

a. For HR (BLK employees). At around 2:45 a.m. WIB when the researchers arrived at the BLK, it turned out that the hall seemed deserted, only a few employees were still in place until the hour came home.
b. Facilities and infrastructure for office space are inadequate. Because one standard size room is occupied by 6 employees. And also in the door computer training room was broken.
c. Programs from BLKs that come into contact with Indonesian labor migrants are still lacking, as there has not been any specific training from BLK for prospective migrant workers.
d. There are no special instructors for each district so that the training activities are slowed down.
e. There is no Manpower Office in Sampang District. The existence of the Employment field is sheltered by the service of micro and labor business cooperatives. This was felt to be out of tune, so that the programs adjusted to the BLK program (There was no specific program in the field of labor).
f. Needs needed at Sampang BLK are instructors. Because of the 4 districts in Madura, there are only instructors in Sumenep. This is as stated by the following Head of BLK:
"Of the 4 regencies in Madura, the only instructor was Sumenep, because he was directly sheltered by Surabaya. Well, one instructor was shared with other districts, so that it was not effective. (interview with Agus Sumarso) "
g. According to the regent. Keteenaga Kerjaan "Bisrul Hafi" Most of the Sampang people, especially in the North, do business in the sale and purchase of illegal drugs. So that makes them lazy to work well.
h. In connection with the revitalization of the Work Training Center in addition to the HR aspect, public awareness about the importance of certificate of expertise is very lacking. Communities that have expertise certificates from the BLK do not function the certificate as well as possible. The following is an excerpt from the official interview with Pak Bisrul Hafil Sampang.
"In addition to human resources, awareness from the community is still lacking. Like TKI, there are still many who are illegal. They just come along when there is an invitation without taking care of the papers and other requirements. Even though they have a certificate of expertise provided by the BLK, they don't function the certificate as well as possible ". (interview with Mr Bisrul Hafil, Sampang)

3.4 Training and management model at BLK

The condition of the BLK in the four districts located on the island of Madura is adequate, both facilities and buildings. In 2016-2017 there is a dormitory as a place for lodging participants during training. Equipment is also adequate, such as sewing, computer, welding, carpentry, automotive etc. There is no program from the Office that relates to improving worker skills for prospective migrant workers abroad because there is already a PT CTKI that handles skills training for prospective migrant workers. (interview with Pak Agus Sumarso).

Initially, the BLK was under the auspices of the Office of Social and Manpower, but starting in 2016, the BLK was under the auspices of the Office of Cooperatives, Micro Enterprises and Employment, which previously were under the auspices of the Office of Social and Manpower (interview with Mr. Agus Sumarso)

3.5 Employment training center institutional strengthening model

To design a model of BLK institutional strengthening, it must be based on strengths and weaknesses (threats)

a. The programs of the BLK did not touch the needs of Indonesian migrant workers. There has been no specific training from BLK for prospective TKI.
b. The program for prospective TKI has only been carried out by PT CTKI
c. There are no special instructors for each Regency. So that it slows down the implementation of training activities.
d. Needs are needed in all three BLKs (outside the Sumenep BLK), namely instructors.
e. Public awareness about the importance of certificate of expertise is very lacking. Communities that have expertise certificates from the BLK do not function the certificate as well as possible.

Based on the previous analysis, it was obtained an illustration of the BLK institutional strengthening model, namely the BLK must coordinate (cooperation) with PT CTKI in terms of training prospective migrant workers. So that the training programs at the BLK synergize with PT CTKI, namely training and improving the skills of prospective migrant workers. In addition, the people who want to register as TKI must have at least a skill certificate obtained from the BLK. This will not only ease the task of PT CTKI in training the community (prospective TKI) as well as preparing the community to be better prepared in the world of work.

The condition of the BLK in the four districts located on the island of Madura is adequate, both facilities and buildings. In 2016-2017 there is a dormitory as a place for lodging participants during training. Equipment is also adequate, such as sewing, computer, welding, carpentry, automotive etc. There is no program from the Office that relates to improving worker skills for prospective migrant workers abroad because there is already a PT CTKI that handles skills training for prospective migrant workers. (interview with Pak Agus Sumarso).

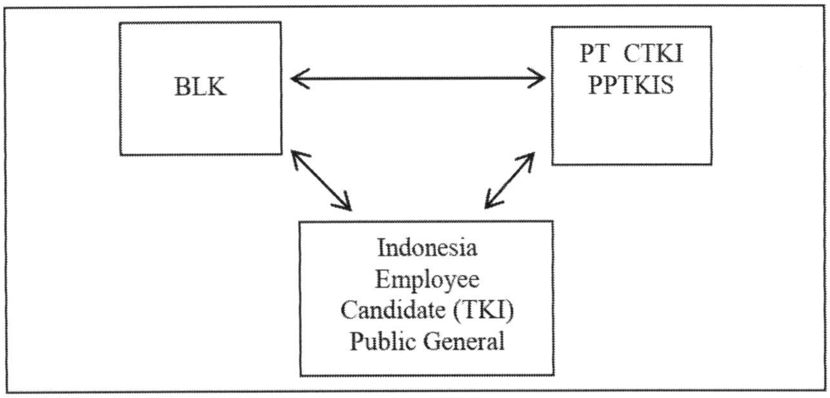

The target of training at the BLK is not to find work for graduates but rather to obtain skills by being trained for about one month on a working day with the duration of the training starting at 08.00 to 16.00 or equivalent training held for 240 hours per class. The training participants at the BLK are not limited to only the residents of Bangkalan (Madura) but may from outside the Bangkalan district as long as the person concerned has a KTP and is around 18 years to 35 years old.

BLK registration is usually opened at the beginning of the year, the requirements are such as a copy of a high school / equivalent diploma, a copy of an SKCK, a photocopy of an ID card, an original doctor's certificate, and a photo that must be completed. By registering at the beginning of the year students who graduate from high school or equivalent who wish to register must wait for 6 months, (it can be said that prospective participants are rested before facing a solid schedule). BLK training also does not require students to become workers because with the skills they have, students can open a business to open employment opportunities. So BLK becomes a motor in reducing unemployment. The lack of a blocking BLK program, which is the lack of information on BLK in high schools, the scheme is actually waiting instead of picking up the ball, the BLK is lagging behind with recruitment which is usually carried out by private schools and universities ahead of national examinations by going to schools - the upper secondary school. Not to mention the selection carried out through written examinations which actually denied the literal meaning of the Vocational Training Center. Because the BLK should be an extension of the regional government to help the government alleviate unemployment through education, the cost of which has already been borne by the government. For this reason, the government, according to the Minister of Manpower, Hanif Dhakiri will continue to strive and encourage competency improvement for workers through various strategies to improve BLK quality through reorientation, revitalization and rebranding of BLK.

First, the reorientation of the BLK related to the priority sector in the BLK, there must be a priority sector including the profession. Second, revitalization of BLK to improve quality so that BLK graduates can meet industrial needs. Third, the rebranding of BLK in order to increase public trust is higher. Thus, unemployed people get the opportunity to have skills and can enter the labor market. While those who have worked, can improve their competencies so that their competitiveness increases.

Especially for revitalization, the Ministry of Manpower continues to revitalize BLK, especially BLK-BLK owned and managed by regional governments (pemda). The Ministry of Manpower's Director General of Training and Productivity Development (Binalattas) Khairul Anwar (March, 2016) said the effort was made so that BLK was able to keep up with technological advances and the needs of the labor market and industry. Some aspects that have been addressed in revitalizing this BLK include infrastructure and training equipment, quantity and quality of instructors, training methods and curriculum, and management of BLK management. Binalattas targets to revitalize 70 BLK in 2016 so that the quality of Indonesian workers will be better so as to accelerate the reduction of unemployment and increase the competitiveness of workers at both the national and international levels.

Those who are trained and educated in BLK besides formal workers are also potential entrepreneurs. Although to educate and train someone to become an entrepreneur is much more difficult than training and educating skilled formal workers. The types of training that are in demand include computer vocational training, electric welding vocational education, vocational education, vocational automotive, sewing vocational, vocational embroidery, and vocational carpentry. Through competency-based job training in BLK, it is expected that Indonesian labor migrants entering the labor market both in the local, national and foreign labor markets can continue to be upgraded and ready to win competition in the MEA era.

Training at BLK generally only takes about one month or is identical with around 240 hours per class. Training with a sufficient duration of time is indeed inseparable from the addition of funds, labor, and space as well as the understanding of the training itself that distinguishes it from education. Training (training) is a short-term educational process that uses systematic and organized procedures so that non-managerial workers learn the knowledge and technical skills for specific purposes. Job training according to Act No. 13 of 2003 Article I Paragraph 9. is the whole activity to give, obtain, improve, and develop work competencies, productivity, discipline, attitudes, and work ethic at a certain level of skills and expertise in accordance with the level and job qualifications and jobs.

According to Marzuki Training is teaching or giving experience to someone to develop behavior (knowledge, skill, attitude) in order to achieve something desired. Training for employees is a process of teaching certain knowledge and skills and attitudes so that employees are more skilled and able to carry out their responsibilities better, according to standards.

The training is more focused on improving the capabilities and expertise of HR related to jabtan or functions that are the responsibility of the individual concerned at this time (current job oriented). The goal to be achieved and a training program is to improve individual performance in the current position or function. In the science of human resource management (HRM) training and / or education is usually a package and is one of the main tasks of HRM in the field of HR development. It is said that development is the second operational function of personnel management, the development of new / old employees needs to be carried out in a planned and continuous Employee development program should be carefully arranged and based on scientific methods and guided by the skills needed by the company or an educational institution now and for the future. Every company personnel / education institution is required to work effectively, efficiently, quality and quantity of work well so that the competitiveness of a company / educational institution is greater.

So the definition of development is an effort to improve the technical, theoretical, conceptual, and moral capabilities of employees in accordance with the needs of work / position through education and training. In addition, development (development) can also be interpreted as preparing individuals to assume different or higher responsibilities in companies, organizations, institutions or educational institutions. Development tends to be more formal, involving anticipation of individual abilities and expertise that must be prepared for future office interests. The targets and development programs involve a broader aspect, namely increasing the ability of individuals to anticipate changes that might occur unplened change or planed change. From this understanding shows that the focus of career development is to increase the mental ability of the workforce. Development is more focused on increasing ability in decision making and expanding human relations for top management and middle level management while training is intended for employees at lower levels (executors).

According to Hasibuan (2012) the notion of practice and development is different. Training is intended to improve the mastery of a variety of specific, detailed and routine work skills and techniques, in which training prepares employees (labor) to do the work now. While the development (Developrnent) has a wider scope in an effort to improve and improve knowledge, abilities, attitudes and personality traits.

According to Mariot Tua Efendi H (2005) that training and development can be defined as a planned effort from the organization to improve the knowledge, skills and abilities of employees. Furthermore, Tua Mariot added that training and development are the same two concepts, namely to improve knowledge, skills and abilities. But, judging by its purpose, generally the two concepts can be distinguished. Training is more emphasized in increasing the ability to do specific jobs at this time, and development is more emphasized in increasing knowledge to do work in the future.

4 Conclusion

BLK is a forum that prepares prospective workers or entrepreneurs with various skills needed by the world of work / industry so that BLK graduates are expected to have good field competencies in terms of knowledge, skills, and attitudes (behaviors) that are tailored to their talents, interests, and work goals desired by prospective graduates. Therefore, the role of BLK needs to be improved with one of them is to revitalize BLK to increase the knowledge, education and skills of prospective workers. The revitalization of the BLK began with an inventory of the problems and constraints faced by the BLK and prospective workers as well as the potential and strength that exists by looking at the opportunities and threats that exist in the labor market both nationally and internationally.

Based on the inventory, needs analysis is then carried out based on the relevance and comparability between the provision of theory and practice provided as long as they take part in education or training at the BLK with real conditions in the workforce / industry. The results obtained were the establishment of a BLK revitalization model for increasing knowledge, education, and skills of prospective workers. The resulting model has also considered institutional strengthening including the support of stakeholders.

REVITALIZATION OF BLK FOR IMPROVING ABILITY MIGRANT WORKERS:
KNOWLEDGE, EDUCATION, AND SKILLS BASED ON LINK AND MATCH

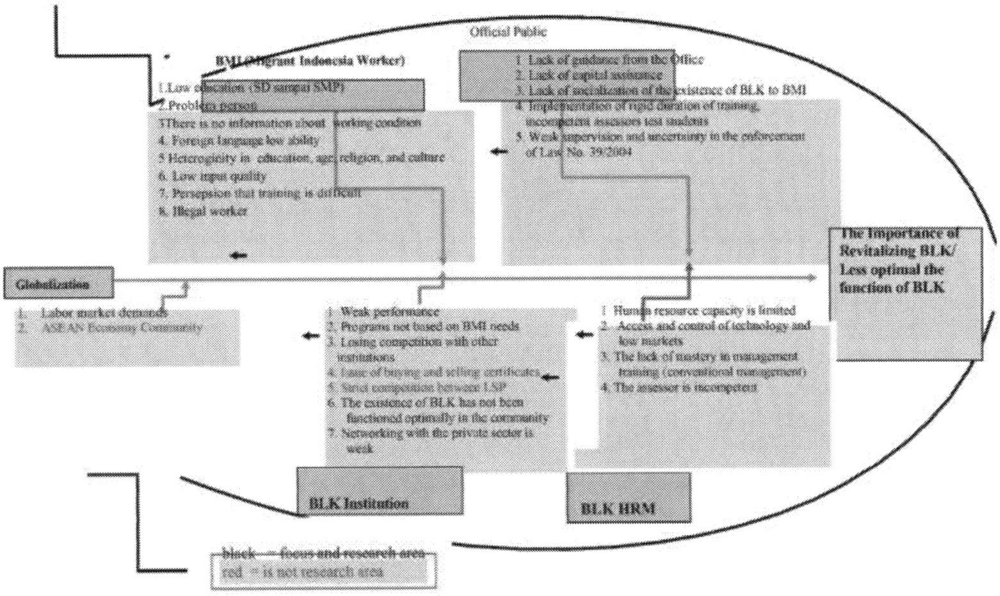

Bibliography

[1] A. Ahmad, E. Journal of Goverment (2014)
[2] A. Mangkunegara and A. Prabu, *Human Resource Management* (PT. Refika Aditama, Bandung, 2006)
[3] A. Pandji, *Business Management* (PT. Asdi Mahastya, Jakarta, 2000)
[4] Data on Disnakertransduk of the East Java Provincial Government, 2015
[5] E. Maesaroh and M. Sulton, E.journal.unikama Method (2015)
[6] H. Fitroh, J. of Public Admin 1, (2010)
[7] I. Jusuf, *Basic Principles of Training Management (From Need Analysis to Training Program Evaluation)*, (Scholar Human, Jakarta, 2001)
[8] Interview with Pak Bisrul Hafil, Sampang BLK. Interview with Mr. Agus Sumarso (53 years old)
[9] M. Sjafri, *Strategic Human Resource Management* (Ghalia Indonesia, Jakarta, 2004)
[10] M.S.P Hasibuan, Human Resource Management Jakarta: Teenagers Rosdakarya Nailasuhada-m.blogspot.com/…/pelatihan-dan-pengembangan-sumberdayamanusia, accessed 20 November 2013
[11] M. S.P Hasibuan, ibid , 86.
[12] M. Tua Efendi H, *Manusa Resource Management: procurement, development, compensating, and increasing employee productivity* (Grasindo Widiasarana Indonesia,Jakarta,2005)
[13] Office of Social Affairs, Manpower and Transmigration of Bangkalan Regency. Profile of Technical Implementation Unit of Office of Work Training Center, 2013
[14] R. Ningsih and S. Rahayu, *Assessment of Developmental Levels of Cooperatives with the DLA Method*. (LSP21, Malang, 2002)
[15] R. Ningsih and S. Rahayu, *Assessment of the Level of Development of Village Unit Cooperatives with the DLA Method* (2004)
[16] R. Ningsih and S. Rahayu, *Strategy for Improving the Performance of Cooperative Identity, UMKM Sampang Service* (2009)
[17] R. Ningsih and S. Rahayu, *Strategic Study of Overcoming Cooperative Bankruptcy, UTM and Sampang Cooperative Cooperative Office* (2009)
[18] R. Ningsih, S. Rahayu, *Performance Assessment of Women's Cooperatives. Office of Cooperatives and MSMEs* (2011)
[19] R. Ningsih and S. Rahayu , *Assessment and Assistance in Women's Cooperative Performance. UMKM and UTM Cooperative Service*. 2015-2019 RPJMN (2012)
[20] T. Eddy, J. Technical 8, (2013)
[21] Wispandono, *PUPT Proposal (Revitalization of Work Training Centers to Improve Skills of Indonesian Migrant Workers in Madura)* (LPPM UTM, 2016)
[22] Zulkifli, et al. , *Quality Cost Reporting to Minimize the Risk of Defective Products to Increase Company Profit* (Self-Reliance, 2013)

What is Behind Green Industry Motive to Maintain Rural Areas?

Ali Akbar Anggara [1*], *Bambang* AgusPramuka [2]

[1] UniversitasJenderalSoedirman
[2] UniversitasJenderalSoedirman

Abstract. The environmental damage is one of the essential factors that determine of disasters. Industrial activities that are mostly in urban areas are contributors to this problem. These activities often exploit resources from rural to an urban area. So that industry is required to be actively involved in maintaining the rural area alike implementing a green industry thatcan be represent through the application of green accounting which is disclose environmental information. The purpose of this paper is to explain the involved an industry in rural areas. Especially, to explain benefit and industry motive. This study uses legitimacy theory and stakeholder theory as a theoretical framework to explain the role of industry. The research method used a quantitative approach and seconder data from Indonesia stock exchange with the LQ45 index as a sample. The data analyzed using multiple regression and path analysis to examine environmental factors that influence the economic performance. The result of this study shows that environmental performance and environmental cost has an influence on environmental disclosure then influence economic performance. Furthermore, it can be identified that the company get the benefit from environmental performance and environmental cost. Also, Economic performance is a reason why an industry involved in maintaining the rural area.

1 Introduction

The environmental damage is the main factors that causes disasters. In 2004, the United Nations High- Level Threat Challenges and Change Panel explain that environmental degradation as one of ten threats to the world.The waste is one of reason the environmental damage. Since 1950, The amount of waste has always increased. There are significant increase from 1950 until 2015. In 1950 the waste was at 2 million tons per year, and in 2015 the world produced 381 million tons of waste per year. The industrial is main contributor on waste production in kind of plastics as much as 146 million tons (Geyer et al., 2017). Then,industrial activities that are mostly in urban areas are contributors on environmental problems. The one of problem that caused by industry is land transition. The activities often exploit resources from rural to an urban area.

*Correspondingauthor:author@email.org

Today, the company must be aware that the supply is limit. The everything of raw material sourced by environment has limited supply so that need action to keep the sustainability. Three bottom line is explain that profit, people and planet has become important element for sustainability. So that industry is required to be actively involved in maintaining the rural area alike implementing a green industry.

Green industry practices are achieved through the application of green accounting in accounting practices(Clarkson et al., 2011).Actually, green accounting practice alike disclose environmental information is still voluntary in Indonesia. Evenenvironmental disclosure is an important part of financial statement (Gray, 1994). Environmental disclosure is can be push the company to improve environmental performance because the everything of environmental information which did by company will be publish to the public. Indonesia has constitution Number 40 of 2007 about company, social and environmental responsibility is an obligation that must be carried out by each company. The regulation made by government to push the company in order to interest with environment.

The public will be judge based on environmental disclosure(Clarkson et al., 2011). Then, environmental disclosure is one kind of method of the company's responsibilities to the public. It is relevant with theory of legitimacy is explain that stakeholder or community will be involve on legitimacy.Of course, the valuation will be influence to company economic performance (Zhongfu et al., 2011). This term will create the paradigm that the company with excellent environmental performance will disclose excellent environmental information (Ingram and Frazier, 1980; Freedman and Wasley, 1990). However, the application of green industry is like two sides of a coin on the one hand, it will bring benefits to the industry, but on the other sideit will cause potential increases in costs and decrease the profit (Zulhaimi, 2015).

This paper will be answer the benefit for the company if do green industry practice and disclose the environmental information. Based on analysis, this research will explain whether the economic performance is main company motive to did green industry practice.

2 Literature review

This study uses theory of legitimacy and stakeholder as a framework to describe the industry role. Grayet al. (1995) explained that the company sustainability depends on stakeholder. Environmental disclosure is the result of considerationfrom the company and stakeholders. The companies have responsibilities to each group or individuals who can or have been affected by the company obligation. Stakeholders can control themselves to create decision to take a part on environmental or not (Mitchell et al., 1997).

Legitimacy theory is a company management system that is orientedtowards alignments with society, government, individuals and community groups. For this reason, as a system that prioritizes public support, the company's operations must be following the expectations of the community.Legitimacy is considered necessary for companies because the legality of society to the company is a strategic factor for the development of the company in the future.Barkemeyer (2007) revealed that the explanation of the strength of the organization's legitimacy theory in the context of corporate social responsibility in developing countries hastwo things, namely the ability to place profit maximization motives to make a clearer picture of the company's motivation to increase its social responsibility. Then, the legitimacy of the organization which can include cultural factors that form different institutional pressures in different contexts.

This theory states that the company and the surrounding community have close social relations because both are bound in a social contract. Legitimacy theory states that the existence of a company in an area because it is supported politically and guaranteed by government regulations and parliament which is also a representation of society. Thus, there is an indirect social contract between the company and the community, which in this case gives the city the cost and benefits for corporate sustainability (Andreas, 2011).
According to the theoretical framework, this study will be answer the benefit for stakeholder who concern on environmental and what kind legitimacy will achieve by company. So that the study focused on testing environmental performance and environmental cost as a factors that influence environmental disclosure. These variables are representative from green industry practice on accounting term. Then, this study examine the influence and mediate function from environmental disclosure on economic performance.

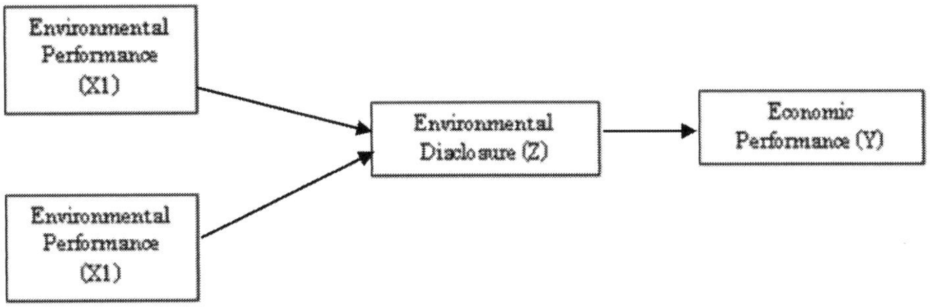

Fig. 1. Research Model

H1: Environmental performance influences environmental disclosure.
H2: Environmental cost influences environmental disclosure.
H3:Environmental disclosure influences economic performance.
H4: Environmental disclosure mediate variable between environmental performance andeconomic performance.
H5: Environmental disclosure mediate variable between environmental cost and economic performance.

3 Research method

Based on the kinds of problems discussed in this study, this type of research is classified as a quantitative approach. The quantitative approach is research carried out by collecting data in the form of numbers and then processed and analyzed to obtain a piece of scientific information behind these numbers (Martono, 2011).This study aims to describe, verify, or test a symptom. (Siswono and Yuli, 2010).
The population in this study is the total number of companies listed on the Indonesia Stock Exchange. This research uses a sample of companies listed as LQ45 companies. Tandelilin (2010) explain that index established by IDX especially for 45 companies who have prospects, financial conditions, and high transaction value.It means, investor and trader are aware on company value which represented by economic value.
Sample selection is made by using a purposive sampling method with predetermined criteria. By using this method, in study expect to get information from specific target groups and to obtain a representative sample according to the specified criteria (Sekaran, 2006). The requirements are as follows:

a. The company listed on LQ45 index Indonesia Stock Exchange
b. Publish their financial statement for 2015, 2016, 2017, and 2018.
c. Stand on LQ45 index at 2015, 2016 and 2017
d. Publish a sustainability report at 2015, 2016 and 2017
e. Following the PROPER program at 2015, 2016 and 2017.
f. Inform environmental cost in financial statement.

3.1 Conceptual and Operational Definition of Variable

3.1.1 *Environmental Performance*

Good environmental performance is one term of the company's concern for the earth. Environmental performance is the company's performance in creating a pleasant environment (Suratno et al., 2006). PROPER is used as an environmental performance measurement tool in Indonesia and one of the compliance instruments since 1995. The author uses measurements using the rating scale according to the Ministry of Environment, namely:
5 = Gold; 4 = Green; 3 = Blue; 2 = Red; 1 = Black

3.1.2Environmental cost

The definition of the environment and environmental cost management are stated in PSAK No. 33 concerning general mining accounting. Firoz and Ansari (2010) provide examples of estimates related to environmental costs, including provision of clean-up costs, provision for rehabilitation in the mining industry, provision of claims for contingencies, provision of environmental costs such as overcoming air pollution, noise pollution, gas, and hazardous waste and provision of purchasing equipment to control pollution. The equation as follows:
Environmental cost = Total environmental cost on rural area (cost).

3.1.3Environmental Disclosure

Reporting or disclosure of socio-environmental accounting information related to aspects of the interaction between company organizations and the social environment and physical environment (Saudagaran, 2011). GRI is reporting, disclosure of standards that are indicators of social responsibility carried out by the company to create/provide the benefits of reporting to corporate stakeholders. This guideline is designed to be used by organizations in size, sector, or location.
Index GRI G4= (Total GRI EN item in the company)/(Total environmental item GRI EN)

3.1.4Economics Performance

Economic performance is the relative performance of a company in the same industry, which is characterized by the annual return of the industry concerned (Suratno et al, 2006). In this study, the authors limit the measurement of the company's economic performance using Tobin's q. The indicators are using for measuring corporate performance variables from an investment perspective that has been tested in various top management situations (Wolfe and Sauaia, 2003).
Tobin's Q=((Total share x Price)+(Total Liabilities+Inventory-Current Assets))/(Total Assets)
The data is analysed by Multiple Regression technique for analysis factor that influence dependent variable and path Analysis to determine mediatefunction of environmental disclosure.

4 Result and discussion

4.1 Result

4.1.1 Result of Descriptive Test Based on Variables

Result of Descriptive Test Based on Variables is described output from companies sampled based on Environmental Performance (EP), Environmental Cost (EC), Environmental Disclosure (ED) and Economic Performance (EcP).

Table 1. Result of Descriptive Test Based on Variable.

	EP	EC	ED	EcC
Mean	3.4737	86,883,564,908	.3632	1.2816
Mode	3.00	3,977,754,008	.20	.90
Std. Deviation	.60345	90,805,557,774	.23644	.55549
Minimum	3.00	3,977,754,008	.10	.30
Maximum	5.00	363,502,070,000	.90	2.70

From Table 1. it can be concluded that the companies that are sampled on average spend an environmental cost of 86.88 billion per year and at least companies spend an environmental cost of 3.97 billion per year. Furthermore, from the data, it can be seen that the environmental performance is good, which is at 3.00, that indicates the company has a blue PROPER rating. Color is explained that companies made environmental management efforts that are required following the provisions and legislation. (kemenlh, 2013).
Environmental Disclosure shows that average of company who publish sustainability is only follow 12 indexes of GRI (0.3632 x 34 item). Even there is company that only follow 3 indexes of GRI (0.10 x 34 item). GRI has made a standard for sustainability report to report information related to three bottom line, namely profit, people, and the planet. The measure has been used internationally. However, there are still many companies that do not provide complete information. It is because the company does not carry out activities as found in GRI Environmental.

4.1.2 Simultaneously Test (F test)

F statistics test shows whether all the independent variables included in the model have a joint or simultaneous influence on the dependent variable. The test is often called testing the overall significance of the regression line to be tested whether X1 and X2 are linearly related on Y(model II) and test X1, X2 and Y on Z (model III).

Table 2. F test Output

Model		Model II		Model III	
		F	Sig.	F	Sig.
	Regression	3.610	.023	2.792	.042
	Residual				
	Total				

Table 2. shows the result of the F test output, which the result is model II is a significant correlation, namely Sig. value is 0.023 < 0.05 also Sig. value is 0.042 < 0.05. Automatically, if Sig. value is significant, then F is higher than F table too.

4.1.3 Partial Test (t-Test)

This analytical tool is used to examine H1, H2, H3, H4, H5, H6, and H7. t-test shows how far the influence of the independent variable on the dependent variable as partial. The test results as follow table 3.

Table 3. . t Test Output

Model	Unstandardized Coefficient B		Sig.	
	Model II	Model III	Model II	Model III
(Constant)	-.042	1.560	.839	.003
Environmental Performance	.185	.141	.008	.417
Environmental Cost	-1.011	-1.876	.020	.088
Environmental Disclosure		1.290		.003

a.Dependent Variable: Environmental Disclosure (Model II)
b.Dependent Variable: Economic Performance (Model III)

According to Table 3 shows Environmental Performance (X1) and Environmental Cost (X2) has significant influence as partial on Environmental Disclosure (Y) are 0.008 and 0.020. The sig value is lower than 0.05 as significant probability standard. It is mean H1 and H2 are accepted

Unstandardized Coefficient B in model II shows the direction of change for each variable. Environmental Performance shows a positive value of 0.185. It is different from the Environmental cost which indicates a negative trend of -1.011. It means that every additional unit of information about environmental performance will also strengthen the influence of environmental disclosure of 0.185. It is different from the Environmental cost which will have a negative influence on environmental disclosure. However, the negative value of coefficient B.While in model III, an independent variable that has a direct influence on economic performance (Z) is environmental disclosure (Y). It is indicated by Sig. Y value is 0.003, which means smaller than 0.05. So, H3 is accepted. Continuously, environmental disclosure has a positive influence on economic performance. It can be seen from the unstandardized value of B coefficient, which is equal to 1.290.

4.1.4 Path Analysis

To answer H4 and H5, the study is used path analysis to find out the direct and indirect effects, which then compared them. Based on the previous model, a complete model can be described, as shown in Figure 2. The model is a combination model of model II and model III. It is needed to compare the drivers, directly and indirectly, to find out whether Y has a function as a mediating variable or not.

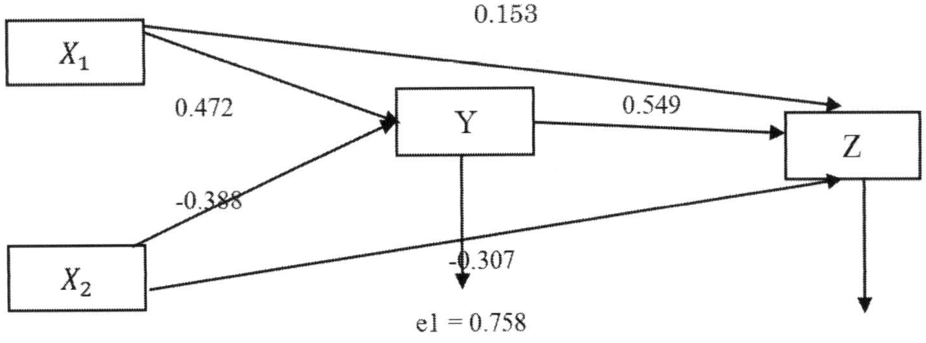

Figure 2. Research Model Output

H4 is environmental disclosure variables can mediate environmental performance with economic performance. The study compares the value of the direct effect of X1 on Z at 0.153, and the indirect effect of X1 on Z through Y is 0.259 (0.472 x 0.549). From these results can be seen that the value of indirect influence is greater than the direct effect. So, it can be sure that Y can mediate X1 and Z.

H5 is environmental disclosure variables can mediate environmental costs with economic performance. The study compares the value of the direct effect of X2 on Z at -0.307 and the indirect effect of X2 on Z through Y, which is equal to -0.213 (-0.388 x 0.549). From these results can be seen that the value of indirect influence is greater than the direct effect. Then it can be sure that Y can mediate X2 and Z.

Table 4. Summary of Hypothesis Test

Hypothesis	Value	Decision	Conclusion
H1	.008	Significant	Accepted
H2	.020	Significant	Accepted
H3	.003	Significant	Accepted
H4	Direct: 0.153 and Indirect: 0.259	Mediate	Accepted
H5	Direct: -0.307 and Indirect: -0.213	Mediate	Accepted

4.2 Discussion

4.2.1 The Influence of Environmental Performance on EnvironmentalDisclosure

Based on the results of the regression test and path analysis explain that environmental performance is influence on environmental disclosure variables (looks table 4. Summary of Hypothesis test). Preston (1981), Al- Tuwaijri (2004), Clarkson et al. (2011) states that there is a positive relationship between environmental performance and environmental disclosure that companies with high environmental performance have incentives to communicate company information by making environmental disclosure reliable. It means, environmental disclosure will be good or bad depend on environmental performance.

One of the grand theories in research is the theory of legitimacy. The theory explains the importance of the legitimacy of the public. Environmental performance from a company can answers this theory. Environmental performance has a positive effect (Unstandardized coefficient B = 0.185) on environmental disclosure. This value means that every increase of one PROPER level will increase the influence up to 18.5%. Environmental disclosure becomes a good medium to get attention from stakeholder. The company can build the brand image through environmental disclosure so that the company can get the public sympathy. It shows that excellent environmental performance can provide value to the company.

4.2.2 The Influence of Environmental Cost to Environmental Disclosure

Environmental costs have an influence although the direction of the negative coefficient results in unstandardized coefficients B = -1.011. When the company reporting environmental cost so much, it will decrease environmental disclosure quality. The company should have report environmental cost but should be consider carefully. It means the company currently considers environmental cost is only an additional expenditure for the company which will eventually become a profit-reducing account for the company application of green accounting to the company is like two sides of a coin, on the one hand it will bring benefits to the industry but on the other hand it will cause potential increases in costs through environmental costs (Zulhaimi, 2015). The situation becomes challenging for management to consider the proportion of environmental costs that will be incurred by the company. The good calculations are needed to control each sides of this coin in order to be optimal. Business is indeed looking for profit, so stakeholders cannot abandon this substance. Actually, it will be good if the company does not report environmental cost because environmental cost has negative influence to environmental disclosure.

4.2.3 The Influence of Environmental Disclosure and Economics Performance

Companies publish information about the environment through environmental disclosure aimed at building an image on the company and getting attention from the public. Based on environmental disclosure, stakeholders will understand how the company's role and concern for the environment. These stakeholders will assess and decide whether they will legitimize or even invest. Companies with high environmental disclosures in their financial statements will be more reliable. Reliable financial statements will affect economic performance, where investors will respond positively to higher stock price fluctuations, and vice versa. If the environmental disclosure of a company is low, then investors will react negatively to the variations in stock prices which decrease from year to year.

Gray (1993) explains that environmental disclosure is an essential part of a company's financial statements. The problem is that environmental disclosure in the annual report in Indonesia is still voluntary even though the Indonesian government has regulated this in regulations, one of which is Republic of Indonesia Law No. 47 of 2012 concerning social and environmental responsibility. Also though environmental disclosure has a positive impact on economic performance

(Zhongfuet al., 2011). Stakeholder theory states that the amount of environmental information disclosed by the company will affect stakeholders so that it affects stock prices and affects the company's annual return. Almilia and Wijayanto (2007) found a relationship between environmental disclosure and economic performance.

4.2.4 4Environmental Disclosure is Intervening Variable Between Environmental Performance and Environmental Cost on Economic Performance

That are influence as directly to economic performance. Both variables are need environmental disclosure as media to show up their value. The variables will be meaningful If the company report environmental performance and environmental cost in special space, that is environmental disclosure. The special space can be highlight for investor and trader who are interest with environmental issues. Based on this study explain that investors and traders are consider environmental performance and environmental costs in investing their money in the LQ45 listed companies. The investors are be aware that taking environmental, social, and corporate governance, or ESG, issues onboard is in the long-term interest of the investments. Not doing so could pose a financial risk to their investments.
Today, market regulators are press the company to disclose information on environmental issues, and given the focus of markets on short-term profit, companies are not always forthcoming with full disclosures on environmental risks even though in Indonesia still as voluntary information. Ceres Network from Socially Responsibility Investment (SRI) explain that today the industry brings an effect to the earth because their activity. It is difficult for the future then need energy and climate change policy. Good environmental disclosure will be to provide complete information to investors. However, excellent environmental performance cannot be separated from the costs that must be incurred to achieve it. During hot issues about environmental campaigns, it is appropriate for stakeholders to start caring about environmental issues. It's time for the Three bottom lines to come true, so it's not just a slogan.

From the discussion, it can conclude that stakeholders in Indonesia have been concerned about environmental issues expressed in environmental disclosure, in this case, through the GRI 4.0 standardized sustainability report. Thus, although environmental disclosure is a type of voluntary disclosure, there is no loss for the company to disclose it to the public. The reason also becomes the basis of environmental performance, and environmental costs do not directly affect economic performance. It is because investors will easily read clearly when the environmental performance and environmental costs are published in environmental disclosure. Besides, from the test results, it can be ascertained that investors are still sensitive to the environmental cost magnitude. It is because the environmental cost will be a burden that also reduces the company's revenue

References

1. Al-Tuwaijri, Sulaiman A., Theodore E. Christensen, and K. E. Hughes. (2004). The RelationsAmong Environmental Disclosure, Environmental Performance, and Economic Performance: A Simultaneous Equations Approach. Accounting, Organizations and Society.29. 447-471.M. Ben Rabha, M.F. Boujmil, M. Saadoun, B. Bessaïs, Eur. Phys. J. Appl. Phys. (to be published)
2. Algifari. (2000). AnalisisRegresi, Teori, Kasus&Solusi. BPFE UGM, Yogyakarta.

3. Bewley, and Yue Li. (2000). Disclosure of environmental information by Canadian manufacturing companies: A voluntary disclosure perspective. DOI: 10.1016/S1479-3598(00)01011-6
4. Burgwal, Van de and Jose Oliviera Vieira. (2014). Environmental Disclosure Determinants in Dutch Listed Companies. Cont Fin USP. 64. 60-78. ISSN 1808-057X
5. Clarkson, Peter M., Overell and Chapple. (2011). Environmental Reporting and its Relation to Corporate Environmental Performance. Journal of Accounting, Finance and Business Studies. 47(1). 27-60
6. Fahmi, Irham. (2012). PengantarPasar Modal. Bandung: Alfabeta
7. Firoz, Mohammad and Aziz Ansari. (2010). Environmental Accounting and International Financial Reporting Standard. International Journal of Business and Management. 5(16)
8. Freedman, M. and C. Wasley. (1990). The association between environmental performance and environmental disclosure in annual reports and 10-Ks. Advances in Public Interest Accounting 3: 183–193.
9. Geyer, Roland, Jenna Jembeck, Kara Lavender Law. (2017). Production, use, and fate of all plastics ever made. Science Advance. 3(7). e1700782. DOI: 10.1126/sciadv.1700782
10. Gray, R. (1994). Corporate Reporting for Sustainable Development Accounting for Sustainability in 2000AD. Environmental Value. 3(1). 17-45.
11. Hughes, Susan,. Allison Anderson and Sarah Golden. (2001). Corporate environmental disclosures: Are they useful in determining environmental performance?. Journal of Accounting and Public Policy. 20(3). 217-240. DOI: 10.1016/S0278-4254(01)00031-X
12. Ingram, and Frazier K. (1980). Environmental Performance and Corporate Disclosure. Journal of Accounting Research. 18(2). 614. DOI: 10.2307/2490597
13. O'Donovan, Gary. (2000). Legitimacy Theory As An Explanation For Corporate Environmental Disclosures. Faculty of Business and Law. Victoria University of Technology. Melbourne, Australia.
14. Patten, D. M. (2002). The Relation Between Environmental Performance and Environmental Disclosure: A Research Note. Accounting, Organizations, and Society. 27. 80.
15. Preston, D. (1981). Environment, Society and Rural Change in Latin America. Rural Sociology. 46(4). 764
16. Roberts, R.W. (1992). Determinants of Corporate Social Responsibility Disclosure: An Application of Stakeholder Theory. Accounting, Organizations and Society. 17. 595-612.
17. Saudagaran, S. (2001). International Accounting: A User Perspective. South Western College Publishing.
18. Sembiring, Eddy Rismanda. (2006). Karakteristik Perusahaan dan Pengungkapan Tanggung Jawab Sosial: Study Empirispada Perusahaan yang Tercatat di Bursa Efek Jakarta", jurnalMaksiUniversitasDiponegoro Semarang. 6. 69-85.
19. Suratno, Darsono, and Siti Mutmainah. (2006). Pengaruh Environmental Performance Terhadap Environmental Disclosure Dan Economic Performance: StudiEmpirisPada Perusahaan Manufaktur Yang Terdaftar Di Bursa Efek Jakarta Periode 2001-2004. SimposiumNasionalAkuntansi IX Padang. 23-26 Agustus.
20. Tandelilin, Eduardus. (2010). PortofoliodanInvestasiTeoridanAplikasi. Edisipertama. Yogyakarta : Kanisius

21. Veronica, Strang. (2009). Integrating the Social and the Natural Sciences in Environment Research: A Discussion Paper. Environment and Sustainable Development. 11(1). 1-18
22. Wardhani, Galuh and Sugiharto T. (2013). Pengaruh KinerjaKeuangan Perusahaan Ukuran Perusahaan, Kinerja Lingkungan Terhadap Intensitas Pengungkapan Pelaksanaan Tanggung JawabSosial Perusahaan Manufaktur Yang Terdaftar Di Bursa EfekIndonesia. Proceeding PESAT (Psikologi, Ekonomi, Sastra, Arsitektur & Teknik Sipil). 5.
23. Wolfe, J., &Sauaia, A. C. A. (2014). The Tobin's Q As A Company Performance Indicator. Developments in Business Simulation and Experiential Learning. 30.
24. Zhongfu, Yu, Jian Jianhui, He Pinglin. (2011). The Study on the Correlation between Environmental Information Disclosure and Economic Performance-With empirical data from the manufacturing industries at Shanghai Stock Exchange in China. Energy Procedia. 5. 1218-1224. DOI: 10.1016/j.egypro.2011.03.213
25. Zulhaimi, Hanifah. (2015). PengaruhPenerapan Green Accounting TerhadapKinerja Perusahaan. JurnalRisetAkuntansi Dan Keuangan. 3 (1). 603-616

An Integrative Model of Place Image, Product Image and Consumer Receptivity

Ratna Roostika[1*], *Muafi*[2], and *Agnès Retno* Permata[3]

[1]Univesitas Islam Indonesia, Yogyakarta, Indonesia
[2]Univesitas Islam Indonesia, Yogyakarta, Indonesia
[3]Univesitas Islam Indonesia, Yogyakarta, Indonesia

Abstract. Place image as an intangible cue has the power to influence the behavior of the buyer. Place in marketing discipline has been analyzed from country level to regional/territorial level. Focusing on a territorial context, the purpose of this study is to examine the place image effects on both the perception of local product and place as destination to visit. A local hand made batik industry from Yogyakarta is chosen to test the model. By analyzing 110 valid sample using SEM-PLS method, some findings from this study are: 1) Place image influences local product beliefs. 2) Product beliefs influence product choice intention. 3) Place image influences place beliefs. 4) Place beliefs influence place visit intention. This study contributes to the place image and product image integration model where majority of previous study were empirically examined as a separate model. By combining place image and its local product image, the local government may design distinct local product marketing strategy align with its local place image.

1 Introduction

Image as an intangible cue has the power to influence the behavior of buyers. This cue has long been recognized by marketers globally. The development of image theory as behavior influencer has developed starting from product image in micro context to place image as more macro context. Similarly, branding theory also developed from product branding into place branding. Currently it is very common practices that government is using marketing approach to market and to brand their region / city / country. For example, in the more intense tourism competition, many destinations promote their region branding such as I Amsterdam, Be Berlin, I love New York, Jogja Istimewa, and Enjoy Jakarta. In the wider geographical scope, country branding and marketing are also well organized such as 100% pure New Zealand, Incredible India, and Your Singapore. As part of successful place branding, positive image building is very critical. Place image is used for geographically defined places and it is considered the most macro level application of marketing and branding practices to date. According to Hawkins, et al. [20], images are mental schemata in human memory, that consist of a complex combination of cognitive and affective

* Corresponding author: ratna.roostika@uii.ac.id

components. In align with this definition, the overall image of a place that is perceived by individual is developed from complex components. In the hospitality literature, the performance of local service sectors, local hospitality, local products quality, local geographical sites and local infrastructures will together make the total image of a place.

The country-of-origin effect on buyers' behavior is particularly because of the positive place image that further affects on product image then further positive attitude for purchase or other positive behaviors. Similarly, country image influences product branding and buyers' positive behaviors. That place-related behavior is influenced by place image has been acknowledged in many tourism researches [25]. At the country level, developed countries have taken many benefits in combining the image power of the country to their products. For example, Korea with its Samsung, Germany with BMW and Mercedes Benz, USA with Ford and General Motor, Italy with Ferrari, Japan with Sony, etc. One country may also powerful in several industries such as Japan with its electronics and automotive and South Korea with its electronics and cosmetics industry. If country-of-origin (COO) works at country level, a narrower context which commonly known as territory-of-origin (TOO) is recently also well practiced and discussed in the marketing literature [30]. The use of TOO is similar to COO perspective, where the image of a territory or region may influence the receptivity of the local products [27]. The TOO concept therefore can be applied by not only well-developed countries but also developing countries as far as they have a strong region image and distinctive products. The marketers can use the region's image to boost its local product sales. In the marketing literature the product country image (PCI) studies have shown that country image helps consumers to evaluate the country's products, which is called as halo effect [18]. The hallo effect is useful and applicable in the situation where consumers are unfamiliar with the foreign products.

However, the study of product country image (PCI) and tourism destination image (TDI) are commonly discussed separately. The studies that relates between place image (destination image) and product image is very limited particularly at the regional or territorial level [12;28]. The present study aims to address this research gap by incorporating place image, product beliefs, destination beliefs and the behavioral consequences. More specifically, this study was designed to address these main objectives: 1) to test the influence of place image on product and destination beliefs. 2) to test the product and destination beliefs influence on product and destination receptivity.

2 Batik Yogyakarta

In order to achieve the objective of this study, batik industry is chosen to test the model proposed that integrate between place image and local product image. Yogyakarta is one of region in Indonesia which is considered as one of the most favorite tourists' destinations. Yogyakarta is known as cultural city, tourism city as well as student city. In the past time, this region is the center of Mataram Kingdom, and recently the culture and heritage are remained strongly preserved. While the region it self already maintained a good reputation for long, Yogyakarta is also supported by some well known products such as batik, silver, bamboo crafts and many other local foods and culinary. Batik industry is chosen since it is a product that 1) Batik has been part of Javanese long cultural history. 2) Yogyakarta is considered as one of the centers of batik production. 3) Many Batik experts reside in this region. 4) Batik from Yogyakarta has its own uniqueness. Thousands of batik designs have been produced and varieties of batik quality are also offered. The price ranges differently from the cheap one to the very expensive one. Hand write batik is the most expensive kind of batik since it is manually drawn. The less expensive is the stamped batik and a combination between hand written-stamped batik. Majority of batik consumers will consider Yogyakarta and the areas surrounding when they need to find good quality batik. For these reasons, this study will test the affect of Yogyakarta, as a city with its well-maintained image to the receptivity of its Batik industry. It is expected that if the proposed model is accepted, a strategy which combine the strength of place image and the potential local product could be developed.

3 Literature review

3.1 Place image

Place image has started to become a popular research in the tourism context [7;22]. Place image is defined as the total of beliefs, ideas, and impressions that someone as individual or as a group have on a destination [14]. It is a subjective and also a complex concept where it can be divided into two components namely cognitive (perceptions of individual attributes, for example the quality of services, the friendliness of staff) and affective (overall impressions, for example the emotion or mood towards the destination) [11]. The majority of place image or destination image studies are commonly focused on destination attributes which are used as quantitative or structured data (e.g. [5;17;31]). The main reason for observing image of a destination is to understand the perception of the potential tourists on the destination as well as to understand how tourists will respond by their attitudes and behaviors [11]. Previous research has identified that tourists prefer to choose a destination that has the most favorable image [23]. Recently, the more studies have evidenced on the destination image direct impact on travel behaviors and destination image also plays a critical role in the process of making choice [16].

3.2 The relationship between place image to product and destination beliefs

The marketing literature concerning on PCI provides the base for the study of COO [24]. This study, argues that the national origin of a product may influence product evaluations by the consumers. This proposition then leads to the theoretical foundation for country of origin (COO) effect on consumer behaviors. After this study, research on perceived COO as a cue for consumers product evaluation become proliferated. For example, some researches drew on empirical evidence which identified the moderating effect of COO on consumer by product category (e.g. [1;3]). With this logic, this study uses the PCI framework to examine the contribution of place image (destination image) to product beliefs at regional level. The recent studies, Stepchenkova [29] and Zhang, et al., [31] found that country image affects destination image. Further, the study also found that this positive image affects tourists' visitor intentions. The work of Elliott and colleagues (e.g. [10;9]) have developed more integrative models that combine PCI and TDI to further see the consumer behaviors. The present study follows DeNisco, et al., [8] and Mody, et al., [24] in evaluating the place image effect on products and destination beliefs. These studies found the relationships between place image and product and destination beliefs. More specifically place image affects beliefs [8;9]. In order to explore how consumer perceptions of place image may affect their attitudes on products and destinations beliefs, therefore this study proposes:

H1. Place image has positive influence on product beliefs.

H2. Place image has positive influence on destination beliefs.

3.3 Intention to choose and to visit

Elliot and Papadopoulos [9] argue that beliefs may affect behavioral intention or receptivity. In the tourism context, this behavior intentions can be translated into several actions such as intention to purchase, intention to choose, intention to recommend, intention to support and intention to visit. Of particular interest here, this study examines the intention to choose the local products as well as the intention to visit the destination. Following DeNisco, et al., [8] and Elliot and Papadopoulos [9] integrative model, this study combined two research streams, 1) consumer perceptions about products, and the other is 2) consumer perceptions on destinations. Previous studies have identified the consecutive relationships starting from place image (cognitive and affective), followed by beliefs and then finally the behavioral action whether to choose or to visit.

3.4 The relationship between belief and intention to choose/visit

Beliefs affect behavioral intention or receptivity [9]. The term receptivity is used to incorporate consumers' willingness to buy, to choose, to visit or to travel. Receptivity is also associated with overall assessment of places as producers and tourism destinations. The relationship between destination image and product image has been confirmed in the studies (eg. [4]). Similarly, the relationships from product and destination beliefs to product and destination receptivity also confirmed in some studies (eg. [4;9]). More specifically, the main contribution from this study is the integrative framework, where previous studies only provided empirical test in separate study. For example, study in destination image was not consider the relationship to product image original from the related destination. Similarly, studies on product image rarely relates to destination image of the related products. By incorporating place/destination image, the beliefs and the intention to choose/visit, the local government or travel agency could design marketing strategy that match between local place image and local potential products. Behavioral intentions such as intention to visit a destination is defined as the likelihood where someone is deciding to visit a place [26]. As a concept, the knowledge on behavioral intention is important by both academic researchers and industry since it can be used as a predictor of actual behavior. Kaplanidou [21] identified some previous researches which already observed antecedent of behavioral intentions such as destination image, trust, quality, value, attitude, etc. Beliefs is an attitude when people are already trust in something. Beliefs according to study from Elliot and Papadopoulos [8] and DeNisco, et al., [9] affect behavioral intentions. Product beliefs may affect intention to choose product. For destination belief, it may affect the intention to visit. Therefore, this study further proposes:

H3. Product beliefs has positive influence on intention to choose product

H4. Destination beliefs has positive influence on intention to visit a place

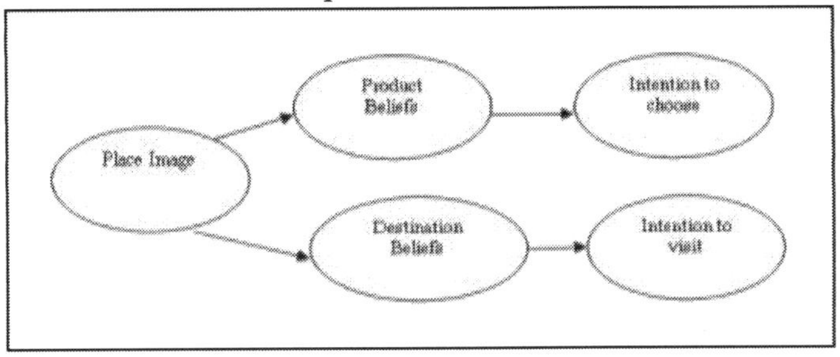

Fig. 1. Research Model

4 Research methods

4.1 Research instrument

This study applied quantitative method in order to achieve the objective of the research. A survey method was taken by developing questionnaires and distribute the paper-based questionnaire to respondents who live outside the region of Yogyakarta. The questionnaires were developed from previous studies in the same context which mainly taken from Elliot and Papadopoulos [8] and DeNisco, et al., [9]. This study applies Likert scale for close questionnaire ranging from 1 very disagree to 5 very agree. Purposive sampling method was chosen where respondents should not be the residents of Yogyakarta.

4.2 Respondents' profile

After three weeks of data gathering, 110 respondents were successfully gathered and valid for further statistical processes. Table 1 provides data of the respondents' characteristics. As seen from table 1, majority of the respondents are female household. They are between 31-40 years old and they understand very well about Yogyakarta and Batik from Yogyakarta. Majority of the respondents' occupation are "other". After gone through further interview, the meaning of "other" is translated as they don't have permanent job. Sometimes they can get side jobs such as food catering/cooking, sewing, farming, planting, etc. Other can also be interpreted as housewife and no specific job. Even though majority are holding bachelor, but they prefer to be a house wife.

Table 1. Demographic Profile of The Respondents

Demographic	Number of Respondents	%	Cumulative %
Gender			
Male	19	17.3%	17.3%
Female	91	82.7%	100%
Age (Years)			
15 – 30	0	0%	0
31 – 40	52	47.3%	47.3%
Older than 40	58	52.7%	100%
Occupational Status			
Students	0	0%	0%
Public Sector	0	0%	0%
Private Sector	1	9%	9%
Self Employed	41	37.3%	38.2%
Others	68	61.8%	100%
Monthly Expenditure			
Less than Rp. 500.000/mth	13	11.8%	7.6%
> Rp. 500.000 – 1.000.000	31	28.2%	40.0%
> Rp. 1.000.000 – 3.000.000	43	39.1%	79.1%
> Rp. 3.000.000	23	20.9%	100%
Education			
High School	20	18.2%	18.2%
Bachelor	90	81.8%	100%
Master	0	0%	100%
Other	0	0%	100%

4.3 Structural equation modelling

Partial least squares structural equation modelling (PLS-SEM) was chosen as statistical tool to test the proposed model. Due to the complexity of the proposed model, PLS-SEM is widely used [2]. PLS enable research with non-normal data distributions as well as data with either formative measure or reflective measure [2]. In the areas of marketing, tourism, strategic management, hospitality and behavioral studies, PLS is considered a prominent SEM technique [2;19].

4.4 Measurement model

PLS assess two procedures which are measurement model and structural model. That is why we called two step approaches. The measurement model is for testing and preparing data reliability and validity. In measuring construct's reliability, composite reliability (CR) is used and the value should be higher than 0.7 [6]. As can be seen from table 2, composite reliability (CR) were all above 0.7 thus satisfied the requirement of data reliability. The item's reliability is also important which can be seen from the item loadings in table 2. The item loadings should be higher than 0.50 according to Hair, et al., [19]. Validity of the measure was checked by testing convergent validity, which can be seen from AVE values. AVE value is accepted when the value is above 0.5. Table 2 shows that all AVE value was above 0.5, or ranging from 0,7023 (destination beliefs) to 0.7776 (intention to visit). The result from the composite reliability can also be translated as the measure for convergent validity for all constructs investigated in the study.

Table 2. Descriptive Statistic of Constructs

	Items	Loading	Mean
Place Image			
CR = 0.9241	CA1	0.8708	0.8665
Cronbach Alpha = 0.8953	CA3	0.9039	0.9058
AVE = 0.713	CK2	0.6028	0.5922
	Ck5	0.9068	0.9005
	CK6	0.8968	0.8903
Product Beliefs			
CR = 0.8765	PP2	0.8849	0.8874
Cronbach Alpha = 0.7877	PP3	0.8831	0.8814
AVE = 0.7042	PP4	0.7414	0.7405
Destination Beliefs			
CR = 0.7854	DP1	0.848	0.8419
Cronbach Alpha = 0.93	DP2	0.7436	0.7392
AVE = 0.7023	DP3	0.9138	0.9137
Intention to Choose			
CR = 0.8868	MMP1	0.7589	0.7672
Cronbach Alpha = 0.8305	MMP2	0.8511	0.8554
AVE = 0.6625	MMP3	0.8447	0.8457
	MMP4	0.7977	0.8088
Intention to Visit			
CR = 0.9129	MMD1	0.8746	0.8689
Cronbach Alpha = 0.8664	MMD2	0.9068	0.9043
AVE = 0.7776	MMD4	0.8634	0.8671

Table 3. AVE Square Root

	Destination Beliefs	Intention to Choose Destination	Intention to Choose Product	Place Image	Product Beliefs
Destination Beliefs	**0.838**	0	0	0	0
Intention to Choose Destination	0.6983	**0.8818**	0	0	0
Intention to Choose Product	0.68	0.8079	**0.8139**	0	0
Place Image	0.6293	0.8	0.7078	**0.8444**	0
Product Beliefs	0.5605	0.8362	0.7475	0.8144	**0.8391**

While convergent validity can be identified via composite reliability, discriminant validity can be assessed using AVE square root. According to the criterion for threshold value, Fornell and Larcker [13] suggested that the square root of AVE of each construct should be higher than the correlation with other construct. Table 3 shows that the square root of AVE of each construct (as identified in the bold number) have value higher than the correlations with other constructs. Since the measurement model have shown satisfactory measure for reliability and validity, thus all remaining items were further analyzed in the structural model. This study dropped some items 1) 4 items from place image, 2) destination belief 1 item, and 3) intention to visit 1 item. The final result is the final measure after deleting some items due to not qualifying the PLS validity and reliability requirement.

4.5 Structural model

After the measurement model was completed, the structural model assessment can be tested. The results from the structural model is shown in Figure 2. There are four R2 can be seen from Figure 2. R2 of product beliefs is 0.663. R2 of intention to choose product is 0.559, R2 of destination beliefs is 0.396, and R2 of intention to visit is 0.488. This suggests that place image is considerably strong in affecting product beliefs (66.3%) and destination beliefs (39.6%). Further, product beliefs explain intention to choose product quite strong with 55.9%. Similarly, destination beliefs also explain intention to visit quite strong with 48.8%, while the other 51.2% is explained by other factors.

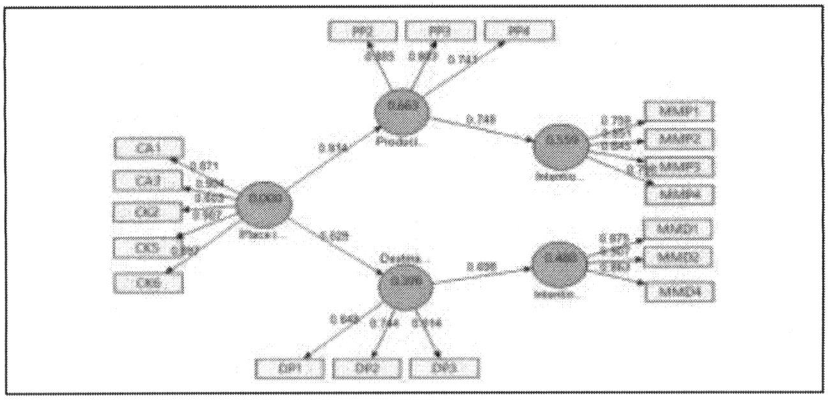

Figure 2. PLS Results

Table 4. Path Coefficients

Independent Variable	Dependent Variable	Original Sample (Standardized estimate)	T Statistics (|O/STERR|)
Destination beliefs	Intention to visit Destination	0.6983	17.0504
Place image	Destination beliefs	0.6293	12.4538
Place image	Product beliefs	0.8144	21.5048
Product beliefs	Intention to choose product	0.7475	14.1675

The structural model from PLS explained that all the relationships proposed were supported. This means that hypothesis 1 to hypothesis 4 were supported. The proposition that place image positively affect product beliefs (H1) was supported with significant $\beta=0.8144$, $p<0.01$ and t statistics at 21.5048. The proposition that place image positively affect destination beliefs (H2) was supported with significant $\beta=0.5293$, $p<0.01$ and t statistics at 12.4538. The proposition that product beliefs positively affect intention to choose product (H3) was supported with significant $\beta=0.7475$, $p<0.01$ and t statistics at 14.1675. The proposition that destination beliefs positively affect intention to visit (H4) was supported with significant $\beta=0.6983$, $p<0.01$ and t statistics at 17.0504. The results of this study have important implications for batik industry and Yogyakarta as destination where the image and reputation of Yogyakarta can be managed to support the local product such as batik.

5 Discussion

This study proposes four hypotheses that combine between tourism destination image (TDI) concept to product country image concept (PCI). The first hypothesis was supported thus place image influences product beliefs. This can be interpreted that positive image of an area may lead to public's positive perception on the product produced by the related area. Batik Yogyakarta has long been preserved and developed particularly in the kingdom family. Public perception on the quality of life in Yogyakarta, quality of education, quality of health, hospitality of the local, peaceful and pleasant area can assist people to think positively about local product produced in Yogyakarta. The positive perception by people that batik Yogyakarta is a good product, is made by expert, innovative and have high historical values can be because of long time efforts made by people of Yogyakarta in building their positive image for their region. The result of this study supports the previous literature in the same context by Elliot and Papadopoulos [9] and DeNisco, et al., [8].

Similarly, the second hypothesis is place image positively affect destination beliefs. Relating to batik industry, Yogyakarta image as a whole has positive impact on public beliefs about Yogyakarta as a destination. The better the image the better the destination beliefs. This means that Yogyakarta image will lead people to think that when in Yogyakarta, they should find many interesting places to visit such as the Kraton (kingdom palace), the temples, the beach, the volcano and the shopping areas. So even though someone has never visit Yogyakarta, but by being exposed to the positive information about Yogyakarta, people will start to build beliefs about Yogyakarta as a place to visit. The result of this study thus supports the previous literature from Elliot and Papadopoulos [9] and DeNisco, et al., [8].

Hypothesis 3 and hypothesis 4 proposing that when customers build beliefs about place and local product, then most possibly that their behavior should show positive response. Beliefs is prerequisite for future behavior. Thus, business managers, government and education institutions should understand and be clear that there are processes to make consumers choose the offerings. These hypotheses give us some views that even though image may lead to direct impact on behavioral intentions, building "beliefs" in product and in place should lead to better response from consumers. The finding from this study supports the previous literature from Elliot and Papadopoulos [9] and DeNisco, et al., [8].

6 Conclusion and recommendation

Place image as an intangible cue may influence the behavior of the consumers. The marketing discipline has recognized the country-of-origin effect on products with brand where the image of country or place may influence consumers' perception on product. By using this logic, narrower area such as the image of territory could also influence the local products receptivity by the market. The objective of this study is to find the influence of place image on product and destination beliefs as well as further on the receptivity or intention to choose the product and to visit the place. Batik industry in Yogyakarta is chosen to test the model. Purposive sampling was chosen where respondents were individual who are not the residents of Yogyakarta but familiar with Yogyakarta. There were four proposed hypotheses built for this research and the PLS-SEM suggests that: 1) The influence of place image on product beliefs is supported. 2) The influence of place image on destination beliefs is supported. 3) The influence of product beliefs on intention to choose product is supported. 4) The influence of destination beliefs on intention to visit the destination is supported.

The theoretical implications are that the combination of tourism destination image and product image could be modeled in one research framework, so that it could provide the theoretical base on place image and product image relationships. Secondly, testing the research model in batik industry provides the wider evidence in different industry on the application of integrative model between tourism destination image and product country image. The practical contributions are that the statistical support to the research model should be translated by the managers, public policy makers as well as academician that marketing strategy on local products should consider the reputation or image of the place where the local product originally comes from. This strategy could add differentiation on the products as well as helps the market to easily recognized and build positive image to those local products.

7 Acknowledgements

The authors are thankful to DP2M Ministry of Research, Technology and Higher Education of the Republic of Indonesia and Universitas Islam Indonesia, Yogyakarta, for supporting this research and providing the necessary facilities.

References

[1] Z. U. Ahmed, J. P. Johnson, X. Yang, C. K. Fatt, H. S. Teng, and L. C. Boon, Int. Mark. Rev. (2004)
[2] F. Ali, S. M. Rasoolimanesh, M. Sarstedt, C. M. Ringle, and K. Ryu, Int. J. Contemp. Hosp. Manag. (2018)
[3] M. Andéhn, F. Nordin, and M. E. Nilsson, J. Consum. Behav. 15, 225 (2016)
[4] A. Beerli and J. D. Martin, Ann. Tour. Res. 31, 657 (2004)
[5] C.-F. Chen and D. Tsai, Tour. Manag. 28, 1115 (2007)
[6] W. W. Chin, in Handb. Partial Least Squares (Springer, 2010), pp. 655–690

[7] A. De Nisco, N. Papadopoulos, and S. Elliot, Int. Mark. Rev. (2017)
[8] Q. Deng and M. Li, J. Travel Res. 53, 69 (2014)
[9] C. M. Echtner and J. R. B. Ritchie, J. Travel Res. 31, 3 (1993)
[10] S. Elliot and N. Papadopoulos, J. Bus. Res. 69, 1157 (2016)
[11] S. Elliot, N. Papadopoulos, and L. Szamosi, Anatolia 24, 5 (2013)
[12] M. Florek, J. Int. Stud. 6, (2013)
[13] C. Fornell and D. F. Larcker, J. Mark. Res. 18, 39 (1981)
[14] M. G. Gallarza, I. G. Saura, and H. C. García, Ann. Tour. Res. 29, 56 (2002)
[15] D. Gertner, Place Brand. Public Dipl. 7, 91 (2011)
[16] J. J. Hahm and K. Severt, J. Hosp. Tour. Insights (2018)
[17] J. Hahm and Y. Wang, J. Travel Tour. Mark. 28, 165 (2011)
[18] J. F. Hair, M. Sarstedt, C. M. Ringle, and J. A. Mena, J. Acad. Mark. Sci. 40, 414 (2012)
[19] C. M. Han, J. Mark. Res. 26, 222 (1989)
[20] C. K. A. Hawkins, Del I, Best Roger J., Consumer Behavior: Building Marketing Strategy, 8th ed. (McGraw-Hill, Boston, 2001)
[21] K. Kaplanidou, J. Sport Tour. 14, 249 (2009)
[22] C. King, N. Chen, and D. C. Funk, J. Hosp. Tour. Res. 39, 3 (2015)
[23] B. Leisen, J. Serv. Mark. (2001)
[24] M. Mody, J. Day, S. Sydnor, X. Lehto, and W. Jaffé, Tour. Manag. Perspect. 24, 139 (2017)
[25] S. Pike and C. Ryan, J. Travel Res. 42, 333 (2004)
[26] R. Ratna, A. Hidayat, and M. Muhajir, Int. J. Res. Stud. Manag. (2019)
[27] R. Roostika, Russ. J. Agric. Socio-Economic Sci. 6, 97 (2019)
[28] S. Stepchenkova, (2015)
[29] H. Zhang, X. Fu, L. A. Cai, and L. Lu, Tour. Manag. 40, 213 (2014)
[30] M. Zhang and D. Merunka, Asia Pacific J. Mark. Logist. (2015)

Research on Public Participation in Recording Intangible Cultural Heritage in Rural Area

Yanling Guo [1*], *Yonghui Wang* [2]

[1] School of Management, Hebei University, Baoding City, Hebei Province,China
[2]School of Management, Hebei University, Baoding City, Hebei Province,China

Abstract. Being an important part of cultural heritage, intangible cultural heritage (hereinafter referred to as "ICH") needs to be passed down from generation to generation. Seemed as the gene of national cultures, ICH is very important for the development and prosperity of human culture, for the development of social economy and civilization, and for the identity of individuals and ethnic groups. Since the publication of the *Convention for the Safeguarding of the Intangible Cultural Heritage* in 2003, ICH safeguarding activities have been fully implemented throughout the world. Recording ICH is a basic and important way to safeguard ICH, this way plays an important role and has been widely approved throughout China and other countries. It has been applied in many ICH item safeguarding, such as Nanjing Yunjin, Qinhuai Lantern Festival. As we all know, ICH resources are very rich in China. There are 39 ICH items have been listed in " the Representative List of the Intangible Cultural Heritage of Humanity", 1,372 ICH items have been listed in "the Representative List of the Intangible Cultural Heritage in China". There are a large number of famous and not famous ICH items scattered in urban areas. To record so many ICH resources, it is not enough to rely on the power of the government. As the creator and inheritor of ICH, the public should play an important role in the safeguarding activities. Attracting the public to participate is a prerequisite of ICH safeguarding. Through questionnaires, interviews and other research methods, this paper introduces the public's awareness and attitude about recording ICH, and their ability and willingness to participate in ICH safeguarding activities. Based on the current status, It analyzes the advantages of public participation in the protection of ICH , and the disadvantages, opportunities and threats are also analyzed. Based on the international principle of "government-led, public participation", this paper proposes to construct a "folk-government" model to optimize the quality of recording ICH. To ensure the quality of public participation, it further points out that some measures should be taken, such as strengthen the government leadership to provide necessary support, identifying their advantages and responsibilities of different participants to joint, expanding the participation content and diversifying the participation method. These measures are all aimed at promotion the safeguarding and inheriting the ICH.

* Corresponding author: yanlingguo02@163.com

1 Introduction

Since the publication of *the Convention for the Safeguarding of the Intangible Cultural Heritage* in 2003, the safeguarding of intangible cultural heritage (hereinafter referred to as "ICH") has been fully implemented throughout the world. In China, the national ICH inheritor recording project has been launched in 2015, and the national ICH recording project has been launched in 2017. A large amount of records are created. ICH records include all kinds of documentation during the process of ICH safeguarding and inheritance, such as text, photographs, sound and video recordings. These records are products of ICH safeguarding and inheritance, and also are the basis for further safeguarding and inheritance. With the development of ICH recording project, a large number of ICH distributed in rural areas that never attracted the attention of national government departments can be effectively protected by records. The public, as the familiar and inheritor of ICH is bound to become the main force in the ICH recording project. Relative to government agencies, public refers to all individuals and organizations, including inheritors, non-governmental organizations, and ordinary persons who maybe participate or interested in ICH safeguarding.

Different members are different in values, knowledge structure, cultural literacy, and social responsibility. These factors determine their different performance in recording ICH. Generally, there are quality problems exited in records in terms of authenticity, integrity, systematicity, security, etc. Based on the status of public participation in the safeguarding of ICH, this paper discusses strategies to optimize public participation in recording ICH

2 Material and Methods

Through questionnaires, interviews and other research methods, this paper introduces the public's awareness and attitude about recording ICH, and their ability and willingness to participate in ICH safeguarding. Based on the current status, it analyzes the advantages of public participation, and the disadvantages, opportunities and threats are also analyzed.

2.1 The realistic foundation of public participation in recording ICH

2.1.1 Wide participation in the safeguarding of ICH

With the public cultural consciousness and cultural self-confidence have been gradually improved, the public has become an important force in the safeguarding of ICH. They have participated in most safeguarding activities, from the application for the list of heritage, the identification of inheritors, the construction of training school, and the construction of the ecological and cultural area. The wide application of high technologies such as mobile internet and new media provides a more efficient platform for public participation in the safeguarding, inheritance, propaganda and innovation of ICH. For example, in "the First Intangible Cultural Heritage Innovation and Design Competition" launched on April 2018, it has attracted more than 20 teams including university students, Internet companies and cultural companies.

2.1.2 Public participation in the safeguarding of ICH has received legal support

Article 15 of the Convention for the Safeguarding of the Intangible Cultural Heritage declares that each State Party shall ensure the maximum participation of community groups and individuals, and permit their active participation in management when carrying out ICH safeguarding activities. "Government-led, Social-participate" is also the general principle of ICH safeguarding throughout the world.Articles 9 and 10 in the Law of the People's Republic of China on Intangible Cultural Heritage clearly declares that "China encourages and supports citizens, legal persons and other organizations to participate in the safeguarding of ICH, and organizations or individuals that make significant contributions should be rewarded." At present, the government pay more attention on the ICH that have been listed in certain safeguarding lists, and the higher the level, the more attention it attracts (for example, listed in " the Representative List of the Intangible Cultural Heritage of Humanity", "the Representative List of the Intangible Cultural Heritage in China"). While scattered in rural areas or undeveloped areas, a lot of ICH are rapidly disappearing because of their low level or not be listed. Public power is playing the key role in safeguarding and inheriting these ICH.

2.1.3 The public may be the important force in recording ICH

Recording is the basic way to safeguard ICH. Especially for ICH which are weak and frangible, recording means a kind of rescue. The public is the familiar, inheritor, and beneficiary of the ICH, the public has an natural advantage in the recording of ICH. In China, from the early stage , folk songs, folk stories and myths are recorded and arrangement in "the Book of Songs", to the ten books of Chinese folk literature and art in the 1980s, to the recording project of ICH nowadays, all these have demonstrated the public power in recording ICH.

2.2 Practical obstacles for the public to participate in recording ICH

2.2.1 Weak participation consciousness

The government and the public are the two main forces in ICH safeguarding. But the two forces are not balanced and coordinated from the perspectives of laws and regulations formulating, ICH survey, and inheritor identification. In the era of "application for the heritage ", ICH are graded by government. Some ICH have been paid enough attention by national governments and international organizations (such as UNESCO), while a large number of ICH are facing vanished with the deaths of inheritors due to low grade. In the process of recording ICH, the public is not familiar with what should be recorded, how to recorded, and related technologies and standards during recording activities. Public participation is also limited to passively obey the government arrangement during the field survey, records collection, and oral recording. The lack of consciousness of participation leads to lack of initiative, which leads to problems existed in the quality of records, such as not complete, not authenticity, not dynamic, etc.

2.2.2 Insufficient capacity for public participation

The traditional long-standing ICH has been seen as the blood and gene of local people. It is an integral part of the local mass's life. However, because lack of necessary knowledge and skills, public participation capacity in recording ICH is often not enough.

Firstly, knowledge about records management is not enough. ICH inheritors and the mass have accumulated and preserved a large number of records during their ordinary lives and activities. However, how to organize these records (including scientific classification, system arrangement, etc.), how to control the storage conditions to extend their life requires a amount of special knowledge. During the author's survey, it is often found that some precious records are lost, mildewed, and the disc cannot be played. Secondly, with the wide application of a large number of technologies, digital devices (such as digital cameras, smart phones, computers, etc.) are needed to record ICH, this is difficult and not reachable for people who are old and live in rural area.

2.2.3 Limited in content and participation ways

Recording enables ICH to be "preserved and passed on". At present, public participation is not enough in the process of investigation, recording, preservation, inheritance and propaganda of ICH. Generally, most of the public participation is limited only in the records collection stage, and the public participation is rarely in other stages, such as records information organization, data mining, publishing, database construction and management, website construction and operation, etc. Most of these activities are considered too professional, which directly prevent the public from participating. As to the participation ways, ICH recording is mostly government-led, and most of products are preserved in library, museum, archives ,intangible cultural heritage centers and other related institutions. Therefore, public participation in ICH recording is controlled by government. The Limited content and participation ways of public participation inevitably lead to the disappear of a large number of traditional cultures.

2.3 Optimization Strategies for Public participation in Recording ICH

2.3.1 Adjust the relationship between government and the public in recording ICH

You are free to use colour illustrations for the online version of the proceedings, but any print version will be printed in black and white unless special arrangements have been made with the conference organiser. Please check whether or not this is the case. If the print version will be black and white only, you should check your figure captions carefully and remove any reference to colour in the illustration and text. In addition, some colour figures will degrade or suffer loss of information when converted to black and white, and this should be considered when preparing them.

"Government-led, Social-participate" is one of the international general principles in the safeguarding of ICH. "Government led" refers that the government should provide corresponding support to mobilize, organize, manage the safeguarding of ICH through management regulation designing, organizational construction, and funds providing. But in fact, many people(including government departments) misunderstand the "government-led", and interpret it as the government should control all the safeguarding activities. This misunderstanding lead the government to too excessive participation in ICH protection, which has also led to the gap (estrangement) between government departments and inheritance groups. Recording ICH means dynamically recording the past, present, and future of all kinds of ICH items through records and so that the following generations can know about the traditional culture through records. The ICH can exist for a long time in form of records even if some traditional culture will disappear in the history. The public is the basic power of creating, thriving, inheriting and innovation of ICH, so we must adjust the relationship between the government and the public in the process of recording ICH.

Firstly, public participation means public participation should not limited in the scope of donation, oral recording, while should participate in the whole process of recording ICH, including collection, storage, arrangement and utilization of records. For example, public should participate in records appraisal, identifying the records authenticity and retention period, records compilation, exhibition, description, website designing and so on.

Secondly, public participation means comprehensive participation, fully participating into the recording activities, regulation designing, organization system and funds support. Public should be fully given the right to know, to supervision and decision-making. Taking the regulation designing as an example, generally, most regulations including management regulation, operational norms as well as standards and specifications are drawn up by the government or relevant departments, the public only implement these regulations and standards. Public comprehensive participation should also participate in the regulations construction through public hearing, discussion, inspection, supervision and so on. The public's will, demands and realistic conditions should be given full respect. Public comprehensive participation ensure the regulations are scientific ,and also ensure they are carried out smoothly.

Thirdly, public participation means all the public members' participation. During recording ICH, the public should include all the groups and individuals that have actual or potential influence on the safeguarding of ICH. Different individuals and groups with different cultural literacy and knowledge have different capability, attitude and interest during recording the ICH, but all of them are indispensable and interrelated forces.In the rural area, the elderly ordinary people think that they are far from the specialty work of ICH safeguarding and recording. In fact, they are the most loyal inheritors and guardians of traditional skills and folk customs, and also the main source for creating the authentic and integral records. The participation of the whole public can make the ICH records "taken from the folk and used for folk"

2.3.2 Strengthening professional guidance to ensure the quality of public participation

ICH listed in safeguarding inventories of all levels are often recorded by the related government departments and recoeds are kept centrally. While for a lot of ICH items that have not been listed in inventories, records are scattered, even not be recorded, local individuals are the main power of recording them. These recording activities are casual, not professional. To improve the records quality, the government departments should provide enough professional guidance on what to record, how to record,etc.

First of all, recording scope should be defined. The recording scope should cover the emergence, development, evolution, protection, inheritance and all activities of the cultural item. So the recording activity should go on constantly. Modern technologies and equipment should be used as much as possible. It is not only beneficial to the preservation and dissemination in the network environment, but also make the factors that can not be reflected by pictures and texts (such as feeling, strength and skills, etc.) can be collected and feedback.

Secondly, records appraisal must be emphasized enough. ICH (such as customs, etiquette, folk music and dance) come from local people's everyday life. Therefore, the value of ICH records must be judged by the local people. On the basis of their daily life, the public understand its value in continuing the national spirits, propagating the national concepts, and maintaining the national emotions. From the perspective of records management, the public needs to know the records appraisal work, including its significance, content and methods. The public should learn to judge the records' value, authenticity and retention period from the factor of content, source, practice and external characteristics. Comparison, content analysis, verification and other different methods should be used in records appraisal.

Thirdly, ICH records need to be systematically managed through specialty methods such as classification, arrangement, serial coding and so on. The public needs to know the classification of ICH items and the classification of records belong to the same ICH item.

These skills and knowledge can help to turn the scattered and disorderly records into systematic records resources.

Finally, records preservation is also an urgent problem to be solved. In our investigation, we found that due to the unsatisfied storage conditions, a large number of records kept in the folk faced problems such as paper yellowing and crisping , bitten by insects and mice, unclear old photos and images, and unable to be played discs. Therefore, the public needs to know the basic preservation skills and knowledge, measures. The public should know the appropriate conditions for keeping the records in proper carriers, and the alternative repair plans and channels for seeking help in case of problems

2.3.3 Optimizing the process of public participation

Firstly, the composition of the public needs to be analyzed. The public can be divided into different parts, such as universities and research institutions, social groups, local elites, inheritors, ICH enthusiasts and ordinary people, etc. Every part has advantages and disadvanges. For example, universities and research institutions have advantages on theoretical literacy and are good at grasping the basic characters and regular patterns through appearances. Experts and scholars from fields of records management, culture, tourism, literature and art can provide adequate theoretical guidance for recording ICH. Most social groups (such as societies and associations) are devoted to a certain field, with relatively complete and standardized organizational structure and management system. Their members with high cultural literacy can undertake the tasks of organization, communication and coordination in the ICH recording work. Local cultural elites have strong influence and appeal in local areas, and often make outstanding contributions to the inheritance, protection and innovation of local culture. The inheritors are the "living soul" of ICH and grasp the essence of ICH, however, their cultural literacy, cultural consciousness and recording capability are very different from each other. ICH enthusiasts are a very energetic group of the public. Many of them record ICH actively, upload and share ICH information through websites, micro-blog, WeChat and other media. Ordinary people is the largest group of the public. ICH is closely related with their daily life. Every member of the public is an important component in the recording of ICH. They must complement and combine with each other.

Secondly, the process of public participation needs to be optimized according to the "Folk-Government" model. ICH is rooted in the local daily life. Recording ICH needs to adopt "Folk-Government" model, that is to say, the public is the main power under the help and support of the government, so that the ICH recording is carried out in the natural space. The public participation process should be optimized as follows: When an ICH item proposed to be recorded (proposed by the government or the public), this proposal will be submitted to the local records management department. After accepting the proposal, the records management departments decompose the recording work into several tasks (include but not limited to records collection, arrangement, preservation, utilization, digitalization etc.) ,interpret every task in detail, provide proper support, including methods, standards, content, and all that need be paid attention to. And then records department should release the tasks and recruit volunteers through television program, website and other new media. After releasing, the public may apply for certain tasks according to their abilities, interests and hobbies. When the number of enrolled volunteers reaches a certain threshold, the volunteer team is established. The volunteer team will be divided into different groups based on the tasks for recording ICH. For instance, there are groups for collection ,for organization, for preservation, for utilization, for propaganda, for digitalization, etc. The records departments will arrange records managers for each group to guide, supervise and assist in the completion of their tasks.

To ensure the recording quality, the recording process should be embedded with three-layer reviews. The first layer review is undertaken by the members of the same task group to find and reserve quality problems such as not implement standards and any other problems during recording. After completing their own tasks, volunteers may submit the results on the platform, so as to facilitate other members to review and find out quality problems. So the second layer review is taken by members of other teams, to find not authenticity, not accurate, non-standard or any other problems. To encourage volunteers to actively participate in the review, proper rewards and honorary titles are needed. The third layer review is undertaken by the experts group, which composed of experts from records management, archival science and other related fields according to the category and characteristic of ICH. The experts group's is responsible for review if the recording process is undertaken scientifically and if the ICH can be safeguarded by those records. In the whole recording process, the decomposition, distribution, implementation, guidance and review are carried out according to the characteristics, existing situation and requirement of the ICH.

A drop of water can reflect the brilliance of the sun, ICH can reflect the style and appearance of local life. The breadth and depth of ICH recording need to be expanded. Scientific and effective public participation is the main source of ICH recording. However, public participation is still in its primary stage. The public participation needs to be optimized through the professional guidance, through redesigning the public participation model and process , through strengthening the government proper support, and through analyzing and developing different member's advantages and disadvantages so as to cooperate together.

References

1. T. Janke and L. Iacovino, Arch. Sci. **12**, 6 (2012)
2. T. Cook, Arch. Sci. **6**, 13 (2013)
3. S. Pietrobruno, Int. J. Herit. Stud. (2014)
4. C. Guccio, M. F. Martorana, I. Mazza, and I. Rizzo, in Cult. Herit. a Chang. World (2016)
5. G. Cozzani, F. Pozzi, F. M. Dagnino, A. V. Katos, and E. F. Katsouli, Pers. Ubiquitous Comput. (2017)

POTENTIAL OF NEW INSTITUTIONAL ECONOMICS FOR RURAL COMMUNITY DEVELOPMENT

P. Eko Prasetyo [1*], *Adryan* Setyadharma [1], and *Nurjannah* Rahayu Kistanti [1]

[1] Department of Development Economics, Faculty of Economics, Univeritas Negeri Semarang, Indonesia

Abstract. Rural economic inequality is a fundamental problem which makes it important to seek its causes and find its policy solutions immediately. The new institutional economic potential development policy is the right choice. The purpose of this article is to explain more broadly the importance of new institutional economic potential in developing rural communities. Especially, to explain various economic dimensions of new institutions in a comprehensive manner in determining the level of community development. This research method used structured survey that has been well-designed and measured for data collection, variable measurement, and data analysis. Quantitative and qualitative data are gathered by using integrated methods between various disciplines; economics-sociology and economics-geography. Further, to interpretation the data, we used economic and cultural concepts; informal economy, new institutional economics, gravitational economics and cultural anthropology. Then, the method of analysis is path analysis using recrusive form of correlation model with multiple path equation systems. The main research results show that the new institutional economic potential is a key factor in developing rural communities and reducing inequality. Besides, the new institutional economics will encourage economic growth and public welfare

1 INTRODUCTION

Currently, rural communities in Indonesia are not only experiencing complex problems, but also interesting challenges to study and develop such as the increase of economic inequality, and the decline of social capital. Other problems that cannot be taken for granted are poor commercial infrastructure, limited social networks, increasingly loss of productive personnel, and the limitation of other facilities. Not only rural development policies have been implemented in various ASEAN member countries, but also have become priorities and industrialization's complementary. These efforts contribute to higher inclusive growth and more employment in rural areas (Thanh, 2016). The aim of rural development in Indonesia is to encourage quality economic growth; to reduce unemployment, poverty and

* Corresponding author: pekoprasetyo@mail.unnes.ac.id

inequality. Many policy programs for rural areas have been carried out such as counseling, education and training, various grants and subsidized soft loans. However, the program has not helped many rural communities to rise and develop independently. The success of each program can vary, even in the same geographic area. To help participate in solving the problem, it is proposed to focus on the role and function of the potential of "institutions" especially through the new institutional economics. The urgency of institutions in the broadest sense, can be at the center of all efforts in developing rural communities. Furthermore, this article proposes a new institutional economics as a new reference that can be used to encourage economic development and improve the welfare of rural communities.

Rural economic development is a part of community development. The economic development of rural communities is as a process of increasing income and living standards and the welfare of rural communities followed by increasing quality of life and quantity of settlements, better quality of social, psychological and cultural institutions in rural communities. Community development is a process where people are together with the government and other stakeholders to improve the socio-economic and cultural conditions of community members (Latopa, 2015). This improvement will determine their better choice and create an environment which make people can use their full potential to lead productive lives (Shaffer 1989). Rural community development is a process carried out by community members. This is a process in which local communities cannot only create more jobs, income, and infrastructure, but also help their communities become fundamentally better and are able to manage change (Cavaye, 2006). The function of economic institutions in society is as an institution that provides security of ownership rights and relatively equal access to economic resources to the wider community (Netar, 2017). The result of Netar's research (2017) explains that there is a large impact from the institutions that exist in the community on livelihoods in rural areas

The economic development of rural communities will certainly involve active participation and many institutional elements of related community development such as social capital, human capital, social networks, economic institutions, entrepreneurial business opportunities or rural industries and so on. The article is to describe the problem in achieving economic development of rural communities and community development. Specifically, it is to describe the role of local wisdom of social capital in the community and new institutional economics potential in the development of rural communities. According to Netar (2017), the role of social and economic institutions in a society will dominate the economic development process. They determine attitudes, motivations and conditions for development. If the institutions are elastic and encourage people to take advantage of economic opportunities and further to lead a higher standard of living and inspire them to work hard, economic development will occur and vice versa. However, this problem has been properly observed by UNO that economic development is not possible without supporting environment (Netar, 2017)

The basic concept of new institutional economics (NIE) is the economic behavior of a person or group which is strongly influenced by certain institutions. This institution is very strong and in this case are interpreted as "rules of the game" in a community group, both formal and informal. Therefore, the key to the success of economic development for the development of rural communities through new economic institutions requires sufficient prerequisites as well as the support of all institutional elements that exist in society and active participation of the community productively. All individuals in society and institutional elements are expecting progress and must make changes that are beneficial to the progress of society. That is, the economic role of the new institution is taken as a facilitator or rules of the game; to overcome common inequality, prosperity and justice. Meanwhile, every behavior of the active participation of individuals and society is a determinant of institutional progress. Because institutions are inanimate objects, institutions cannot develop if there are no individuals or groups in the community who develop them, and conversely individuals and community groups will be difficult and cannot develop well if existing institutions do not facilitate the individual or group to be actively participate and take action for the progress. In other words, the new institutional economics is proposed as a rule of the game in encouraging the active behavior of the community to create business opportunities, efficiency, influenceiveness, productivity, economic growth and competitiveness and shared prosperity.

The research argues that both novelty and empirical institutional performance of the Indonesian central government are ininfluenceive and inefficient, and are unable to reach rural areas, so that the inequality in rural areas increases. This argument is supported by the result of a research (Yildirim, 2016) which concludes that developing country institutions are generally less influenceive. In developing countries the quality of service is bad as well as its bureaucracy. It is high cost, weak human resource capacity, unskilled, corrupt, and the existence of manipulation in the justice system. So that, the level of social trust is low and the existence of institutions does not have adequate legal regulations and sanctions. Hence, it is proposed that social capital, human capital, entrepreneurship, and new institutional economics are the pillars of sustainable economic development that are expected to be able to increase rural economic growth and reduce inequality. This argument is supported by the result of a research (Raja, 2014) which stating that if natural capital, physical capital and human capital are the wealth of the nation, social capital contributes more to harmonious growth. Raja, (2014) emphasizes that with the new institutional economics framework, social capital is a driver of sustainable development. The aim is to introduce to potential institutions; new social capital, human capital, entrepreneurship and institutional economic framework in Indonesia which can be the main sources of sustainable development and reduction of inequality through rural community development efforts

2 Literature review

Institutional theory has been increasingly used in research and entrepreneurial implementation, (Bruton, 2010). Based on literature studies, the emergence of the term "institutional economics" (institutional economics) was first introduced by Walton Hamilton in 1919. The well known institutional school of economics are Thorstein B. Veblen, Wesley Mithell and John R. Commons (Rutherford, 2001). However, this institutional school of economics did not develop and was inferior to the Neo-Classical School with econometric tools and welfare economics by J.M. Keynes. Further, in 1970s, this institutional economy reappeared with a more standard concept and as a center to discuss ideas and had a greater economic impact, so that since this emergence, it is often known as New Institutional Economy (NIE) and in 1970s was considered Old Institutional Economics (OIE). The OIE person who is considered the most powerful and influential and can also be called the father of institutional economics is Thorstein Bunde Veblen (1857-1929). Whereas, the famous followers of Veblen are like; Joseph Schumpeter, Gunnar Myrdal, and Kenneth Galbraith. However, in various literatures, they are often referred to the flow of Islamic Institutions.

The emergence of NIE is a success story which have been through many stages (Menard, 2018). The emergence of the NIE as a center for new ideas, there are not only many institutional research programs running, but also many competing ideas (Williamson, 2000; Richter, 2005). There are famous NIE figures such as Douglas C. North, Ronald Coase; Elinor Ostrom; and Oliver E. Williamson (Richter, 2005). Also, there are several important functions and important roles that NIE stands for as arguments which are used as the rationale. First, NIE is an effort to incorporate institutional theory into the economy. However, NIE modifies and expands the NCE theory to be able to overcome and deal with various problems (North, 2003). What is built and maintained is the basic assumption of scarcity and competition as the basic theory of competition in microeconomics. The NCE assumption is instrumental rationality. Second, NIE is more flexible because it emphasizes "rules" or economic policies and not the dominance of free market roles. Where the institutional framework determines the types of skills and knowledge that are considered to have maximum results (North, 2003). Third, NIE is important because it is built humanistically by formal and informal forces to improve economic growth, competitiveness and shared equity (Acemoglu, 2005, Prasetyo, 2019). Another argument is that the new institutional economics approach is more multidisciplinary in nature, namely placing economics in the framework of a unity with other social sciences such as; sociology, anthropology, geography, history, politics, law and real life culture of the local community.

There is small difference between OIE and NIE namely; that the emergence of OIE tends to be a reaction to the feeling of dissatisfaction with the flow of Neo-Classical (NCE) which is actually a continuation of the flow of classical economics. Whereas, the emergence of NIE tends to be more constructive, perfect and complete. There are four basic theoretical concepts used by the NIE school in discussing the economic behavior of society, namely; (1) transaction cost theory, (2) property rights theory, (3) public choice theory, and (4) game theory, (Williamson, 1985, 2000; Kim, 2005). The first three bases are often referred to as the NIE golden triangle (Menard, 2018). Meanwhile, another difference between OIE and NIE is that the NIE school still uses two basic Neo-Classical assumptions that are important namely; rational individual behavior and the existence of a clear individual preference function. Based on these differences, the NIE is often considered not to create conflict with NEC and its nature as development, but it is refining and completing. Williamson (2000) has introduced the evolution of the NIE theory through four levels of "social analysis", and the first level is considered the lowest level is; social theory namely; as informal rules that are inherent in the local community such as tradition, norms, customs and religion. This level moves very slowly. Then, the second level is; economics of property rights, the third level is transaction cost economics, and the fourth level as the highest level is resource efficiency and incentive structure originating from neoclassical economics or agency theory.

Institutions are humanly designed boundaries forming interactions among communities, both formally and informally and their enforcement characteristics (North, 1994). Institutions are the center of many rural development efforts and offer a place of reference for community members, development researchers, and practitioners to develop a sense of norm and common understanding to work (Meador, 2017). Meanwhile, social capital is the main source of contribution to harmonious economic growth (Raja, 2014). Therefore, it is important to gain insight into the relationship between social capital and rural welfare (Ali, 2006). The Estrin Research Result, (2016) states that it is necessary to strengthen the importance of human capital's ability to enter entrepreneurship, and institutions act as important contingencies. This phenomenon strengthens the argument of the human capital theory into social entrepreneurship. Furthermore, Nielsen (2007) asserts human and social capital in a guideline for intervention in policy implementation. It is very useful to link with systemic approaches such as; system innovation, business systems, and also social capital theory as an intermediary theory in the context of intermediate applications. Whereas, at the high level of adaptation and flexibility in the global market, it needs to be combined with high social benefits and high wage rates, as a balance of human capital. In addition, entrepreneurial incentives and individual responsibilities must be also strengthened (Nielsen, 2007).

Economic development is a part of community development aiming to build five important things, namely; physical, financial, human, social and environmental (Cavaye,

2006). According to Cavaye (2006), the development of rural local industries involves relatively small industrial or institutional group facilities such as market alliances, and this is part of economic development. Through participation and social networks, they build social capital and new skills to build human capital. Furthermore, they develop new economic options in the form of institutions to build physical and financial capital and to improve their environment. Based on all the literatures discussed above, the framework of this research is that; the existence of the potential of social capital, human capital and social networks owned, then they build entrepreneurial business opportunities in the rural area and market alliances as a forum for rules of the game. All of these activities are processes of economic development as part of developing rural communities. Meanwhile, the role and function of the new socio-economic institutions dominate the process of economic development

3 Methodology

3.1 Structural equation model.

This research is an exploratory research using a quantitative approach. The main types of explanatory variables are; social capital, networking and human capital. This main explanatory variable is used to explain the efforts to establish entrepreneurial business opportunities, and new institutional economics, as a way to produce productive economy for the development of rural communities. Further, the main key variable is the new institutional economics variable and it refers to endogenous variable. In the path analysis research model, the explanatory explanatory variable can be considered as an exogenous variable that will determine the formation of endogenous variables. There are two endogenous variables, namely the main endogenous variables, namely the new institutional economics, and entrepreneurial business opportunity variables as the second endogenous variable as well as exogenous variables or moderators of the new institutional economics variables. Based on this information, the structural equation model is as follows

$$Y = \rho_{YX1}X1 + \rho_{YX2}X2 + \rho_{YX3}X3 + \mathcal{E}_1 \quad (1)$$

$$Z = \rho_{ZX1}X1 + \rho_{ZX2}X3 + \rho_{ZY}Y + \mathcal{E}_2 \quad (2)$$

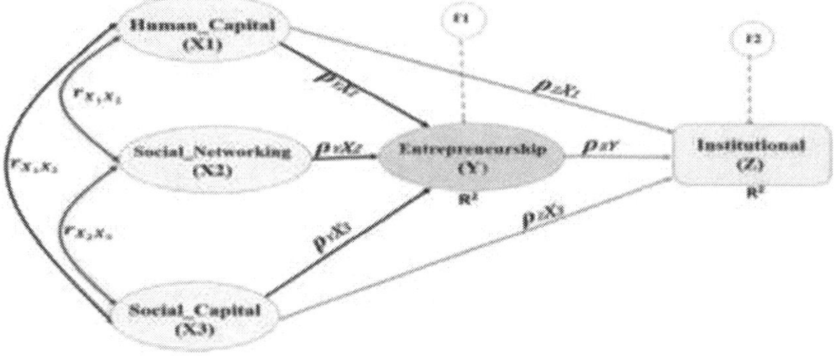

Fig.1. Path analysis of the recrusive form correlation model with multiple path equations
The acknowledgements should be typed in 9-point Times, without title.

Based on figures-1, this research aims to analyze the direct influence, indirect influence and total influence on the new institutional economics. In quantitative analysis of the path analysis model is used to strengthen the argument and to show that standard coefficients can be used and the case of multicollinearity can be reduced

3.2 Research design and explanation

Based on the use of the level of depth of analysis and its type of data, this research tends to be exploratory research designed by the use of an explanatory quantitative descriptive research approach. The main purpose is to explore and analyze deeper the problems of economic development as an effort in developing rural communities. Various issues in the village that have been revealed above are especially; the problem of low economic growth, and high unemployment, poverty and inequality. These must be resolved through economic development so that rural communities are able to develop independently to improve their standard of living and welfare. As economic development is part of community development, the main institutional variables of the new institutional economics potential are proposed as a solution through rural economic development. Because the purpose of the research is to explore, analyze deeply the various factors or other variables that are related and influence the main variables of the institution, the other variables used are more referring to the model (1) and (2). The position of the institutional variable (Z) is an endogenous variable. While some other variables as explanatory (exogenous) are; Human capital (X1), Social networking (X2), Social capital (X3), and entrepreneurship, (Y).

Social capital variables, social networks and human capital as the main key variables used in this research refer to the basic model of Wiliamson theory (1985, 2000). In the initial stage (level 1), this research was designed with the basic concept of "social theory" as described in the Williamson model. That is, collaboration between the main exogenous variables will encourage the formation of entrepreneurial business opportunities in a rural area. Furthermore, in order for the existence of the entrepreneurial business to develop institutional factors which are good, safe and fair. One of the institutional factors referred to is the new institutional economy as a platform and a solution. For the needs of in-depth analysis, primary data are needed and secondary data as supplementary one. The method used is the socio-economic and economic-geographic science approach. While the analytical method tends to be a science approach; economics and economics-sociology. Furthermore, for the purposes of quantitative-descriptive-exporatory analysis, all variables are quantitative and the data tend to be measured through an economic value approach using the gini ratio index method.

4 RESULTS AND DISCUSSION

4.1 Descriptive results and discussion

Economic development is regarded successful if it is able to improve the standard of living and welfare of the community to the rural area. Community welfare is an economic and social dimension. This means that there is a relationship of interdependence between economic growth, economic development and social development. Thus, to achieve good economic development in rural communities, it is not enough to only create economic growth, but it also its growth has to be high, human quality, humanistic principles are maintained and environmentally friendly and social condition is improved. Achieving a high level of economic growth is indeed important, but it is more significant it also improve the quality and the benefits of economic growth itself. If the high economic growth can be achieved, but economic development is not able to reduce the problem of unemployment, poverty and inequality, then this economic development cannot be said successful (Seers, 1979, Chibba, 2008). According to Dudley Seer (1979), the goal of economic development is to reduce poverty, unemployment and inequality. According to Seers (1969), if one or two or even three of these conditions are not available, then development has not succeeded, even though the income per capita has increased in large numbers (Chibba,

2008). Figure 2 shows conditions in Indonesia. In the last 15 years (the era of President SBY and Jokowi), it is found that the economic growth is able to reduce poverty and unemployment, but inequality remains stagnant at a high level, above (0.39), even in rural level of inequality tends to increase. If you refer to the opinion of Dudley Seers, then economic development in Indonesia cannot be said successful.

To make the process of economic growth give a positive influence, it has to be also accompanied by the increase of life quality per capita and a policy of national resource utilization, to achieve individual and social progress in each country (Haller, 2012). Quality of life is a multi-dimensional concept based on a complicated causal relationship of reciprocal relationships of variables that are difficult to measure because they are created from two subjective personal and psychological dimensions (Murgas, 2015). Therefore, the most important part of life quality is the phenomenon of achieving a better life. Thus, Indonesia must strengthen the economic and social institutional system, as well as good and right economic and political governance. The government's policy program in providing village funds is a good program, but unfortunately, its governance is not implemented properly making it have no positive impact on reducing poverty, unemployment and inequality in the rural areas. This problem is generally caused by the problems of economic and political institutions even though the socio-cultural institutions are strong. However, the impact of governance on development is still subjective, complex and not well understood and correct. For example, there is still misuse of village funds. This is proven by the condition of economic and social development in rural areas which is still challenging and it is suspected that the quality of government institutions are still severe. For instance, corruption cases and also inefficient administrations from the national to the local level

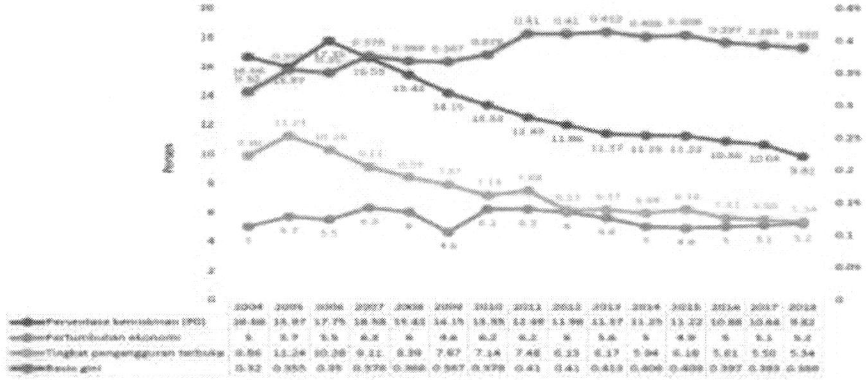

Fig.2. Description of economic quality growth, unemployment, poverty, and inequality.

Based on the above figures and explanations, it can be stated that the unsuccessful implementation of central government policies in rural areas is due to the fragility of economic and political institutions in the local as a reflection of the weakness of national political institutions. To conclude, there is less awareness of economic and social policies in which they are the key to the success of humanist and sustainable economic development. The quality of economic growth in Indonesia as a sufficient condition in the success of economic development because of human capital (Prasetyo, 2008). Meanwhile, social capital is the main source of contribution to harmonious economic growt (Raja, 2014). Therefore, economic, social, political and cultural institutions in the rural areas must be formed and strengthened by the factors of human capital and social capital in the local community as the strength of the local wisdom. If the institution formed is not supported by the strength of local wisdom from both human capital and social capital factors, it will still be easily fragile. This is due to human capital and social capital are the main investments in driving the success of economic growth and social economic development.

Indonesia's vision states that people are the main source of development because the goal of development in Indonesia is to develop Indonesian humanity as a whole and to develop the entire Indonesian community. So that, persons are the main subject and not as objects in the development. With the improved human and social capital, the creativity, innovation and productivity will be stronger and the quality of economic growth will be increasingly high. In my opinion, economic growth in Indonesia is not yet qualified because of economic growth has not been able to encourage the success of achieving the objectives of economic and social development. The high level of income inequality is a proof that even equity-economic achievement can still be declared unsuccessful especially the prosperous social equity is still not fulfilled. I suggest that there are two main reasons namely the quality of human capital and social capital and also the in accurate measurement. Economic growth as measured by Gross Domestic Product (GDB) in Indonesia is inaccurate and can be misleading, preferably economic growth is more real if measured by the Gross National Product (GNP).

It is argued that the attention of Indonesia today and in the future is not only given to to achieve high economic growth as measured by DGP, but also economic growth should be focus on quality. It should be immediately known who are the biggest dominant players and tthose taking the benefits of the achievement of GDP; what kind of mechanism is formed in achieving economic growth as measured by GDP? What mek the rural communities cannot enjoy the results of these economic growth achievement. In other words, if only a high economic growth program could be achieved, it would still be difficult to reduce development problems, especially it is still difficult in reducing income inequality. Therefore, it takes courage and commitment of all elements of the Indonesians to build the capacity of quality human resources and social capital, and the courage to change the mechanism of how to measure economic growth to a more real thing. Without commitment, the biggest bonus demographic that will occur in 2030 can only be a burden for government, and become an object of development and instead of the main subject and capital in development.

4.2 Quantitative results and discussion

The purpose of using the path analysis model is to explore, detect and strengthen the description of the dependence between the various economic phenomena described above. Basically there are three forms of the path analysis model namely; (1) correlated path model, (2) mediated path model, and (3) independent path model. This is because the exploratory research design referred to is to multiply more in the role of variables in terms of the level of relations and at the same time the degree of influence of exogenous variables on endogenous variables both; direct, indirect and total influence. The meaning of the recrusive path analysis model is the causal relationship between endogenous and exogenous variables is in the same direction. Although it is assumed that the direction of the relationship can be of two-way causality, but to make it shorter and easier to understand, they were only selected and presented in the form of a recrusive model only. Furthermore, there are two models form of structural equations, then the form of the model figure chosen is a double lane model. Then, based on the research method above, the steps to present the path analysis results can be presented as follow.

The results of the study in table-1 for model-1 are to explore and examine how the influence of explanatory variables on entrepreneurial business opportunity variables in rural areas. While for model 2 is to explain the most influencing factors on the new institutional economy. Model-1 shows that social capital factor gives the first largest positive influence of 55 percent and are significant towards entrepreneurial business opportunities. In model-1, significant human capital factor is only able to make the second largest contribution that is 33.5 percent of entrepreneurial business opportunities. Meanwhile, the social network factor significantly only contributed a third of 14.1 percent to entrepreneurial business opportunities. In social theory, empirically, social capital is the basic values of local wisdom that have been inherent in almost all communities in rural areas in Indonesia. Although at present, the social capital value tends to decrease. The social capital values in this article are measured by the ratio between dimensions of the level of trust, volunteerism, social participation, tolerance and reciprocity as well as a sense of sympathy and empathy, have provided the main strength and entrepreneurial business opportunities. Furthermore, these social capital factors will form a social network within the community and strengthen shared knowledge and skills to increase the growth capacity for entrepreneurial business. Thus, the results of this research support previous research conducted by Raja (2014) stating that social capital is the main source of contribution to harmonious economic growth and encourages sustainable development. The results of this research also support Madriz's research (2018) which mention that human capital and social capital are drivers for entrepreneurship

Model		**Unstandardized Coefficients**		**Standardized Coefficients**	**t-stc**	**Sig.**
		B	Std. Error	Beta		
1	(Constant)	-.046	.029		-1.575	**.118**
	Human_capital (X1)	.352	.056	.335	6.230	**.000**
	Social_networking (X2)	.168	.066	.141	2.537	**.012**
	Social_capital (X3)	.633	.069	.550	9.219	**.000**
Model		Unstandardized Coefficients		Standardized Coefficients	t-stc	**Sig.**
		B	Std. Error	Beta		
2	(Constant)	.198	.020		9.749	**.000**
	Human_capital (X1)	.262	.048	.323	5.407	**.000**
	Social_capital (X3)	.157	.066	.177	2.392	**.018**
	Entrepreneurship (Y)	.368	.066	.476	5.552	**.000**

a. Dependent Variable: Entreprenurship (Y)
b. Dependent Variable: Institutional (Z)
c. All exogenous variables are significant for endogenous variables at level-1 (99%)

Table 1. Multiple regression results against; entrepreneur and new institution economics.

In social analysis, social networks is a social structure that is formed by individual relations in a social organization because of the similarities in goals, vision, ideas, friends, harmony, and closeness. In social analysis, this social network is bound in which it also has rules of the game that must be maintained wiht full of awareness. Social relations formed through the spirit of entrepreneurship which is based on the uniqueness of social interaction "tuna satak bathi sanak". It is a new innovation to form a new economic and social

 https://doi.org/10.1051/shsconf/20208601015

institution that is believed to improve economic growth and welfare of the community (Prasetyo, 2019). In model-2, the result shows that the entrepreneurship (Y) factor has significantly been able to provide the first largest contribution of 47.6 percent to the formation of institutional (Z). It indicates that increased capacity of entrepreneurs has encouraged the formation of new institutions in rural communities. Thus, the emergence of new social-economic institutions means that there has been an increase in capacity and capacity for social memory (Shvedosky, 2016).

In model-2, the human capital factor has significantly contributed to the second largest amount of 32.3 percent to the formation of a new institutional economy for playing an important role in improving its business. Meanwhile, social capital factors are proven to have declining contribution even though there was still a significant influence on the formation of a new institutional economy. Empirically, using the perspective of socio-economic analysis, this phenomenon is natural and often occurs in real life. This already began since the process of social network interaction became increasingly widespread. Through participation and social network, they build social capital and new skills to improve human capital. The existing role of social capital might be still strong, but in the next process, when they began to develop new economic options as alternatives to new businesses and began to form new institutions, their role began to decline. Actually, the option for new institutions to be formed is to expand their business and simultaneously build joint physical and financial capital and to improve their environment. However, when economic business principles emerge and are dominant, then, the elements of the values of the formation of a social capital are slightly diminished. However, the results of this research are in line with previous research conducted by Cavaye (2006). Cavaye states that the development of local industries, including entrepreneurship in rural areas, involves relatively small industrial or institutional groups. Then, through participation and social networks, they build human capital to form larger institutions. However, the results of Cavaye's research did not explain quantitatively the impact on the social capital itself. In fact, the emergence of new social-economic institutions is built from the social capital in which, in economics, it can be as part of an influenceive knowledge management system although in the social system it can cause disruption of levels of resilience and vulnerability in social life.

Table 2. The partial correlation of the Pearson product moment matrix.

Model-1		Entrepreneurship	Human_capital	Social_netwoking	Social_capital
Pearson Correlation	Entrepreneurship (Y)	1.000	.671	.580	.800
	Human_capital (X1)	.671	1.000	.390	.510
	Social_networking (X2)	.580	.390	1.000	.559
	Social_capital (X3)	.800	.510	.559	1.000
Model-2		NIE	Human_capital	Social_capital	Entrepreneurship
Pearson Correlation	NIE	1.000	.733	.723	.835
	Human_capital (X1)	.733	1.000	.510	.671
	Social_capital (X3)	.723	.510	1.000	.800
	Entrepreneurship (Y)	.835	.671	.800	1.000

a. Dependent Variable: Entreprenurship (Y)

b. Dependent Variable: Institutional (Z)

c. All exogenous variables are significant with endogenous variables at level-1 (99%)

Based on the partial matrix correlation values in table-2, both in model-1 and model-2, it shows the value that are still consistent with the results in table-1. In model-1, the value of social capital correlation still shows the strongest correlation with the level of entrepreneurial business opportunities in rural areas showing significant and strong correlation value of 80 percent. It means that social capital factors not only offer a large and significant level of influence on entrepreneurial business opportunities, but also strong and significant correlations between social capital factors and entrepreneurial business opportunities. Further, the problem of entrepreneurial business opportunities also significantly still has the first strongest level of correlation with the new institutional economy. In model-2, it is also interesting to see the level of correlation between social capital factors and the level of institutional economy. Although the magnitude of the level of influence of social capital has declined to the existence of a new institutional economy, the magnitude of the level of partial social capital correlation is still relatively strong and significant with a new institutional economic level of 72.3 percent, and only slightly with a strong and significant correlation value between level of human capital with a new institutional economic level of 73.3 percent. This phenomenon proves that the level of social capital factors has not been able to contribute to the increase in new institutional growth. However, it still has a strong and significant level of correlation. This means that the resilience of social capital factors is still good for the existence of new institutions produced.

There is an interesting thing to describe further in model-2, both in table 1 and 2. In model-2 in table-1, it shows the greatest and significant level influence between entrepreneurship and new economic institutions. Then, in table-2 there is also the strongest correlation between entrepreneurship and the new economic institution. The result shows that there is a causal relationship between entrepreneurship and new economic development. Thus, it supports previous research conducted by Elert (2017). The result of Elert's research (2017) show that it has contributed to the existing theory by passing an in-depth analysis of entrepreneurial responses to institutions and finding a causal relationship between the two. Meanwhile, the results of Foss's research (2016), stated the opposite. The institutions will influence entrepreneurial activities more through the conditions of entrepreneurial action itself.

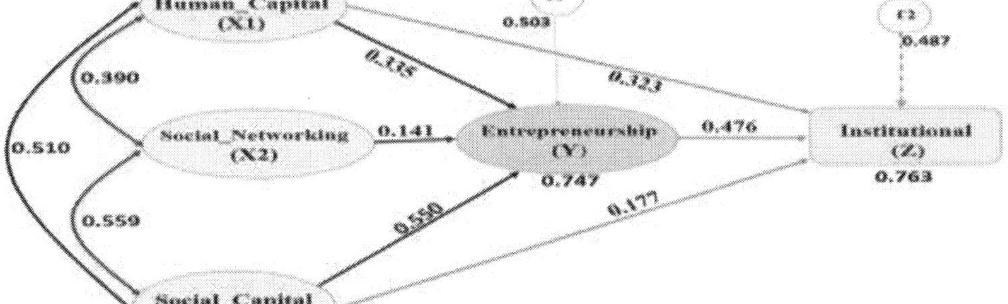

Fig. 3 The results of path analysis of the recrusive form correlation model with multiple path equations

Based on the results of the research in table-1, and table-2, a basic framework for path analysis can be generated as illustrated in figure-3 and tables-3. Figure-3 describes the form of multiple path analysis model correlation lines. Based on figure-3, it can be seen that the direction of the arrows of each exogenous variable regarding the new institutional endogenous economic vaiebel has a direct and indirect influence. Furthermore, the results of multiplication of values in figure-3 will be obtained in the angular form values as in table-3. Figure-3 shows that that the direct influence of human capital and social capital factors is greater than the influence of human capital and social capital factors on institutions. However, the decline in the value of social capital appears to be greater, from 55.0 percent to 17.7 percent, compared to the decline in value in human capital, namely from 33.5 percent to 32.3 percent. The phenomenon of the results of this research supports the previous research conducted by Bosma, (2018). Bosma's research (2018) found that when entrepreneurial channel for quality, and variations in national economic growth increase, the importance of human capital is proven to be reduced

Table 3. The results of direct influence, indirect influence and total influence

Variabel	Direct influence	Indirect influence				Indirect influence	Total influence
		HC	SN	SC	EO		
Human_capital (HC)	0.104		0.009	0.029	0.052	0.090	0.142
Social_networking (SN)	0.020	0.009		0.007	0.003	0.010	0.013
Socal_capital (SC)	0.031	0.029	0.007		0.046	0.082	0.128
Entrepren_opport (EO)	0.226	0.052	0.003	0.046		0.101	0.327
Total							**0.610**

Then, various processes of in-depth study of the level of partial correlation and the level of influence partially and the path analysis model are performed as in figure 3, the value of the results of path analysis can be obtained as in table-3. Table-3 illustrates obvious degree of direct influence, indirect influence and total influence of exogenous variables on new institutional endogenous economic variables. Table-3 shows that entrepreneurial business opportunity factors are able to provide the first biggest influence either direct, indirect and total influence, each of which is equal to; 22.6 percent; 10.1 percent; and 32.7 percent. Further, social capital factors are also interesting issues. Social capital factors are still able to give a total influence on institutions at 12.8 percent. However, it turned out that the indirect influence was even greater at 8.2 percent, rather than the direct influence of only 3.1 percent. Whereas, the human capital factor is able to give a total influence on institutions which is equal to 14.2 percent, slightly higher than the influence of social capital. However, the human capital factor also still offers a greater level of direct influence of 10.4 percent and an indirect influence of only 9.0 percent. The phenomenon of the magnitude of the influence of social capital turns out to occur at the initial level, when business begins to expand, the social capital role decreases

5 Conclusion

The author has not only examined the role of social capital, human capital and networking on entrepreneurship, but also the role of the existence of entrepreneurial businesses on institutions. The results show that there were strong and significant positive relationships, and significant positive influences between social capital, human capital and networking on entrepreneurship. Further, there is also a strong positive relationship and significant; human capital, social capital and entrepreneurship on with a new institutional economics. I suggest that it is important to further study how the causality relationship and the influence of institutions on entrepreneurship and economic growth. Also, it is necessary to examine how the new institutional economic quality of the institution is able to encourage productive entrepreneurship and encourage economic growth. This is due to good economic development should be able to create social equity. Development must also consider humans as the subject of development and not as objects. Through the development of entrepreneurship and institutions, the level of economic growth can still be significantly improved towards a better one. Hence, it is recommended that a serious policy commitment should be made to improve human capital capacity and strengthen social capital. If entrepreneurship and institutions are strong the improved capacity of human and social capital, it can increase human resource quality, better understanding on humanistic principles, and sustainable economic growth

6 Acknowledgement

We are very grateful to the Directorate General of Strengthening for Research and Development, Ministry of Education of the Republic of Indonesia (DRPM DIKTI), who has provided financial assistance in conducting research, especially in the PDUPT decentralization research scheme for the years 2019-2021 with contract numbers: 192 / SP2h / LT / DRPM / 2019

References

1. D. Acemoglu, S. Johnson, and J. Robinson, Am. Econ. Rev. **95**, 3 (2005)
2. A. K. Ali and A. E. Mansur, Potential Soc. Cap. Community Dev. Eds. Yokoyama, S. Al., Asian Product. Organ. (2006)
3. N. Bosma, J. Content, M. Sanders, and E. Stam, Small Bus. Econ. (2018)
4. G. D. Bruton, D. Ahlstrom, and H. Li, Entrep. Theory Pract. **34**, 421 (2010)
5. J. Cavaye, (2006)
6. M. Chibba, World Econ. **9**, 1 (2008)
7. N. Elert and M. Henrekson, Found. Trends Entrep. (2017)
8. S. Estrin, T. Mickiewicz, and U. Stephan, J. Bus. Ventur. **31**, 449–467 (2016)
9. C. Bjørnskov and N. J. Foss, in *Acad. Manag. Perspect.* **20**, 3 (2016)
10. J. Kim and J. T. Mahoney, Manag. Decis. Econ. **26**, 223–242 (2005)
11. A.-L. A. LATOPA and S. M. Bashir, Conf. Pap. Present. 2015 (n.d.)
12. C. Madriz, J. C. Leiva, and R. Henn, Small Bus. Int. Rev. **2**, 29-42 (2018)
13. J. E. Meador and S. Skerratt, J. Rural Stud. **53**, 144-155 (2017)
14. C. Ménard, RAUSP Manag. J. **53**, 3 (2018)
15. F. Murgaš and H. Böhm, 15th Int. Multidiscip. Sci. GeoConference SGEM 2015. **3**, 213-220 (2015)
16. T. Netar, (2017)
17. D. C. North, Int. Libr. Crit. WRITINGS Econ. **84**, 3 (1994).
18. D. C. North, New Institutional Econ. Third World Dev. **21**, 31-40 (2003)
19. P. E. Prasetyo, Int. J. Econ. Financ. Issues **9**, 3 (2019)
20. P. Raja, Pertanika J. Soc. Sci Hum **22**, 97-110 (2014)
21. .R. Richter, Eur. Bus. Organ. Law Rev. **6**, 161-200 (2005)
22. M. Rutherford, J. Econ. Perspect. **15**, 3 (2001)
23. D. Seers, Dev. Theory Four Crit. Stud. 9 (1979)

24. D. S. Kraybill, (1990)
25. V. Shvedovsky, A. Standrik, and Y. Bilan, Econ. Sociol. **9**, 2 (2016).
26. V. O. T. Thanh12 and N. A. Duong, (2016)
27. O. E. Williamson, J. Econ. Lit. **38**, 3 (2000)
28. O. E. Williamson, The Economic Institutions of Capitalism: Firms, Markets, Relational Contracting (Free Press, 1985).
29. A. Yildirim and M. F. Gokalp, Procedia Econ. Financ. **38**, 347-359 (2016)

CREATIVE ECONOMIC DEVELOPMENT IN ECONOMIC EDUCATION PERSPECTIVE

Daryono[1*], Sugeng Wahyudi[2], Suharnomo[3], Indi Djastuti[4], Mahfudz[5], Amie Kusumawardhani[6] , Sri Supriyati[7]

[1]Faculty of Economics and Business, Universitas Jenderal Soedirman
[2]Faculty of Economics and Business, Diponegoro University.
[3]Faculty of Economics and Business, Diponegoro University
[4]Faculty of Economics and Business, Diponegoro University
[5]Faculty of Economics and Business, Diponegoro University
[6]Faculty of Economics and Business, Diponegoro University
[7]Faculty of Economics and Business, Diponegoro University

Abstract. This study aims to determine the effect of entrepreneurial knowledge on creative attitudes of creative economy sub-sector Small Medium Enterprises in Kendal sub-district in the context of creative economic development. The population in this study were 129 Micro Small Medium Enterprises in the creative industry sub-sector. The total sample in this study was as many as 33 Micro Small Medium Enterprises in the creative industry subsector. The results of this study indicate that entrepreneurial knowledge has a positive and significant influence on the creative attitude of creative economy sub-sector Micro Small Medium Enterprises in Kendal sub-district, Central Java City, Indonesia.

1 INTRODUCTION

The informal sector has a strategic role in economic development, especially in overcoming poverty and unemployment. This also relates to the role of economic education in the informal sector through the transformation of the values and character of entrepreneurship in business activities. The informal sector is dominated by micro, small, and medium enterprises (MSMEs). Some research results show that business actors in the informal sector are dominated by MSMEs, both in developing countries and in developed countries (Gurtoo and Williams, 2009, Schneider and Williams, 2013, Webb et al., 2009, Williams, 2009a, Williams, 2009b, Williams, 2009c, Williams, 2010).

The informal sector is dominated by MSMEs. The development and contribution of the MSME sector is inseparable from the growing growth of new entrepreneurs in the community. The trend of the emergence of entrepreneurship is due to economic pressure, so that someone is forced to create their own jobs. In addition, there are other factors, namely unemployment, frustration of previous jobs, and the need for decent living. Current trends in the development of MSMEs in Indonesia are mostly directed at the creative economy sector which is largely determined by human capital.

The most important asset in the global era is human capital. A nation that has comparative advantages in natural resources will not do much in the arena of global compe*tition without being supported by the excellence of human capital. Only a nation

* Corresponding author: daryonojvc@gmail.com

that excels in human capital will be able to take part and master knowledge and its application in the real form of problem solvers encountered in life (Thomas et al., 2013).

The concept of creative economy is an economic concept based on human abilities and skills. Human civilization is divided into three economic waves. A paradigm shift in economic development cannot be separated from changes in global scale. This change was marked by a shift in economic development from the agricultural, industrial, and information sectors to the creative economy sector. The development of the creative economy sector of a nation will compete and have an impact on social life if managed properly. Therefore, the rise of a new economic wave requires innovation and creativity of the community, so that the quality of human resources is needed as creative economic actors. This condition must be linked to the ability to manage potential. Therefore, the concept of creative economy must be in accordance with the ability of innovation and creativity in managing existing local potential.

The development of the creative economy cannot be separated from the development of the creative industry because the development of the creative economy requires the support of the economic industry. The development of the creative economy is believed to be a way for developing countries to follow the development of the global economy. This is because the creative economy sector relies more on creativity and intellectual community in developing existing local potential. On the other hand, the development of a creative economy in a particular-place depends heavily on the quality of human resources in developing the creativity of an intelligence. In other words, the development of creative economy is an alternative economic activity for the community in developing their potential (United Nations Development Programme, 2013).

This explanation can show that economic change is determined by the quality of human resources in exploring creativity as innovation as the main impetus in developing a creative economy. Therefore, if the community has innovation and creativity, the acceleration of economic development will be faster. Therefore, the choice in developing a creative economy will increase competitive business. The development of the creative economy sector must also be oriented to the cultural aspects of society. Therefore, the development of profitable economic sectors will have an impact on the rise of community-driven business opportunities. Therefore, structured and systematic development will encourage the development of this sector in the region. According to the explanation, it shows that creativity arising from cultural wealth will create jobs. Therefore, with the development of the creative economy, it will increase the income of the community (Hamilton, 2009, Cunningham and Higgs, 2008).

One of the sub-districts in Central Java City which is identical with the creative economy is Kendal sub-district. Kendal sub-district is the centre of Central Java's creative economy area. With a variety of potential, such as the number of hotels that are ready to accommodate tourists, as well as malls and markets as a strategic location to market creative products, it becomes a strength for Kendal sub-district in empowering the economy of the community through creative economic development programs. In Kendal sub-district, there are 71 hotels that are ready to become supporting facilities for the marketing of creative products, so that they can build a positive image for Kendal sub-district through tourism facilities. There are 129 creative zones in Kendal sub-district in 2016.

Table 1
Creative Zone Data for 2016

Creative Industries Subsector	Number of Business Actors
Interior design	5
Clothes	55
Art	28
Performing Arts	1
Publishing and printing	4
Culinary	36
Total	129

Source: Creative Zone of Kendal Sub-district in 2016

Empirical facts show that the majority creative products in the Kendal sub-district are produced by MSMEs included into the informal category. The informal sector MSMEs still have number of problems, some of which include low levels of creativity and innovation because of lack of knowledge. The lack of creative attitudes results in low innovative attitudes, achievement motivation, orientation in profit, perseverance and fortitude, determination of hard work, having a strong drive, energetic, and initiative.

Knowledge aspects are important in the era of creative economy. The study results of Hasan (2012) regarding the characteristics of small-scale industrial labour shows that workers in small-scale industries in Central Java City have the characteristics of utilizing their economic knowledge, both obtained formally and informally in carrying out production activities. Other findings from Hasan (2014) related to productivity and elasticity of industrial sector employment opportunities show that knowledge and education are one of the most important aspects in determining economic behaviour, especially those related to employment opportunities and other economic decisions.

Based on this, the development of the creative economy in the perspective of economic education is a study that needs to be done because economic education is one of the efforts to foster a creative economy in Kendal sub-district. The development of the creative economy in the perspective of economic education is carried out in the form of knowledge transfer, both formally, non-formally, and informally. Transferred knowledge as an effort to increase innovation and creativity of MSMEs engaged in the creative economy is divided into two, namely tacit and explicit. The character of tacit knowledge is constructed from individual experience and is a basic form of explicit knowledge. Tacit knowledge represents knowledge possessed by a person, attached to the mind, and cannot be separated from the person who owns it. As consequence, tacit knowledge is difficult to transfer. Explicit knowledge is knowledge that is easily changed in formal and systematic language so that it is easier to transfer than tacit knowledge.

Based on this, it can be stated that the transfer of knowledge in the development of the creative economy carried out is realized in the aspect of entrepreneurial knowledge related to creative attitudes. Entrepreneurial knowledge consists of the variable of knowledge of entrepreneurship with the indicators are knowledge of business planning, knowledge of decision making, knowledge of managing the company, knowledge of managing finances, knowledge of production techniques/processes, knowledge of designing products, knowledge of marketing techniques, knowledge of distributing goods, knowledge of accounting administration, and technical knowledge of controlling the company.

While the variable of creative attitude, referring to the creativity level theory of McGregor (2008) is a creative process dimension whose indicators are fluency in thinking, originality in thinking, and flexibility in thinking. Based on this, the hypothesis of this study is knowledge of entrepreneurship has a positive effect on creative attitude.

2 METHOD

This research is quantitative research. The data used in this study were sourced from the SME industry subsector in the creative industry in Central Java. The types of data in this study include primary data and secondary data. Primary data was obtained through questionnaires and direct interviews with respondents regarding economic education and creative economic development, especially those related to the variable of knowledge of entrepreneurship and creative attitudes. Whereas secondary data was obtained from the Creative Economy Agency, namely the number and distribution of creative economy subsector MSMEs in Kendal sub-district as well as data in the form of other documentation relating to research problems.

The population in this study are all business actors who are engaged in the creative economy in Kendal sub-district as many as 129 SMEs in the creative industry sub-sector. Sampling was determined by using purposive sampling technique, the aim was to capture a truly representative sample with the study, so obtained as many as 33 creative industries sub-sector MSMEs in the sub-district of Kendal. The data analysis technique used in this study is multiple linear regression through the SPSS version 21 program.

3 RESULT AND DISCUSSION

A summary of research results can be seen in Table 1. Based on the results of data processing through the SPSS Version 21 program, the regression equation Y = 2,861+ 7,199 X1 + e is obtained. This equation shows that entrepreneurial knowledge has a positive influence on creative attitudes. This means that if knowledge increases, it will also increase creative attitudes. The contribution of entrepreneurial knowledge to the creative attitude of creative industry sub-sector MSME actors in Central Java City was 54.2 percent, while the remaining 45.8 was the contribution of other variables not included in the research model.

Table 2
Estimation Results

Independent Variable and Constant	Regression Coefficient	Standard of Error	t_{count}	Significance
Constant	2.861	5.371	1.701	0.041
Entrepreneurial Knowledge	7.199	0.142	5.504	0.041
*Level of confidence α = 5%				
R Square = 0.542				

The results of the significance test through the t test using SPSS Version 21 can be seen in Table 2. The results of data processing show that entrepreneurial knowledge has a significant influence on the creative attitude of the creative industry sub-sector MSMEs in Central Java City.

Table 3
Significance Test

Independent Variable	t_{count}	Significance	Significant Level	Conclusion
Constant	1.701	0.041	0.05	Significant
Entrepreneurial Knowledge	5.504	0.041	0.05	Significant

The findings of this study indicate that entrepreneurial knowledge has a significant influence on the creative attitudes of SMEs in the creative industry sub-sector in Central Java City. Entrepreneurial knowledge refers to an individual's appreciation of the concepts, skills, and mentality expected of an entrepreneur. This knowledge can be obtained and developed through entrepreneurial activities, which in this case is an entrepreneurial activity carried out by creative industry sub-sectors in the city of Central Java. The findings of this study are in line with studies identifying types of entrepreneurial knowledge that complement each other in producing creative attitudes. This type of knowledge relates to an individual's ability to recognize entrepreneurial opportunities that involve discovery and innovation.

The ability of creative industry sub-sectors MSMEs in the city of Central Java to achieve certain successes is influenced by entrepreneurial factors. The implications of the application of the concept of entrepreneurship itself in business organizations stated that if entrepreneurs want to succeed in the face of open competition in the current global era must have a creative and innovative entrepreneurial spirit, especially seeing the current conditions of course it takes modern entrepreneurs, who have more insight, thinking ahead, always following developments, and open to new concepts and ideas. With a far-sighted outlook, always working and being intentional, creating something new, open to new experiences, including being dissatisfied with what is done now is the side of success in carrying out its business activities related to its competence. The determinants of success are entrepreneurial competencies, in this case, creativity and innovation as determinants that are the characteristics of entrepreneurship which are implied from the entrepreneurial knowledge gained.

In connection with the relationship with entrepreneurial knowledge, creative attitude is the unification of knowledge from various fields of different experiences to produce new and better ideas. Creativity is also a producer of new ideas and innovations as translators of new ideas into new companies, new products, new services, new processes, or new methods of production. Research findings that show the presence of influence between entrepreneurial knowledge on creative attitudes are in line with some previous literature and research. Creative attitude is the core of entrepreneurial knowledge (Wood and McKinley, 2010). Creative entrepreneurs will always work hard and continuously to improve ideas and solutions, by making gradual changes and improvements to their work.

4 CONCLUSSIONS AND RECOMENDATIONS

The results show that entrepreneurial knowledge has a significant influence on the creative attitudes of the MSMEs in the creative industry sub-sector in Central Java City. The results of this study have implications for the development of creative industry sub-sector MSMEs in Central Java City need to consider aspects of entrepreneurship education, both formal, non-formal, and informal, so that through education is formed entrepreneurial knowledge. Through this knowledge, it is expected that creative attitudes can be formed to realize the business success of the creative industry subsector MSMEs in Central Java City

5 REFERENCES

1. Cunningham, S. D. & Higgs, P. L. Creative Industries Mapping: Where Have We Come From and Where Are We Going? *Creat. Ind. J.* **(1)**, 18-22 (2008).
2. Gurtoo, A. & Williams, C. Entrepreneurship and the Informal Sector: Some Lessons from India. *Int. J. Entrep. Innov.* **(10)**, 55-62 (2009).
3. Hamilon, L. Building the creative economy in Nova Scotia. *The Research Committee of the Nova Scotia Cultural Action Network.* 27-12 (2009).
4. Hasan. M. Characteristics of Small Indutrial Labor. *J. Develop. Agric. Ec.* **(2)** (2012).
5. McGregor, D. Thinking: developing learning. *A guide to thinking skills in education, McGrawHill.* (2008)
6. Schneider, F. & Williams, C. C. The Shadow Economy (Instiute of Economic Affairs, London, 2013).
7. Thomas, H., Smith, R. R. & Diez, F. Human Capital and Global Business Strategy (Cambridge University Press, New York, 2013)
8. Isar, Y. R., UNCTAD & UNDP. *Creative Economy Report 2013: Widening local development pathways. United Nations Development Programme and the United Nations Educational, Scientific and Cultural Organization* (2013).
9. Webb, J. W., Tihanyi, L., Ireland, R. D. & Sirmon, D. G. You say illegal, i say legitimate: Entrepreneurship in the informal economy. *Academy of Management Review* (2009).
10. Williams, C. C. Entrepreneurship and the off-the-books economy: Some lessons from England. *Int. J. Manag. Enterp. Dev.* **(7)**, 429-444 (2009a)
11. Williams, C. C. The motives of off-the-books entrepreneurs: Necessity- or opportunity-driven? *Int. Entrep. Manag. J.* **(5)**, 203-217 (2009b).
12. Williams, C. C. Beyond Necessity-Driven Versus Opportunity-Driven Entrepreneurship . *Int. J. Entrep. Innov.* **(14)**, 44-60 (2008).
13. Williams, C. C. Spatial variations in the hidden enterprise culture: Some lessons from England. *Entrep. Reg. Dev.* (2010), 22, 403-423.
14. Wood, M. S. & McKinley, W. The production of entrepreneurial opportunity: a constructivist perspective. *Strateg. Entrep. J.* (2010), 4,66-84.

RURAL WOMEN ENTREPRENEUR AND CAPACITY BUILDING: ROLE OF NGOS IN EMPOWERING WOMEN IN INDIA

Dr.P.Maduraiveeran [1] , Dr. G.P.Sudha [2]

[1] Professor, Anna Centre for Public Affairs, University of Madras, Chennai – 600 005, India
[2] Guest faculty, Anna Centre for Public Affairs, University of Madras, Chennai – 600 005, India

1 Introduction

After the Second World War participation of women in industrial and service sector has increased due to evident need for producing more and providing services. In developed countries, the numbers of self employed women have also phenomenally increased. It is true that women were most attached with kids, kitchen and knitting. This was replaced by powder, pappad and pickles. In due course of time women became both small and medium entrepreneurs. However, women have been facing lot of challenges and barriers. As far as India is concerned, women in rural areas had been confined themselves to small business and micro level or cottage industries found in the village side. Majority of rural women entrepreneurs are concentrated in low-paid, low-skilled and low-productivity jobs. They have basic indigenous knowledge, skill and potential to establish and manage enterprises.

"Any rural woman or group of rural women who innovates, imitates or adopts an economic activity may be referred as a rural women entrepreneur. Similarly an adult women who creates, who owns and runs an enterprise in rural area can be defined as rural women entrepreneur."

As far as India is concerned, women's participation in trade, industry and commerce, requiring entrepreneurship is still not up to mark because of the various problems and stigma associated with their gender roles. Thus promotion of entrepreneurship and economic empowerment of women pose a challenge to the government funding agencies and non-governmental organizations. Dealing with government agencies, knowing the government regulations, and the taxations, time management, balancing life and family ,negotiating skills, networking and stress management skills are necessary for them.

Some of the issues challenging the women entrepreneurs particularly rural area are shortage of finances, limited mobility, male dominance, lack of education, performing motherly duties and lack of achievement motivation.3 Another important finding is [Rizvi and Gupta, (2009)], that government sponsored entrepreneur development activities have benefited only a small section of urban middle class women because of their level of education, access to information and their family support.

Despite a new work culture and increased participation of women in workforce, women in entrepreneurial roles are very limited. The probable reason may be their conflicts between competing work demands, personal and family needs. However, entrepreneurship provides more opportunities for self employment more income and enhanced economic status to the rural women.

Along with the women empowerment processes, development of entrepreneurship, training of women as entrepreneur, capacity building for women entrepreneurs and bank linkages and inculcating skill training have also simultaneously taking place. It is in this context the research entitled "Rural Women Entrepreneurs and Capacity Building: Role of Non-Governmental Organizations in Empowering Women in India" has been undertaken.

2 Limitation of the study

This study mainly emphasizes the role of non-governmental organizations in empowering women in rural areas of Villupuram district of Tamilnadu, India for understanding the crucial role of nongovernmental organizations. Three nongovernmental organizations from Villupuram district, particularly which are providing capacity building, training to women entrepreneurs have been chosen. They are :

1. Auroville Village Action Group (AVAG)
2. Community Action for Rural Development Society (CARDS)
3. Rural Education Action For Liberation (REAL)

According to World Bank, investing more in business of women rather in men leads to greater development of a nation. Empowering women in entrepreneurship leads to break the inequalities and reduces poverty. The government has realized the importance of women entrepreneurship. As a result, it offers a variety of programmes for women entrepreneurs. Such entrepreneurs are helped by Tamil Nadu Corporation for Development of Women, which in turn are implemented by various non- governmental organizations

3 Objective of the study

The main objective of the study is to analyze the role of non- governmental organizations engaged in capacity building activities for women entrepreneurs in Villupuram district of Tamilnadu.

1. To study the socio-economic background of the rural women entrepreneurs and their interest in entrepreneurship.
2. To enumerate the various strategies adopted by non- governmental organizations for empowering women through entrepreneurship.
3. To analyse the adequacy of training and capacity building activities undertaken by non-governmental organizations particularly for women entrepreneurship.
4. To suggest various measures for improving the rural women enterprisesEquations should be centred and should be numbered with the number on the right-hand side.

3.1 Challenges Faced by Rural Women Entrepreneurs

1. Personal Challenges
2. Social Challenges
3. Financial Challenges
4. Marketing Challenges
5. Managerial Challenges
6. Technological Challenges
7. Skill Challenges

3.2 Problems faced by Women Entrepreneur

1. Capability training
2. Finding new customers because of instability of demand
3. Obtaining financing
4. Dealing with the employee problems
5. Product pricing
6. Planning for market expansion
7. Handling legal problems
8. Determining and maintaining product quality
9. Collaborating agency
10. Diversification and decentralization
11. Women entrepreneurs needs differ

A strong desire to do something positive is an inbuilt quality of women entrepreneur who are capable in contributing values in both family and social life. With the advent of media especially digital media, women have become more aware about their trades, rights and also working culture. The opportunities and challenges available to women in the digital era has made them job providers instead of jobseekers by becoming women entrepreneurs. They are capable of becoming innovative and produce more new products. They can establish their own identity and status in society. They can economically support their family and become role models for others.

They are been motivated by success stories, of friends and peer groups. Through their economic empowerment they create future for their wards. They also are supported by government through subsidy, bank loans and marketing guidance, etc. They become economically independent and generate more employment for others. Moreover, the new challenges and opportunities provide them self - esteem and self- fulfillment.

It is found that majority 56 per cent of the rural women entrepreneur are in the age group of 40-50. More than 90 percent women are married and nearly half of the rural women have more than secondary grade and many of them are graduates. Education remains the most important factor - to get loan, to understand the dynamics of entrepreneurship, manufacturing and marketing process, human resources, etc.,

More than 60 per cent rural women entrepreneurs face financial challenges.

4 Conclusion

Development of women Entrepreneurship particularly in rural areas is an indication of social transformation and breaking of traditional taboos. The capacity building programmes provided by non- governmental organizations have transformed the middle class women into innovative, hardworking, achievable women entrepreneur. Though the non-governmental organizations and governmental support is limited in many ways but play a crucial role in empowering women in rural areas.

Another important barrier to women entrepreneur is geographical locations and the availability of raw materials and marketing opportunities particularly in rural areas and for rural women entrepreneur it is a big challenge. They may find raw material and produce a lot of goods and services but they would not find marketing agencies. Here the non-governmental organizations help the rural women entrepreneurs to locate marketing opportunities for their products, services because the non- governmental organizations operates at national and international level or they can use their networking power beyond the geographical location. For instance the coconut shell products produced in Kerala are marketed in the time square of New York through non- governmental organizations supported online marketing with eco friendly leveling.

Yet another most valuable input provided by non- governmental organizations is introduction of innovative techniques and providing technical knowhow from international collaborating agencies to illustrate several dry fishing women entrepreneurs uses solar technology in coastal areas of Kanyakumari and Gujarat - with the help of international non- governmental organizations. Similarly, preservation and food processing women entrepreneurs use such solar heaters for their entrepreneurship in Villupuram district in cashew nut processing, export industries which provide considerable foreign exchange

5 Suggestions

Women who underwent some type of entrepreneurial training must given another chance when particular women or group want to switch over to a new avenue or diversification of their enterprises to produce new products and services. Non- governmental organizations found in the Villupuram district have identified a number of new entrepreneurial ventures for women entrepreneurs mobilized by Self Help Groups.

Among the rural entrepreneurs technology adoptions secure to be very minimal despite several training and awareness programme use of internet and e-commerce as well as online transactions are not encouraging therefore further emphasize must be made into this area for digitally empowering entrepreneurship among rural women. Branding, common branding of the products of the entrepreneurs promoted by the SHGs still remain a unfinished task because through geographically women entrepreneurs were federated among themselves wither they produce different products based on the availability of raw materials in their area. Therefore, common branding still remains an issue. Rural women entrepreneur are not able to compete with mega malls and multinational companies, in the range of products, investment and exhibiting their products in a mega malls and halls, public relations activities, etc.,

The number of women entrepreneur is growing. Research has identified that woman entrepreneur are less likely to grow. Women entrepreneurs face a wide variety of challenges both in starting and developing their business ventures. Considering these challenges, entrepreneurial advising, and training and education services found to play a positive role in venture success.

It is true that needs of entrepreneur service/care needs differ based upon the enterprises and the entrepreneurial stage. But the government gives training on a monotonous manner here non- governmental organizations differ and provide selective and needed training to bridge the gap between existing and expected needs of the entrepreneurs.

References

1. R. Sivanesan, Int. J. Res. Manag. Bus. Stud. (2014)
2. Ibid.
3. D. K. VYAS, C. P. PATEL, F. H. SHAH, and M. S. BHATT, An Int. Peer Rev. Multidiscip. Res. J. 19 (2006)
4. P. Adinarayana Reddy, Women Work Particip. Status Train. Oppor. (2011)
5. Jhunjhunwala and M. Jhunjhunwala, *Indian Approach to Women's Empowerment* (International Specialized Book Service Incorporated, 2004)
6. A. Prasad, *Women Empowerment and Globalisation: A Modern Perspective* (Swastik Publications, 2012)

Financial Literacy of Economics and non-Economics Student

Ascaryan Rafinda[1,2], Timea Gal[2]

[1]Faculty of Economics and Business Universitas Jenderal Soedirman, Indonesia
[2]University of Debrecen, Károly Ihrig Doctoral School of Management and Business, Hungary

Abstract. The aim of this research is to identify the level of financial literacy of students in economic faculties and non-economic faculties in Indonesia. This research was conducted to determine the understanding of students in the economic and non-economic faculties on the level of personal financial literacy. The survey was conducted to test the level of student financial literacy in various University in Indonesia. In total there were 206 participants from students in Indonesia who took the survey. An independent sample T-Test was conducted to determine differences in the level of understanding of students in the two groups. The results of this research indicate that there are no differences in the level of financial literacy in economic and non-economic students. This result shows that economics student does not guarantee they have higher personal financial knowledge.

1 INTRODUCTION

The results of a financial literacy survey of the Indonesian in 2016 explain that the average financial literacy score from Indonesian was only 29.6% well literated. The capital city of Jakarta has the highest position with 40% of its population being well literated while the farthest province of West Papua has the lowest literacy rate, only 19.3% of its population being well literated (OJK), 2017).

Financial knowledge can be obtained from various several sources. The surrounding environment such as family and friends is one of the source of financial knowledge (Bowen, 2002; Chen & Volpe, 2002; Norvilitis & MacLean, 2010; Shim, Serido, Bosch, & Tang, 2013; Xiao, Tang, Serido, & Shim, 2011)). The role of formal financial education also has a contribution to financial maturity and financial behaviour (Alsemgeest, 2015; Norvilitis & MacLean, 2010; Urban, Schmeiser, Michael Collins, & Brown, 2018).

Various studies examine the level of financial literacy in tertiary institutions (Anthes, 2004; Godfrey, 2006; Murphy, 2005). Higher education is the highest formal educational institution in Indonesia, the majority of Indonesian people will start working after graduating from higher education institutions. This makes universities the last opportunity to help the community in increasing their financial knowledge (Harrington & Smith, 2016).

Because after work, no matter how to literate their financial knowledge, they have to deal with financial issues and decisions.

The issue regarding the low level of financial literacy of Indonesian generally occurs in tertiary institutions. Some research findings identify that students in Indonesia have low levels of financial literacy (Margaretha & Pambudhi, 2015; Mendari & Kewal, 2013). Previous studies have identified a positive relationship between financial education in tertiary institutions and the level of one's financial knowledge. The role of universities in increasing financial literacy in Indonesia needs to be explored further. The strategic position of tertiary institutions in improving public financial literacy in Indonesia needs to be identified in current conditions and evaluated for future improvement.

This research identifies the condition of student financial literacy in tertiary institutions to determine the current level of financial literacy in Indonesian tertiary institutions. Universities that have various disciplines in this research will be classified into two groups, namely students from the economic and business group and non-economic and business group. This is to facilitate the identification of gaps between the two of them.

Research that examine the level of student financial literacy in university is important because it can determine the current conditions and the role of universities to improve financial literacy level in Indonesia. This research will explain how the condition of financial literacy in Indonesian universities.

2 LITERATUR REVIEW

Various studies have examined the impact of financial education on a person's life. Some interesting research results to study are as follows. Basic financial knowledge education is needed for day-to-day financial management but investment and retirement can be left to professionals (Alsemgeest, 2015). The results of the research found that cognitive aspects in financial education were still lacking to improve one's financial knowledge. There is no one solution for all problems in financial education found (Alsemgeest, 2015).

Some financial literacy research at universities also found the effects of financial education on a financial behaviour. Credit card behaviour was found to be different between students who took financial education and students who did not participate in financial education (Norvilitis et al., 2006).

Students on economics and business at universities get courses related to finance. Although not directly studying management of personal financial, financial courses in general can provide financial perspectives that are quite important in financial decision making. So students in the field of economics and business have a higher understanding of finance than students from non-economics and business. Based on these arguments, the hypothesis proposed in this research is that non-economic and business students have a lower understanding of the level of financial literacy than economics and business students.

3 RESEARCH METHOD

The data in this research were taken using a questionnaire. Online and paper based questionnaires were distributed to participants in the form of students in Indonesia. The questionnaire instrument was preceded by informed consent that explained the willingness to participate in this research. Respondents who are willing will proceed to the next question, who are not willing cannot proceed to the next question.

The sample in this research focuses on undergraduate students on several universities in Indonesia. A total of 206 people participated in this research from public and private universities in Indonesia. This research takes two months to collect all the data. This research classifies the sample into two groups namely economic and business group and non-economic and business groups. Participants from the economic and business study

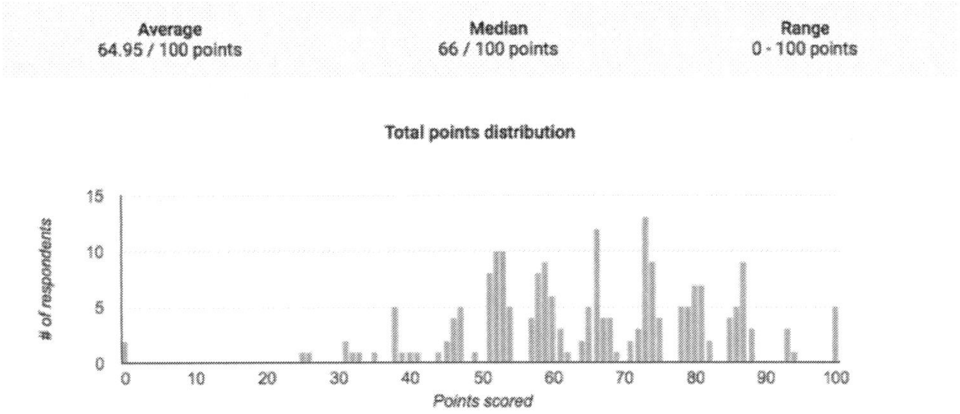

Figure 1 : Statistics Descriptive

group were 87 people and from non-economic and business people 116 people. Here the descriptive statistics from the research sample: From all the research sample, the average financial literacy score is 64,95. The median is 66 out of 100 points. The maximum score is 100 out of 100 and the lowest is 0 out of 100 points.

Independent sample unequal variance T-Tests are performed to examine the differences test of financial literacy level. Unequal variance t-test was done because the two groups have different variances. The financial literacy instrument measure financial literacy in three areas, namely personal investment, credit transactions and personal financial management. A total of 15 multiple choice questions were given to participants to find out the level of financial literacy. Each participant in this research fills in the instrument within 10-20 minutes depending on the speed of the participants answering the question.

The results of the different tests show that there are statistically significant differences between the two groups. The first group is the non-economic and business group, the second group is the economic and business study group. The first group had an average score of 61.26, while the second group had an average score of 69.44. The following is a summary of the results of the statistical tests:

t-Test: Two-Sample Assuming Unequal Variances

	First Group	*Second Group*
Mean	61,26437	69,43966
Variance	176,10372	262,66589
Observations	87,00000	116,00000
Hypothesized Mean Difference	0,00000	
Df	199,00000	
t Stat	-3,94774	
P(T<=t) one-tail	0,00005	
t Critical one-tail	1,65255	
P(T<=t) two-tail	0,00011	
t Critical two-tail	1,97196	

Hasil Uji Beda T-Test Unequal Variance

4 RESULT

Based on the data taken and tested with statistics, the data shows that the average of the first group was 61.26 and the average of the second group was 69.44. The results of the unequal variance T-Test show that there are significant differences between the two groups. This significant difference shows that the first group has an average value of financial literacy lower than the average value in the second group.

The first group which is a group of students from non-economic and business programs has an average value of financial literacy lower than the group of students from economic and business programs. Students in the first group have the lowest score of 38 out of a total of 100 points, and the highest score of 85 out of a total of 100 points. Whereas the second group had the lowest score of 45 and the highest score of 100. In total there were 4 participants who received perfect scores in answering questions in this study.

5 CONCLUSION

Students from the economic and business study group have higher financial literacy scores compared to the non-economic and business study group. Although these students do not get formal courses on personal finance, but they have sufficient basic financial knowledge in three fields (investment, credit and personal finance). Optimizing students' personal financial knowledge could be done by giving special courses in the form of personal finance. So students can more specifically understand personal financial management.

This research limits financial literacy to only three areas (investment, credit and personal finance) so it does not measure financial knowledge in other fields. Further research can be optimized by measuring other fields in personal finance to make it more comprehensive.

6 REFERENCES

1. Alsemgeest, L. Arguments for and against financial literacy education: Where to go from here? *Int. J. Consum. Stud.* (2015).
2. Anthes, W. L. Financial illiteracy in America: a perfect storm, a perfect opportunity. *J. Financ. Serv. Prof.* **58(6)**, 49 (2004).
3. Faulcon Bowen, C. Financial knowledge of teens and their parents. *J. Financ. Couns. Plan.* **13(2)**, 93-102 (2002).
4. Chen, H. & Volpe, R. Gender Differences in Personal Financial Literacy among College Students. *Financ. Serv. Rev.* **11(3)**, 289-307 (2002).
5. Godfrey, N. S. Making Our Students Smart about Money. *Education Digest: Essentials Reading Condensed for Quick Review 71*(**7**), 21–26 (2006).
6. Harrington, C. & Smith, W. College Student Interest in Personal Finance Education. *SSRN Electron. J.* **25**, 351-372 (2016).
7. Kewal Anastasia Sri, S. S. M. Tingkat Literasi Keuangan Di Kalangan Mahasiswa Stie Musi (Financial Literacy Level among Students Stie Musi). *J. Econ.* **9(2)**, 130-140 (2013).
8. Margaretha, F. & Pambudhi, R. A. Tingkat Literasi Keuangan Pada Mahasiswa S-1. *Tingkat Literasi Keuang. Pada Mhs. S-1 Fak. Ekon.* **17(1)**, 76-85) (2015).
9. Murphy, A. J. Money, Money, Money: An Exploratory Study on the Financial Literacy of Black College Students. *College Student Journal*, **39(3)**, 478 (2005).

10. Norvilitis, J. M., & MacLean, M. G. The role of parents in college students' financial behaviors and attitudes. *J. Eco. Psycho.* **31(1)**, 55–63 (2010).
11. Norvilitis, J.M. *et al.* Personality Factors, Money Attitudes, Financial Knowledge, and Credit-Card Debit in College Students. J. App. Soc. Pshycho. **36 (6)**, 1395-1413 (2006).
12. Shim, S., Serido, J., Bosch, L., & Tang, C. Financial identity-processing styles among young adults: A longitudinal study of socialization factors and consequences for financial capabilities. *J. Cons. Aff.* **47(1)**, 128–152 (2013).
13. Urban, C., Schmeiser, M., Michael Collins, J., & Brown, A. The effects of high school personal financial education policies on financial behavior. *Economics of Education Review* (2018).
14. Xiao, J. J., Tang, C., Serido, J., & Shim, S. Antecedents and Consequences of Risky Credit Behavior Among College Students: Applicatin and Extension of the Theory of Planned Behavior. *J. Pub. Pol & Mark.* **30(2)**, 239–245 (2011).

The Identification of Mangrove Ecosystem Services for Decision Making

Suharno[1*], *Emmy* Saraswati[2]

[1]Economics and Development Studies, Faculty of Economics and Business, Jenderal Soedirman University
[2]Economics and Development Studies, Faculty of Economics and Business, Jenderal Soedirman University

Abstract. Directly or indirectly, goods and services available in natural ecosystems will contribute to human welfare. The human ability to calculate the economic value of ecosystem goods and services is an important thing to do for integrated environmental decision making, sustainable business practices, and land use planning with its geographical scale, and the level of local-social wisdom. For this reason, a comprehensive review and study are needed by analyzing mangrove ecosystem services. Discussions with in-depth descriptive methods are applied to evaluate mangrove ecosystem services specifically, and focus methods and techniques are used for data analysis, and further to understand their potential and disadvantages.

1 Introduction

Mangroves are intertidal ecosystems that are located along tropical and subtropical coasts. Mangrove swamps supply food needs for ecosystems such as sediment traps with various nutrients absorbed and prevent erosion on shorelines (Jusoff, 2013). Mangrove forests also supply various types of raw materials, such as food, fuelwood, wood, and tannins (De Groot et al. 2002). It is noted, among others, the types of fish, shrimp, and crabs using mangrove ecosystems as ground nurseries and feeding ground (Robertson & Duke, 1987; Beck et al. 2001). Unfortunately, the indication of global data records mangrove forests has dropped dramatically (Giri et al. 2008). Observation findings from 1980, had a 25% decline in mangrove forests (FAO 2007). The main contributors are a land conversion for shrimp ponds, illegal logging, freshwater pollution, and utilization (Barbier et al. 2011).

As part of the Segara Anakan estuary ecosystem and the largest mangrove forest area, the mangrove forest in Segara Anakan Lagoon, Cilacap Regency once triumphed in Java in the past. The area of mangroves in the Segara Anakan area reaches 21,500 ha (Sasaki & Sunarto 1994). Note that in various periods, several studies reported that environmental stress in the Segara Anakan mangrove forest continued. The situation with various pressures reduced mangrove forest in 2013 with the remaining land area of 6,716 ha (Purwanto et al. 2014). The process of ongoing development that continues to increase such

* Corresponding author: suharno@unsoed.ac.id

as settlements, industries, rice fields, fish ponds causes the conversion of mangrove areas that have surpassed regenerative ability to live mangroves (Sukarjo, 2016). Continued efforts to pressure human activities have caused an increase in land surface with mangrove status being critical. For this reason, an assessment of the critical status of the mangrove forest is needed. How to measure the critical indicators of land for mangroves can be done by assessing the density of mangroves with the vegetation index. Calculation of the vegetation index of this model is done by utilizing remote sensing technology (Adam et al. 2010). The vegetation index is very important because it can be used as an indicator in biomass estimation (Foody, 2003), estimated leaf area index (Kovacs et al. 2004) and estimation of primary productivity (Komiyama et al. 2008). Patterns of leaf spectral characteristics can distinguish radiation intensity from reflected electromagnetic power (Jusoff, 2006).

Mangrove forests are the main ecosystems that support life that is important in coastal and marine areas. Every ecosystem has a benefit value, both direct and indirect benefits in the form of ecosystem services. Ecosystem services are ecological processes or ecosystem components that have the potential to provide a flow of benefits to the community. The Millennium Ecosystem Assessment (2005) classifies ecosystem services in four main categories, namely provisioning services, regulating services, and cultural services. Therefore, to realize sustainable development with the concept of Ecosystem-Based Management and ecosystem services in Cilacap District especially in Kampung Laut Subdistrict, it is necessary to analyze mangrove ecosystem services in some coastal areas of Kampung Laut District so that the government and the community can know the benefits or services owned by the mangrove forest so that the forest can continue to be preserved and used as a fortress in disaster risk reduction, especially coastal disasters and can be a reference in disaster management for regions with similar characteristics. The purpose of this study was to analyze the services of mangrove ecosystems in the mangrove area of Ujung Alang Village, Kampung Laut District, and calculate the value of mangrove ecosystem services.

2 Analytical method

The data collected is divided into two types, namely primary data and secondary data. Primary data includes all results of interviews and direct observations in the field. Secondary data in the form of data collected from the results of literature studies, literature review, and institutional data.

Identifying ecosystem services is based on the Millennium Ecosystem Assessment (2005), where identification is divided into three services, namely provider services, regulatory services, and cultural services. Assessment of ecosystem services in Ujung Alang Village uses a short approach to economic valuation by calculating the total economic value of mangrove ecosystem services. The assessment is based on the value of each activity identified in ecosystem services.

Analysis of research data was carried out using descriptive analysis techniques. Descriptive analysis is used to describe the results of the calculation of mangrove ecosystem services.

3 Results and discussions

Segara Anakan Lagoon is a national strategic area in accordance with Indonesian government regulation no 26/2008. The total area of the Segara Anakan lagoon and its environmental area is 24,968 ha (Ludwig 1984). The lagoon region has coordinates of 7 °

37'22 "-7 ° 47'37" south latitude and 108 ° 45'11 "-109 ° 2'54" east longitude. Segara Anakan Lagoon has three rivers that supply high sediment, namely; Citanduy River, Cibeureum River, and Cikonde River, which have an impact on increasing the expansion of the characteristics of mangroves in Segara Anakan Lagoon and its surroundings. This condition is in accordance with the opinion of Setyawan (2010) which states that geologically, the Segara Anakan lagoon area tends to decrease due to sedimentation and mangrove expansion.

The opening of rice fields encouraged the decline of the Segara Anakan mangrove area. The opening of rice fields has happened for quite a long time and still continues to this day. In addition, until now there is still found illegal logging due to market demand, wood products from mangrove logging can be used as firewood for cooking needs, and home pillar materials. Illegal logging has reduced the diversity of mangrove species (Yuwono et al. 2007) although outreach efforts on the importance of preserving mangrove ecosystems by government activists and conservation have always been made to inhibit illegal logging. Changes in land use from Segara Anakan and the decline in mangrove areas can be seen in Figure 1.

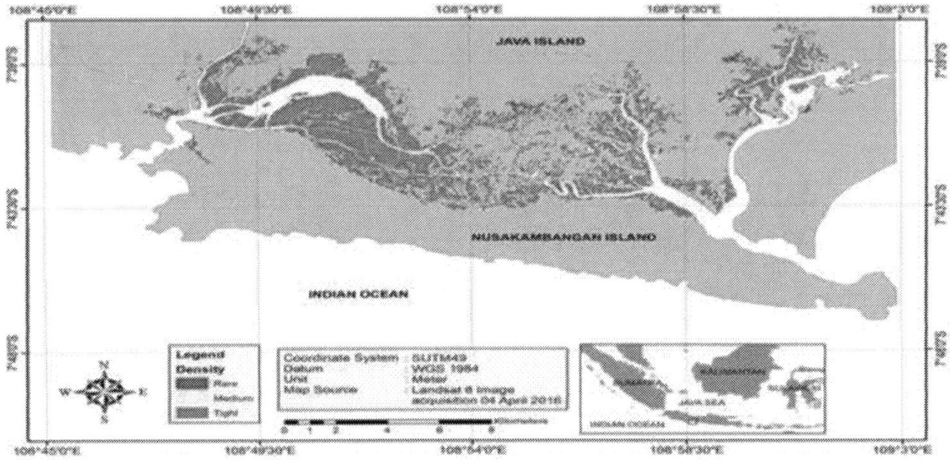

Fig 1. Segara Anakan Land Mangrove Area

The identification of mangrove ecosystem services aims to facilitate the grouping of benefits provided by the ecosystem.

1. Provider / Production Services

Provider/production services are services/benefits provided by mangrove ecosystems in the form of goods and services. The resulting activities are in the form of food providers for the surrounding community in the form of tothok fish, shrimp, crab, and shellfish producers. Data from BPS (2017), the people of Ujung Alang Village are predominantly working as fishermen so that not many people use mangrove forests to meet their daily food needs.

The Department of Environment, Water, Heritage, and the Arts (2009) explained that provider services do not always talk about the use of organisms living in an ecosystem. In addition to utilizing organisms that live in the mangrove ecosystem, the utilization of parts of mangrove trees can also be categorized in the activities of the service provider. However, based on the results of interviews, processing of parts of mangrove trees such as stems, leaves, and seeds is in the experimental stage so that the processed products cannot be commercialized.

Table 1. Summary of Services and Activities

Types of Services	Example Activities
Provider / Production Services	a. Food providers (fish, shrimp, crabs, and tothok shells)
Regulatory / Regulatory Services	a. Benefits of agriculture b. Protective ponds from coastal erosion c. Provider of biodiversity d. Carbon producer
Cultural Services	a. Provision of tourism opportunities

2. Regulatory or regulatory services

The second service is regulatory or regulatory services found in Ujung Alang village. Regulatory/regulatory services are the result of the ability of ecosystems to regulate climate, water cycles and biochemistry, land surface processes, and various biological processes (Department of Environment, Water, Heritage, and the Arts, 2009). The form of activity from regulatory services or regulators in Ujung Alang village is in the form of activities of ponds, providers of biodiversity, and carbon producers.

Providers of biodiversity can be seen from the research conducted by Sari et al. (2016). Based on the research, there are several types of coastal plants that can be found in Ujung Alang Village. Regulatory services or regulation of the mangrove ecosystem, one of which is a carbon producer. However, the services of mangrove ecosystems in regulating carbon in this study were not examined in more detail because of limited data.

3. Cultural services

Cultural services are services related to benefits that humans get through entertainment, the development of reasoning, relaxation, and spiritual reflection (La Notte et al., 2017). Cultural services can also be said as a bonus that comes along with public awareness of the sustainability of an ecosystem. Based on observations in the field, tourism opportunity service providers are only found in Ujung Alang Village. Utilization of tourism by the people of Ujung Alang Village began to be active in 2010. Based on the results of interviews with tourism managers, Ujung Alang Mangrove tourism began with the creation of a mangrove ecosystem restoration in 2001, which until now continues to grow. The growth of mangrove ecosystem tourism in Ujung Alang Village is also followed by the emergence of new jobs that can improve the economy of surrounding communities such as lodging houses, warungs, sales of mangrove seedlings, etc.

Value of ecosystem services

Assessment of ecosystem services aims to determine how important an ecosystem is. The greater the value obtained from the ecosystem, the more important an ecosystem is for the sustainability of the surrounding community. Products produced by an ecosystem have different values according to people's needs. Assessment of ecosystem services is carried out using an economic valuation approach. Economic valuation is one method of valuing a resource with an economic approach (De Groot, 2006).

1. Value of fishery provider/production services

The mangrove ecosystem as a food provider means that the ecosystem becomes a producer of food sources for the surrounding community. The value of ecosystem services as food providers is derived from the purchase price or market price of fish, shrimp, and crabs. These products were chosen because they are the dominant products that are usually sought after by the surrounding community both as daily food items or for sale. The value of service providers from the mangrove ecosystem is presented in Table 1.

Tabel 2. Fisheries Services

Commodities	Price (Rp)	Estimated Value of Direct Benefits (Rp)
Fish	8,000-15,000	150,345,590,321
Tiger shrimp	50,000-70,000	668,584,049,585
Shrimp krosok	8,000-20,000	58,985,980
Mangrove Crab	25,000-55,000	24,950,578,987
Jumlah		912,866,199,880

2. Value of Regulatory Services or Regulations

Erosion and siltation are one of the major threats to people living in lagoon areas. Erosion can cause damage to the lagoon ecosystem, but silting for some people is used as rice-producing agricultural land. Awareness is needed to maintain the mangrove ecosystem in Kampung Laut Subdistrict, one of the pictures of the adaptation of coastal communities in Kampung Laut District to erosion and siltation that occur. Based on the results of interviews and field identification, the regulating services owned by the mangrove ecosystem in Kampung Laut District are divided into three benefits. The three benefits include agricultural production, protection of ponds from coastal erosion, and maintaining biodiversity.

a. Value of agricultural benefits

The total area of agricultural land in Kampung Laut District is an area of 979 hectares. With sedimentation in the Segara Anakan lagoon, which tends to increase agricultural area (Reichel et al., 2009). Then it can be seen the estimated total value of direct agricultural benefits of Rp. 1,455,786,980 per year.

b. The protective value of ponds from coastal erosion

Pond cultivation in Kampung Laut Subdistrict is polyculture or more decreasing. As explained from one commodity cultivated in one pond area where the selected commodity is mutually beneficial and does not interfere with the growth and development of each commodity. Aquaculture is dominated by two villages in Kampung Laut Subdistrict, namely Klaces Village and Ujung Alang Village. Aquaculture is concentrated in Bondan. Traditional system cultivators have land for cultivation ranging from 0.5 to 5 ha depending on the capital of each farmer.

c. Value of biodiversity

Biodiversity is all living things on earth (plants, animals, and microorganisms) including the genetic diversity they contain and the diversity of ecosystems they form (Kusmana, 2015). Biodiversity values are obtained from calculations that refer to the Ruitenbeek (1992) biodiversity value study, which is the US $ 15 ha/year, which is then applied to each study area. Adjustment to the study area begins with compound values so that the calculations can be close to accurate. The value of direct benefits from the utilization of mangrove resources for firewood with a total benefit of Rp. 3,768,889,760 per year.

3. Value of cultural services

The cultural services of the mangrove ecosystem that are developing in some sub-districts of Kampung Laut are tourism activities. Cultural service valuation methods can use measurements using tourist travel costs or use the income from tourism activities. The cultural service assessment at Ujung Alang uses an approach to income from ongoing tourism activities. Such tourism activities include withdrawal of boat rentals, sales of mangrove seedlings, and the results of rental of residents' houses around tourist attractions. Assessment of cultural services through tourism activities shows that these activities can increase income for villagers. Based on the calculation of cultural service values, the value of cultural services in Ujung Alang is estimated at Rp. 107,920,000 / year.

4 Conclusion and Implications

Based on the results of this research, it can be concluded several things regarding the analysis of mangrove ecosystem services. Types of the utilization of Segara waters resources Puppies carried out by the community are the utilization of fisheries, agriculture, and utilization of mangrove wood as fuel. Whereas the total direct benefit value for these activities is Rp. 918,198,786,620 per year. This value illustrates the considerable economic benefits of the Segara Anakan water resources. However, the high sedimentation problem in Segara Anakan can threaten the sustainability of utilization in these waters.

Policies to overcome problems due to the sedimentation process require a systematic action by adhering to ecological principles. The solution to the sedimentation problem must be carried out from upstream to downstream because without a comprehensive solution it can disrupt all the activities in Laguna.

Bibliography

[1] E. Adam, O. Mutanga, and D. Rugege, Wetl. Ecol. Manag. **18**, 281 (2010)
[2] E. B. Barbier, S. D. Hacker, C. Kennedy, E. W. Koch, A. C. Stier, and B. R. Silliman, Ecol. Monogr. **81**, 169 (2011)
[3] M. W. Beck, K. L. Heck, K. W. Able, D. L. Childers, D. B. Eggleston, B. M. Gillanders, B. Halpern, C. G. Hays, K. Hoshino, and T. J. Minello, Bioscience **51**, 633 (2001)
[4] R. de Groot, Landsc. Urban Plan. **75**, 175 (2006)
[5] R. S. de Groot, M. A. Wilson, and R. M. J. Boumans, Ecol. Econ. **41**, 393 (2002)
[6] Department of the Environment, Water, Heritage, *Ecosystem Services: Key Concepts and Applications, Occasional Paper No 1.* (Canberra: Department of the Environment, Water, Heritage, and the Arts, 2009)
[7] G. M. Foody, Int. J. Remote Sens. **24**, 4035 (2003)
[8] C. Giri, Z. Zhu, L. L. Tieszen, A. Singh, S. Gillette, and J. A. Kelmelis, J. Biogeogr. **35**, 519 (2008)
[9] Ismail, Sulistiono, S. Hariyadi, and H. Madduppa, AACL Bioflux (2018)
[10] K. Jusoff, J. Sustain. Sci. Manag. **1**, 27 (2006)
[11] K. Jusoff, Polish J. Environ. Stud. (2013)
[12] A. Komiyama, J. E. Ong, and S. Poungparn, Aquat. Bot. **89**, 128 (2008)
[13] J. M. Kovacs, F. Flores-Verdugo, J. Wang, and L. P. Aspden, Aquat. Bot. **80**, 13 (2004)
[14] La Notte, D. D'Amato, H. Mäkinen, M. L. Paracchini, C. Liquete, B. Egoh, D. Geneletti, and N. D. Crossman, Ecol. Indic. **74**, 392 (2017)
[15] P. E. Purwanto AD, Asriningrum W., Winarso G., in *Di Hear. Semin. Nas. Penginderaan Jauh*, Kartasasmita et al (ed)(Bogor, Indonesia, 21 April 2014)
[16] C. Reichel, U. U. Frömming, and M. Glaser, Reg. Environ. Chang. **9**, 335 (2009)
[17] A. I. Robertson and N. C. Duke, Mar. Biol. **96**, 193 (1987)

[18] H. Ruitenbeek, *Mangrove Management: An Economic Analysis of Management Options with a Focus on Bintuni Bay, Irian Jaya* (EMDI/KLH, Jakarta, 1992)
[19] L. K. Sari, L. Adrianto, K. Soewardi, Atmadipoera, A. S., and E. Hilmi, in *AIP Conf. Proc.* (AIP Publishing, n.d.), p. Vol. 1730, No. 1, 080002
[20] Sasaki Y., Sunarto H., hutan Mangrove. Segara Anakan Lagoon. Dalam: Takashima F, Soewardi K, Editor. Penilaian ekologi Untuk Perencanaan Pengelolaan Segara Anakan Lagoon, Cilacap, Jawa Tengah. NODAI Pusat Program Internasional. Tokyo University of Agriculture. JSPS-DGHE Program, pp. 95-106 (1994)
[21] S. Sukardjo, Malaysian J. Sci. **35**, 73 (2016)
[22] B. P. Statistik, *Kecamatan Kampung Laut Dalam Angka Tahun 2017* (Badan Pusat Statistik, Cilacap, 2017)
[23] C. Kusmana, in *Pros. Semin. Nas. Masy Biodivers. Indones. Vol. 1 Nomor 8 Hal 1747* (2015)
[24] L. HF, *Ulasan Sastra, Pemantauan Lingkungan Segara Anakan Dan Penggunaan Proyek Yang Optimal* (Institut Teknik Hidrolik, Badan Penelitian dan Pengembangan, Departemen Pekerjaan Umum, Indonesia, 1984)
[25] M. E. Assessment, *Ecosystem And Human Well-Bein: Current State and Trends* (Islandpres, Wasington, 2005)
[26] W. B. Setyawan, in *Semin. Nas. (Biodiversitas Dan Bioteknol. Sumberd. Akuatik)* (2010)
[27] E. Yuwono, T. Jennerjahn, I. Nordhaus, E. Riyanto, M. Sastranegara, and R. Pribadi, Asian J. Water, Environ. Pollut. **4**, 61 (2007)

THE STAKEHOLDER ANALYSIS FOR FISHERIES MANAGEMENT

Suharno[1*], Agus Arifin[2], and Ary Yunanto[3]

[1]Economics and Development Studies, Faculty of Economics and Business, Jenderal Soedirman University
[2]Economics and Development Studies, Faculty of Economics and Business, Jendera*l Soedirman University
[3]Management Department, Faculty of Economics and Business, Jenderal Soedirman University

Abstract. The south sea of Central Java as a fishing ground for south coast fishers makes its resources common-pool resources, and the utilization of ownership rights becomes unclear. As a result, fisheries resources experience excessive utilization, which indicates the phenomenon of the tragedy of the common. Based on these conditions, appropriate rules and policies are needed in the institutional management of fisheries resources to remain sustainable. The purpose of this study is: to identify the roles of stakeholders in fisheries management. The analysis used is stakeholder analysis. The results of the study show that: Stakeholder groups involved in fisheries management are subjects, bystanders, actors, and players.

1 INTRODUCTION

Causes of failure of fisheries resource management due to complex problems. Emerging fisheries problems are usually multi-sector and various aspects such as ecology, economy, social, ethics, and management (Suharno et al., 2019). Campbell et al. (2012) suggested that the effectiveness of the watershed is highly dependent on each relevant stakeholder to submit and obey existing rules. According to Salayo et al. (2006), Merino et al. (2008) conflicts that often occur in the use of fisheries resources, because they have two different objectives, namely for the purpose of protecting fisheries resource ecosystems and ecological processes, and another aim is to exploit fisheries resources used in meeting economic needs and welfare of the population. Further conflicts that develop are conflicts of interest between sectors and between agencies involved in watershed management and conservation (Irnawati et al. 2013, Tampubolon and Satria 2013).

Fisheries management is an obligation mandated by Law No. 31/2004 and Law No. 45/2009 concerning fisheries. Fisheries management is defined as all efforts, including integrated processes in information gathering, analysis, planning, consultation, decision

* Corresponding author : suharno@unsoed.ac.id

making, allocation of fish resources, and implementation and law enforcement of laws and regulations in the field of fisheries, carried out by the government or authority others that are directed at achieving the sustainability of the productivity of aquatic biological resources and agreed objectives.

Since the implementation of decentralization in Indonesia has implications for natural resource management, it is the active participation of local communities in the process of managing the area. Until now, the role of the community in the management of fisheries resources is still very limited as a subject in activities. It is still necessary to develop partnership relationships between stakeholders based on the concept of mutual need relationships and equality, sustainable and mutually beneficial relations. Some of the negative impacts of development activities are still found, for example, the emergence of elimination of communal rights, environmental degradation, the lack of rights of local communities to access public needs, and the emphasis on political bargaining positions.

The problem of this research is that each role in the overall fisheries management has not been identified. Community-based fisheries resource management that is suitable to be applied today is accommodating community involvement in fisheries resource management based on participatory fisheries / co-management, Pomeroy and Williams, (1994) mentioning a central element of co-management is a partnership between government, society, and users other resources. Implementing co-management as a joint management concept means that various stakeholders (stakeholders) agree to share roles in management, rights, and responsibilities, over a region or natural resource as intended by the main goal so that management is more precise, efficient, fair and equitable (Susilowati, 1999; Nikijuluw, 2002). Discussion of fisheries resource management is inseparable from coastal and marine areas and stakeholders so that collaborative fisheries resource management is needed which involves small-scale fishers and stakeholders.

This condition is purified about the need for research on sustainable management of fisheries resources, especially in Cilacap waters with mapping based on parties / stakeholders so that they can accommodate the interests of various related parties, namely: protection of fisheries resources, preservation of the surrounding environment, and involving human elements that cannot be separated as social and economic beings.

2 ANALYTICAL METHOD

The function of fishery production is a link between the use of production input factors (fisheries as a composite index of ships/boats, fishing equipment, fuel, labor and ability to manage) with the level of production (output). A large number of fishing gear that is operated in waters and limited fishing areas will create utilization conflicts between fishers and the threat of damage to fisheries resources.

In-depth observations are needed to analyze the condition of the fishery through a study of fisheries resource management by collaborating full involving small-scale fishers and stakeholders. Current conditions for managing fisheries resources have not collaborated fully involving small-scale fishers and stakeholders. The potential of fisheries resources must be managed well by looking at the various aspects involved.

The research was carried out by directly observing the problem (by collaborating fully with fisheries management involving fishers and stakeholders) that exist and are being faced in the waters of Cilacap.

This study uses data to be collected in the form of primary data and secondary data. In general, the types of data collected from both primary and secondary data are Data on the role of fisheries stakeholders (fishers, government, and the private sector) in the management of small-scale shrimp fisheries.

Secondary data was collected through data related to research problems obtained in the form of documents and files originating from reports of activities, books, legislation, and journals related to the management of shrimp fisheries in the waters of Cilacap and elsewhere with similar problems. Secondary data, among others, comes from Government agencies (Ministry of Marine Fisheries / MMF; Fisheries Service; Central Statistics Agency, and related agencies).

Primary data was collected by the technique of distributing questionnaires to respondents, in-depth interviews, direct observations in the field, and focus discussion groups. This data can also be in the form of references from expert opinions. Reference to expert opinion is collected from expert opinions during interviews and when opinions or ideas are conveyed through print and electronic media about a problem related to research. These experts can come from Government agencies (Ministry of Maritime Affairs / Fisheries and Fisheries Services), higher education, research institutions, and fisheries observers/practitioners. The collection of primary data through questionnaires is carried out with the stages of sampling group selection, identification/selection of respondents, and data collection to respondents.

Stakeholder analysis is needed due to the overlapping of each role whose status is not clear in the stakeholder. Reed et al. (2009) explain the stakeholder analysis model that can be described to be used as a model. Stakeholder analysis requires the following steps:

1. Identify stakeholders and their roles
2. Identify expectations for stakeholders about the program
3. Benefits-benefits / benefits that will be obtained for stakeholders
4. Categorizing and differentiating stakeholders based on their interests and influence.

3 RESULTS AND DISCUSSIONS

The definition of a stakeholder is a person, or group, or institution because its role can be influenced positively or negatively due to the existence of a program/project activity, or vice versa a person, or group, or institution can influence a program/project activity. Stakeholders are mapped into a matrix of stakeholder analysis according to the value of importance and influence.

The amount of interest is assessed from:

1. Stakeholder involvement in fisheries management
2. Reliance of stakeholders on fisheries management
3. The role of each stakeholder related to fisheries management
4. Benefits obtained by stakeholders from fisheries management
5. Stakeholders' interests in sustainable fisheries management

The amount of influence is assessed from:

1. Instruments and sources of strength (power) at each stakeholder
2. Position of the degree of stakeholders in decision making
3. HR support for the management of fisheries resources
4. Funding and management capabilities for fisheries resource management
5. Interaction with other stakeholders

Assessment of the importance and influence of stakeholders using the Likert scale, namely the value of 4: very high, 3: high, 2: sufficient, 1: less, 0: low. The amount of value obtained by each stakeholder is a point for the magnitude of interest and points for the magnitude of influence. After knowing the value of importance and influence, each stakeholder is mapped into a matrix of interests and influences using Mactor software.

Data retrieval is done by direct interviews and questionnaire to representatives of each party identified from the results of actor analysis, processing qualitative data from interview results is quantified by referring to tiered data measurements.

Sustainable shrimp fisheries management in Cilacap waters has several stakeholders involved in its management. Stakeholder analysis is carried out to identify and map actors (level of importance and influence) in the management of sustainable shrimp fisheries in Cilacap waters. Based on the results of interviews and deepening, 11 stakeholders were involved in the management of shrimp fisheries. In Table 1. a profile of the roles of each stakeholder involved is presented.

Table 1. Stakeholders for Sustainable Fisheries Management in Cilacap Waters

No	*Stakeholders*	Role
1	Fisheries & Marine Services (District and Province)	Regional authorities that protect the integrity of their territory, improve the welfare of the local community, & preserve the environment of their area
2	Indonesian Fishermen Association of Cilacap	Community activity groups that participate in fisheries management together with fishers and fishing communities
3	Non-governmental organization	Community activity groups that control the compliance of rules and local wisdom in the local environment
4	Unsoed	Educational institutions, community practitioner groups, academics, researchers, observers in fisheries management
5	Cilacap Fisheries Association	Community activity group that connects fishers with other related parties
6	Traders	Supplier group of fishermen's goods and services, & users of fishermen's catch products
7	Village	Village authorities that have the authority to regulate, serve, and manage the rights and obligations of citizens in their administrative areas
8	KUD Minosaroyo	Fisheries goods and services supplier group, & fish auction process authority at Fish Auction Place
9	Local Community	Community groups in various groups: age not yet productive, productive age and unproductive age
10	Ministry of Maritime Affairs and Fisheries	National fisheries concession authority
11	Fishers	Groups that have the skills and knowledge to run operational fisheries management

Source: Primary Data Processed, 2019.

In order to manage fisheries resources, several objectives must be determined that is related to all stakeholders. The objectives to be achieved are, 1. sustainable shrimp fisheries production, 2. increasing fisherman income, and 3. increasing fisherman catches.

After knowing the role of each party, then categorize stakeholders based on their interests and influence. Stakeholders are mapped into stakeholder analysis matrices based on their importance and influence.

Assessment of the importance and influence of stakeholders using Likert scale is value 4: very high, 3: high, 2: enough, 1: less high, 0: low. An assessment of the magnitude of stakeholder interests in the management of shrimp fisheries in Cilacap waters can be seen in Table. 2.

Table 2.
Assessment of Stakeholder Interests

MDI Criteria	Fisheries Service	HNSI	NGO	Unsoed	Association	Traders	Village	KUD	Local Community	MMF	Fishers
Fisheries Service	0	3	1	1	3	3	1	3	1	3	4
HNSI	1	0	3	1	3	3	1	3	1	3	3
NGO	1	1	0	1	1	1	1	1	1	1	1
Unsoed	1	1	1	0	1	1	1	1	1	1	1
Association	1	2	1	1	0	1	1	2	1	1	2
Traders	1	1	1	1	1	0	1	1	2	2	2
Village	2	2	2	2	2	2	0	2	2	2	2
KUD	2	1	1	1	3	3	2	0	3	3	3
Local Community	1	1	1	1	1	1	1	1	0	1	1
MMF	3	2	1	1	2	1	1	1	1	0	3
Fishers	3	1	1	1	2	1	1	2	2	2	0

Source: Primary Data Processed with Mactor Software, 2019.

Assessment of the importance and influence of stakeholders based on the objectives to be achieved using a Likert scale, namely the value of +4 or -4: very high, +3 or -3: high, +2 or -2: enough, +1 or -1: less high, 0: low. Assessing the magnitude of stakeholder interests by basing their objectives on the management of shrimp fisheries in Cilacap waters can be seen in Table 3.

Table 3.
Assessment of Stakeholder Interests based on Purpose

2MAO Criteria	Sustainable Fisheries Production	Increase in Fishers's Income	Increase in Fishers's Catches
Fisheries Service	3	3	3
HNSI	2	3	3
NGO	1	1	1
Unsoed	1	1	1
Association	-1	2	2
Traders	-1	1	1
Village	2	2	2
KUD	-1	1	1
Local Community	-1	1	1
MMF	3	3	3
Fishers	-2	3	3

Source: Primary Data Processed with Mactor Software, 2019.

After identification, we are mapping the interests and influence of stakeholders in fisheries management in the waters of Cilacap, then mapping into the stakeholder grid analysis matrix. This matrix consists of 4 quadrants, namely spectators (Spectator), Actors (Actors), Subjects (Subject) and Players (Player), the position of each stakeholder will be mapped according to the results of the assessment of importance and influence.

Based on the stakeholder grid analysis in Figure 1. it was found that the fisheries service, KUD Minosaroyo, Ministry of Maritime Affairs and Fisheries, and fishers were in the position of players (player). The Cilacap district fisheries office and the Central Java fisheries department get a mandate to manage sustainable fisheries activities which are commanded by the Ministry of Fisheries and Marine Affairs. The conditions in its operation, fisheries management in Cilacap also involved the KUD Minosaroyo and fishers to share their roles in carrying out their activities.

Villages and HNSI have the role as subject (subject) having a high enough interest in fisheries management in Cilacap waters, while the influence they have is not as big as their interests. Village and HNSI as social control of the operation of fisheries management following applicable rules, the orderliness of the social environment of the community, and the need for local labor. The welfare of the community and fishers, improving the quality of life (religion, education, health, public facilities, culture), the orderliness of the social environment of the community are closely related to sustainability in fisheries management. Villages and HNSI can shift their roles into players by increasing their influence, especially in interacting with other stakeholders, which have been weak, considering that other influences are difficult to improve.

Local communities, associations, and traders are at the position as an actor. Local communities, associations, and traders have a high interest in fisheries management in Cilacap waters, in this case, related to the availability of shrimp supply, fluctuations in fisheries commodity prices and selectivity rules for fishing gear.

Unsoed and NGOs have the lowest interests because it is more coordination, not executor or policymaker. Unsoed and NGOs have influence in coordination and policymaking in fisheries management.

Figure 1.
Assessment of Stakeholder Interests

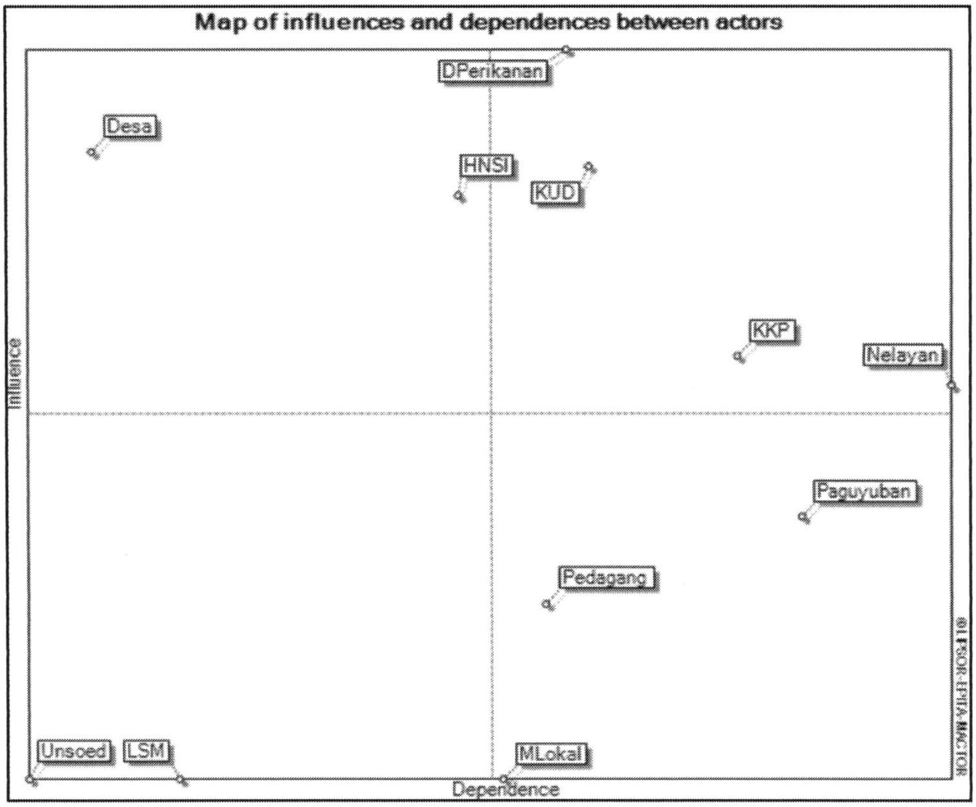

Source: Primary Data Processed with Mactor Software, 2019.

4 CONCLUSION AND IMPLICATIONS

The result of the analysis of the choice of shrimp resource management in Cilacap waters from key-persons is cooperative co-management management that is through the initial exchange of information, the views of the community begin to be included in the agenda and issues as input in the management of fisheries resources, based on:

The findings of the stakeholders' analysis show: fisheries service, KUD Minosaroyo, Ministry of Maritime Affairs and Fisheries, and fishermen as key players / players in managing Cilacap fisheries; Villages and HNSI act as subjects; Local communities, associations and traders in positions as actors; while Unsoed and NGOs only have low roles as spectators.

This research has evaluated the roles of stakeholders in current fisheries management. Furthermore, it can be recommended by setting up a new institution that is not currently able to strengthen fisheries management and is able to run a comprehensive and collaborative fisheries management model by accommodating the existence of elements of local institutional wisdom in the fisheries sector of Cilacap Regency.

5 BIBLIOGRAPHY

1. Campbell, S. J. *et al.* Weak Compliance Undermines the Success of No-Take Zones in a Large Government-Controlled Marine Protected Area. *PLoS One* **7**, 11 (2012).

2. Irnawati, Ririn. *et al.* Teknik Interpretative Structural Modeling (ISM) Untuk Strategi Implementasi Model Pengelolaan Perikanan Tangkap Di Taman Nasional Karimunjawa. *Jurnal Ilmu Pertanian dan Perikanan*. **2 (1)**, 75-86 (2013).
3. Merino, G., Morales-Nin, B., Maynou, F. & Grau, A. M. Assessment and bioeconomic analysis of the Majorca (NW Mediterranean) trammel net fishery. *Aquat. Living Resour.* **(22)**, 75-56 (2008) .
4. Bambang, S. & Bowo, F. Dosen Fak. Hukum Universitas Wahid Hasyim Alumni MIH UNDIP Nikijuluw, V.P.H. 2002. Rezim Pengelolaan Sumberdaya Perikanan . Jurnal Ilmiah Ilmu Hukum QISTI 38. *J. Ilm. Ilmu Huk. Qisti* (2002).
5. Pomeroy, R. S. & Williams, M. J. Fisheries Co-management and Small-scale Fisheries: A Policy Brief. *International Center for Living Aquatic Resources Management* (1994).
6. Reed, M. S. *et al.* Who's in and why? A typology of stakeholder analysis methods for natural resource management. *J. Environ. Manage.* **(90)**, 1943-1949 (2009).
7. Salayo, N. D., Ahmed M., Garces, L. & Viswanathan, K. An Overview of Fisheries Conflicts in South and Southeast Asia: Recommendations, Challenges and Directions. *Naga WorldFish Cent. Q.* **(29)**, 1-2 (2006).
8. Suharno, Anwar, N. & Saraswati, E. A technique of assessing the status of sustainability of resources. in *IOP Conference Series: Earth and Environmental Science* **(250)**, 1 (2019).
9. Susilowati, I. An Analysis of Co-Management Fisheries In West Sumatra Province, Indonesia: A Case Study of Ikan Larangan. *Reserach Report, International Center For Living Aquatic Resource Management*, Manila, Philipines. (1999).

Traditional Agroforestry Ecosystem for Rural Prosperity

Carolina Carolina [1*], *and Febtri* Wijayanti[2]

[12] Research Center for Science Technology and Innovation Policy and Management
Indonesian Institute of Sciences

Abstract. Pursuance of prosperity has always been the aim of rural development program. A strategy to vitalizing rural economy to ensure equitability is achieved through optimal utilization of local resources. It is an inclusive approach, widely implemented which also applied in Indonesia. In order to examine the concept, a qualitative study was conducted in Sumba Barat Daya in East Nusa Tenggara Province, in which agriculture is the leading sector of regional development. Assessment is focussed toward the potency of natural resource, specifically traditional agroforestry system, producing variety of estate crops which are sold in local, regional and international market as well. Representing different type of market system, assessment on cashew and coffee crops were discussed. The analysis indicates that potency to achieve prosperity seems more promising when the strategy implemented is local driven. We encounter that farming activity in managing coffee based agroforestry generates local innovation, not just in ecosystem management, but also in post harvest handling and local market system establishment. Therefore, management of coffee agroforestry leads better to inclusivity in comparison with cashew. However, to ensure sustainable support of traditional agroforestry ecosystem to rural livelihood, understanding socio-ecological dynamics of rural Sumba should be acquired in order to generate supportive relevant policies.. yamu,to,

1 Introduction

Amidst many obstacles found in the policy enactment (Long, 1987; Knickel et al., 2017; Soleh, 2017) rural development has been implemented as one of strategy to reduce inequality in many countries. Indonesia declare the importance in Law no 6/2014 on Village Development programs in which prosperity is stated as the goal. Similar in goal 6 of Sustainable Development Goals, indicating that it is not just a local quest but a globally significant pursuit. The fact that approximately 45% of Indonesia's population live in rural areas, where most of the poor abide is a strong motivation to always strive for betterness thorugh rural development strategy. The policy accomodates diversity of ideas being implemented as every regions may have different characters to cater, implying that

* Corresponding author: carolina.4q@gmail.com

understanding rural sosio-ecological system is necessary in order to obtain appropriate approach of rural development (Schouten et al., 2009)

Rural development to achieve people's prosperity has undergone transformation from considering economy as sole factor of progress and modernisation to present nuance of balancing economic, social and environment consideration as sustainable procurement (Rivera, et al., 2018) Nevertheless it indeed requires multidimensional thinking to contextualize prosperity as goal of rural development, a strategy that should be designed for equal distribution of wealth within society and areas (Pouw & Gupta, 2016). As a long known environmental management practice, agroforestry provides wide range of opportunities for farmers to gain sustainable benefits (Nair, 1991; Liu et al., 2014; Leakey, 1996) recognize it as a globally important ecosystem that will remain important in the long run due to its capacity to support livelihood, and provide environmental benefits as well. Being traditional, its sustainable optimal ecosystem service is known to be supported by local knowledge (Nair, 1991), an important assets to vitalize rural economy.

As agrarian base region, Indonesia's strategy of rural development consider local resources as important assets to achieve the mission. In this study, we explore the strength of traditional agroforestry system in support to rural economic activity on the assumption that utilization of local resources will be the most approachable strategy.

2 Methodology

2.1 Study Area

The study was conducted in Sumba Barat Daya in East Nusa Tenggara Province, an agrarian region as most of many areas in Indonesia. Ecologically, it is a part of Sumba Island in East Nusa Tenggara Province Indonesia (Figure 1). About 331.894 people lives in an area of 1.445,32 km2, confirming the region as the most populated district in the island. Among them, 70 % work at agriculture sector to include agroforest management (Statistics, 2018). Landuse can be chategorized into wet land and dryland, which are 12.3% and 46.2% of total 72,057 Ha. respectively. In agroforestry ecosystem management, they plant various species of trees for subsistence and also commercial purpose. Sumba Barat Daya region consists of 11 subdistricts i.e. Kodi Bangedo, Kodi Balaghar, Kodi, Kodi Utara, Wewewa Selatan, Wewewa Barat, Wewewa Timur, Wewewa Tengah, Wewewa Utara, Loura, and Kota Tambolaka. Socio-culturally the area is divided into three different parts i.e. Kodi, Wewewa and Loura. Kodi and Wewewa differs in ecosystem and language spoken; whereas Laora is occupied by mixture of cultures to include immigrants from outer island.

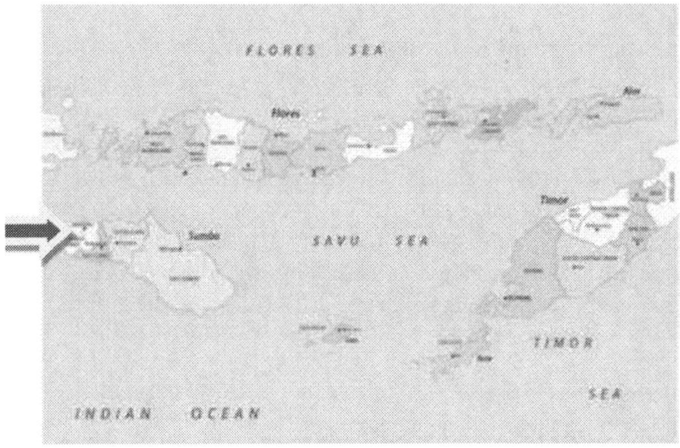

Fig. 1. Map of East Nusa Tenggara Province
(Source:www..shutterstock.com)

2.2 Research Method

Agroforestry is recognized as an important traditional land use practice, due to its high levels of biodiversity, traditional knowledge systems, livelihood support system, diversified traditional cultures and its undoubtly economic value (Leakey, 1996; Mercer et al., 2014; Njurumana, 2014; Rivera, et al., 2018). This study focuses on the role of traditional agroforestry as livelihood support to rural community under the assumption that the ecosystem develop overtime by the support of local knowledge (Nair, 1991). The years experience in managing the landuse currently resulted as a system capable of supporting not only subsistence need but also a steady source of family income. Ecosystem management and socio-economic situation interactively connected (Hou et al., 2014), in reality establishment of local market could be resulted by landuse system productivity (Beder, 2011).

To examine the relevance of agroforestry management in rural prosperity, we conducted this study which is designed as a qualitative research to allow understanding of complex reality (Cresweell, 2014) and further analysed amenably. Collection of primary data is conducted by means of in-depth interview to informants i.e head of village, senior farmers, government officers from agriculture sector. We also conducted 2 focus group discussions with coffee and cashew farmers. Secondary data is obtained from local and national statistical office; and also from relevant local government office.

We also use Location Quotient analysis in order to determine comparative advantage of commodities. Commodity with $LQ > 1$ indicates that the commodity is considered as basic commodity and has comparative advantage which is supported by natural resources, a lack in other area (Jawang, 2018). Value of $LQ < 1$ means the commodity can not be considered as basic commodity. LQ equation is as follows :

$$LQ = Pi/Pt \ / \ P.i/P.t$$

Notes :
Pi = commodity production share at sub-district level
Pt = commodity production share at district
P.i = total commodity at sub-district level
P.t = total commodity at district level

3 Result and Discussion

3.1 Traditional agroforestry ecosystem : distribution and productivity

As a traditional land use practice in various socio-ecological system, agroforestry land management is also observed in Sumba Barat Daya. Approximately 47% of agricultural land is dedicated to polycultures of trees and food crops which are planted on the same area at the same period, or differ in time. In general, 13 species of trees grow in agroforest and produce significant quantity of yield. Among those, production of cashew (Anacardium occidentale L.) coconut (Cocos nucifera L.), coffee (Coffea canephora), areca-nut (Areca catechu L.), and Cocoa (Theobroma Cacao) are the five highest (Table 1)

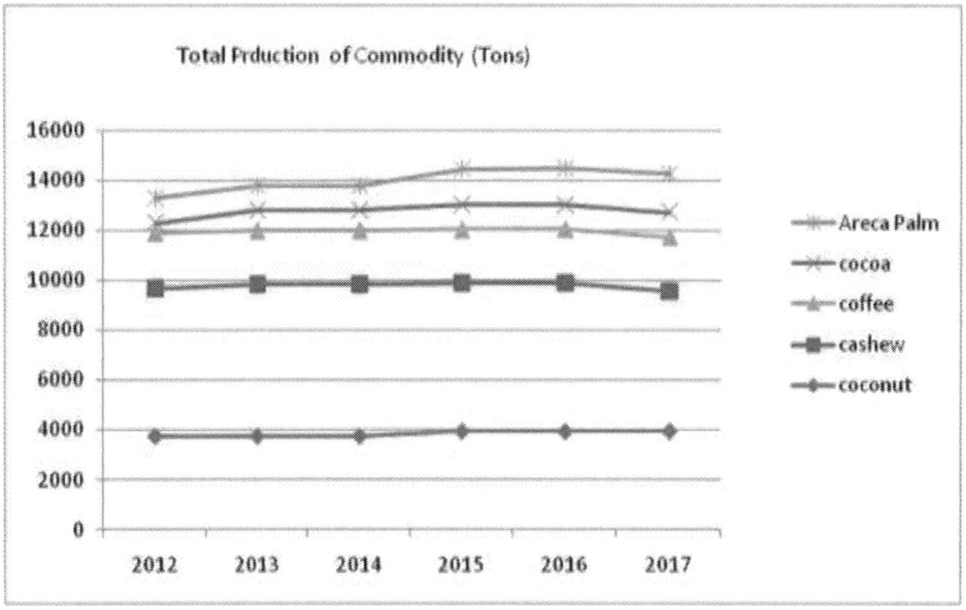

Fig. 2. Total Production of Commodity (2012-2017)

As indicated by Figure 1, there is no significant fluctuation of yields experienced during the period of 2012-2017. It does show stability of productivity although climatic change is not to be ignored. Uniquely distributed, the plants grow in agroforest cropping system in many areas which could be characterized by socio-ecological condition. Except cashew and coconut that are cultivated in plane dry area of Kodi, the other crops grow in higher altitude areas of Wewewa. What is similar is farmer's consideration to the land use as livelihood support they could rely upon. To some extent, the high productivity gives brand to the village and allure buyers to come fetching the produce.

In confirmation of the productivity data above, we performed Location Quotient Analysis . It is necessary to show the strength of certain subject in an area in relevant with a larger area (Tarigan, 2007). In our study, the subject is commodity planted in agroforestry system. The higher LQ, the more comparative the subject is. The analysis usually utilized as a means to identify Produk Domestik Regional Bruto because choosing the right basic commodity will help economic growth of a region. Table 1 show the results of analysis and confirm that cashew, coconut, coffee, arecanut and candlenut are categorized as basic commodity.

https://doi.org/10.1051/shsconf/20208601021

Table 1. Location Quotient Result of Analysis

Kecamatan	coconut	cashew	coffee	cocoa	candlenut	kapok	clove	Areca Palm	Vanilla	Jathropa	Cotton	Tobacco	Betel
Kodi bangedo	1.1850	1.1639	0.2599	2.8483	0.3413	1.3105	0.1219	0.3455	0.8737	3.8974	0.5868	0.9826	0.7446
Kodi Balaghar	0.4527	2.5382	0.0168	0.3252	0.0838	0.0101	0.0000	0.0168	0.0000	0.0268	0.2146	0.4191	0.0134
Kodi	1.8590	1.7191	0.3248	0.0672	0.0093	0.0075	0.0075	0.0709	0.0000	0.0019	0.0000	0.0261	0.0243
Kodi Utara	0.6786	3.1318	0.0166	0.2134	0.0038	0.0077	0.0000	0.0115	0.0000	0.0051	0.0000	0.0460	0.0038
Wewewa Selatan	1.4284	0.0951	1.1272	0.2021	0.6439	0.0020	0.0357	0.5646	0.0059	0.0000	0.0000	0.0028	0.0099
Wewewa Barat	1.0265	0.0384	1.5955	0.3153	0.7458	0.0038	0.0461	0.2345	0.0038	0.0308	0.0000	0.0327	0.0442
wewewa Timur	0.7899	0.0000	1.3396	0.5182	1.1627	0.0253	0.0316	0.1769	0.0126	0.0190	0.0000	0.0038	0.0379
Wewewa tengah	0.5701	0.0000	0.5998	0.2076	0.4416	0.0066	0.0231	1.9872	0.0132	0.0000	0.0000	0.0013	0.2669
Wewewa Utara	0.6615	0.1405	0.5256	0.2129	0.7068	0.0045	0.2175	1.5495	0.0136	0.0036	0.0000	0.0136	0.0680
Loura	0.6503	2.5755	0.1545	0.0579	0.0000	0.0129	0.0032	0.5280	0.0000	0.0322	0.0000	0.0000	0.1030
Kota tambolaka	0.7562	3.2597	0.0000	0.0000	0.0890	0.0127	0.0000	0.0000	0.0000	0.0000	0.0000	0.0000	0.0000

The result also shows that basic commodities in every area is different from one another, and there are similarities among regions. In summary, it is presented in Table 2 as follows.

Table 2. Summary of Basic Commodities in Sub Districts

Sub District	Basic Commodities (LQ>1)
Kodi Bangedo	cashew, coconut, cocoa, kapok, jathropha
Kodi Balaghar	cashew
Kodi	cashew, coconut
Kodi Utara	cashew
Wewewa Selatan	coffee, coconut
Wewewa Barat	coffee, coconut
Wewewa Timur	coffee, coconut, candlenut
Wewewa Tengah	areca-nut
Wewewa Utara	areca-nut
Loura	cashew
Kota Tambolaka	-none-

In summary, results of LQ analysis confirms that Kodi is high producing cashew nuts and also coconut, suggesting that further development could be started from those commodities, whereas Wewewa agroforest is predominantly cultivated by coffee and areca nut.

Table 3. Characters of Kodi and Wewewa Area in Sumba Barat Daya

	Kodi	Wewewa
Topography	Mostly plane	Slope
Altitude	0-300 m above sea level	350-800 m above sea level
Cropping System	monoculture	Polyculture
	Intercrop with foodcrop in rainly season	No additional crop temporally
Dominant crops	cashew, coconut	coffee, areca nut, candlenut

This study was conducted in 2 villages to represent coffee and cashew based areas, i.e. Kadi Roma village in Central Wewewa and Kalaki Kambe village in West Wewewa respectively. Moreover, the areas were selected because of accessibility reason, and also availability of all aspects needed to be explored in order to attain representative situation

3.2 Local Innovation of Land Management

This study examines the role of traditional agroforestry as livelihood support to rural community. Agro forestry is an old land use practice worldwide which has shown strong resilience over centuries and attains global importance (Liu et al., 2014). It is an indispensable land management system due of its high levels of biodiversity, local knowledge systems, livelihood guarantee and diversified traditional cultures. Local farmer-driven innovations becomes an important support of sustainability, especially since farmers incline to efficiently using the resources available to them. The years experience in land use management resulted as an ecosystem capable of supporting not only subsistence need but also a steady source of family income. Management system of dry land as agroforestry can be distinguished as two distinctive ecosystem i,e, traditional agroforestry in Wewewa, called kalio, situated in higher altitude of 350-800 m above sea level and agroforestry system in Kodi, called mango, which is located in plane area. The distinctiveness as indicated in Table 2 show that dominant tree species differs. Kalio in Wewewa is dominated by coffee, whereas mango in Kodi is by cashew.

Although the two main species planted in similar system, i.e agroforestry, cashew and coffee planting have different ecological history. Cashew was introduces through IFAD program in the 1980's, altering high diversity forest to become almost monocrop system. It was part of rural poverty alleviation. Cashew was preferred due to its ability to grow optimally in semi-arid land. Short rainfall season as occur in Sumba is not a limitation for cashew to grow healthily. Over the time, local people adapt the presence of cashew plantation by planting food crops during rainy season. After crop harvest time, they use empty space for herding pigs and chickens, and let them find food in the environment. With the strategy, they will have corn, peanuts, and in many areas, dryland paddy, and healthy pigs. Thus by planting cashew based agroforestry, it enables farmers to develop silvopastoral system.

Unlike cashew plantation, coffee was introduced around 5 decades ago by Timorese, not intentionally introduced by a project. Starting as one of many kinds of tree species planted in kalio, it is now become dominant tree in most of agroforestry system managed. At present coffee plays important role as source of family income, also as a commodity with high social value. Unlike cashew planting system, coffee in kalio is planted with many other tree species with variety of purposes (Carolina, 2016). Fruitcorps and also starchy crops planted in the same area at the same period. They implement a low external input strategy in maintaining the plots without any plant rotation system. Quantitatively, their sole dependence on local innovation over time resulted in present condition that is low productivity of coffee in comparison with national average; however further exploration on ecosystem management need to be performed in order to gain better understanding of the phenomenon.

In general, it is safe to say that the dynamics of agroforestry management predominantly planted with cashew and coffee is to some extent maintained by the establishment of local market, where produce will be readily exchanged with currency. Rural people take the most advantage of existing market whilst caring for the crops to continuously producing as expected.

3.3 Marketing System of Agroforestry Products

Most of produce sold by farmers follows traditional market system. Farmers direct connections is developed with local buyers who provides cash as exchange to their harvest. Regardless of the fact that coffee and cashew have global value, local market system is still the one they rely upon. Coffee market system is depicted in Figure 3.

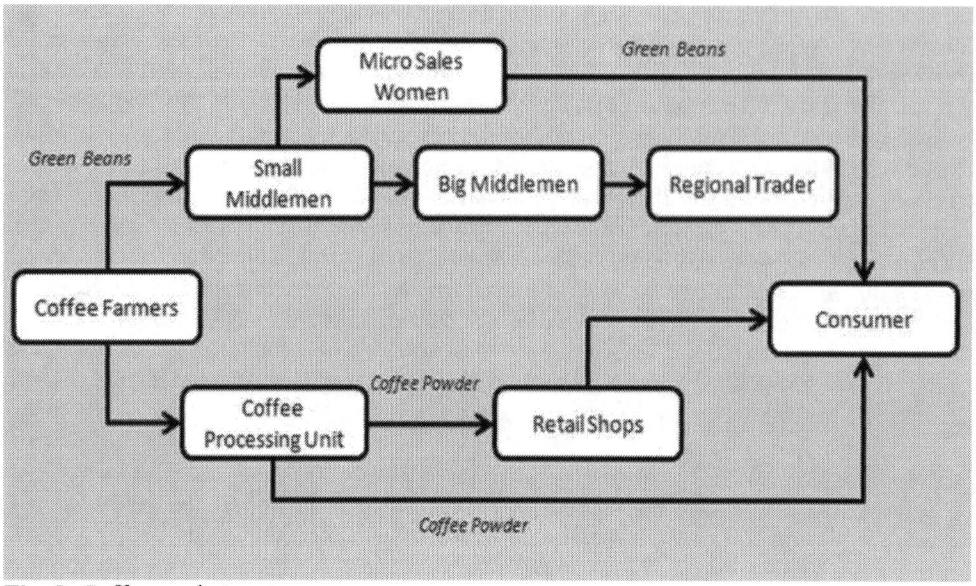

Fig. 3. Coffee market system

Harvest time is the period of farmers processing their coffee bean to become green bean. They will store the green bean for sometimes if not cashing it all directly when currency is needed. Coffee creates many types of sales, all integrated in local market system. From small middlemen (called "popolele"), who are mostly establish direct relation with farmers, to midsize middlemen who collect coffee greanbeans from several small buyers, they also have microsaleswomen called "mamamok" who sells greenbeans in mugs. It is an interesting market system that provide opportunity for many people to take "a pinch" of profit from their business. They call it "profit sharing for many". There are indeed bigsize buyer or middlemen or "popolele besar" who establish connection to send green beans to other region.

Coffee farmers also have oopportunity to sell their product to coffee processing units. Although the quantity needed is small compare to a village productivity, the unit has potency to become an effective part of village market system of local coffee.

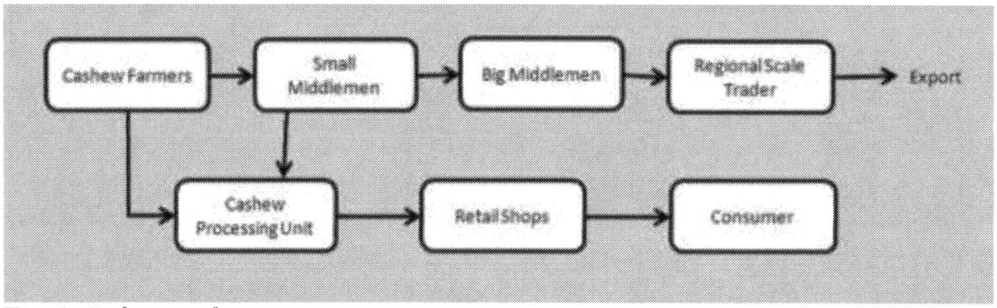

Fig. 4. Cashew market system

Cashew market system is less complicated because it is not a product they consume daily. A complicated processing technology and also the high sales price of raw cashew motivate them to sell rather than consume. Middlemen in cashew market system work based on free competition, although there are different levels of middlemen actively taking part in the supply chain. Small middlemen reside in villages, whereas bigger one cover more than one village. They collect the product up to certain quantity before selling it to commodity trader which directly connected with big companies dealing with export goods. Unlike the case of coffee market system, cashew market chain confirms that farmers as producers is an isolated entity who may loose opportunity to gain fair share of their work.

Although coffee and cashew are considered as global commodities, farmers have no direct access. Being smallholders, their traditional management is not only applied in caring for the plant and plantation, but also in marketing the product. As smallholder farmers, fulfilling subsistence need is their main concern. It is expressed by the way they sell the product. Unaware of the high value of crops, farmers usually sell in small portion, enough to have money for couple of days life. Only at certain occassion, they do sales in bigger quantity, although by global-scale business standard, it is still considered far from sufficient. By doing so, farmers only receive small amount of margin.

In reference on how farmers establish capability to manage the ecosystem, from crop planting to harvest the produce, and successfully partaking on the market system, we envisage a better scheme to increase not only the margin received but also bargaining power in the market. In order to meet global standard, a strategy need to be developed aiming at supporting and protecting the well being of rural community. However the idea should be carefully implemented not to cause certain parties being left out. One approach to accomplishing this for the small farmer is through the development and running of cooperatives and appropriate supply chains (Jang & Klein, 2011). Direct sales to small middlemen place the farmers in uncertain situation. They do not have access to price information, and give in to market fluctuation scenario played by the middlemen. Creation of off farm economic activities based on local products will be a advantageous. However, both crops retain different characters to be carefully referred. Based on series of field activities performed with coffee and cashew farmers in two different villages, we could derive a conclusion that is not difficult to transfer processing technology, what is challenging is to establish an attitude of technology application to produce standardize product.

However, it is a challenge worth taking when we aim at achieving rural prosperity. Appropriate supports should be provided, not only in part of increasing land productivity through strengtheing local knowledge, but also providing support in technology advancement and marketing system (Molua, 2005). The efforts should be designed to attain value add shared in equity.

3.4 Rural Development Program : In reality

The potential of agroforestry as support to livelihood is found in many areas such as Malawi, Bangladesh (Ndalama et al., 2015; Hanif et al., 2018) although socio-ecological hindrances to success are unavoidably found (Hanke et al., 2017). Evidence from the case studies conducted in two villages namely Kadi Roma and Kalaki Kambe, representing predominantly coffee and cashew agroforestry are potential to be reffered as local natural resources assets for rural development program. However, the idea of promoting local natural resource is not in the agenda. Farmers, who are the managers of ecosystem, seems to think, work, and act alone. There seems no connection between village authorities, nor local government officers from relevant sector. Crop cultivation in rural ecosystem obviously an expression of local-driven-innovation materializing from long experience in managing the environment. Crop handling and processing technology implemented are mostly traditional, sufficient to meet the local market standard. Efforts to enhance level of technology acquired was initiated by local government program but lacking in continuous assistance to establish sustainable capacity.

The fact that Act 6/2014 recommend optimal use of local resources by means application of appropriate technology is not well translated to reality. Although the concept is confirmed by LQ analysis, acknowledgement to potency of local resource specifically the local driven innovation i.e. agroforestry ecosystem is not a concern. Although specialization in one of the main resources can boost inequality of development between regions (Hidayat, 2014), as local economy and social assets, it worth promoting (Zega et al., 2013). Inequality can be avoided by establishment of a strategy to enhance value of local products and create linkage among sectors to ensure shared benefits across rural regions (Hidayat, 2014).

The starting points of major mismatch between visions and reality may be due to limited capacity of local village rulers to interpret Act 6-2014. Socialization and assistance from their superior direct them to focus on infrastructure development such as rural roads and the like. Its lack of thorough understanding of the opportunity provided by rural development policy measures will lead them to missing the prospect of achieving condition of just prosperity

4 Conclusion

In reference to the ecosystem potency, it is concluded that socio-ecological innovation stemmed from agroforestry ecosystem in pursue of rural prosperity might be achieved by means of nurturing local innovation system as valuable assets to development i.e. technology of agroforestry ecosystem management to include seed selection, plant, maintenance, harvest; institutionalizing local market system to ensure fair share of benefit accross rual community; promotion of collective action in many aspects of activities i.e. agroforestry resource management, post harvest technology, sales and marketing; enhancing the role of government to provide necessary supports to ensure sustainable rural development

4.1 Acknowledgement

This study is part of the work supported by National Priority Program "Appropriate Technology for Rural Economy Development based on Coffee Agroindustry in Sumba Barat Daya District 2019" conducted by Center for Appropriate Technology Indonesian Institute of Sciences. We thank our colleagues Mirwan A. Karim, Arie Sudaryanto who have provided great support to our field research and also Hari Siswoyo Aji who has constantly provided valuable insights.

References

1. S. Beder, Environ. Conserv. (2011)
2. Carolina. (2016). Peran strategis agroekosistem kopi rakyatdi Kabupaten Sumba Barat Daya Provinsi Nusa Tenggara Timur (The strategic role of smallholders coffee agroecosystem for South West Sumba District of East Nusa Tenggara Province). Jurnal Inovasi, 13, 65-74.
3. W. J. Creswell, *Research Design. Qualitative, Quantitative, and Mixed Methods Approach*, 4th ed. (United States: SAGE Publications, Los Angeles, 2014)

4. M. A. Hanif, R. M. Roy, M. S. Bari, P. C. Ray, M. S. Rahman, and M. F. Hasan, Small-Scale For. **17**, 505 (2018)
5. H. Hänke, J. Barkmann, C. Coral, E. Enforskaustky, and R. Marggraf, Ecol. Soc. (2017)
6. M. Hidayat, Econ. J. Emerg. Mark. (2014)
7. Y. Hou, S. Zhou, B. Burkhard, and F. Müller, Sci. Total Environ. (2014)
8. W. Jang and C. M. Klein, Ann. Oper. Res. (2011)
9. U. P. Jawang, J. Pengelolaan Sumberd. Alam Dan Lingkung. (Journal Nat. Resour. Environ. Manag. **8**, 396 (2018)
10. K. Knickel, M. Redman, I. Darnhofer, A. Ashkenazy, T. Calvão Chebach, S. Šūmane, T. Tisenkopfs, R. Zemeckis, V. Atkociuniene, M. Rivera, A. Strauss, L. S. Kristensen, S. Schiller, M. E. Koopmans, and E. Rogge, J. Rural Stud. (2018)
11. R. Leakey, Agrofor. Today **8**, 5 (1996)
12. R. W. Long, Publius J. Fed. **17**, 15 (1987)
13. F. W. Mercer, D. E., Frey, G.E. & Cubbage, *Handbook of Forest Economics* (Earthscan from Routledge, New York, 2014)
14. E. L. Molua, For. Policy Econ. (2005)
15. P. K. R. Nair, For. Ecol. Manage. (1991)
16. E. Ndalama, G. Kamanga-Thole, and E. Missanjo, Int. J. For. Hortic. **1**, 5 (2015)
17. G. N. Njurumana, D. Marsono, Irham, and R. Sadono, J. Mns. Dan Lingkung. (2014)
18.. N. Pouw and J. Gupta, Curr. Opin. Environ. Sustain. (2017)
19. M. Rivera, K. Knickel, I. de los Rios, A. Ashkenazy, D. Q. Pears, T. Chebach, and S. Šūmane, J. Rural Stud. (2018)
20. M. Schouten, M. Van der Heide, and W. Heijman, EAAE 113 Semin. Belgrade, Repub. Serbia, December 9-11 (2009)
21. A. Soleh, J. Sungkai (2017)
22. R. Tarigan, Ekonomi Regional: Teori Dan Aplikasi (Bumi Aksara, Jakarta, 2007)
23. L. Weiwei, L. Wenhua, L. Moucheng, and A. M. Fuller, J. Resour. Ecol. (2014)
24. S. B. Zega, A. Purwoko, and T. Martial, Peronema For. Sci. J. **2**, 157 (2013)

The Development of MSMEs International Networks through Internationalization

M. ELFAN KAUKAB

Sains Al-Quran Jawa Tengah in Wonosobo, Indonesia, Faculty of Economics Universitas

Abstract. This article studies the effects of relationship bonding, entrepreneurial cognition, and market dynamism towards internationalization and their impacts in MSMEs network growth. This article examines hypotheses derived from theory and research employing structural model analysis in 150 exporting-MSMEs in Central Java province, Indonesia. This research reveals that relationship bonding influences internationalization while entrepreneurial cognition and market dynamism do not. Internationalization affects the MSMEs network growth. This leads to the finding that MSMEs internationalization in Central Java, Indonesia is promoted more by social factors instead of individual and macro environment, while network growth is stimulated by individual and macro environment instead of social factors. Social determinant directs internationalization and individual and macro determinant regulates network expansion. The implementation of network theory and ecosystem management theory to comprehend MSMEs internationalization and network growth. This research provides supports to encourage educators and policy makers upgrade the skills of MSMEs practitioners in developing international business. This research compares social perspective, individual perspective, and macro environment in an empirical study in a developing country. This research offers solution to comprehend internationalization and network growth involving market dynamism and entrepreneurial cognition as well as bonding relationship as determinants.

1 INTRODUCTION

Internet development benefits MSMEs in marketing their products overseas. Approximately 75% of economic impact provided by internet is taken up by MSMEs traditional companies that only sell their products and services on the internet by employing its high productivity (Manyika and Roxburgh, 2011). The MSMEs vast development in a country is essential as MSMEs are tools to build entrepreneurship, innovation route, industrial dynamic source, and work creator instrument (Ahmed, 2015). These are the reasons for broader intensive MSMEs development. This development can be achieved through internationalization. Internationalization is state boundaries crossing in an organization's growth process (Child and Rodrigues, 2005).

There are many variables proven to influence MSMEs internationalization. They are grouped into strategic variables (like opportunities) (Dunning, 2001), economic variables (Ghoshal, 1987), and non-economic variables (George, Wiklund and Zahra, 2005).

Nevertheless, the previous studies on non-economic factor are divided between whether MSMEs internationalization is influenced by cognitive or behavioral factor only, and whether it is also influenced by social and individual factor (Dew, Grichnik, Mayer-haug, Read, and Brinckmann, 2014; Rae, 2005). Those are still unanswered and require further studies. Some research have attempted to comprehensively unravel the question, but indirectly employing cognitive construct (Kyvik, Saris, Bonet, and Felício, 2013; Zahra, Korri, and Yu, 2005) and behavioral construct (Moen and Servais, 2002; Reihlen and Apel, 2007). Research involving individual factors (Zahra et al, 2005; Kyvik et al, 2013) and social factors (Ellis, 2011; Arenius, 2002; Chetty and Agndal, 2006) have also been conducted, although without comprehensive construct such as entrepreneurial cognition representing individual and social bonding factors as collective factors. Therefore, more comprehensive and inclusive research are needed when testing factors influencing internationalization and MSMEs network development.

Furthermore, internationalization theory has not been broadened to consider three things at once: social factors, rational factors, and contingency factors. As far as the writers know, these come from three bigger theories namely network development (Johanson and Mattson, 1987), FDI (Foreign Direct Investment) theory (Caves, 1971), and dynamic capability theory (Teece, Pisano, and Shuen, 1997). Employing structural survey, this article aims to explore the correlation of bonding, entrepreneurial cognition, market dynamism, internationalization, and network development in MSMEs that have internationalized in Central Java province, Indonesia. There are three contributions emanated from this research. First, this research contributes in internationalization literature by showing how variables focusing on individuals, groups, and environment affect MSMEs internationalization work. This research is also instrumental in literature by providing empirical validation for the correlation of social factors, rational factors, and contingency factors in internationalization.

Second, this research operates network theory in bonding variable, FDI theory in entrepreneurial cognition, and dynamic capability in market dynamism. Third, based on Indonesian internationalized MSMEs samples, this research explores the effects of those variables to internationalization and network development, which leads to validating findings obtained in industrialized countries (for example Knight and Liesch, 2016; Vatamanescu, Zbuchea, Pinzaru, and Andrei, 2016; Veugelers, 2008). This eventually provides description of internationalization behavior in developing countries. It is expected that this research can act as a base for future research exploring MSMEs internationalization, especially in developing countries.

2 THEORETICAL FRAMEWORK

The social aspect of entrepreneurial construct related to internationalization generally employs social capital variable (Arenius, 2002; Chetty and Agndal, 2007; Ellis, 2011). It is generally understood that good social capital generates positive effect to internationalization. One form of social capital is bonding. Bonding is a behavioral variable as it is derived from the application of some techniques such as customization, information sharing, and managerial bonding development (Davcik and Sharma, 2016). Bonding can also be a collective variable because it involves perspective of an existing social network.

Literature in marketing classify three types of bonding namely economic bonding, social bonding, and structural bonding (Harrison-walker and Neeley, 2012). Economic bonding is the amount of money and time spent to build bonding, social bonding is virtual interactions among the members to build connections, and structural bonding is commitment given to the network that makes it difficult to leave the bonding (Quinton and Harridge-march,

2010). The bonding applied in this research is social bonding, to make it in line with social capital theory.

Bonding capital is a part of social capital that better suits MSMEs, as MSMEs have relatively weak bargain position that forces them to rely on bonding mechanism in aligning interest in behavioral-based contract (Arenius, 2002). Nevertheless, there is only one research investigating the correlation of bonding capital and internationalization (Vatamanescu et al., 2016). The research reveals that bonding capital affects positively towards MSMEs internationalization. Overall literatures support the notion that social capital positively correlates to internationalization speed (Arenius, 2002).

Apart from using social capital theory, bonding can also be analyzed by network theory. In accordance with network theory, internationalization is strongly affected by network of formal and informal bonding (Coviello and Martin, 1999). Bonding is a form of formal and informal network meant by network theory. Thus, the following hypothesis is proposed:

Hypothesis 1: Bonding will show positive correlation with MSMEs internationalization

Hypothesis 2: Bonding will show positive correlation with MSMEs network development.

The concept of entrepreneurial cognition is a new one developing in entrepreneurship studies (Mitchell et al., 2002). Studies in entrepreneurial cognition have emerged in 1976, but this was still implicit in nature that it faced conceptual challenge (Grégoire, Corbett, and Mcmullen, 2011). Phenomena as learning failure and problem solving in business are clearly cognitive phenomena in entrepreneurship (Corbett, Neck, and Detienne, 2007). Other similar things are how one learns how to become an entrepreneur and correlates his knowledge with his business intention (Pihie, Bagheri, and Sani, 2013). However, the concept of entrepreneurial cognition has just been developed to include all cognitive aspects that have been learned in the previous entrepreneurship studies.

A basic question in entrepreneurial cognition is how an entrepreneur thinks (Tang, Kacmar, and Busenitz, 2012). In this case, several constructs have been proposed like entrepreneurial perception, entrepreneurial intention, entrepreneurial efficacy, faith in entrepreneurship, entrepreneurship study, entrepreneurial capacity, and so forth (Acs and Audretsch, 2010; McKenzie, Ugbah, and Smothers, 2007). It is still in debate whether this cognition is formed individually or collectively (Dew et al., 2014) and whether cognition, without including behavior, can lead to internationalization (Rae, 2005).

Research in entrepreneurial cognition have revealed that there is correlation between entrepreneurial cognition and internationalization (Knight and Liesch, 2016). Entrepreneurial cognition is regarded as how an entrepreneur proactively takes steps in internationalization, especially one who has previously had experience in international business be it personally or as an employee in multinational company (Aspelund, Madsen, and Moen, 2007; Luostarinen and Gabrielsson, 2006; Mcdougall, Oviatt, and Schrader, 2003; Weerawardena, Sullivan, Liesch, and Knight, 2007). Entrepreneurial cognition can be seen as individual characteristic determining if an MSME takes internationalization step from the beginning (Aspelund et al., 2007; Luostarinen and Gabrielsson, 2006; Mcdougall et al., 2003). Apart from experience, specialized knowledge seems to play role in assisting internationalization (Fan and Phan, 2014; Rialp, Rialp, and Knight, 2005; Weerawardena et al., 2007). Learning orientation, as one cognitive aspects, has also known to affect as internationalization booster (Baum, Schwens, and Kabst, 2011; Freeman, Edward, and Schroder, 2006; Kuivalainen, Sundqvist, and Servais, 2007; Mudambi and Zahra, 2007; Rialp et al., 2005). Entrepreneurial cognition should be seen as capability in supporting internationalization (Rialp et al., 2005). As it is recognized, capability and competence are parts of performance determinants, superior in resource-based view perspective. Moreover, entrepreneurial cognition provides superior market knowledge for entrepreneur, which eventually gives benefits triggering internationalization (Aspelund et al., 2007; Fan and Phan, 2014; Freeman and Cavusgil, 2007; Garcia-Canal, Duarte, Criado, and Llaneza,

2002; Mathews and Zander, 2007; Mcgaughey, 2007; Mudambi and Zahra, 2007; Rialp et al., 2005). Manager with entrepreneurial cognition will be able to understand, recognize, and exploit opportunities existing in international market (Zahra et al., 2005). This ability is highly necessary as rushing internationalization will ruin MSMEs performance (Mudambi and Zahra, 2007). However, some research have indicated that early internationalization does increase capability that boosts MSMEs performance (Autio, Sapienza, and Almeida, 2000; Zahra, Ireland, and Hitt, 2000; L. Zhou, Barnes, and Lu, 2010).

The correlation of entrepreneurial cognition and internationalization can be elaborated by employing FDI theory (Coviello and Martin, 1999). According to this theory, internationalization is a result of rational decision made by entrepreneur in developing his business. Often times this is a gradual decision in which company first runs in local market, then moves to international one. The research in entrepreneurial orientation reveals that this concept positively affects MSMEs development (Wiklund, 1998). A study conducted by Wiklund and Shepherd (2005) found that environmental dynamic and entrepreneurial orientation have roles in promoting MSMEs development. This is in line with resource-based view that human resource is the most important resource in MSMEs development (Wiklund, Patzelt, and Shepherd, 2009).

It can be concluded then, from entrepreneurial cognition studies and FDI theory mentioned earlier, that entrepreneurial cognition will affect MSMEs internationalization. Thus, the following hypotheses are proposed:

Hypothesis 3: Entrepreneurial cognition will show positive correlation with MSMEs internationalization

Hypothesis 4: Entrepreneurial cognition will show positive correlation with MSMEs network development

Market dynamism represents contingency factor which a determinant in MSMEs internationalization and network development. This factor is employed in dynamic capability theoretical model (Atanassova and Clark, 2015), in which a company's dynamic capability is determined by market dynamism (Eisenhardt and Martin, 2000). Dynamic capability is organizational and managerial process of creating, coordinating, integrating, reconfiguring, and altering resources and skill to maintain business and increase competitiveness by adapting to environmental changes (Atanassova and Clark, 2015; Eisenhardt and Martin, 2000; Teece, 2007; Wang and Ahmed, 2007).

Competition creates market dynamism(Stone and Badawy, 2011). Market dynamism is among of external triggers to generate dynamic capability of an organization (Atanassova and Clark, 2015; Cavalcante, Kesting, and Ulhøi, 2011; Du and Banwo, 2015). It is generally known that market changes create more growth opportunities (Wiklund et al., 2009; H. Zhou and de Witt, 2009). The growth opportunity is utilized by MSMEs to introduce new product/service in the market, therefore prompting competitive advantage and yielding better financial result (Wiklund, 1998). This opportunity is also used to expand the market to sell the products and services and therefore creating new networks in many places (Ratten and Dana, 2007). Thus, market dynamism is found in various research as a growth determinant for high profile company (Moncada-paternò-castello and Cincera, 2012; Veugelers, 2008).

Based on vast accepted correlation between market dynamism and MSMEs development, we presume that:

Hypothesis 5: Market dynamism will show positive correlation with MSMEs internationalization

Hypothesis 6: Market dynamism will show positive dynamic with MSMEs network development

Unlike FDI theory, the network theory holds that social bonding contributes role and large influence towards business transactions in internationalization (Johanson and Mattsson,

1987). This theory opposed FDI theory that states internationalization as of rational strategic decision making process (Coviello and Martin, 1999). According to network theory, internationalization is the result of members' behavior pattern inside the network that creates formal and informal networks. Bonding interaction and development establish internationalization of a business. The company in time uses the available network bonding to proactively execute its business (Boojihawon, 2004).

Network theory has been applied in bonding variable. The yielded social capital, as a source of competitive advantage, should be developed to generate more capital. This means that when internationalization has been achieved, there will be more efforts to widen the networks into more areas. Therefore, MSMEs develop into international market. This has led to the following hypotheses:

Hypothesis 7: MSMEs internationalization will show positive correlation with MSMEs network development

3 METHODS

The population of this research is MSMEs in Central Java province that have exported their products. There are 973 MSMEs involve in this research. Their data are obtained from Dinas Koperasi dan UMKM (Cooperative and MSMEs Service) Central Java. Samples are determined by considering the minimum required sample in SEM (Structural Equation Modeling) analysis, which is the data analysis method applied in this research. Hair, Eingle and Sarstedt (2011) state that minimum sample of SEM analysis is ten times the questions in the variable with the most questions. Variable with the most questions here is entrepreneurial cognition variable, consisting of 20 questions. Hence the minimum sample is 20 x 10 = 200.

The researchers managed to obtain 150 valid questionnaire results. Although this seems low, we argue that our samples represent 15,4% of the total exporting-MSMEs population in the location where the study is conducted. This means that the number is quite representative. The detailed samples characteristics are shown in Table 1.

Table 1 MSMEs Characteristics

Characteristics	%	Characteristics	%
The number of foreign partners		Industry	
1-10	93,3	Laminated wood	12,7
11-20	6,7	Rattan craft	0,7
Experience in international market		Bamboo craft	2,7
1-3 years	18,0	Batik craft	19,3
4-7 years	40,7	Eyelash craft	4,7
8-10 years	20,7	Wood craft	54,0
More than 10 years	20,7	Ceramic craft	4,7
The number of export-destination countries		Masks	0,7
1-2	40,0	Traditional hats	0,7

Characteristics	%	Characteristics	%
3-4	37,3	Location	
5-6	16,7	Rural	79,3
More than 6	6,1	Urban	20,7

Data are collected in three different techniques namely email (if the MSMEs have it), direct field visits, and fellow researchers' assistance. When responses have reached 200 or when the entire population has filled out the questionnaires, we begin the data analysis. Different test is carried out to examine the way three methods fill out the questionnaires in order to seek for potential bias. If significant different results present, data collecting method variable needs to be controlled by incorporating it as variable in analyzing the data.

Bonding relationship is measured by using instrument comprises of three related items in connection to MSMEs and their overseas partners. The instrument is adapted from Woodside and Baxter (2015). Those items are strong social bonding among the people involved in business partners' organizations overseas, highly advantageous bonding, and the many shared goals possessed by MSMEs and their overseas partners. They have a 5-option scale ranging from "strongly disagree" (coded 1) to "strongly agree" (coded 5).

Entrepreneurial cognition employs questionnaires by Mitchell, Smith, Seawright, and Morse (2000) developed by Urban (2008), which consist of willingness dimension and ability dimension. Total 20items are used to measure entrepreneurial cognition in this research. They are valued in binary option scale. The first option indicates high entrepreneurial cognition while the second one does not. The more options which are made for the first option means that the higher the entrepreneurial cognition is.

Market dynamism is a two-indicator variable. The specific items applied are adapted from Zhou and de Wit (2009). Dynamic market indicators are measured in 5-point Likert scale, where '1' means 'strongly disagree' and '5' means 'strongly agree.' Meanwhile internationalization indicators are valued in the same 5-point Likert scale but with different options for each item which generally provides composite view for the option taken (for example 1 means 1-20%; 2 means 21-40%, and so on).

MSMEs internationalization is measured in four indicators, including (i) percentage of foreign market profits, (ii) percentage of sales revenue of foreign market, (iii) the number of employees working for foreign market, and (iv) the ability of organization structure to explicitly accommodates international activities (Asugman, Johnson, and Mccullough, 1997).

Meanwhile international network utilizes instrument of Mudambi and Zahra (2007). MSMEs network development variable is measured in three items: (i) the number of export destination country growth, (ii) the number of companies cooperating with the MSMEs growth, and (iii) MSMEs export intensity growth.

Respondents are also asked to provide information related to age (under 35 years old; between 35-55 years old; more than 55 years old); the number of foreign partners (1-10; 11-20; 21-30; 31-40; more than 40); education (elementary school; junior high school; high school/vocational school; diploma; graduate; post graduate); field of study (economics; others); and experience in international business (1-3 years, 4-7 years, 8-10 years, more than 10 years). Data analysis of this research is carried out by employing multivariate statistical techniques Structural Equation Modeling (SEM).

4 RESULT

Confirmatory analysis is conducted to check the reliability of all the indicators using their loading. As good reliability and validity are hard to reach for entrepreneurial cognition variable, it is reduced into three items with the biggest factor loads. The same thing happens to market dynamism variable, which is not reliable. Since it contains only two items, we separate this variable into two single items. Internationalization variable is also reduced into one item out of four previously planned. Relationship bonding and growth variables are not reduced because they are valid and reliable. Table 2 shows that all indicators have reached the accepted value as they have passed the defined threshold. VIF values of each construct are less than 5, indicating the absence of multicollinearity problem (see Table 2).

Table 2 CR, Cronbach's alpha, AVE, and VIF for research constructs

Variables	CR	Cronbach's alpha	AVE	VIF
Relationship bonding	0,761	0,751	0,522	1,104
Entrepreneurship cognition: willingness script	0,847	0,843	0,651	1,216
Entrepreneurship cognition: ability script	0,737	0,685	0,515	1,166
Market dynamism: customer needs	1,000	1,000	1,000	1,171
Market dynamism: products change	1,000	1,000	1,000	1,113
Internationalization	1.000	1.000	1,000	1,083
Network growth	0,746	0,719	0,510	1,285
Number of foreign partners	1.000	1.000	1.000	1,032
International market experience	1.000	1.000	1.000	1,066

Figure 1 displays path coefficient (β), p value, and R2 value of hypothesized relationships in this research. Model suitability values of the research are RMSEA = 0,058; GFI = 0,909; AGFI = 0,841; CFI = 0,931; NFI = 0,829; and TLI = 0,893. Three (AGFI, NFI, and TLI) out of the six parameters are less than the minimum threshold of 0,900. Nevertheless, this number is still marginal as it is in the range of 0,800 – 0,900. Thus, the model is accepted.

Figure 1 Structural model

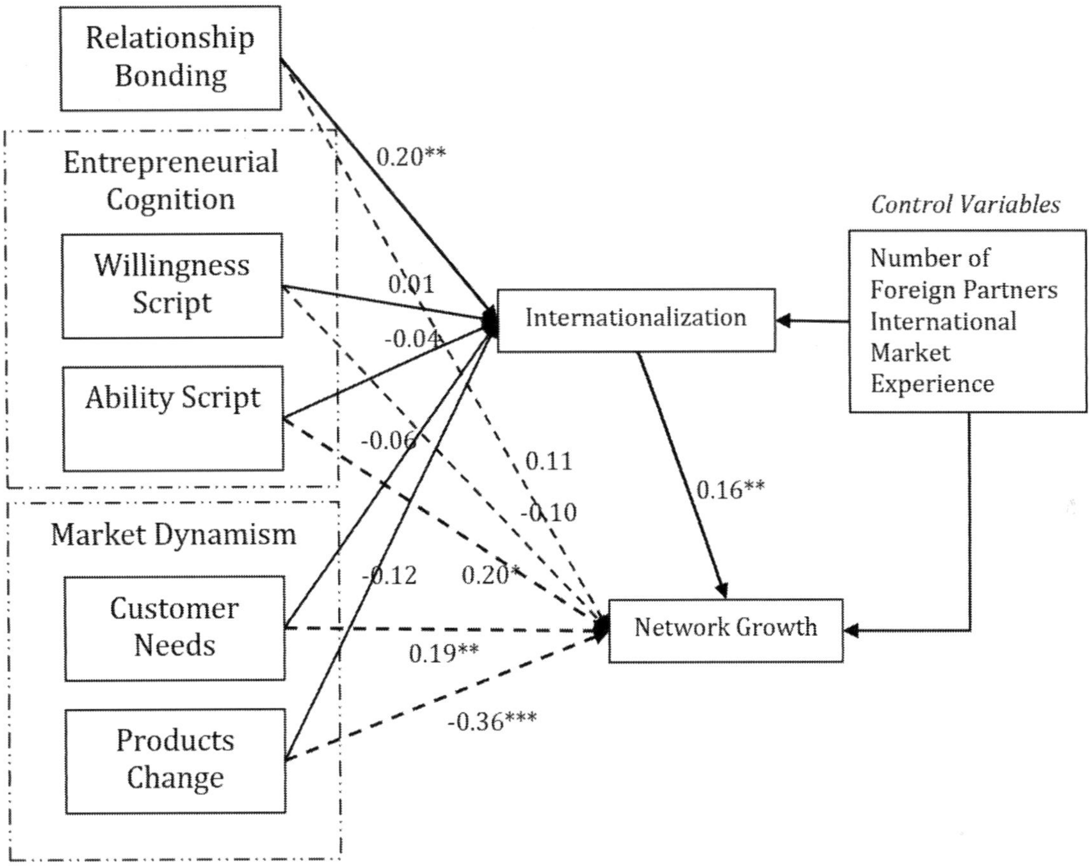

*** p ≤ 0.01; ** p ≤ 0.05; * p ≤ 0.10

This research examines seven hypotheses to predict MSMEs internationalization and growth by employing relationship bonding variable, entrepreneurial cognition variable, and market dynamism variable. In line with Figure 1, the result accepts hypothesis 1 and hypothesis 7. Hypothesis 4 and 6 are relatively mixed due to two accounted variables: entrepreneurial cognition and market dynamism, which each consists of two dimensions and only one dimension shows relationship in the hypothesized direction. Meanwhile the result rejects hypothesis 2, 3, and 5.

Bonding relationship, entrepreneurial cognition, and market dynamism variables are proven to predict 6,9% variation in MSMEs internationalization. Meanwhile these variables together with internationalization predict 30,6% variant of network growth.

To have knowledge on the model's predictive validity, we calculate the value of Stone-Geisser Q2. A less than 0 Q2 value indicates that the model has relevant predictive value(Hair Jr, Hult, Ringle, and Sarstedt, 2016). The Q2 value of this model is 0,354 and thus proves big predictive relevance.

5 DISCUSSION

For MSMEs samples in Central Java, Indonesia, the relationship bonding variable is the strongest variable influencing internationalization. On the other hand, entrepreneurial cognition and market dynamism variable play roles in network growth. The insignificant correlation between relationship bonding and network growth does not accord to the view that social capital will speed up internationalization (Arenius, 2002). Nevertheless, relationship bonding does create positive correlation to internationalization, as revealed by Vatamanescu et al (2016).

The plausible explanation of the insignificant effect of relationship bonding towards network development is that MSMEs operate in close network which focuses only on the existing relationships (Behari, 2016). Since the consumers' focus is on this, the network development will depend on the consumer-triggered growth instead of on the MSMEs' themselves. There has been internal growth in the existing network which happens qualitatively. This confirms the finding in management study where tight relationship bonding results in mutually relationship of the parties involved (Nuryakin, Aryanto, and Setiawan, 2018). That is worth seeking as it promises large advantages in business relationship (Paduraru et al., 2016). In the meantime, when network growth is on the go, a commitment needs to be made which means that relationship bonding should be developed with new partners. There is a need for cultural investment (Billore and Borg, 2014) in it and this does not depend on the relationship with the existing partner. This is related more to the internationalization, entrepreneurial cognition, and market dynamism, which show significant correlation in network growth in this research. It will be difficult for MSMEs to manage complex network in line with the development of their networks (Boojihawon, 2004), while the existing relationship bonding has guaranteed well growth.

This research has also found that internationalization is only significantly influenced by relationship bonding which is a social factor. Cognitive variable in terms of entrepreneurial cognition does not affect internationalization. Market dynamism variable, which is a contingency variable, does not correlate significantly to internationalization. This means that of the three factors: social, rational, and contingency, only the social factor has been proven to promote MSMEs internationalization, at least in MSMEs sample in Central Java. This finding reaffirms social perspective of MSMEs internationalization (Ellis, 2011; Arenius, 2002; Chetty and Agndal, 2006).

Although market dynamism does not affect internationalization, it does influence the network growth. However, the two variables employed yield different results. Product change variable negatively affects the network development, unlike customer needs variable that positively influences it. The reasonable clarification of the negative effect towards market dynamism in product change dimension is that product change disrupts the existing pattern of MSMEs products. This in turn will discourage MSMEs to develop their networks. Research conducted in Indonesia also reveal that MSMEs prefer to maintain the existing profit instead of do expansion (Ikhsan, Almahendra, and Budiarto, 2017). MSMEs respond by reducing customers and focusing on costumers who demand their products in traditional form which is their specialty. In other words, MSMEs are unable to cope with product change requested by dynamic market. This confirms the finding that the bigger the company is, the more adaptive it is to cope with dynamic environment(Yu, Lumpkin, Parboteeah, and Stambaugh, 2019). The research shows that highly dynamic environment may negatively influence company's performance (Davis, Eisenhardt, and Bingham, 2009). Product change is different from customers' taste change, because customers might want something which is not totally new – a variation of the old product. This is why the other variable of market dynamism, namely customer needs, positively and significantly influences network development.

6 CONCLUSION AND IMPLICATION

Overall, the result of this research supports the network theory because relationship bonding significantly influences internationalization. Nevertheless, since international network growth can also be seen as internationalization variable, supports are also provided for FDI theory and dynamic capability theory owing to the finding that entrepreneurial cognition and market dynamism affect the MSMEs' international network growth.

This finding has important implication both theoretically and practically. First, this research adds to the existing literature that market dynamism contributes in MSMEs development in complex manner (Davis et al, 2009). Dynamic market does not necessarily encourage MSMEs development as the MSMEs may opt to hold out, withdraw, or follow the changes by innovating. It means that market dynamism is a strategic bait for MSMEs (Seo and Chae, 2016). This explains why this research as well as other research in developing countries (Donkor, Donkor, and Kwarteng, 2018) find that market dynamism will not affect MSMEs performances (Donkor et al, 2018). This can be assessed from ecosystem management theory perspective which affirms the strategic role of market dynamism in shaping internal dynamic of a company to manage their resources (Crielaard, Wubben, and Omta, 2018).

The second theoretical implication relates to theoretical contestation in MSMEs internationalization. Since internationalization in this research is influenced solely by relationship bonding, the finding supports the network theory more (Coviello and Martin, 1999), than FDI theory (Caves, 1971) or dynamic capability theory (Teece et al, 1997). Eventually, empirical validation can only be provided for social factor and not for rational factor of FDI perspective and contingency factor of dynamic capability perspective.

Third, this research encourages review towards entrepreneurial cognition. It shows that only ability aspect can promote international entrepreneurial activities. Willingness aspect alone cannot generate either internationalization or international network growth. This means that sharp theoretical difference is necessary to separate willingness from ability as two different constructs in entrepreneurial studies. The gap between willingness and ability is a common thing found in economic practice (Ashraf, Glaeser, and Ponzetto, 2016). Some entrepreneurs have willingness but have no ability, while the other have the ability but without willingness(Richards, Kammerlander, and Zellweger, 2019). In this case, entrepreneurial cognition as a variable with willingness and ability dimension might find it hard to produce valid construct. Entrepreneurial cognition variable can only be formed as a unity when the motivation gap separating willingness and ability (Kotlar, de Massis, Frattini, and Kammerlander, 2016) can be closed.

In the meantime, the practical implication emerged is that people of Central Java should build a belief that they are able to develop their MSMEs businesses to international level. Just willingness is not enough to create MSMEs business growth. This research reveals that it is the ability script, not the willingness script, which creates MSMEs network growth internationally. To develop ability, one needs to take concrete steps by expanding experience or by taking training and education. Therefore, MSMEs administrators should spare their times to join entrepreneurial trainings held by government, especially the ones focus on ability development, instead of just be aware of it. This kind of training helps to develop their ability in effectively handling international business (Vahidnia et al., 2019).

This means that other stakeholders like educators and policy makers should also develop training and education program that promote MSMEs practitioners' learning activity(Keane, 2016). These trainings and educations have been implemented in various ministry programs such as PKBM (Community Learning Activity Center) from the Ministry of Education, center of business incubator from Ministry of Industry, national entrepreneurship movement from the Ministry of Cooperative and MSMEs, and also in

ministry of specific sector such as young generation of agriculture from the Ministry of Agriculture (Hadi, Suardi, and Cahaya, 2015; Mirzanti, Simatupang, and Larso, 2015). However, these programs are not comprehensive yet and require more concrete policy in minister level to generate productive entrepreneurial ecosystem as an effort of internationalization (Hermanto and Suryanto, 2017).
Finally, this research has some limitations. First, the finding is limited for MSMEs in Central Java. Most of the MSMEs engaged in this study are those working on handicraft sectors. This kind of MSMEs largely depends on tourism sector to attract consumers. Second, research involving market dynamism aspect are complex ones owing to the connection to strategic management (Donkor et al, 2018). The mechanism joining entrepreneurial cognition, relationship bonding, and network growth might be more complex and can only be reached when the research design employs qualitative method. Next research should use mixed method to reveal the mechanism as well as causal relationship involving mediator and moderator variables in the model, especially for variables in relation to strategic management. Third, this research might have been bias as the data used are those collected from MSMEs practitioners themselves (Ikhsan et al, 2017). Moreover, a cross-sectional design does not facilitate the belief on the causality hypothesized in this research. Thus, the subsequent research needs to use longitudinal design and employ secondary data such as the company's reports.

7 REFERENCE

Acs, Z. J., and Audretsch, D. (2010). Introduction to the 2nd Edition of the Handbook of Entrepreneurship Research. In Z. J. Acs and D. Audretsch (Eds.), Handbook of Entrepreneurship Research (2nd ed., pp. 1–22). New York: Springer. https://doi.org/10.1007/978-1-4419-1191-9

Ahmed, M. (2015). Entrepreneurship and Economic Development : A Discourse on Selected Conceptual Issues Entrepreneurship. Daffodil International Journal of Business and Economics, 9(1), 187–193.

Arenius, P. M. (2002). Creation of Firm-Level Social Capital, its Exploitation, and the Process of Early Internationalization. Helsinki University of Technology Institute of Strategy and International Business.

Ashraf, N., Glaeser, E. L., and Ponzetto, G. A. M. (2016). Infrastructure, Incentives and Institutions (No. 21910). Massachusetts.

Aspelund, A., Madsen, T. K., and Moen, O. (2007). A review of the foundation , international marketing strategies , and performance of international new ventures. European Journal of Marketing, 41(11/12), 1423–1448. https://doi.org/10.1108/03090560710821242

Asugman, G., Johnson, J. L., and Mccullough, J. (1997). The Role of After-Sales Service in International Marketing. Journal of International Marketing, 5(4), 11–28.

Atanassova, I., and Clark, L. (2015). Social Media Practices in SME Marketing Activities: A Theoretical Framework and Research Agenda. Journal of Customer Behaviour, 14(2), 163–183.

Autio, E., Sapienza, H. J., and Almeida, J. G. (2000). Effects of age at entry , knowledge intensity , and imitability on international Growth. Academy of Management Journal, 43(5), 909–924.

Baum, M., Schwens, C., and Kabst, R. (2011). International as opposed to domestic new venturing : The moderating role of perceived barriers to internationalization. International Small Business Journal, 31(5), 536–562. https://doi.org/10.1177/0266242611428343

Behari, D. B. (2016). Engagement Pattern of Users in Digital and Social Media Marketing DSMM A Service Industry Perspective. KIIT University.

Billore, G., and Borg, J. E. (2014). Cultural entry barriers for SMEs: An exploratory study of cultural entry barriers for foreign SMEs entering Japan. Linnaeus University.

Boojihawon, D. (2004). International entrepreneurship and network relationships: the international marketing communications sector. In M. Jones and P. Dimitratos (Eds.), Emerging Paradigms in International Entrepreneurship (pp. 217–248). Cheltenham: Edward Elgar Publishing.

Cavalcante, S., Kesting, P., and Ulhøi, J. (2011). Business model dynamics and innovation : (re) establishing the missing linkages. Management Decision, 49(8), 1327–1342. https://doi.org/10.1108/00251741111163142

Chetty, S., and Agndal, H. (2007). Social Capital and Its Influence on Changes in Internationalization Mode Among Small and Medium-Sized Enterprises. Journal of International Marketing, 15(1), 1–29.

Child, J., and Rodrigues, S. B. (2005). The Internationalization of Chinese Firms : A Case for Theoretical Extension ? Management and Organization Review, 1(3), 381–410.

Corbett, A. C., Neck, H., and Detienne, D. R. (2007). How Corporate Entrepreneurs Learn from Fledgling Innovation Initiatives: Cognition and the Development of a Termination Script. Entrepreneurship Theory and Practice, 829–852.

Coviello, N. E., and Martin, K. A. (1999). Internationalization of Service SMEs : An Integrated Perspective from the Engineering Consulting Sector. Journal of International Marketing, 7(4), 42–66.

Crielaard, J. P., Wubben, E. F. M., and Omta, O. S. W. F. (2018). The Fit Between Employee Roles and Market Dynamism: An Ecological Management Perspective. Revista de Administracao Mackenzie, 19(2). https://doi.org/10.1590/1678-6971/eRAMR180114

Davcik, N. S., and Sharma, P. (2016). Marketing resources , performance , and competitive advantage : A review and future research directions. Journal of Business Research, 69(12), 5547–5552.

Davis, J. P., Eisenhardt, K. M., and Bingham, C. B. (2009). Optimal Structure, Market Dynamism, and the Strategy of Simple Rules. Administrative Science Quarterly, 54, 413–452. https://doi.org/10.2189/asqu.2009.54.3.413

Dew, N., Grichnik, D., Mayer-haug, K., Read, S., and Brinckmann, J. (2014). Situated Entrepreneurial Cognition Situated Entrepreneurial Cognition. International Journal of Management Review. https://doi.org/10.1111/ijmr.12051

Donkor, J., Donkor, G. N. A., and Kwarteng, C. K. (2018). Strategic planning and performance of SMEs in Ghana The moderating effect of market dynamism. Asia Pacific Journal of Innovation and Entrepreneurship, 12(1), 62–76. https://doi.org/10.1108/APJIE-10-2017-0035

Du, J., and Banwo, A. (2015). Promoting SME Competitiveness : Lessons from China and Nigeria. American Advanced Research in Management, 1(1), 1–12.

Eisenhardt, K. M., and Martin, J. A. (2000). Dynamic Capabilities: What Are They? Strategic Management Journal, 21, 1105–1121.

Ellis, P. D. (2011). Social ties and international entrepreneurship : Opportunities and constraints affecting firm internationalization. Journal of International Business Studies, 42, 99–127. https://doi.org/10.1057/jibs.2010.20

Fan, T., and Phan, P. (2014). International New Ventures: Revisiting the Influences behind the "Born-Global" Firm. Journal of International Business Studies, 38(7), 1113–1131.

Fornell, C., and Larcker, D. F. (1981). Evaluating Structural Equation Models with Unobservable Variables and Measurement Error. Journal of Marketing Research, 18(1), 39–50.

Freeman, S., and Cavusgil, S. (2007). Toward a Typology of Commitment States Among Managers of Born-Global Firms : A Study of Accelerated Internationalization. Journal of International Marketing, 15(4), 1–40.

Freeman, S., Edward, R., and Schroder, B. (2006). How Smaller Born-Global Firms Use Networks and Alliances to Overcome Constraints to Rapid Internationalization. Journal of International Marketing, 14(3), 33–63.

Garcia-Canal, E., Duarte, C., Criado, J., and Llaneza, A. (2002). Accelerating international expansion through global alliances : a typology of cooperative strategies. Journal of World Business, 37, 91–107.

Grégoire, D. A., Corbett, A. C., and Mcmullen, J. S. (2011). The Cognitive Perspective in Entrepreneurship : An Agenda for Future Research. Journal of Management Studies, 48(6), 1443–1478. https://doi.org/10.1111/j.1467-6486.2010.00922.x

Hadi, C., Suardi, I., and Cahaya, A. (2015). Entrepreneurship and Education : Creating Business Awareness for Students in East Java Indonesia. Procedia - Social and Behavioral Sciences, 177(July 2014), 459–463. https://doi.org/10.1016/j.sbspro.2015.02.396

Hair Jr, J. F., Hult, G. T. M., Ringle, C. M., and Sarstedt, M. (2016). A primer on partial least squares structural equation modeling (PLS-SEM) (2nd Edition). Thousand Oaks: Sage Publications.

Harrison-walker, L. J., and Neeley, S. E. (2012). Customer Relationship Building on the Internet in B2B Marketing: A Proposed Typology. Journal of Marketing Theory and Practice, 12(1), 19–35.

Hermanto, B., and Suryanto. (2017). Entrepreneurship Ecosystem Policy in Indonesia. Mediterranean Journal of Social Sciences, 8(1), 110–115. https://doi.org/10.5901/mjss.2017.v8n1p110

Ikhsan, K., Almahendra, R., and Budiarto, T. (2017). Contextual Ambidexterity in SMEs in Indonesia: A Study on How It Mediates Organizational Culture and Firm Performance and How Market Dynamism Influences Its Role on Firm Performance. International Journal of Business and Society, 18, 369–390.

Johanson, J., and Mattsson, L. (1987). Interorganizational relations in industrial systems: a network approach compared with the transaction-cost approach. International Studies of Management and Organization, 17(1), 34–48.

Keane, S. (2016). Entrepreneurial Self-Efficacy (ESE): A Business Model Approach. National University of Ireland.

Knight, G. A., and Liesch, P. W. (2016). Internationalization : From incremental to born global. Journal of World Business, 51, 93–102. https://doi.org/10.1016/j.jwb.2015.08.011

Kotlar, J., de Massis, A., Frattini, F., and Kammerlander, N. (2016). The Paradoxical and Time-Varying Effects of Family Ownership on Absorptive Capacity. Academy of Management Proceedings, 1, 12592.

Kuivalainen, O., Sundqvist, S., and Servais, P. (2007). Firms ' degree of born-globalness , international entrepreneurial orientation and export performance. Journal of World Business, 42, 253–255. https://doi.org/10.1016/j.jwb.2007.04.010

Kyvik, O., Saris, W., Bonet, E., and Felício, J. A. (2013). The internationalization of small firms: The relationship between the global mindset and firms' internationalization behavior. Journal of International Entrepreneurship, 11(2), 172–195. https://doi.org/10.1007/s10843-013-0105-1

Luostarinen, R., and Gabrielsson, M. (2006). Globalization and Global Marketing Strategies of Born Globals in SMOPECs. Thunderbird International Business Review, 48(6), 773–801.

Manyika, J., and Roxburgh, C. (2011). The great transformer : The impact of the Internet on economic growth and prosperity.

Mathews, J., and Zander, I. (2007). The international entrepreneurial dynamics of accelerated internationalization. Journal of International Business Studies, 38(3), 387–403.

Mcdougall, P. P., Oviatt, B. M., and Schrader, R. (2003). A Comparison of International and Domestic New Ventures. Journal of International Entrepreneurship, 1, 59–82.

Mcgaughey, S. L. (2007). Hidden ties in international new venturing : The case of portfolio entrepreneurship. Journal of World Business, 42, 307–321. https://doi.org/10.1016/j.jwb.2007.04.011

McKenzie, B., Ugbah, S., and Smothers, N. (2007). "Who is an Entrepreneur?" Is it Still the Wrong Question? Academy of Entrepreneurship Journal, 13(1), 23–44.

Mirzanti, I. R., Simatupang, T. M., and Larso, D. (2015). Mapping on Entrepreneurship Policy in Indonesia. Procedia - Social and Behavioral Sciences, 169, 346–353. https://doi.org/10.1016/j.sbspro.2015.01.319

Mitchell, R. K., Busenitz, L., Land, T., Mcdougall, P. P., Morse, E. A., and Smith, J. B. (2002). Toward a Theory of Entrepreneurial Cognition: Rethinking the People Side of Entrepreneurship Research. Entrepreneurship Theory and Practice, 93–104.

Mitchell, R. K., Smith, B., Seawright, K., and Morse, E. (2000). Cross-Cultural Cognitions and the Venture Creation Decision. Academy of Management Journal, 43(3), 974–993.

Moen, O., and Servais, P. (2002). Born Global or Gradual Global ? Examining the Export Behavior of Small and Medium-Sized Enterprises. Journal of International Marketing, 10(3), 49–72.

Moncada-paternò-castello, P., and Cincera, M. (2012). Enterprises ' Growth Potential in the European Union : Implications for Research and Innovation Policy. The IUP Journal of Entrepreneurship Development, 9(4), 7–40.

Mudambi, R., and Zahra, S. A. (2007). The survival of international new ventures. Journal of International Business Studies, 38, 333–352. https://doi.org/10.1057/palgrave.jibs.8400264

Nuryakin, Aryanto, V. D. W., and Setiawan, M. B. (2018). Mediating effect of value creation in the relationship between relational capabilities on business performance. Contaduría y Administración, 63(1), 1–21. https://doi.org/10.22201/fca.24488410e.2018.1178

Paduraru, T., Vatamanescu, E.-M., Andrei, A. G., Pînzaru, F., Zbuchea, A., Maha, L. G., and Boldureanu, G. (2016). Sustainability in Relationship Marketing: An Exploratory Model for the Industrial Field. Environmental Engineering and Management Journal, 15(7), 1635–1647.

Pihie, Z. L., Bagheri, A., and Sani, Z. (2013). Knowledge of cognition and entrepreneurial intentions : Implications for learning entrepreneurship in public and private universities. Procedia - Social and Behavioral Sciences, 97, 174–181. https://doi.org/10.1016/j.sbspro.2013.10.219

Quinton, S., and Harridge-march, S. (2010). Relationships in online communities : the potential for marketers. Journal of Research in Interactive Marketing, 4(1), 59–73. https://doi.org/10.1108/17505931011033560

Rae, D. (2005). Entrepreneurial learning : a narrative-based conceptual model. Journal of Small Business and Enterprise Development, 12(3), 323–335. https://doi.org/10.1108/14626000510612259

Ratten, V., and Dana, L. (2007). Internationalisation of SMEs : European comparative studies Mary Han Isabell Welpe. International Journal of Entrepreneurship and Small and Medium Enterprise, 4(3), 361–379.

Reihlen, M., and Apel, B. A. (2007). Internationalization of professional service firms as learning – a constructivist approach. International Journal of Service Industry Management, 18(2), 140–151. https://doi.org/10.1108/09564230710737790

Rialp, A., Rialp, J., and Knight, G. A. (2005). The phenomenon of early internationalizing firms : what do we know after a decade (1993 – 2003) of scientific inquiry ? International Business Review, 14, 147–166. https://doi.org/10.1016/j.ibusrev.2004.04.006

Richards, M., Kammerlander, N., and Zellweger, T. (2019). Listening to the Heart or the Head ? Exploring the "Willingness Versus Ability" Succession Dilemma. Family Business Review, 1–24. https://doi.org/10.1177/0894486519833511

Seo, Y. W., and Chae, S. W. (2016). Market dynamics and innovation management on Performance in SMEs : Multi-agent simulation approach. Procedia Computer Science, 707–714. https://doi.org/10.1016/j.procs.2016.07.060

Stone, A., and Badawy, L. T. (2011). SME Innovators and Gazelles in MENA - Educate, Train, Certify, Compete! MENA Knowledge and Learning, (43), 1–4.

Tang, J., Kacmar, K. M., and Busenitz, L. (2012). Entrepreneurial alertness in the pursuit of new opportunities. Journal of Business Venturing, 27, 77–94. https://doi.org/10.1016/j.jbusvent.2010.07.001

Teece, D. J. (2007). Explicating Dynamic Capabilities: the Nature and Microfoundations of (Sustainable) Enterprise Performance. Strategic Management Journal, 28, 1319–1350. https://doi.org/10.1002/smj

Urban, B. (2008). Entrepreneurial Cognitions: Linking Willingness and Ability Expert Scripts to Self-Efficacy and New Venture Creation Decision. SA Journal of Industrial Psychology/SA Tydskrif Vir Bedryfsielkunde, 34(3), 22–28.

Vahidnia, H., Mitchell, J. R., Smith, J. B., Assaf, A. M., Mitchell, R. K., and Araci, Ö. (2019). Journal of Business Venturing Insights Further exploring international entrepreneurial cognitions : The case of the Middle-East. Journal of Business Venturing Insights, 11, 1–8. https://doi.org/10.1016/j.jbvi.2019.e00112

Vatamanescu, E.-M., Zbuchea, A., Pinzaru, F., and Andrei, A. (2016). The Impact of Relational Capital on SMEs Internationalization: Leveraging Online versus Offline

Business Networking. In European Conference on Knowledge Management (p. 926). Academic Conferences International Limited.

Veugelers, R. (2008). The Role of SMEs in Innovation in the EU: A Case for Policy Intervention? Review of Business and Economics, 53(3), 239–262.

Wang, C. L., and Ahmed, P. K. (2007). Dynamic capabilities : A review and research agenda. International Journal of Management Reviews, 9(1), 31–51. https://doi.org/10.1111/j.1468-2370.2007.00201.x

Weerawardena, J., Sullivan, G., Liesch, P. W., and Knight, G. (2007). Conceptualizing accelerated internationalization in the born global firm : A dynamic capabilities perspective. Journal of World Business, 42, 294–306. https://doi.org/10.1016/j.jwb.2007.04.004

Wiklund, J. (1998). Small Firm Growth and Performance: Entrepreneurship and Beyond. Internationella Handelshögskolan.

Wiklund, J., Patzelt, H., and Shepherd, D. A. (2009). Building an integrative model of small business growth. Small Business Economics, 32, 351–374. https://doi.org/10.1007/s11187-007-9084-8

Wiklund, J., and Shepherd, D. (2005). Entrepreneurial orientation and small business performance : a configurational approach. Journal of Business Venturing, 20, 71–91. https://doi.org/10.1016/j.jbusvent.2004.01.001

Woodside, A. G., and Baxter, R. (2015). Imprinting, Honeymooning, or Maturing: Testing Three Theories of How Interfirm Social Bonding Impacts Suppliers' Allocations of Resources to Business Customers. Australasian Marketing Journal, 23(2), 96–106.

Yu, A., Lumpkin, G. T., Parboteeah, K. P., and Stambaugh, J. E. (2019). Autonomy and family business performance : The joint effect of environmental dynamism and national culture. International Small Business Journal: Researching Entrepreneurship, 37(2), 153–185. https://doi.org/10.1177/0266242618811893

Zahra, S. A., Ireland, D., and Hitt, M. A. (2000). International Expansion by New Venture Firms: International Diversity, Mode of Market Entry, Technological Learning, and Performance. Academy of Management Journal, 43(5), 925–950.

Zahra, S. A., Korri, J., and Yu, J. (2005). Cognition and international entrepreneurship : Implications for research on international opportunity recognition and exploitation. International Business Review, 14(2), 129–146. https://doi.org/10.1016/j.ibusrev.2004.04.005

Zhou, H., and de Witt, G. (2009). Determinants and dimensions of firm growth. Zoetermeer.

Zhou, L., Barnes, B. R., and Lu, Y. (2010). Entrepreneurial proclivity , capability upgrading and performance advantage of newness among international new ventures. Journal of International Business Studies, 41, 882–905. https://doi.org/10.1057/jibs.2009.87

Value of innovation to enhance marketing performance (clarification)

Suliyanto Suliyanto[1*], *Rahab* Rahab[1], and *Mansur* Chadi Mursid[2]

[1]Faculty of Economics and Business, Jenderal Soedirman University, Indonesia
[2]Faculty of Islamic Economics and Business, Institut Agama Islam Negeri (IAIN) Pekalongan, Indonesia

Abstract. The purpose of this paper is to analyze value of innovation as the role mediating in the relationship between NPD Innovation and NPD Performance, and also to clarified in Exploratory Factor Analysis and Confirmatory Factor Analysis. We proposed value of innovation basis on shariah as mediating role in this study developed from diffusion of innovations theory. The questionnaires will be given to only the managers of the board of shariah micro finance in Pekalongan and Banyumas, Indonesia with total 171 responden from 27 LKMS (Baitul Maal wa Tamwil, Baitul Tamwil, Kospin Jasa Shariah, and KSPP Shariah). Value of innovation basis on shariah significantly affect to marketing performance. Research limited in first order, future research can examined in second order research. LKMS adapted value of innovation basis on shariah to get product launch easily. Value of innovation basis on shariah as the new variable and theory concept. Indirect, NPD Innovation positive significantly affect to marketing performance with value of innovation basis on shariah and product launch success are as mediating role.

1 Introduction

Studies have provided various perspectives on the performance effect of product innovativeness, and several scholars have argued that product innovativeness positively affects new product performance (Mishra, Kim, & Lee, 1996; Atuahene-Gima, 1996; Hultink & Robben, 1995; Fang, 2008; Akroush, 2012; Millson, 2013; Bicen, Kamarudin & Johnson, 2014; Santos, Basso, Kimura & Kayo, 2013; Huang & Tsai, 2014; Vinayak & Kodali, 2014; and Warren, 2017). Conversely, several studies have indicated that product innovativeness is negatively associated with new product performance. Buyers may be averse to new products with a high degree of innovativeness because of a heightened potential of social, performance, or financial risks that accompany the purchase of such products (Sethi, 2000). Several studies have also observed that product innovativeness does not influence new product performance (Calantone et. al., 2006), unidentified (Santos, et. al., 2013), whereas other evidence supports the hypothesis that a negative effect occurs (Cooper, 1979; and Fu & Jones, 2008).

* Corresponding author: author@email.org

An idea that is not compatible with the prevalent values and norms of a social system will not be adopted as rapidly as an innovation that is compatible (Rogers, 1983). The adoption of an incompatible innovation often requires the prior adoption of a new value system. An example of an incompatible innovation is the use of contraception in countries where religious beliefs discourage use of birth-control techniques, as in Moslem and Catholic nations. This research explore and examine value of innovation that is compatible with microfinance shariah environment in Indonesia.

2 Literature review

2.1 NPD performance

Godener and Soderquist (2004) identified seven areas of measurement that were related to NPD which are financial performance measurements, customer satisfaction measurements, process management measurements, innovation measurements, strategic measurements, technology management measurements and knowledge management measurements. Sherman et. al., (2005) utilized six performance variables in their investigation. Of these six variables, product prototype development proficiency, product launch proficiency and design change frequency were process-oriented performance variables. While variables like market forecast accuracy and technological core competency fit were grouped as performance competencies, the only standalone variable was product development cycle time. Ledwith and O'Dwyer (2009) reported that new product performance is measured in terms of market-level measures, financial measures, customer acceptance measures, product-level measures and timing measures. Liu et. al., (2005) identifed three performance measures are as follows, (1) new product life cycle; (2) new product sales and profits; and (3) time to market for new product. We use marketing performance measures proposed by Akroush (2012) in the present study by exogen aksen method (Suliyanto, 2018).

2.2 NPD innovation

Innovation is vital to the survival of modern corporations (Ko, To, Zhang, Ngai, & Chan, 2011). Rogers (1983) defined an innovation as an idea, practice, or object that is perceived as new by an individual or other unit of adoption. A product, service, or process can be the subject of innovation (Bhoovaraghavan and Vasudevan 1996). Schumpeter (1934) classifies innovations in two types: (1) radical and (2) incremental. Radical innovations are those originating from the process of creative destruction, a term coined to explain technological or market paradigm break throughs, shifting to something completely new and that can be represented by a product or a process. Product innovativeness refers to the level of perceived newness, originality, and uniqueness of a product (Garcia & Calatone, 2002). For any organization, NPD innovation is crucial in achieving the success. Vinayak & Kodali (2014) proposed six elements NPD innovation, (1) product innovation; (2) process innovation; (3) market innovation; (4) service innovation; (5) behavioral innovation; and (6) managerial innovation.

Product innovation is often referred to as the novelty and meaningfulness of new products introduced to the market in a timely fashion (Wang and Ahmed, 2004). Product innovation is critical to product success which in turn is highly related to sustainable business success, providing great opportunities for businesses in terms of growth and expansion into new areas (Cooper, 2000; Henard and Szymanski, 2001). Process innovation refers to the introduction of new production methods and new technology that can be used to improve production processes (Wang and Ahmed, 2004). Maravelakis et. al., (2006) emphasized that process innovation may result in product innovation and likewise product innovation may force process innovation, an inference that product innovation and process innovation are strongly correlated. In our study, process innovation constructs have been taken as those related to product development process issues concerning production methods and the distribution cycle.

Market innovation is the newness of approaches that organization adopts to enter and exploit the targeted market, i.e. innovation related to market research, advertising and promotion as well as identification of new market opportunities and entry into new markets (Wang and Ahmed, 2004). Market innovation is central to product innovation and likewise, product innovation maintains a central focus for product newness. Here, we refer to market innovation in the context of novelty of market-oriented approaches. Similarly, service innovation refers to the differences and novelties that can be built into the dimensions of intangible service offerings (Zolfagharian and Paswan, 2008). In service innovation, activities are undertaken to deliver core services so as to attract more consumers (Oke, 2007), which in turn tend to create a new revenue streams. Behavioral innovation in the organization is directly related to the people and their practices. Here, innovation is brought in to the social system of an organization like focusing on the innovative practices, culture, the overall internal receptivity to new ideas and innovation adapted by individuals and teams in the organization. Managerial innovation practices focuses more on leadership/senior management's role in building the organizational structure, administrative processes and enabling the human resources toward an innovative culture. In the present study, management strategy on innovation, administration or leadership innovation, focus on feasibility studies or risk-taking attitude of management, support for knowledge management, organization's characteristics and motivation of people to innovate were taken as constructs of managerial innovation.

2.3 Value of innovation basis on shariah

Service science is the study of service systems and of the co-creation of value within complex constellations of integrated resources (Spohrer et. al., 2007, 2008). Service is the application of competences (knowledge and skills) by one entity for the benefit of another (Vargo and Lusch, 2004, 2006). This definition provides a fresh perspective for understanding economic phenomena, by implying that value is created collaboratively in interactive configurations of mutual exchange. It centers on the participants, processes, andresources that interact to create value in service systems. So value and value creation are at the heart of service and are critical to understanding the dynamics of service systems and to furthering service science. But value is an elusive term.

Marketing basics highlight the importance of understanding different consumer segments and relating to their needs (Kotler and Armstrong, 2006). Mirroring the holistic view of Islamic legislation through its prime underpinning objectives (maqasid ash-shari'ah) and seeing Islam not only as a culture but as creed ('aqidah), worship ('ebadat), interactions (mu'amalaat), and morality (akhlaq), El- Bassiouny (2014) attempted to set a humble precedent aimed at presenting a macro-marketing view (see p. 46) of the potential implications of Islamic marketing according to the macro-level and integrated approach of the spirit and heart of Islam, namely the sources and goals of Islamic shari'ah. Overall depiction of the transcendental values integration model discussed in El-Bassiouny (2014).

The "maqasid ash-shari'ah" of preservation of self, intellect, posterity, wealth, and faith represent the soul of Islamic legislation that permeate its inherent value system, and offer a broad framework for actions and deeds consistent with its morals, priorities, and ideals (El-Bassiouny, 2015). Based service dominant logic (Vargo & Lusch, 2006) and Islamic religiusity (El-Bassiouny, 2015), we proposed measure in the present study as value of innovation basis on shariahwith the following five statements, (1) create values religiusity to stakeholders; (2) use a spiritual brand character; (3) prioritization of values religiusity; (4) co-creation business based shariah values; and (5) create values religiusity system, preservation of self, intellect, posterity, wealth, and faith.

2.4 Product innovation advantage

Quality of the new product is perhaps the most important factor affecting success (Cooper and Kleinschmidt, 1987). We mentioned previously that quality control is more challenging for the service firm than for the manufacturing firm. Unfortunately, attempts by service firms to ensure consistency in quality through standardization decouple the service personnel from customers, encourage the domination of NSD by operations, reduce the influence of marketing and the understanding of customer needs (Lovelock, 1984, 1983; Mahajan, et. al., 1994).

Further, in comparison with their manufacturing counterparts, service firms are less efficient in the innovation process, face greater customer and competitor uncertainties, and greater incompatibility of innovations with customer requirements and firm competencies (Carman and Langeard, 1980; Brentani, 1993; Easingwood, 1986; Edget, 1993). These factors are predicted to result in lower competitive advantage/quality for new services in comparison with new products. In addition, services firms have relatively greater difficulty in sustaining advantage compared to manufacturing firms.

Consequently, whereas perceived advantage is the number one factor affecting new product success, its effect on success of new services is less profound (Cooper, et. al., 1991). This research use differen, unique, innovativeness, and quality as indicators variable product innovation advantage.

2.5 Product launch success

A prime innovation success factor is the proficiency of process activities such as idea screening, market, technical and financial assessments, and launch (Cooper, et. al., 1991; Cooper, et. al., 1987; and Dwyer, et. al., 1991). Although, the greater interaction between service providers and customers should lead to better awareness and understanding of customer requirements in new service development (NSD), other factors militate against this. Patent protection of intellectual property in services is not nearly as effective as in products. New services can be copied as easily and quickly by competition.

Managers of services firms do recognize the need for launch activities. such as market tests, but dispense with them because of the high cost, service being a copy of competitors, difficulty of producing test market conditions, and the need to beat competition to market (Morone and Berg, 1993).

We use five indicators as measurement product launch success. They are easily, quickly, interest, intention, and use. Customer easily and quickly to adopt new product launch, then customer interest and intention to use toward new produk. Finnally, customer use new product in the early launching.

3 Proposed model and hypotheses

Figure 1 introduces the paper's proposed model and its related hypotheses.

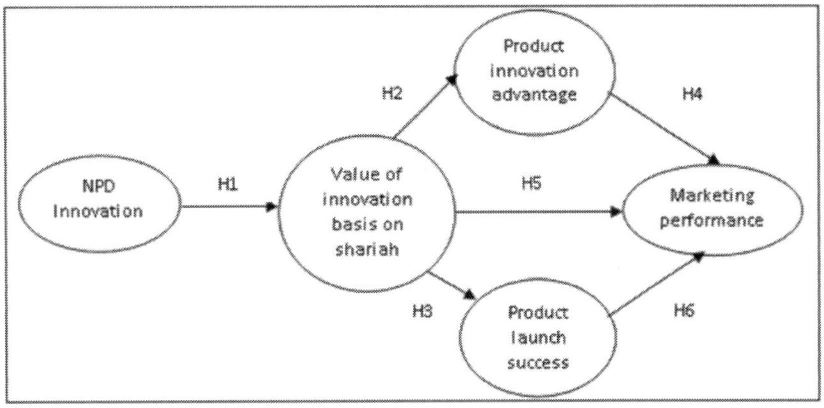

Fig. 1. Propose research model.

According to the model, we proposed six hypotheses, (1) NPD Innovation is positively associated with value of innovation basis on shariah; (2) Value of innovation basis on shariah is positively associated with product innovation advantage; (3) Value of innovation basis on shariah is positively associated with product launch success; (4) Product innovation advantage is positively associated with marketing performance; (5) Value of innovation basis on shariah is positively associated with marketing performance; (6) Product launch success is positively associated with marketing performance.

4 Research methods

Table 1. Construct and indicator construct.

No	Construct	Indicator construct
1	NPD Innovation	Newness/novelty/originality/uniqueness Administration/leadership innovation Customer focus/customer relationship management After-sales support services Employees individual innovativeness New transaction methods
2	Value of Innovation Basis on Shariah	Mu'amalat (special worship) 'Ebadaat (general worship) Innovativeness Creativeness Human well-being
3	Product Innovation Advantage	Differen Unique Innovativeness Quality
4	Product Launch Success	Easiliy Quickly Interest Intention Use
5	Marketing Performance	Product launch on time New product speed to market Sales new product Market share Marketing benefit

Purposive sampling is applied to this study. The questionnaires will be given to only the managers of LKMS (the board of shariah micro finance) in Pekalongan and Banyumas, Indonesia with total 171 managers from 27 LKMS (Baitul Maal wa Tamwil, Baitul Tamwil, Kospin Jasa Shariah, and KSPP Shariah).

5 Result

With regard to construct validity, as recommended by Hair et. al. (2012), exploratory factor analysis (EFA) and confirmatory factor analysis (CFA) were used to assess constructs validity. All the research items were subjected to EFA. An index of Kaiser's measure of sampling adequacy (overall MSA = 0,867) and Bartlett's test of sphericity Sig. 0,000 suggested that factor analysis is appropriate for analyzing the data. Based on eigenvalue greater than 1, the results of EFA indicate that the research items loaded on five factors, four relective and one formative (NPD Innovation). To validate the findings that emerged from using EFA, the four factor model was evaluated by CFA using smartPLS 3.0.

Table 2. VIF values dan outer weights NPD innovation.

NPD Innovation	VIF	Outer Weights
Newness/novelty/originality/uniqueness	1,855	0,055
Administration/leadership innovation	1,707	0,252*
Customer focus/customer relationship management	1,911	0,087
After-sales support services	2,577	0,047
Employees individual innovativeness	2,601	0,597**
New transaction methods	2,577	0,237*

** p < 0,01 * p < 0,05

Evaluate outer model for NPD Innovation shown in Table 2. VIF values is not between 5 – 10, administration/leadership innovation and new transaction methods was significant (p<0,001), and employees individual innovativeness was significant (p<0,005).

Table 3. R^2 values.

Constructs	R Square	Adjusted R Square
Product launch success	0,698	0,694
Product innovation advantage	0,027	0,014
Marketing performance	0,667	0,653
Value of innovation basis on shariah	0,736	0,732

To evaluate inner model using R2, we calculate Q2, and Goodness of Fit (GoF). Value of Q2 = 0,974 and GoF = 0,575. Accordingly, research model is fit and robust to examine hypotheses (Tenenhaus, 2004).

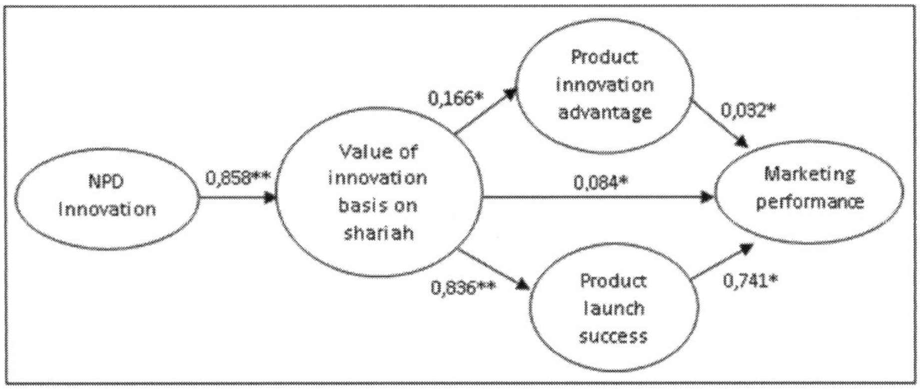

Fig. 2. Result from the structural equation analysis. Note: *p<0,05 **p<0,01.

6 Direct, indirect, and total effect

NPD Innovation significantly affect toward value of innovation basis on shariah, and value of innovation basis on shariah indirect significantly affect toward marketing performance. Value of innovation basis on shariah affect toward product launch success, and product launch success affect toward marketing performance.

Table 4. Direct, indirect, and total effect

Relationship	Direct	Indirect	Total	T Statistics	P Values
NPDI → VoIBoS	0,858**	-	0,858**	41,861	0,000
VoIBoS → PIA	0,109*	-	0,109*	3,050	0,029
VoIBoS → PLS	0,836**	-	0,836**	25,404	0,000
PIA → MP	0,049*	-	0,049*	2,427	0,041
VoIBoS → MP	0,084*	0,624**	0,709**	3,026	0,037
PLS → MP	0,741**	-	0,741**	5,410	0,000
NPDI → PLS	-	0,717**	0,717**	20,199	0,000
NPDI → PIA	-	0,142	0,142	1,036	0,301
NPDI → MP	-	0,608**	0,608**	9,962	0,000

** p < 0,01 * p < 0,05
NPDI = New product development innovation
VoIBos = Value of innovation basis on shariah
PIA = Product innovation advantage
PLS = Product launch success
MP = Marketing performance

Table 5. Exploratory factor analysis and confirmatory factor analysis.

Constructs	Exploratory Factor Analysis		Confirmatory Factor Analysis		
	eigen values	loading factor	composite reliability	outer loading	AVE
Value of innovation basis on shariah	9,570		0,874		0,587
Mu'amalat (special worship)		0,726		0,571	
'Ebadaat (general worship)		0,779		0,703	
Innovativeness		0,755		0,878	
Creativity		0,651		0,858	
Human well-being		0,690		0,779	
Product innovation advantage	2,392		0,821		0,544
Differen		0,893		0,824	
Unique		0,885		0,905	
Innovativeness		0,894		0,606	
Quality		0,916		0,557	
Product launch success	1,404		0,932		0,663
Easily		0,619		0,846	
Quickly		0,568		0,790	
Interest		0,606		0,794	
Intention		0,558		0,853	
Use		0,593		0,800	
Marketing Performance	1,076		0,917		0,689
Product launch on time		0,684		0,819	
New product speed to market		0,774		0,860	
Sales new product		0,795		0,791	
Market share		0,792		0,851	
Marketing benefit		0,759		0,825	
Kaiser-Meyer-Olkin Measure of Sampling Adequacy.					0,867
Bartlett's Test of Sphericity		Df			210
		Approx. Chi-Square			1056,012
		Sig.			0,000

Table 6. Examine hypotheses.

Relationship	Coefisien	p values
H1: New product development innovation → Value of innovation basis on shariah	0,858	0,000**
H2: Value of innovation basis on shariah → Product innovation advantage	0,109	0,029*
H3: Value of innovation basis on shariah → Product launch success	0,836	0,000**
H4: Product innovation advantage → Marketing performance	0,049	0,041*
H5: Value of innovation basis on shariah → Marketing performance	0,084	0,037*
H6: Product launch success → Marketing performance	0,741	0,000**
SRMR	0,097	
d_ULS	3,525	
d_G1	2,301	
d_G2	1,906	
Chi-Square	605,797	
NFI	0,952	
rms Theta	0,179	

* p < 0,05 ** p < 0,01

7 Discussion

The NPD Innovation that formed from six dimensions as a whole, positively affects toward value of innovation basis on shariah, but only managerial innovation, behavioral innovation, and process innovation are significantly. This indicates that indirectly NPD Innovation affect toward marketing performance. There is a positive and significant effect of NPD Innovation toward value of innovation basis on shariah, value of innovation basis on shariah positively affects toward product launch success, and product launch success positively affects toward marketing performance. Shariah values are referred to the theory Islamic Religiosity, developed by El-Bassiouny (2015) in the concept of transcendental values integration, states that the process of developing shariah value, especially in the service of a company based on shariah involves value co-creation activities that have a positive impact on marketing activities (in this case supporting research results).

Vargo (2006) argues that, there is a value creation process in instilling trust in customers when marketing or selling new products resulting from innovation and product development. With regard to mediation variable, as recommended by Baron and Kenny (1986), in the first order shown that both value of innovation basis on shariah and product launch success mediated between NPD Innovation and marketing performance.

8 Conclusion

The conclusions of the research are the creation of value of innovation basis on shariah values and product launch success mediating the relationship between NPD Innovation and NPD Performance in the board of sharia microfinance, it must be corroborated by the trust of customers in the name of products that have a spiritual character. Second order research as recommendation in future, and explore compatibility value innovation in the large objects.

References

1. M. N. Akroush, Compet. Rev. **22**, 343 (2012)
2. K. Atuahene-Gima, J. Prod. Innov. Manag. **13**, 35 (1996)
3. R. M. Baron and D. A. Kenny, J. Pers. Soc. Psychol. **51**, 1173 (1986)
4. S. Bhoovaraghavan, A. Vasudevan, and R. Chandran, Manage. Sci. **42**, 232 (1996)
5. P. Bicen, S. Kamarudin, and W. H. A. Johnson, J. Bus. Res. **67**, 2877 (2014)
6. R. J. Calantone, K. Chan, and A. S. Cui, J. Prod. Innov. Manag. **23**, 408 (2006)
7. J. M. Carman and E. Langeard, Strateg. Manag. J. **1**, 7 (1980)
8. R. G. Cooper, J. Mark. **43**, 93 (1979)
9. R. G. Cooper and E. J. Kleinschmidt, J. Prod. Innov. Manag. **4**, 169 (1987)
10. R. G. Cooper, Res. Technol. Manag. **43**, 38 (2000)
11. R. G. Cooper and U. de Brentani, J. Prod. Innov. Manag. **8**, 75 (1991)
12. L. Dwyer and R. Mellor, J. Prod. Innov. Manag. **8**, 39 (1991)
13. N. El-Bassiouny, J. Bus. Res. **69**, 569 (2016)
14. N. El-Bassiouny, J. Bus. Res. **67**, 42 (2014)

15. E. Fang, J. Mark. **72**, 90 (2008)
16. F. Fu, E. Jones, and W. Bolander, J. Pers. Sell. Sales Manag. **28**, 351 (2008)
17. R. Garcia and R. Calantone, J. Prod. Innov. Manag. **19**, 110 (2002)
18. A. Godener and K. E. Söderquist, R D Manag. **34**, 191 (2004)
19. J. F. Hair, M. Sarstedt, C. M. Ringle, and J. A. Mena, J. Acad. Mark. Sci. **40**, 414 (2012)
20. C. T. Huang and K. H. Tsai, Ind. Mark. Manag. **43**, 1407 (2014)
21. E. J. Hultink and H. S. J. Robben, J. Prod. Innov. Manag. **12**, 392 (1995)
22. K. K. B. Ko, C. K. M. To, Z. M. Zhang, E. W. T. Ngai, and T. L. K. Chan, J. Bus. Res. **64**, 1327 (2011)
23. P. Kotler and G. Armstrong, Principles of Marketing (Prentice-Hall, Upper Saddle River: New Jersey, 2006)
24. A. Ledwith and M. O'Dwyer, J. Prod. Innov. Manag. **26**, 652 (2009)
25. P. Lo Liu, W. C. Chen, and C. H. Tsai, Technovation **25**, 637 (2005)
26. C. H. Lovelock, J. Mark. **47**, 9 (1983)
27. R. F. Lusch and S. L. Vargo, in *Serv. Log. Mark. Dialog, Debate, Dir.* (2006), p. 449
28. J. Mahajan, A. J. Vakharia, P. Paul, and R. B. Chase, Int. J. Res. Mark. **11**, 1 (1994)
29. E. Maravelakis, N. Bilalis, A. Antoniadis, K. A. Jones, and V. Moustakis, Prod. Plan. Control **17**, 283 (2006)
30. M. R. Millson, Eur. J. Innov. Manag. **16**, 317 (2013)
31. S. Mishra, D. Kim, and D. H. Lee, J. Prod. Innov. Manag. **13**, 530 (1996)
32. J. Morone and D. Berg, J. High Technol. Manag. Res. **4**, 123 (1993)
33. A. Oke, G. Burke, and A. Myers, Int. J. Oper. Prod. Manag. **27**, 735 (2007)
34. E. Rogers, Difussion of Innovations (The Free Press, 1983)
35. D. F. L. Santos, L. F. C. Basso, H. Kimura, and E. K. Kayo, J. Bus. Res. **67**, 527 (2014)
36. J. A. Schumpeter, Harvard Econ. Stud. **46**, 404 (1934)
37. R. Sethi, J. Mark. **64**, 1 (2000)
38. J. D. Sherman, D. Berkowitz, and W. E. Souder, J. Prod. Innov. Manag. **22**, 399 (2005)
39. J. Spohrer, P. P. Maglio, J. Bailey, and D. Gruhl, Computer (Long. Beach. Calif). **40**, 71 (2007)
40. J. Spohrer, S. L. Vargo, N. Caswell, and P. P. Maglio, *The service system is the basic abstraction of service science*, in Proceedings of the 41st Annual Hawaii International Conference on System Science, January, (2008)
41. Suliyanto, Metode penelitian bisnis untuk skripsi, tesis, dan disertasi (ANDI, Yogyakarta, 2018)
42. M. Tenenhaus, S. Amato, and E. V. Vinzi, XLII SIS Sci. Meet. 739 (2004)
43. S. L. Vargo and R. F. Lusch, J. Mark. **68**, 1 (2004)
44. S. L. Vargo, P. P. Maglio, and M. A. Akaka, Eur. Manag. J. **26**, 145 (2008)
45. K. Vinayak and R. Kodali, Meas. Bus. Excell. **18**, 39 (2014)
46. C. L. Wang and P. K. Ahmed, Eur. J. Innov. Manag. **7**, 303 (2004)
47. N. L. Warren and A. Sorescu, J. Mark. **81**, 64 (2017)

48. M. Zolfagharian and A. Paswan, J. Serv. Mark. **22**, 338 (2008)

RELATIONSHIP BETWEEN TRAITS: COGNITIVE ABILITIES, GENERAL SELF-EFFICACY AND NEUROTICISM TO CAREER SUCCESS MEDIATED BY SPECIFIC SELF-EFFICACY

Wiwik Priyatin[1*]

Department of Human Resources Management, Faculty of Management, University of Jenderal Soedirman

Abstract. This research is an associative correlational type of research with aim to analyze the correlation between traits of cognitive ability, general self-efficacy and neuroticism with career success mediated by specific self-efficacy. The data sample were 31 lecturers which were chosen by using random sampling technique of 68 lecturers of Diploma of Nursing in Banyumas and Cilacap Areas as sample size. The data was analyzed by employing multiple regression analysis and Pearson product moment correlation. The results showed that cognitive ability and neuroticism has no positive relationship to career success mediated by specific self-efficacy. Meanwhile, general self-efficacy has positively strong relationship to career success with specific self-efficacy as the mediator.

1 Introduction

Career success poses as main study in Human Resource Management Study. It is something that is very much needed and becomes important for every individual because the achievement of it is a form of success in his/her work and reflects personal satisfaction (Steyrer et al., 2001). Success in a career is everyone's dream as well, because a career is something that someone wants to achieve in the future as a measure of his/her success. This study entitled "Relationship between Traits: Cognitive Ability, General Self-Efficacy and Neuroticism on Career Success mediated by Specific Self-Efficacy which aimed to explain the relationship between traits: cognitive ability, general self-efficacy and neuroticism with career success among lecturers of Diploma of Nursing in Banyumas and Cilacap Areas. The population target in this study was lecturers of Diploma of Nursing in Banyumas and Cilacap Areas.

---

* Corresponding author: [1]wiwikaura428@gmail.com

2 Methodology

This research was an associative correlational type of research which aimed to analyze the correlation. The data sample of this study was 31 respondents from 68 lecturers by using random sampling as the sampling technique. Questionnaires were used to collect the data, Multiple Regression Analysis using SPSS was used for data analysis technique, and Pearson Product Moment Correlation was used to measure the strength of relationships between variables. The hypothesis framework is shown in Figure (1) and described as follows: H1(+) Cognitive Ability is positively related to Specific Self-Efficacy (SSE), H2(+) Specific Self-Efficacy has a positive relationship to Career Success, H3(+) General Self-Efficacy (GSE) is positively related to Specific Self-Efficacy, H4(-) Neuroticism has negative relationship to Specific Self-Efficacy, and H5(-) Neuroticism is negatively related to career success. Researcher used questionnaire from Shockley, at al., (2015) as a tool to measure career success and cognitive abilities, while for measured the general self-efficacy researcher used questionnaire from Schwarzer, R., & Jerusalem, M. (1995), and to measure the specific self-efficacy for assignments difficulty level researcher used questionnaire from Bandura, (1998).

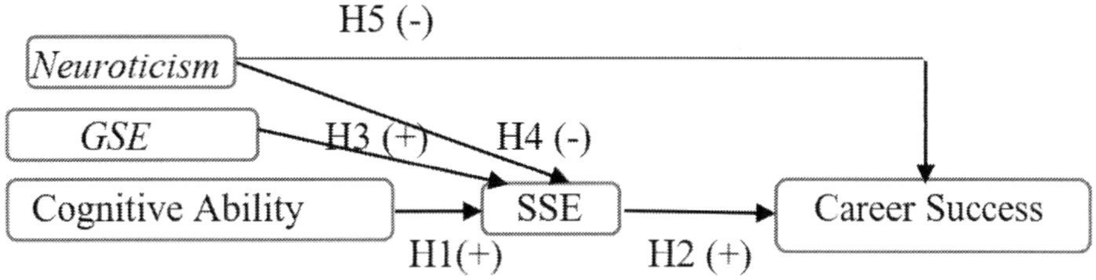

Figure. 1. A figure of a relationship between variables in this study

3 Results

Based on the analysis, this study found that cognitive ability had no positive relationship to career success mediated by specific self-efficacy, and Neuroticism had negative relationship to career success mediated by specific self-efficacy. Meanwhile, there was a positively strong relationship between general self-efficacy and career success mediated by specific self-efficacy. So, H1 and H2 were denied whereas H3, H4, and H5 were fulfilled. The results of the correlation test sig of cognitive abilities with specific self-efficacy was equal to 0.181> 0.05 so that there was no relationship, with Pearson Correlation value was 0.246 degrees means that the correlation is weak. However, the result of correlation test of sig neuroticism value with specific self-efficacy is 0.009 <0.05 means there is a relationship, with Pearson Correlation value was - 0.460 degrees means that it had moderate correlational relationship. Meanwhile, the results of the test of correlation sig value of general self-efficacy with specific self-efficacy is 0,000 < 0.05 which means there was a relationship, with Pearson Correlation value was 0.796 degree which means it had strong correlation. By looking to the result, this study concluded that Neuroticism was negatively related to specific self-efficacy with a degree of moderate correlation. It had to be underlined that the respondents' demographic in this study were male and female whom mostly hold the master degree of nursing.

4 Conclusion

A person with good cognitive abilities is able to increase confidence in his/her ability to mobilize his/her motivation to perform certain tasks (Wood & Bandura, 1989). This is supported by the existence of human capital theory which states that an individual's progress and career success depend on the quantity and quality of human assets brought to the labor market (Becker, 1964). Supported by the research results including cognitive abilities have been proven to be important predictors of performance on various different tasks and it related to task performance through their positive relationship with specific self-efficacy (Stanley M. Gully, Robert, Giled Chen, et al ., 2000). In fact, this study found that there was no positive relationship between cognitive abilities and career success mediated by specific self-efficacy, so there was a gap between the theory and the results of this conducted study.

5 Implication

Further research is needed with a larger number of respondents. Quantitative research methods require a relatively larger sample size (over 30 respondents) in empirical observation (Jogiyanto, 2017) so that they can dig deeper into the relationship between cognitive abilities and career success mediated by specific self-efficacy.

References:

Adiyono, Ratno Purnomo,Wiwiek RA. (2017). Pengaruh Kepribadian Lima Faktor terhadap Kesuksesan Karir. Jurnal Bisnis dan Ekonomi (JBE). Vol.24, No.2, 162-176.

Abele E Andrea & Spurk Daniel. (2009). How Do Objective & Subjective Career Succees Interrelate Over Time. Laser, Labor and Socio-Economic Research Center University of Erlangen Nuremberg. Vol.85 No.35, 803-820.

Arnold Gesell, et. al., (2011). Maturation and Incomplete Man Test. Journal International Cognitive Ability.

Ballout.I Hassan. (2007). Carrer Success The Effects of Human Capital, Person-Environment Fit and Organizational Support. Journal of Managerial Psychology.Vol.22, No.8, 741-765.

Bridgstock, R .(2008). Follow your bliss or show me the money ? career orientation, career management competence and career success in Australian creative workers, Norwich Business Sshool ;University of East Anglia.

Coopersmith.S.(1967). Coopersmith Self- Esteem Inventory Form A. Self Esteem

Institute.

Gully M.Stanley. (2000). Examination of Relationships Among Trait-Like Individual Differences, State-Like Individual Differences, and Learning Performance. Journal of Applied Psychology.Vol.85. No.6, 835-847.

Guthrie P.James.(1998).Career Management Strategies: The Role Of Personality.Journal of Managerial Psychology.Vol. 13 No.5/6, 371-386.

Jogiyanto. (2017). Filosofi dan Metodologi Penelitian. Cetakan Pertama. Yogyakarta: BPFE

Le Huy. (2011). Too Much of a Good Thing: Curvilinear Relationships Between Personality Traits and Job Performance. Journal of Applied Psychology.Vol 96. No.1, 113 – 133.

Luthans Fred. (2011). Oragizational Behavior An Evidence Approach.12th Edition

Nursalam. (2013). Metodologi Penelitian Ilmu Keperawatan. Penerbit Jakarta Salemba Medika. Edisi 3.

Pownall Ian. (2012). Effective Management Decision Making. Ian Pownall & Ventus Publishing Aps.

Shockley, Heather, Laura, Timothy.(2016). Development of a New Scale to Measure Subjective Career Success: A Mixed-Methods Study.Journal of Organizational Behavior, J.Organiz Behavior. Vol .37,No., 128-153.

Song & Day .(2016). Subjective Career Success: A Literature Review & Prospect. Journal of Human Resource and Sustainability Studies. Vol 4, No., 238- 243.

Steyrer Johannes. (2000). Subjective & Objective Career Success – Empirical Findings From the Vienna Career Panel Project Based on The Nation of Career Habitus.

Srivastava & O.P John. (1999). The Big – Five Trait Taxonomy: History, Measurement, and Theoritical Perspectives. Pevin & O.P John (Eds), Handbook of Personality : Theory and research Vol.2 No., 102 – 138.

Suliyanto. (2018). Ekonometrika: Analisis Regresi variabel Dummy, Uji Asumsi Klasik, Validitas dan Reliabilitas, Path Analysis, Structural Equation Modeling (SEM). Tidak dipublikasikan.

Thomas, Kelly, Feldman. (2005). Predictor Of Objectif and Subjective Career Success: A Meta – Analysis. Personel Psychology. Vol 58, No. ,367 – 408.

Wang Mo & Wanberg R Connie. (2017). 100 Years of Applied Psychology Researc on Individual Careers: From Career Management to Retirement. Journal of Applied Psychology.Vol.102, No.3, 546-563.

Scharzer, R,.& Jerusalem, M. (1995). Generalized Self-efficacy Scale. In J. Weinman, S.Wright, & M.Johnston, Measures in health psychology: A user's portofolio. Causal and conrtol beliefs.vol.,No.,35-37.

DiTenun, Smart Application Producing Ulos Motif

Arlinta C. Barus [1*], *Marianna* Simanjuntak [2], and *Verawati* Situmorang [3]

[1,3]Teknik Informatika, Institut Teknologi Del, Laguboti Sumatera Utara
[2] Management Rekayasa, Institut Teknologi Del, Laguboti Sumatera Utara

Abstract. Indonesia is a country which is rich of various traditional cultures and values. One of its representation is traditional woven clothes (well known in Indonesian as kain tenun) which is wide spread throughout Indonesian regions. To support the traditional woven industry, as a relevancy to the industry 4.0 era, we develop DiTenun which is a multiplatform application that is able to produce new motifs of traditional woven intelligently using machine learning approach. The presence of the apps aims to support the growth of traditional weaving industry particularly the small and medium scale ones. The dissemination of the apps is very challenging as traditional woven centers are mostly located in rural area where the digital world has been rarely accessed. In this paper, we present "Ulos" as a case study in the utilization of DiTenun. The implementation of the sustainability of the Ulos industry by DiTenun needs to be adjusted to the development of the industrial era 4.0. Ulos is a traditional woven cloth from Batak tribe, which is located in several rural regions surrounding Toba highland in North Sumatera Utara province. The workflow for producing an item that is marketable is to produce woven fabrics with motifs that have been produced by smart devices. The results of DiTenun can have an impact on the technology produced and on the social life and culture of the weavers. The study shows how DiTenun is designed to support Ulos weavers in creating new motifs of Ulos and to support the economy of relevant small and medium scale industry of Ulos.

1 Introduction

Indonesia is known as a country that is rich in art and culture. One element of the culture is weaving. Weaving is a handicraft in the form of fabric made from yarn (cotton, silk, etc.) by entering threads transversely on a tool and made in areas such as Sumatra, Java, Palembang and Kalimantan. Related to weaving products from various regions, this study examines the weaving results of the North Sumatra region, Ulos.

Ulos is a woven fabric typical of the Batak people who show affection between parents and children, or between someone and other people (Aritonang, 2015). Ulos has a variety of types and motifs that describe its own meaning according to the nature, circumstances, functions, and traditional ceremonies performed. Besides being used for traditional events,

* Corresponding author: author@email.org

Ulos are also used as assets such as table cloth, wall decoration, etc. (Barus, 2015). In addition to traditional ceremonies, Ulos has been produced into fashion products and motifs that can be used into several daily supplies.

The Ulos motif used by weavers today is the Ulos motif that was discovered many years ago and is continuing to be used today. Thus the motif poured on Ulos are the same motif and are produced repeatedly. The same Ulos motif reduces the image and creativity in making Ulos motifs themselves. The various and many Ulos motifs will certainly produce more ideas that can be poured into various weaving products.

This paper presents our work to develop apps DiTenun which has a main function to generate Ulos motif automatically. The generation makes use machine learning concepts and image processing algorithms to learn about Ulos motifs and to produce new Ulos motifs but still have cultural values contained in the Ulos itself. In details, this paper discusses how DiTenun is produced and knows the role of DiTenun in the lives of industrial weavers in industrial 4.0

2 Literature Survey

2.1 Ulos

Ulos is a type of traditional woven fabric originating from the Batak tribe which consists of various types, patterns or motifs and sizes. Ulos is a Batak woven scarf, usually used in traditional ceremonies (marriage, entering a house, death, etc.) (Wikipedia).

Ulos was used by the Batak people as body warmers, but now Ulos cloth is used for other aspects, for example, it is used as Batak ceremonial clothing, Batak community performing arts, and so on. Ulos Batak community consists of various types, namely Toba Ulos Batak, Simalungun, Karo, Pakpak-Dairi and Mandailing-Angkola. Ulos motifs are very diverse, beautiful, unique, and rich in historical and cultural values. Ulos fabric has not been widely developed in the industry because of its traditional and manual workmanship. In addition, the development of the design still revolves around certain weaving motifs for traditional ceremonies. Even though there is so much economic potential that can be developed from Ulos cloth (Pilialang, 2011).

2.2 Rural Area

There is no single best defenition that measures theoretical construct of Rural. The variability in rural defenition is not inherent limitation of rural research, failure to adequate define and describe rural makes it difficult, if not impossible, to confidently make comparisons, and interpatation (Gustavo Anríquez, 2007). Rural areas are more integrated into the world economy, as a result both of liberalisation and improved infrastructure.

2.3 Industry 4.0

The implementation of the sustainability of the Ulos industry by DiTenun needs to be adjusted to the development of the industrial era 4.0 (Mohelsska & Sokolova, 2018). States that Industry 4.0 is based on a boom in digitalization, robotics and automation, and is characterized by a blend of information technology and production processes and in intelligent ways that are characteristic of autonomous machines are called series Internet of Things (IoT) (Toma & Popa, 2018). IoT cannot be separated from the industrial era 4.0, the fourth industrial revolution with intelligent digitalization and automation of value chain products and processes. This is largely based on that the Internet has brought a number of changes to manufacturing and maintenance in the industry, from reducing the production cycle to the automation of maintenance of machinery and equipment, including real and virtual world production integration, where products, production facilities, people and objects are joined by devices software embedded into intelligent and distributed systems (Wilhelm et al, 2018).

2.4 Machine Learning

Machine learning has developed from learning computational theory and pattern recognition. The most effective method is used in data analysis to predict things based on the model and algorithm. This model makes researchers and developers manage patterns to get results and good decision making. For some patterns that have been lost can appear by studying the history of data. The model is based on results collected from training data and it is very difficult to get accurate predictions based on the algorithm developed (Sheena Angra, 2017).

The machine learning technique implemented in DiTenun is image processing. The image can be defined as a two-dimensional function f (x, y) where x and y are the coordinate fields, and the amplitude of f in the x coordinate pair, y is called the intensity of the image at that point. When the value of x, y and the amplitude value are known we can call it a digital image. Digital image processing refers to the processing of digital images, meaning by using a computer (Gonjales).

3 Analisys and Design

3.1 Art and Technology Collaboration

Traditional crafts in this technological progress are slowly because there are not many researchers and innovators who see the potential for technological development, industrial design, preservation of cultural history, and economics in traditional craft industries. In addition, the lack of access to information and access to technology that can be reached by artisans who live in isolated areas far enough from the city center is not helpful at all in utilizing traditional textile craft innovations, especially Ulos. In this field, multidisciplinary collaboration is needed. Collaborative arts and traditional Ulos weaving skills combined with technological sophistication can help facilitate the design process and develop weaving designs.

The development of the industrial revolution, Ulos can be seen as an element of production undergoing transformation, called that the globalization process which penetrated all aspects of life so that the cultural elements between one society and other communities increasingly culturally similar, starting from the motif type, their motif and designations. DiTenun prepares ourselves in the era of industry 4.0 by trying to create or produce Ulos motifs, motif adapted from fractal formulas, involving weaver resources, fostering the resources of weavers or un empowered communities so that weavers can broaden income and sales of production.

3.2 Application Analysis and Design

Image processing with machine learning studies the motifs of weaving and produces new motifs with a way as shown in Figure 1.

SHS Web of Conferences **86**, 01025 (2020) https://doi.org/10.1051/shsconf/20208601025
ICORE 2019

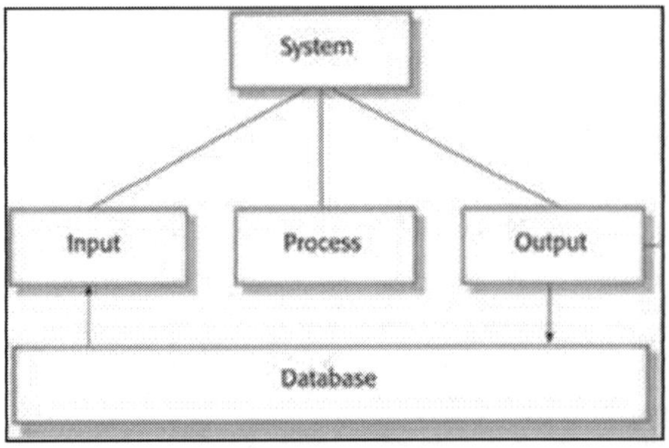

Fig. 1. DiTenun Design

Ulos motif data will be stored in the database. The Ulos motif that already exist today will be sent as output to the system and processed with image processing to produce output in the form of new motif. There are several motif that can be produced. To get other motif, motif that have been successfully generated can be saved to the database. Motif will be sent as input in the system and do image processing and produce and other motif.

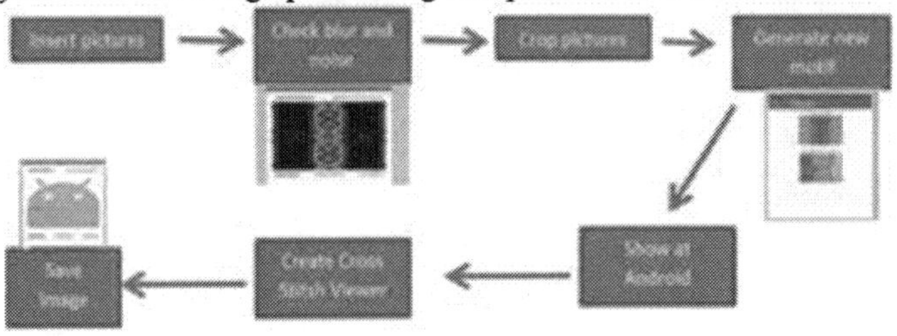

Fig. 2. Generating Motif Process

As depicted in Figure 2, motif images can be edited to get the image the weaver wants. The image will be processed to produce a new motif. The results of new motifs can bedisplayed on Android applications. New motifs generated by the system will be given to weavers as new ideas to produce new Ulos motifs. Weavers are educated to weave Ulos neatly in accordance with predetermined standards so that they are suitable for commercialization.

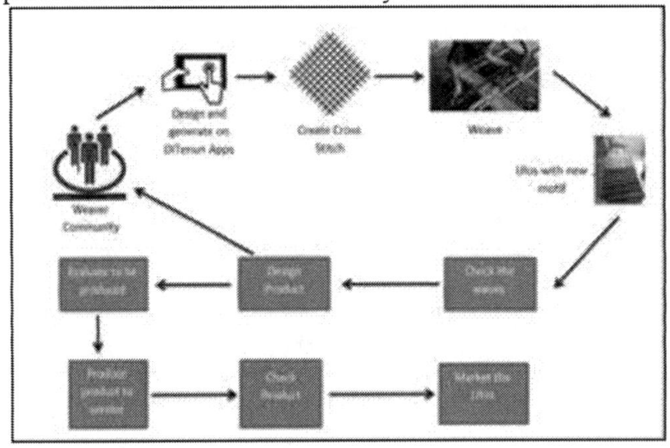

Fig. 3. DiTenun Program

To produce Ulos, a more creative item requires creativity and effort from the community of craftsmen themselves. The workflow for producing an item that is marketable is to produce woven fabrics with motifs that have been produced by smart devices. The new woven fabric will be examined for details of neatness and conformity with the framework. Crafters will design products that want to be made with these fabrics such as bags or clothes. The design and woven fabric will be evaluated to check the feasibility of production before being given to the vendor. After passing the feasibility evaluation, the fabric design can be given to the vendor for processing. After completion of production, the product will be re-examined for feasibility before being sold. Figure 3 presents the flow of tasks to generate new motifs using DiTenun.

3.3 DiTenun Business Analysis

As mentioned, DiTenun is created to support small and medium scale industry, particularly woven clothes industry across Indonesia. By using DiTenun to have new motifs, weavers are expected to able more various designed of clothes that can be adjusted to the market preference. We also provide online system www.ditenun.com that having features to generate and choose your own motif, and then ordering products using the new motif. The online system is also designed as a market place for the weavers to commercialize their products.

We have disseminated DiTenun to Ulos weavers from several regions around Toba area through workshops and training. Based on the meetings, we develop a SWOT analysis of the possible development and implementation of the business continuity of the motif of DiTenun in the industrial era 4.0, the results of SWOT, as follows.

Strengths:

a. Districts on the outskirts of Lake Toba are familiar with Ulos craftsmen (weavers).
b. Mothers and young women usually choose to be weavers rather than farmer in Toba lake region.
c. Weaver communities have used smart phones
d. Weaver communities welcome the development of various motif.
e. New motif training is easy to implement.
f. The local government supports the training program and strengthening the Ulos weaving organization with the motif of DiTenun.

Weaknesses:

a. The price of community-woven Ulos is relatively very cheap.
b. A lot of agents trapping the weavers.
c. The monotonous Ulos motif is only for customary purposes.
d. Allotment of Ulos woven products is only for ceremony cultures, Batak people in general.
e. Lack of connection to the broad market.

Opportunities:

a. Penetration of telecommunications networks is getting better, can reach remote locations of weavers.
b. Production of Ulos with the development of motifs of DiTenun is easily expanded.
c. Contribution of Local Government in supporting village training movements.
d. The industrial organization of post-weave Ulos fabrics is easy to obtain.
e. Broader target market achievement through sharing various functions and Ulos motifs.
f. Ulos are known worldwide.
g. Technology influences the people rapidly, example smartphone and internet.

Threats:

a. Industrial development from motif is not Ulos
b. Some people's views about Ulos is idealism

c. Teenagers are more likely to not like living in the village but rather migrate and do other work. Using the Template

From the analysis, we have to work hard to alleviate weakness and threats and make use the strengths and opportunities to disseminate DiTenun to empower Ulos weavers around Toba. This analysis can also then be used in replicating the use of DiTenun to other types of traditional woven clothes.

4 RESULT AND DISCUSSION

The results of Weaving can have an impact on the technology produced and on the social life and culture of the weavers.

4.1 DiTenun Technology

4.1.1 DiTenun Mobile Application

DiTenun is a smart device using image processing techniques that learn about image dots and take patterns so that they are able to generate new motifs from existing motifs. The development of smart devices already exists in the mobile version. This application has an existing weaving image browsing feature, and runs module pattern generation with image pattern with constant parameters. Nowadays smart devices can run the image pattern with various different parameters (users can customize function parameters) so that from an original image can be produced a variety of different motif using the three techniques that different.

In addition, the interface development process runs iteratively to get frequent and continuous feedback. In addition, a technical method for teaching these smart devices is also needed to crafters. This teaching is needed so that the knowledge of how to use this smart device can be realized properly. The appearance of this application is shown in the Figure 4.

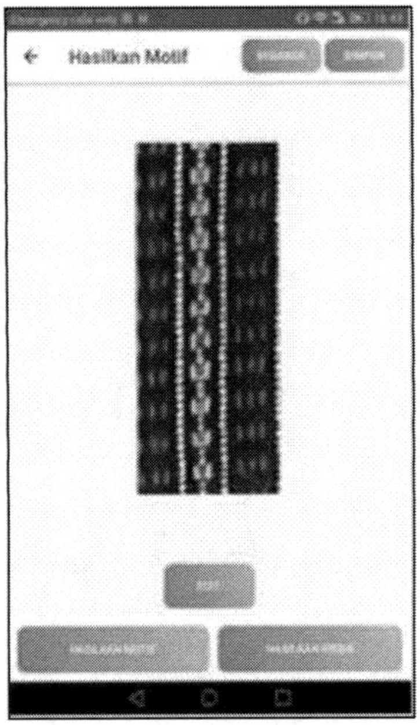

Fig. 4. DiTenun Mobile Application

The application is not only able to produce new motifs but can be poured in a cross stitch form according to the calculation. The resulting cross stitches are then used by weavers as a base for weaving.

4.1.2 DiTenun Web

Besides being available in the mobile application, DiTenun is available on web applications that can be accessed on the link: http://DiTenun.com/. Web development is a medium for exchanging information. Not only as a media that produces information about Ulos, the DiTenun Web is also designed as a medium for buying and selling products produced by the Weaving application.

Fig. 5. DiTenun Website

There are several products that are ready for commercialization, including:

a. Motif Collection Book (hardcopy & softcopy)
b. One Unit Motif (hardcopy & softcopy)
c. Souvenirs (Local and IT Del Souvenirs)
d. Corporate Merchandises (pre order)
e. Fashion Collection (seasonal products)
f. Home Décor Collection (seasonal product)

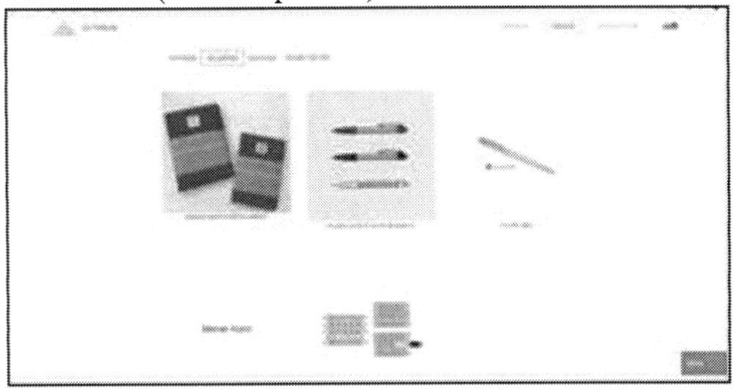

Fig. 6. DiTenun Product

4.2 Weaver Cultures and Social

Implementation of sustainability of a production, starting from: Establishing industrial boundaries, whose purpose focus more, make standards, determine locations and strengthen human resources (Moisa, 2017). In line with Filip (2018), the implementation of sustainability starts from communication and cooperation between technology users, information providers, and information technology systems that can be used in real time. Enables flexible production at lower costs and designs business models. Creating a global network, with the Environmental Management System (EMS) (Yusoff & Rudolf, 2015).

The presence of DiTenun has had the following positive impact.

a. The development of Ulos cultivation business is a prospective livelihood and has a legality principle.
b. Increased technological capacity that will have a huge impact on developing their business going forward Impacts and benefits for partners.
c. Ensure the survival of Ulos artisans through a permanent and sustainable financial income.
d. Ensure weaving skills Ulos is not lost and can develop continuously from generation to generation

5 CONCLUSION

DiTenun is a smart application that can produce Ulos motifs by using image processing with image pattern techniques. Ulos motifs are generated from existing motifs so that they still contain the values contained in Ulos. Ulos motif that has a value is possible to be generated using DiTenun application. The new motif enables flexible production at lower costs and designs business models. There are several products that are ready for commerlization and increase the weaver income. The presence of DiTenun is very precisely in the industrial 4.0 era by utilizing technology to enhance creativity, to increase economy and to keep culture values.

References

1. S. Angra and S. Ahuja, in *Proc. 2017 Int. Conf. Big Data Anal. Comput. Intell. ICBDACI 2017* (2017), pp. 57–60
2. G. Anríquez and K. Stamoulis, *Rural Development and Poverty Reduction* (2007)
3. S. Aritonang, J. Online Mhs. Fak. Ekon. Univ. Riau **2**, 33710 (2015)
4. A. C. Barus, R. Panggabean, A. Sinaga, and Y. Manik, Lemb. Ris. Inst. Teknol. Bandung (2015)
5. W. Bauer, S. Schlund, T. Hornung, and S. Schuler, Logforum **14**, 341 (2018)
6. C. M. Bishop, *Machine Learning and Pattern Recognition* (Springer-Verlag New York Inc., New York, United States, 2006)
7. R. C. Gonzalez, New Jersey 07458 Prentice Hall (n.d.)
8. C. O. Moisă, Valahian J. Econ. Stud. **8**, 2017 (2017)
9. Y. A. Piliang, *Dunia Yang Dilipat: Tamasya Melampaui Batas-Batas Kebudayaan* (Matahari, Bandung, 2011)
10. J. W. G. Putra, *Pengenalan Konsep Pembelajaran Mesin*, 9th ed. (Institut Teknologi Bandung, Bandung, 2017)
11. R. Sathya and A. Abraham, Int. J. Adv. Res. Artif. Intell. **2**, 34 (2013)
12. C. TOMA and M. POPA, Inform. Econ. **22**, 46 (2018)
13. S. Yusoff, R. Nordin, and H. Yusoff, Issues Soc. Environ. Account. **9**, 18 (2015)13. S. Angra and S. Ahuja, in Proc. 2017 Int. Conf. Big Data Anal. Comput. Intell. ICBDACI 2017 (2017), pp. 57–60

Primary Personal Finance Problem in Indonesia

Agus Suroso [1*], *Ascaryan* Rafinda [2], *Timea* Gal [3], and *Osama* Alhendi[4]

[1] Faculty of Economics and Business Universitas Jenderal Soedirman, Indonesia
[2] University of Debrecen, Károly Ihrig Doctoral School of Management and Business, Hungary
[3] The Islamic University of Gaza, Palestine
[4] University of Debrecen, Károly Ihrig Doctoral School of Management and Business, Hungary

Abstract. The main objective of this research is to find out primary personal finance problem in Indonesia. A lot of people do not realize that they need financial knowledge to make financial decision. It leads to bounded rationality when they do the financial decision. Financial problem commonly not different years by years, then by collection financial problem this research could spread and warn other people not to fall into same problem. This research collecting data by questionnaire. Total 360 respondent fill in the questionnaire. The result successfully identify there are several topics in personal finance that become main problem. There are, have no finance plan, debt management and excessive spending. This research imply that we should focus on that topic on educating people in personal finance.

1 Introduction

National survey results show that only 29.66% of Indonesian people financially well litarated (Otoritas Jasa Keuangan, 2016). The total population of Indonesia is more than 220 million people, indicating less than 66 million people have good level of financial literacy. The low level of financial literacy makes a person unable to maximize the potential in financial decision making (Mandell & Klein, 2007). Further impact when a person is unable to make financial decisions to the fullest is the poor planning of pension funds and the low accumulation of assets (Lusardi, 2005), the amount of borrowing costs (J. Hastings, Mitchell, & Chyn, 2010), mistakes in choosing personal investment instruments (M. van Rooij, Lusardi, & Alessie, 2011).

Financial education important do reduce the possibility of poor financial decision making. The Indonesian government is expected to build the foundation of financial education. Through the financial services authority, the Indonesian government promotes the importance of financial education. Various financial modules and financial literacy guidelines have emerged in Indonesia. Various agencies and non-profit institutions participate in increasing public financial literacy. But it is unfortunate because the financial module explain the general of finance and not developed based on real financial problems or cases in the community.

* Corresponding author: roso.fe2014@gmail.com

This research identifies various real financial problems that faced by Indonesian society. In the future data of this research can be used as a database of financial problems in Indonesian society. Focusing on people at vulnerable ages between 23-45 years old this research collected data from a total of 360 participants, collected financial problems faced by people in Indonesia. The results of this research found that there are three topics that are the main problems of the people in Indonesia.

2 Literature Review

Previous research in the field of personal finance has focused on antecedents and consequences. Various antecedent topics from personal finance discuss the factors that can enhance personal finance knowledge. Various factors has been found such as family, environment and friends are factors that can increase personal finance knowledge (Bowen, 2002; Chen & Volpe, 2002; Norvilitis & MacLean, 2010; Shim, Serido, Bosch, & Tang, 2013; Xiao, Tang, Serido, & Shim, 2011).

Previous research in the field of personal finance that focuses on consequences focuses on discussing financial behaviour. The research focuses on the impact of someone who has personal finance knowledge on his financial behaviour. Various studies have found that low personal finance knowledge has an impact on poor financial decision making (J. Hastings et al., 2010; Lusardi, 2005; Mandell & Klein, 2007; Opletalová, 2015; M. C. J. van Rooij, Lusardi, & Alessie, 2012).

There are three most common factors that can increase a personal finance knowledge. Family, friends and the environment are factors that has been found could influence a person's financial knowledge. But not everyone has the same opportunity to get family, friends and environment that support financially. Therefore financial education is one of the alternative factors that can be used to improve people financial knowledge.

Research on financial education is usually related to financial literacy and financial outcomes. The results of a review on various empirical studies found that there are no clear cut results between the three of them (Álvarez-franco et al., 2017). Other research findings found that there is no way to conclude the relationship between financial education, financial literacy and financial outcomes (Fernandes, Lynch, & Netemeyer, 2014; J. S. Hastings, Madrian, & Skimmyhorn, 2013; Kaiser & Menkhoff, 2017; Miller, Reichelstein, Salas, & Zia, 2015). Various problems of the relationship between financial education with financial literacy and financial outcomes can occur because of the non-specific financial training material provided. So that participants do not get material in accordance with expectations and real financial problems encountered.

This research was conducted to identify the real financial cases in Indonesia. The identification of various financial cases makes it easy for educators to develop real case-based financial education. In the future participants who get a real case of financial problems, they are expected to be able to anticipate these problems in the future.

3 Research Method

The data collection was carried out using an online questionnaire. The questionnaire was distributed conveniently through the closest relatives and colleagues. Various social media and chatting applications were used in the process of disseminating research instruments. Social media like Instagram and Facebook were used to process information dissemination, as well as WhatsApp chat applications.

The instrument consists of 20 questions covering 1) demographics (8 questions), 2) financial background (4 questions) and 3) financial education (6 questions). There are two types of questions, namely multiple choice and description. Each participant needs 10-20 minutes to complete filling in the research instrument.

Before collecting the real sample, pilot test was conducted to check the research instrument. A pilot test was conducted on 37 people to identify the level of understanding and the possibility of multi perception of each question asked. The results of the pilot test, carried out minor revisions in several terms to make it easier for participants to understand.

The data collection process was carried out for 40 days, instruments prepared online were disseminated through the closest relatives and colleagues. In an effort to increase participation, selected participants will get incentives with a predetermined category. A total of 10 participants were selected to receive incentives as appreciation for the answers given. A total of 360 participants had heterogeneous respondent demographics.

This research examine using content analysis to understand various financial problems in Indonesia. Content analysis was done by understanding the answers regarding financial problems faced by respondents. This research uses keywords to facilitate the classification of existing problems. For example, a problem falls into the category of debt management when keywords appear, such as debt, credit, instalments, etc. Classification is carried out following the problems that arise from the results of research. There are no restrictions in the process of grouping financial problems.

After classifying based on financial cases, the next step is to determine priorities. Total 5 cases of financial problems arise using keywords. This research counts the number of cases that come out based on data obtained. For example, there are cases of debt management, personal cost management and, personal investments. This research calculate the number of that three cases. The more people fill in regarding personal investments, then the topic will raise and become priority case and vice versa.

4 Result and Discussion

Content analysis was carried out to classify financial problems. From a total of 360 participants, classification was carried out into 5 topics. The following is a list of topics and keywords used in classification:

Table 1. Keywords that used to classify the problem

No	Topic	Keywords
1	Debt Management	Credit, Loans, Installments
2	Saving	Reserve Funds, Idle Money, Emergency Funds
3	Investment	Side Income, passive income, maximize idle funds
4	Expense Management	Difficulties manage money, run out before end of months
5	Financial Planning	Have no plan, have no financial target.

These five topics are the main focus of this research. The next step is calculate the number of cases that arise from each topic of financial problems. The following are the number of cases that emerged based on data collected:

Table 2. Number of cases on Financial Topics

No	Topic	Number of Cases
1	Debt Management	136
2	Saving	72
3	Investment	48
4	Expense Management	54
5	Financial Planning	50
	Total	**360**

Debt management is the most common case based on the results of this research. This shows that many debt problems must be resolved by the Indonesian. In this debt topic, there are various sub-topics of debt that are very specific such as the management of mortgage payments, calculation of vehicle debt instalments, etc.

Identification of these sub-topics is important for further research. This is because the sub-topic database will be used as financial education media in the future based on real cases. Financial education based on real cases will make education participants understand well what the actual conditions in Indonesia.

5 Implication and Limitation

This research has identified various financial problems that faced by Indonesian society. Having discovered various specific financial problems, further implications are the potential for the development of real case-based financial education media. Real cases collected from the results of this research become a database in the development of financial education. So it is hoped that financial education participants can benefit directly.

References

[1] P. B. Alvarez Franco, A. M. Muñoz Murillo, and D. A. Restrepo Tobón, Cuad. Adm. **30**, (2017)

[2] C. Faulcon Bowen, J. Financ. Couns. Plan. (2002)

[3] H. Chen and R. Volpe, Financ. Serv. Rev. (2002)

[4] D. Fernandes, J. G. Lynch, and R. G. Netemeyer, Manage. Sci. **60**, 1861 (2014)

[5] J. Hastings, O. S. Mitchell, and E. Chyn, in *Financ. Lit. Implic. Retire. Secur. Financ. Marketpl.* (Oxford University Press, 2011)

[6] J. S. Hastings, B. C. Madrian, and W. L. Skimmyhorn, Annu. Rev. Econom. **5**, 347 (2013)

[7] T. Kaiser and L. Menkhoff, World Bank Econ. Rev. **31**, 611 (2017)

[8] A. Lusardi, in *Pension Des. Struct.* (Oxford University Press, 2004)

[9] L. Mandell and L. S. Klein, Financ. Serv. Rev. (2007)

[10] M. Miller, J. Reichelstein, C. Salas, and B. Zia, World Bank Res. Obs. **30**, 220 (2015)

[11] J. M. Norvilitis and M. G. MacLean, J. Econ. Psychol. **31**, 55 (2010)

[12] A. Opletalová, Procedia - Soc. Behav. Sci. **171**, 1176 (2015)
[13] Otoritas Jasa Keuangan, Surv. Rep. (2016)
[14] S. SHIM, J. SERIDO, L. BOSCH, and C. TANG, J. Consum. Aff. **47**, 128 (2013)
[15] M. C. J. van Rooij, A. Lusardi, and R. J. M. Alessie, Econ. J. **122**, 449 (2012)
[16] M. van Rooij, A. Lusardi, and R. Alessie, J. Financ. Econ. **101**, 449 (2011)
[17] J. J. Xiao, C. Tang, J. Serido, and S. Shim, J. Public Policy Mark. **30**, 239 (2011)

The Effect of Profitability Ratio, Liquidity Ratio, Leverage Ratio, and Company Size on Sukuk Rating Corporation During 2014-2017 Periods

Suratmi [1*], *Ika* Yustina Rahmawati [2]

[1]Economics and Business Faculty, Universitas Muhammadiyah Purwokerto
[2]Economics and Business Faculty, Universitas Muhammadiyah Purwokerto

Abstract. The purpose of this study was to destermine the effect of profitability ratios, liquidity ratios, leverage ratios, and company size toward sukuk ratings in Indonesia during 2014-2017. The independent variables in this study is profitability ratios, liquidity ratios, leverage ratios, and company size, while the dependent variable is sukuk rating. The object of this study is Corporations for the period 2014-2017, this study used 78 samples. The technique sampling method in this study used purposive sampling. The data analysis techniques used in this study is descriptive statistical tests, and logistic regression tests. The results of the analysis show that simultaneously profitability ratios, liquidity ratios, leverage ratios and company size have a significant effect on sukuk ratings. Partially the size of the company has a significant positive effect on the sukuk rating, while the profitability ratio, liquidity ratio, and leverage ratio have no significant positive effect on the sukuk rating.

1 Introduction

The development of the Islamic capital market is characterized by various indicators including the increasingly widespread sharia capital market players that issue sharia securities in addition to stocks in the Jakarta Islamic Index (JII). Sharia securities that can be traded in the Islamic capital market, one of which is sharia bonds, now known as sukuk (Nurhayati and Wasiah, 2105). Islamic bonds (sukuk) in accordance with the Fatwa of the National Sharia Council No.32 / DSN-MUI / IX / 2002 are long-term securities based on sharia principles issued by issuers to holders of Islamic bonds that require issuers to pay income to sharia bond holders in the form of yield / margin / fee and repay and bond at maturity.

Sharia Capital Market Statistics issued by Otoritas Jasa Keuangan (OJK) in 2015 stated that based on total value and total sukuk emissions and outstanding sukuk show a significant number from year to year where at the beginning of 2010 the total value of sukuk emissions reached 7,815.00 billion later increased in 2015 to 14,483.00 billion. This is in line with the value of outstanding sukuk, where in 2010 the achievement of outstanding sukuk was 6,121.00 billion and then increased in 2015 to 8,444.40 billion

* Corresponding author: arlinta@del.ac.id

rupiahs. Even though in several years the outstanding sukuk had experienced a decline, as in 2011 and in 2014, but in the years after the outstanding sukuk conditions continued to increase (https://analisis-perkembangan-sukuk-di-indonesia/).

In general, bonds are divided into two ratings, namely Investment Grade (Bonds that ratings fall into the top four rating categories) and non-investment grade (Bonds that rating is not included in the four top rating categories). The grade improvement has AAA, AA, A, and BBB ratings. Whereas for non-investment grade with BB, B, CCC, and D. The main purpose of the bond rating is to provide accurate information about financial performance, industrial business position, the company that issues bonds (bonds) in the form of ratings to investors. Investors who are interested in buying bonds should pay attention to bond ratings because the bond rating gives a signal about the probability of a company's failure to pay interest and principal loans. Sudana (2009: 127) stated that one of the considerations of investors in choosing bonds issued by various companies is information about bond ratings. In Indonesia there are three rating agencies namely PT. PEFINDO (Rating Agency of Indonesia), Fitch Indonesia and ICRA.

Brigham and Houstan (2012) stated that the factors that influence bond ratings consist of qualitative and quantitative factors. One of the quantitative factors is financial ratios. A good bond rating is reflected in the company's financial statements, where if the company's performance improves, the rating will increase so investors are interested in investing (Sihombing and Rahmawati, 2015). One way to measure company performance in sukuk rating is profitability, liquidity, leverage, and company size.

2 Literature Review

Signaling theory explains why companies provide capital market information. This theory suggests the importance of companies on giving signals to interested parties or potential investors is the issuance of financial statements by the company. This is done to convey the condition of the company to the wider community because through the delivery of information it can convince the public to invest their capital in the company. One positive signal given by the company to attract potential investors is the bond rating. Bond rating provide information on the security of bonds issued by the company.

According to the Fatwa Majelis Ulama Indonesia No.31 / DSN-MUI / IX / 2002 sukuk is a long term securities based on sharia principles issued by issuers to holders of Islamic bonds. Sukuk requires issuers to pay income to holders of Islamic bonds in the form of profit sharing margin/fee, as well as paying back the bond funds at maturity.

Some of the measurements used are: (1) profitability ratio is this ratio measures the overall effectiveness of management as shown by the size of the level of profits obtained in relation to sales and investment. (2) Liquidity is the ability of a company to fulfill its short-term obligations in a timely manner. A high level of risk indicates that the company has the ability to pay for all its current liabilities and shows the strong financial condition of the company. (3) Leverage is measuring how much the company is financed by debt. The use of debt that is too high will endanger the company because the company will be in the category of extreme leverage, namely the company is caught in a high debt level and it is difficult to release the debt burden. (4) The size of the company is a scale that classifies the size of the company that is related to financial companies (Juliantari and Rasmini, 2013). Company size is the size of the company that can be measured by total assets or sales (Sartono, 2010: 128).

The framework for the formation of hypotheses is as follows:

2.1 The Effect of Profitability on Sukuk Ratings.

The influence of profitability on sukuk ratings according to Mardiah et al. (2016) stated that profitability has a significant effect on sukuk ratings, Kurniawan et al. (2017) stated that profitability has a significant positive effect on sukuk ratings, Tensia et al. (2015) stated that profitability has a significant positive effect on bond ratings, Praptiningisih (2015) stated that profitability has a significant effect on bond ratings, Astuti (2017) stated that profitability has a significant positive effect on sukuk ratings. Pebruari (2016) stated that profitability has a significant positive effect on sukuk rating, the high or low value of profitability or profitability of a company will affect the sukuk rating. This gives rise to the hypothesis:
H1: partially profitability variable has a positive and significant effect on sukuk ratings.

2.2 The Effect of Liquidity on Sukuk Ratings.

Liquidity according to Mardiah et al. (2016), liquidity ratios have a significant effect on sukuk ratings, Kustiyaningrum et al. (2016) stated that liquidity has an effect on bond ratings, Ma'arij (2014) proved that liquidity has a significant effect on bond ratings, Praptiningsih (2015) stated that liquidity ratios have a significant effect towards bond ratings, Astuti (2017) stated that liquidity has a significant positive effect on sukuk ratings, and Pebruari (2016) stated that liquidity has a significant positive effect on sukuk ratings. The higher the liquidity ratio means the greater the company's ability to fulfill short-term financial obligations. Then the second hypothesis is:
H2: Partially liquidity variables have a positive and significant effect on sukuk ratings.

2.3 The Effect of Leverage on Sukuk Ratings.

Leverage in the study of Mardiah et al (2016) stated that leverage has a significant effect on sukuk rankings, Kurniawan et al (2017) stated that leverage has a significant negative effect on sukuk ratings, Tensia et al (2015) stated that leverage has a significant negative effect on bond ratings, February (2016) stated that leverage has a significant negative effect on sukuk rating. High leverage describes a company having a high interest expense and this will have an impact on the low ranking of sukuk. Then the third hypothesis is:
H3: Partially the leverage variable has a negative and significant effect on the sukuk ratings.

2.4 The Effect of Company Size on Sukuk Ratings.

Ma'arij (2014) stated that company size has a significant effect on bond ratings, Tensia et al. (2015) stated that company size has a positive significant effect on bond ratings, Martınus (2014) stated that company sıze has an ınfluence on bond ratıngs, Praptınıngsıh (2015) stated that company size has a significant effect on bond ratings, and according Sihombing (2015), company size has a significant effect on bond ratings. Company size has a positive direction, meaning if there is an increase in company size, it will increase bond ratings, as well as decreasing company size it will lower the sukuk rating. The fourth hypothesis is formed as follows:
H4: partially company size variables have a positive and significant effect on sukuk ratings.

2.5 The Effect of Profitability, Liquidity, Leverage, and Company Size on Sukuk Ratings.

Praptiningsih (2015) stated that simultaneously profitability, liquidity and company size significantly influence bond ratings, Kurniawan and Suwarti (2017) stated that simultaneously variable Profitability, liquidity and leverage have a significant effect on sukuk ratings. , Winanti et al. (2017) stated that simultaneous Profitability and liquidity variables have a significant effect on sukuk ratings, Martinus (2014) stated that company size, debt to equity ratio and current ratio have a significant effect on bond ratings, Kustiyaningrum et al. (2016) stated that profitability, liquidity, leverage and company size have a significant effect on bond ratings, Mardiah et al. (2016) stated that simultaneous liquidity, profitability, and company size influence bond ratings. Astuti (2017) states that simultaneously Liquidity and Profitability have a significant effect on sukuk ratings. Then the fifth hypothesis is formed as follows:
H5: simultaneously variable profitability, liquidity, leverage, and company size have a significant effect on sukuk rating.

3 Study Method

This type of study uses quantitative study, with secondary data. The secondary data used are from 2014 untill 2017 which includes the company's fundamental data that issues sukuk in that year. The population in this study are companies that issues sukuk and are rated by PEFINDO. The sample studied is only corporate companies. The sample selection was done by using purposive sampling to obtain 78 data samples from 2014-2017 periods.

3.1 Data Analysis Method

Hypothesis testing in this study using logistic regression. This is because the dependent variable in this study is a dummy variable, which is a variable that has two alternatives. Logistic regression is generally used if the assumption of multivariate normal distribution is not fulfilled (Ghozali, 2006 in Pandutama, 2012). This means that the independent variable does not have to have a linear normal distribution or has the same variant in each group. In this study logistic regression was used to test the variables of profitability, liquidity, leverage, and firm size on sukuk ratings. The logistic regression used is as follows (Ghozali, 2011 in Kuswara, 2014):

$$Ln \frac{RATING}{1-RATING} = \alpha+\beta 1X1+\beta 2X2+\beta 3X3+\beta 4X4+e \quad (1)$$

$Ln \frac{RATING}{1-RATING}$: Sukuk Ratings, 1 for sukuk in *investment grade* rating and 0 for sukuk in *non-investment grade* rating
α : Constants
β1-β4 : Regression Coefficient for each factor
X1 : Profitabily
X2 : Liquidity
X3 : Leverage
X4 : Company Size
e : Error

4 Operational Definition of Variables

4.1 Dependent Variable (Y)

Dependent variable in this study is Sukuk Rating. This sukuk rating is a rating given by PT. PEFINDO. Based on the ranking definition issued by PT PEFINDO, AAA, AA, A, and BBB ratings are included in the investment grade category while BB, B, CCC and D are included in the non-investment grade category.

The measurement scale used is an ordinal scale, where measurement was made by giving a value of 1 to the investment grade category and a value of 0 to non-investment grade sukuk because the dependent variable in this study is a dummy variable.

4.2 Independent Variable (X)

4.2.1 Profitability Ratio (X1)

This ratio measures the effectiveness of management as a whole which is shown by the size of the level of profits obtained in relation to sales and investment. Profitability is proxied by Return on Assets (ROA).

$$ROA = \frac{Total\ Profit}{Total\ Assets}$$

4.2.2 Liquidity Ratio (X2)

Liquidity is the ability of a company to meet its short-term obligations in a timely manner. Liquidity in this study is proxied by Current Ratio (CR).

$$CR = \frac{Current\ Assets}{Current\ Liabilities}$$

4.2.3 Leverage Ratio (X3)

Leverage is measured to know how much a company is financed by debt. The use of debt that is too high will endanger the company because the company will be in the category of extreme leverage, namely the company is caught in a high debt level and it is difficult to release the debt burden. Leverage ratio is proxied by Debt Equity Ratio (DER).

$$DER = \frac{Total\ Liabilities}{Total\ Share\ Holder\ Equity}$$

4.2.4 Company Size (X4)

The size of the company is a scale that classifies the size of the company that is related to the financial company.

Company size: Ln total assets

5 Results

5.1 Descriptive Statistics

Table 1. Descriptive Statistics Output

	N	Minimum	Maximum	Mean	Std. Deviation
RATING	78	0	1	,95	,222
ROA	78	,00042	,11022	,0272292	,02530688
CR	78	,42299	13,95212	2,4778474	2,38677574
DER	78	,28631	9,01184	2,9326524	2,33465793
SIZE	78	28,88012	34,82768	31,0070206	1,19446212
Valid N (listwise)	78				

5.2 Assessing the Feasibility of the Regression Model (Model Fit)

Based on the results of the table, it is known that the significance value or probability of receiving H0 is 1,000 which indicates a probability value > 0.05 so that H0 is accepted or the model can be said to be fit with the data.

5.3 Assessing the Overall Feasibility of the Model (Overall Model Fit)

This test is used to assess models that have been hypothesized to be fit or not with data. Based on the data in table 4.6 it is known that the results show a decrease in the value of -2logL, which is from block number = 0 of 38,337 to 36,254 on the block number = 1. Therefore, it can be concluded that the model hypothesized is fit with data so that it can be used for study.

5.4 Nagelkerke's R Square

Based on the data in table 2, the value of Nagelkerke's R Square is indicated at 0.499 which means the ability of the independent variable is 49.9% while the remaining 50.1% is influenced by other variables not discussed in this study.

5.5 Regression Coeffiecient

This study examines the effect of variable profitability, liquidity, leverage and firm size on sukuk ratings. Based on the testing of the regression coefficient in table 2, the logistic equation model is obtained as follows:

Table 2. Regression Coeffisient Output

Variables in the Equation

		B	S.E.	Wald	df	Sig.	Exp(B)	95% C.I.for EXP(B)	
								Lower	Upper
Step 1[a]	ROA	-34,277	24,183	2,009	1	,156	,000	,000	499139,918
	CR	-,640	,613	1,088	1	,297	,527	,159	1,755
	DER	-,138	,590	,055	1	,815	,871	,274	2,769
	SIZE	4,603	2,179	4,463	1	,035	99,736	1,394	7134,715
	Constant	-132,282	62,905	4,422	1	,035	,000		

Source: processed data, 2019

$$\text{Ln}\,\frac{RATING}{1-RATING} = -132{,}282-34{,}277ROA-0{,}640CR-0{,}138DER+4{,}603SIZE+e$$

6 Discussion

6.1 First Hyphotesis

Based on the results of hypothesis testing it can be concluded that profitability partially does not have a significant positive effect on sukuk ratings. This is due to the proportion of the use of profits not used in funding cash flows or short-term debt (bonds) but rather to pay dividends in larger amounts. The results of this study are in accordance with the study of Kustiyaningrm et al. (2016), Winanti et al. (2017), Sihombing and Rachmawati (2015) which stated that profitability does not have a significant positive effect on sukuk ratings.

6.2 Second Hyphotesis

Based on the results of hypothesis testing it can be concluded that liquidity partially has no significant positive effect on sukuk ratings. Liquidity is not a consideration in determining bond ratings because the liquidity of each company is different between one company sector and another company sector. This study is in accordance with the study of Kurniawan et al. (2017), Winanti et al. (2017), Martinus and Rosita (2014), Tensia et al. (2015) which stated that liquidity does not have a significant positive effect on sukuk ratings.

6.3 Third Hyphotesis

Based on the results of hypothesis testing, it can be concluded that leverage partially has no significant effect on sukuk ratings. This is because some companies in this study have collateral or guaranteed by their parent company so that the sukuk rating is not based on financial ratios but rather from the company that guarantees it. If the company's debt is weak, it will be strengthened by the guarantee company, so that the bond will be given the same rating as the guarantee company. This study is consistent with the study of Kustiyaningrum et al. (2016), Ma'arij (2014), Sihombing and Rachmawati (2015), Martinus and Rosita (2014) stated that leverage does not significantly influence sukuk ratings.

6.4 Fourth Hyphotesis

Based on the results of hypothesis testing it can be concluded that the size of the company partially has a significant positive effect on sukuk ratings. Based on these results it can be concluded that investors and other information users can use company size as the basis for determining investment in corporate sukuk. The results of this study are in accordance with Tensia et al. (2015), Praptiningsih (2015), Sihombing (2015), Martinus and Rosita (2014) stated that firm size has a significant positive effect on sukuk ratings.

6.5 Fifth Hyphotesis

Based on the results of hypothesis testing it can be concluded that profitability, liquidity, leverage and company size simultaneously have a significant effect on sukuk ratings. This is due to the large liquidity in this study. With high liquidity, it will have an impact on the company's profits. Profit earned is not only used as an operational activity, but rather as payment of dividends and debt financing.

In addition, leverage, liquidity and profitability is one of the financial elements included in PEFINDO's sukuk rating assessment, which is in financial risk. With the existence of good financial risk it is considered less risky in investing so that it will provide a good company rating and the greater the total assets of the company, the greater the size of the company that has more complex operations than smaller companies.

7 Conclusions

Based on study and data analysis, the following conclusions can be drawn. First, partially Profitability does not have a significant positive effect on sukuk rating. Second, Liquidity does not have a significant positive effect on sukuk rating. Third, Leverage has no significant effect on sukuk rating. The study also can be clonclude that based on simultaneous test, the variables of profitability, liquidity, leverage, and company size have a significant effect on sukuk ratings simultaneously.

8 Recommendations

The next study can add other variables that can potentially be used to influence sukuk ratings. The sample only uses corporate companies ranked by PT. PEFINDO during the 2014-2017 period, so that the grace period needs to be added longer in order to get more robust study.

References

1. A. Juliantri and Rasmini, Udayana J. Account. (2013)
2. A. Purwaningsih, Kinerja **12**, 85 (2017)//Purwaningsih, Anna. 2008. *Pemilihan rasio keuangan terbaik untuk memprediksi peringkat obligasi: studi pada perusahaan manufaktur yang terdaftar di BEJ*, Jurnal:Kinerja, Volume 12, No. 1, Tahun 2008.
3. A. T. Kurniawan and T. Suwarti, Pros. Semin. Nas. Multi Disiplin Ilmu &Call Pap. Unisbank Ke-3 **3**, 435 (2017)
4. Brigham & Houstan, *Dasar-Dasar Manajemen Keuangan*, 11th ed. (Salemba Empat, Jakarta, 2013)
5. D. F. J. 2011 KEOWN, ARTHUR, MARTIN, JHON D, PETTY, J. WILLIAM, SCOTT, J. Chem. Inf. Model. 53, 1689 (2011)
6. D. Kustiyaningrum, E. Nuraina, and A. L. Wijaya, Assets J. Akunt. Dan Pendidik. **5**, 25 (2017)
7. E. Pranoto, R. Anggraini, and E. Takidah, J. Wahana Akunt. **12**, 13 (2017)
8. E. S. Utami, D. Anitasari, and T. Endhiarto, Rev. Manag. Entrep. **1**, 27 (2017)
9. E. W. S. N. Kartika Hendra Titisari, J. Akunt. DAN PAJAK **18**, (2017)
10. FATWA DSN-MUI No.32/DSN-MUI/IX/2002, *tentang Obligasi Syariah*
11. H. J. Sihombing and E. N. Rachmawati, J. Ekon. Manaj. Dan Akutansi **Vol 1** No 1, 95 (2015)//Sihombing dan Eka Nuraina Rachmawati. 2015. *Faktor-faktor yang mempengaruhi peringkat sukuk pada perusahaan yang terdaftar di bursa efek Indonesia*. Jurnal Ekonomi. Manajemen dan Akuntansi 1 Vol 24 no 1

12. I Made Sudana, *Manajemen Keuangan Teori Dan Praktik* (Airlangga University Press, Surabaya, 2009)

13. I. Fahmi, *Analisis Laporan Akuntansi*. Bandung: ALFABETA. (2011)

14. I. Ghozali, *Aplikasi Analisis Multivariate Dengan Program IBM SPSS 21 :Update PLS Regresi,* Edisi 8. (Badan Undip, Semarang, 2013)

15. I. Wijayanti and M. P. Priyadi, J. Ilmu Dan Ris. Akunt. **3**, (2014)

16. J. Hartono, Sol. Energy Res. (2013) // Hartono, J. 2013. *Teori Portofolio dan analisis investasi*, Edisi kedelapan, BPE, Yogyakarta. http://studyclubiainkendari.wordpress.com/2016/12/23/analisis-perkembangan-sukuk-si-indonesia/)

17. J. Jusuf, *Analisis Kredit: Untukcredit (Account) Officer* (PT Gramedia Pustaka Utama, Jakarta, 2014)

18. K. O. Tensia, R. Yaya, and E. Supriyono, JBTI J. Bisnis Teor. Dan Implementasi **Vol. 6**, 184 (2015)

19. M. A. A. Elhaj, N. A. Muhamed, and N. M. Ramli, Procedia Econ. Financ. **31**, 62 (2015)

20. M. Martinus and R. Suryaningsih, J. Ultim. Account. **6**, 56 (2014)

21. Ma'arij, Arinutry, Zulbahridar, dan Al Azhar. 2014. *Analisis Faktor Akuntansi Dan Non Akuntansi Yang Mempengaruhi Prediksi Peringkat Obligasi Pada Perusahaan Non Keuangan Yang Terdaftar Di BEI Dan Di Peringkat oleh Pefindo periode 2009-2013.*Jom FEKON (online) Vol. 1 No. 2

22. Mardiah, Gina. Malik, Zaini Abdul dan Nurdin. 2016. *Pengaruh Likuiditas, profitabilitas dan Leverage terhadap Peringkat Sukuk (Penelitian Terhadap perusahaan-perusahaan yang Menerbitkan Sukuk Terdaftar di Bursa Efek Indonesia Selama periode 2009-2014).*Prosiding Keuangan dan Perbankan Syariah. 2460-6561

23. Nurhayati dan Wasiah, *Akuntansi Syariah Di Indonesia* (Salemba Empat, Jakarta, 2015)

24. Praptiningsih. 2015. *Pengaruh profitabilitas dan likuiditas serta ukuran perusahaan terhadap peringkat sukuk.* Bina Widya . Volume 26 Nomor 4, Edisi Oktober 2015,199-208.

25. R. K. Sari, S. Nurlaela, and K. H. Titisari, 2nd International Conference on Technology, Education and Social Science 2018, 418 (2018)

26. R. M. E. Sabeni, Arifin Dwiyanti, Diponegoro J. Account. **3**, 716 (2014)

27. Sugiyono, *Metode Penelitian Manajemen* (Alfabeta, Bandung, 2014)

28. T. Darmadji and Fakhruddin, *Pasar Modal Indonesia : Pendekatan Tanya Jawab*, 3rd ed. (Salemba Empat, Jakarta, 2012).

29. Yohanes, Andrey, 2012. *Pengaruh rasio keuangan dan karakteristik obligasi terhadap rating obligasi korporasi di Indonesia*. Artikel penelitian publikasi:http://bati.imtelkom.ac.id/pustaka/files/16237/jurnal/pengaruh-rasio-keuangan-dan -karakteristik-obligasi-terhadap-rating-obligasi-korporasi-di-indonesia.pdf,diakses pada 20 juni 2015

30. Yulianingsih, G 2013. " *Analisis faktor-faktor yang mempengaruhi prediksi peringkat obligasi pada perusahaan yang listing di BEI*", Jurnal Manajemen S1 Universitas Andalas, Vol. 1, No, 1, halaman 1-11.

Mixed-Model Assembly Line Balancing in The Process of Assembling Trimming Area to Minimize Workstation Using RPW-MVM Method

Muhamad Ali Yusuf [1*], *Dida* Diah Damayanti [2], and *Murni* Dwi Astuti [3]

[1,2,3]Industrial Engineering, Telkom University

Abstract. The development of an industry that continues to move forward coupled with global competition and openness demands that the company continues to evolve and always make improvements in improving the performance of its production process. XYZ Corp is an automotive company based in Germany that produces premium cars, the car assembly process groove at XYZ Corp is a trimming line, mechanical line and finishing line, XYZ Corp produces A-model, B-model and C-model cars. At this time XYZ Corp not achieving the production target due to the car assembly cycle time on the trimming Line 1 area exceeds the specified takt time. Assembly line balancing is required in the trimming area using Mixed-Model Assembly Line Balancing Problem (MALBP) approach to minimize the number of workstation, in the Trimming area assembly line balancing study using the Ranked Positional Weighted with Moving Target (RPW-MVM) method. Alocation constraint should be added due to machine restrictions that cannot be moved. After assembly line balancing, there was a decrease in the number of workstations to 14 workstations with a line efficiency of 86% and balancing efficiency of 97%.

1 Introduction

The development of the industrial that continues to move forward coupled with competition and global openness to meet consumer demand makes companies, especially automotive companies continue to grow and always make improvements. The development of the sales market is very competitive and oriented to customer satisfaction. For example, in the automotive industry, products are made based on several different models and features that are oriented to the customers wants. In an effort to fulfill the demand by building and maintaining assembly lines, the company made a single assembly line for several mixed models. In a situation like this, the problem of the balance of the assembly line will arise because of the demands for different models and features that vary with each product produced (Haq, Jayaprakash, & Rengarajan, 2006).

XYZ Corp is a German automotive company that produces premium cars with high performance. Cars manufactured by XYZ Corp has several types of car products such as model A, model B, and model C. The flow of the car assembly process at XYZ Corp is a

* Corresponding author: aliyusuf.ay30@gmail.com

trimming line and a mechanical line. While the engine, door, tire, and dashboard sub-assembly process is an assembly sub-process on the assembly line in the production department, according to Kumar and Mahto (2013) the assembly line model at XYZ Corp has the characteristics of a mixed-model assembly line because the company produces several types of products on 1 track and there is no setup time for changes in the types of products assembled.

The company can only meet the production target of 90%, the processing time for each work station in trimming is different, Figure 1 is a graphical representation of the processing time in the trimming area. Based on the graph data in Figure 1, it can be seen that at some work stations in the trimming area there is a processing time that exceeds the specified takt time, resulting in an unbalanced trimming assembly line which results in some work station bottlenecks and idle time so that the production target is not reached. Therefore it is necessary to balance the workload in the assembly area. This study aims to design a proposed assembly line with an optimal number of work stations with better line efficiency and balancing performance index.

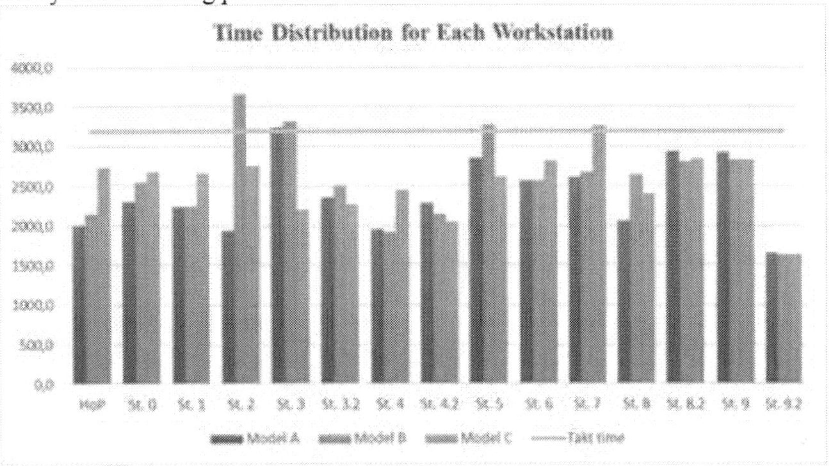

Fig. 1. Time Distribution for Each Workstation

This study focuses on the imbalance of assembly lines. The problem of the imbalance of the assembly line is due to the uneven allocation of workload and the time at several stations that exceed takt time as found in Figure 1. In this study, the balancing of assembly lines is a mixed-model, where assembly lines can produce different models, with adaptability to demand with high levels of efficiency (Reginato et al., 2016). The main purpose of balancing assembly lines according to Eryuk et al (2008) is to minimize the number of work stations with predetermined cycle times or minimize cycle times with the number of work stations that have been determined.

Research "The Design of PC300 Type Hydraulic Excavator Assembly Line System with Genetic Algorithm Method" conducted by Andri Amir (2008) found an increase in productivity from 12 units/day to 13 units/day. Research conducted by Eryuk et al. (2008) regarding "Assembly Line Balancing in a Clothing Company", found that balancing assembly lines using the Ranked Positional Weight (RPW) method resulted in a more optimal number of work stations and line efficiency compared to line balancing using the method Probabilistic Balancing Line. Research conducted by Amen (2001) regarding the solution of assembly-line problems using the heuristic method, found that the heuristic ranked positional weight method has stable results when faced with problems in the real world. Balancing mixed-model assembly lines conducted by Reginato et al. (2016) examines and balances workloads with 7 product models using ranked positional weight with moving target (RPW-MVM) method, the results obtained after workload balancing is increased capacity, line efficiency and balancing efficiency.

Balancing mixed-model assembly lines conducted by Reginato et al. (2016) examines and balances workloads with 7 product models using ranked positional weight with moving target (RPW-MVM) method, the results obtained after workload balancing is incrThe development carried out by Reginato et al. (2016) developed the RPW method in solving problems in the assembly line development, solving assembly line problems using the RPW method which is purely regulated with predetermined cycle times, so that accumulated imbalances can produce assembly line performance. less well, in the RPW-MVM method, there is a limitation of moving target value (MVM) which is used as a reference in balancing and allocating work elements at work stations so that the assembly line is maintained more smoothly and provides a better performance value. eased capacity, line efficiency and balancing efficiency.

This study aims to determine the optimal number of work stations so that it can increase the value of line efficiency and balancing efficiency in the trimming area.

2 Research Method

2.1 Assembly Line Balancing

Assembly line balancing Is a task allocation procedure, where tasks are distributed equally for each assembly work station so that the workload of operators at each work station is balanced (RAJA, 2015). Balance assembly lines are used generally to improve the assembly process for serial work stations. Assembly line balancing is a technique that is often used by industries that want to implement lean manufacturing systems (Kumar and Mahto, 2013). Measures of efficiency (operational performance) of assembly lines can be measured based on (Damayanti and Toha, 2012):

1. Total work stations, each work station will require operating costs, called operational costs.
2. Idle time is a time that is avoided in the context of the efficiency of the assembly line and will result in idle time costs.
3. Waiting time, waiting time in the process is calculated as waiting time cost.

The purpose of balancing assembly lines according to Scholl (1995) is, as follows:

a. Reducing the number of work stations (m) with a predetermined cycle time.
b. Reduce cycle time with the number of work stations that have been determined.
c. Improve line efficiency.
d. Look for a feasible solution with specified cycle time and number of work stations.

2.2 Ranked Positional Weight with Moving Target (RPW-MVM)

Solving line balancing problems on assembly lines using the pure RPW method is set with a predetermined cycle time, so that in allocating work elements to the workstation based on a fixed cycle time, this will result in allocating work elements at each workstation having accumulated imbalances which usually produces poor assembly line performance, to anticipate accumulated imbalances with fixed target limits / fixed cycle times the allocation of work elements must be allocated anywhere and anytime so that line balancing methods are developed on assembly lines based on weighting with moving targets or Moving-Target (MVM). Moving-Target (MVM) calculations on assembly lines are carried out at each work station and balancing assembly lines based on the number of work stations that are then Moving-Target (MVM) values are used as a reference for balancing the allocation of work elements at work stations. Moving-Target (MVM) calculations in the assembly line balancing process will make it easier to configure work stations by allocating work elements in a balanced manner to a predetermined work station (Reginato et al, 2016).

2.3 Performance Indicator

The balancing analysis of assembly lines generated from the RPW-MVM method used is as follows:

1. Line Efficiency (LE)
   Line Efficiency (LE) is an indicator to measure the use of assembly lines that will have an impact on aspects of economic evaluation.

$$LE = \frac{\sum_{i=1}^{k} ST_i}{K.CT} .100\% \quad (1)$$

LE : line efficiency
ST_i : time of the i workstation
K : number of work stations
CT : cycle time

2. Balancing Efficiency
   Balancing efficiency is an indicator that measures the quality of the allocation of work elements to workstations which has an impact on increasing production levels.

$$BE = (1 - \frac{\sum_{j=1}^{W} | \overline{Sj} - Sav|}{W \ x \ Sav}) \ x \ 100 \quad (2)$$

$\overline{Sj}$: Total station time weighted average
W : Number of workstations
Sav : Average station time

2.4 Alocation Constraint

In the assembly process in the trimming area there are several allocation constraint, this limitation is due to machine restrictions that cannot be moved, some work processes or elements that have allocation constraint including:

a. Work elements C22, C21, C20, C19, C18, C17, C14, C11, C12, C16, C15, C13, C23, C10, C2, C8, C7, C6, C1 must be done at station 3.
b. K2 and K12 work elements must be worked on the same station.
c. Work element A2 must be worked on station 1

3 RPW-MVM Model

3.1 Assembly Line Model

The following model assembly lines at PT. XYZ has the characteristics of mixed-models (Raja, 2015), namely assembly lines making products with several variants, where the operation of work for similar product variants for different variants.

The measured performance indicators are:

a. Number of work stations
b. Line efficiency

3.2 Balance of the Existing Assembly Line

The following is the existing performance index information:

1. The number of existing assembly work stations is 15 work stations.

2. The assembly line working time is 25500 seconds/day, cycle time is 3662 seconds/unit, so the capacity is 7 units/day.
3. The assembly line efficiency is 81% and balancing efficiency is 90%.

3.3 Calculation procedure using RPW-MVM

The following are the calculation procedure steps using the RPW-MVM method to balance workload:

1. The first step is to make a combined precedence diagram for all work elements in each model that is worked on the existing work station, for details can be seen on Figure. 3.
2. Next, calculate the proportion of demand. In balancing assembly lines with many models, each model has a different processing time. The average processing time of each model needs to be defined by determining the proportion of demand.

$$pdm = dm/D \tag{3}$$

Where, dm is product demand in period p, where model m = 1, ... M; and D is the total demand for all models produced. The proportion of demand for each model, the proportion of demand for model A models is 43.5%, model B is 27.4% and model C is 29.1%.

3. Calculate cycle time (TC) / takt time based on total production demand. Takt time is 3187.5 seconds/unit, the allocation of work elements must be under takt time.

$$Tc = \frac{waktu\ tersedia\ pada\ periode\ p}{total\ permintaan\ pada\ periode\ p} \tag{4}$$

4. Calculate the average process time weight ($\bar{tk}$) and Average Station Time Total ($\bar{Sj}$) to allocate work elements to RPW-MVM because this method takes into account all models of a mix-model assembly line.

$$\bar{tk} = \sum_{m=1}^{N} pd_m tk_m \tag{5}$$

$$\bar{Sj} = \sum_{k \in j} \bar{tk} \tag{6}$$

5. Perform RPW calculations for each work element by summing the tk of the preceding process based on the combined precedence diagram.
6. Sort according to the weighting of RPW.
7. Calculate the minimum number of work stations (MinW).

$$CTT_m = \sum_{k=1}^{N} tk_m \tag{7}$$

$$MinW = \frac{CTT_m}{Tc}, \; m = 1, ..., M \tag{8}$$

8. Determine the number of work stations j=w.
9. Calculate the moving target number (MVM) for each work station (MVMj,m=1,....,M).

$$CTA_{j,m} = CTT_{j+1,,m} + S_{j,m} \tag{9}$$

$$MVM_{j,m} = \frac{CTT_m - CTA_{j+1,m}}{MinW-(Min\ W-j)} \tag{10}$$

10. Allocating work elements for each model based on RPW weighting values by taking into account the combined precedence diagram and the average weight of the work station ($\bar{Sj}$), allocation of work elements for each work station for all models does not exceed cycle time / takt time (SJ,m = 1, ... , M ≤ Tc
11. Repeat the allocation of work elements until the work elements for each model cannot be moved back.
12. Determine (j = j-1) and do reconsideration MVM j,m=1,...,M.
13. The next step is to validate the level of inequality, that is if ((major MVM j, m = 1, ..., M) ≤ Tc) then do the next step, if ((major MVM j, m = 1, ..., M) ≥ Tc) then repeat step 8 by calculating (MinW = MinW + 1) and repeating the allocation of work elements and then repeating calculations from steps 10 to 13 until all elements of the work are distributed.

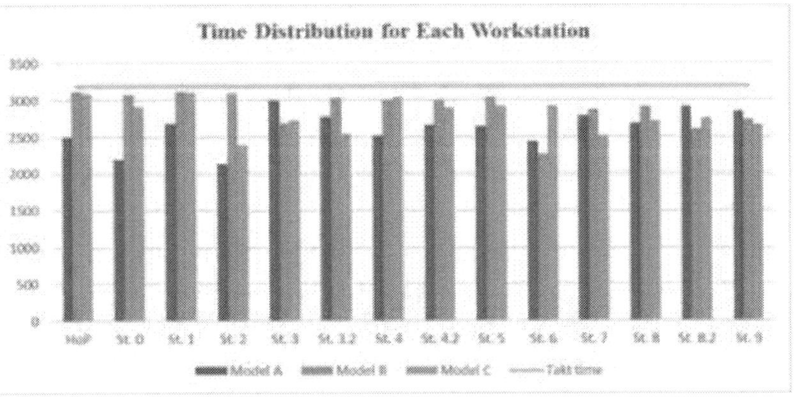

Fig. 2. Time Distribution for Each Workstation after Balancing

4 Conclusions

Based on the results of the calculation and reallocation of work elements for each work station in the trimming area, the optimal number of work stations is found in 14 work stations, with the proposed performance assembly line performance better than the performance index of the existing assembly line. A comparison of proposed and existing performance indexes can be seen in Table 1. After reallocation for each work element, the graph of time distribution for each work station can be seen in Figure 2.

Table 1. Performance Index Comparison

		Current Balancing	New Balancing RPW- MVM
Variable	Time Available in the period (sec)	25500	25500
	Demand (parts/day)	8	8
	Cycle time (sec)	3187,5	3187,5
	Tg (min)	3662	3119
Indicator	Amount of AL workstations	15	14
	Line Efficiency	70%	88%
	Balancing efficiency (BE)	90%	97%

References

1. A. Amir, Peranc. Sist, *Keseimbangan Lini Perakitan Hydraul. Excav. Tipe PC300 Dengan Metod. Algoritm. Genet.* (2008)
2. A. N. Haq, K. Rengarajan, and J. Jayaprakash, Int. J. Adv. Manuf. Technol. **28**, 337 (2006)
3. D. D. Damayanti and I. S. Toha, in *IEEE International Conference on Industrial Engineering and Engineering Management*. (2012), pp. 568–572
4. G. Reginato, M. J. Anzanello, and A. Kahmann, Gest. E Prod**.** **23**, 294 (2016)
5. M. Amen, Int. J. Prod. Econ. **69**, 255 (2001)
6. N. Boysen, M. Fliedner & M. Scholl, ISSN, *Assembly line balancing: Which model to use when?* (2006), pp. 1611-1311
7. N. Kumar and D. Mahto, Glob. J. Res. Eng. **13**, 807 (2013)
8. R. RAJA, Assem. Line Des. Balanc. (n.d.)

9. S. H. Eryuruk, F. Kalaoglu, and M. Baskak, Fibres Text. East. Eur. 16, 93 (2008) pp. 66
10. T. Baroto, *Perencanan Dan Pengendalian Produksi* (Ghalia Indonesia, Jakarta, 2002)

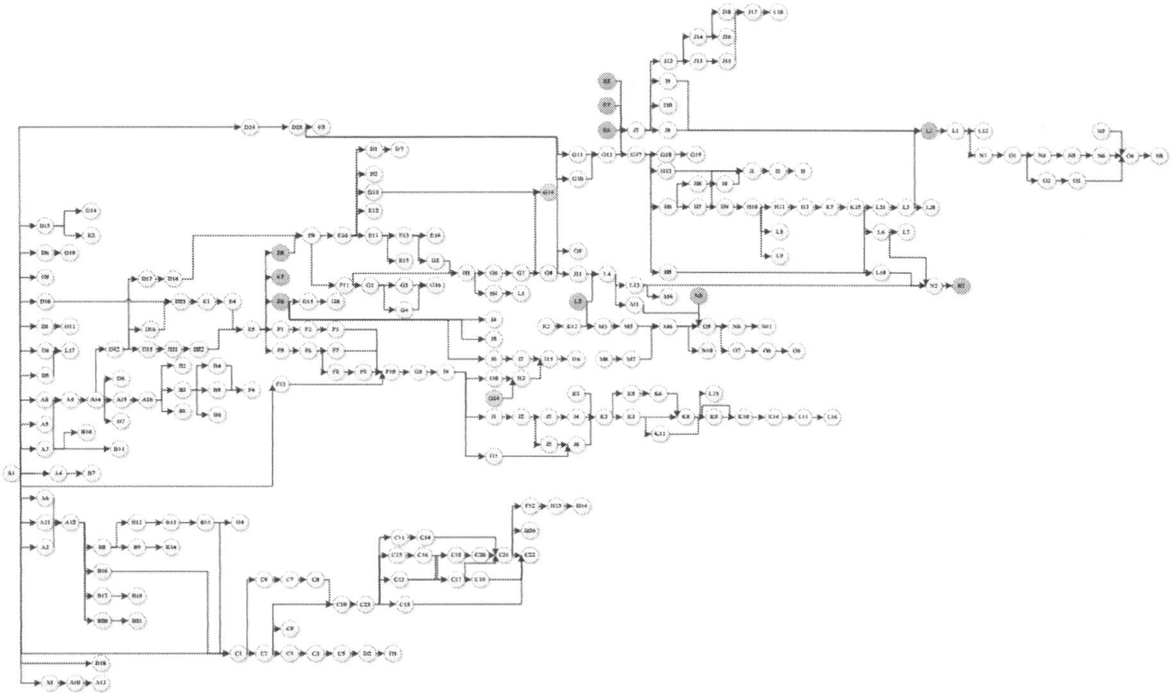

Figure 3. Precedence Diagram

Table 2. Work Element

No	Predesesor	Model			Tk	RPW
		A	B	C		
A1		35	35	35	35	35
A2	A1	350	360	762	473	508
A3	A1		80		22	57
A4	A1	30	40		24	59
A5	A1	33	60	80	54	89
A6	A1	72	55	60	64	99
A7	A1	105	105	95	102	137
A8	A1	58	58	80	64	99
A9	A8, A5, A7	134	140	140	137	393
A10	A1, A3	458	458	458	458	515
A11	A1			80	23	58
A12	A6, A11, A2	125	145	262	170	765
A13	A10	45	45	45	45	560
A14	A9	244	244	244	244	637
A15	A14	280	280	280	280	917
A16	A15	43	43	43	43	960
B1	A16	42		42	30	990
B2	A16	40	40	40	40	1000
B3	A16	90	50	80	76	1036
B4	B3	90			39	1075
B5	B3	93	93	100	95	1131
B6	B3	57	48	80	61	1097
B7	A4	135	120	110	124	183
B8	A12	120	120	160	132	897
B9	B8	75	75	170	103	999
B10	A7	155		110	99	237
B11	A7	300	370	340	331	468
B12	B8	120	120	95	113	1009
B13	B12	480	480	550	500	1510

No	Predesesor	Model			Tk	RPW
		A	B	C		
C13	C23	10	10	10	10	2590
C14	C11	17	17	17	17	2981
C15	C23	33	33	33	33	2613
C16	C15	120	120	120	120	2733
C17	C12, C16	70	70	80	73	2986
C18	C12, C16	135	135	160	142	3055
C19	C17	120	120	120	120	3106
C20	C18	150	140	60	121	3176
C21	C14, C20, C17	360	360	320	348	3999
C22	C21, C13, C19	40	40	40	40	4169
C23	C10			388	113	2580
D1	A1	96	131	96	106	141
D2	C5	60	95	80	75	2603
D3	D2	24	65	30	37	2640
D4	A1	34	69	34	44	79
D5	A9	40	85	50	55	448
D6	A14	122	157	166	144	781
D7	A14	140	175	140	150	787
D8	A1	50	85	80	68	103
D9	A1	75	110	75	85	120
D10	A1	28	63	28	38	73
D11	D1	24	59	24	34	174
D12	A14	80	115	80	90	727
D13	A1	326	361	326	336	371
D14	D13	50	85	70	65	436
D15	D12		115		32	758
D16	D12		253	120	104	831
D17	D12		169	134	85	812
D18	D17	150	245	280	214	1026

No	Predesesor	Model			Tk	RPW
		A	B	C		
B14	B9	60	60		43	1042
B15	B13	445	445	505	462	1972
B16	A12		350		96	861
B17	A12		110		30	795
B18	A1		15		4	39
B19	B17		50		14	809
B20	A12			100	29	794
B21	B20			152	44	838
C1	B16, B15	120	120		85	2068
C2	C1	237	237	237	237	2305
C3	C4	207	207	207	207	2517
C4	C2		18		5	2310
C5	C3	15	15		11	2528
C6	C1	100	110		74	2142
C7	C6	50	50		35	2177
C8	C7	60	60		43	2220
C9	C2	24			10	2316
C10	C2, C8	40	40	40	40	2467
C11	C23	275	275	650	384	2964
C12	C23	180	180	180	180	2760

No	Predesesor	Model			Tk	RPW
		A	B	C		
D19	D8	160		100	99	202
D20	C21	190	245	230	217	4216
D21	D15	170	205	200	188	946
D22	D21	146	181	160	160	1106
D23	D16, D10		85	50	38	906
D24	A1	70	105		59	94
D25	D24		535	240	216	311
E1	D23	70	55	70	66	972
E2	D13	115	100	115	111	481
E3	D25	145	130		99	409
E4	E1	30	25	20	26	998
E5	D22, E3, E4	705	705	600	674	2052
E6	E5	180	165	50	138	2190
E7	E5	130	115		88	2140
E8	E5	240	265	200	235	2287
E9	D18, E5, D9	520	505	450	496	3166
E10	E9	360	370	250	331	3497
E11	E10	265	265		188	3685
E12	E10, E2	90	75	90	86	4029
E13	E11, E12	350	335	300	331	4549

The Impact of Market Value Added to Shareholder Value: Evidence from Indonesia Firms.

Rio Dhani Laksana[1*], *Viviana* Mayasari[2]

[12] Faculty Economics and Business, Universitas Jenderal Soedirman, Indonesia

Abstract. Presently financial experts developed the new concept as a value-based measure of performance in the creation of shareholder value. Market value for shareholders are among the most important goals of firms and owners The purpose of this study is to obtain empirical evidence of Economic Value Added (EVA) affects shareholder value by the method of Market Value Added (MVA) The research using non-financial companies listed on the Stock Exchange from 2013-2017. The result is support that the Economic Value Added (EVA) positive effect on shareholder value or in other words, the higher EVA is the higher shareholder value. The larger the company greater shareholder value and Leverage negative effect on shareholder value as measured using the method of MVA.

1 Introduction

In the recent years, value-based measures have been received a lot of attention during the last two decades, value-based economic value added (EVA) is paid close attention by several accounting, business and nance researchers, corporate professionals, and consultant firms and thus they accept the limitations of traditional measures of performance (Bhasin, 2013). Thus, many scholars criticize traditional measures. Their findings suggest that traditional measures are not proper guidance to make a strategic decision (Laksana et al., 2017). Therefore, we can say that traditional measures still play an important role as instruments for assessing the economic and accounting performance of firms. However, traditional measures only provide information from judgments on past performance (Pinto and Santos, 2011). The purpose of this study is to investigate the impact of profitability on market MVA.

EVA is an indicator of the value creation of an investment. Positive EVA indicates the company managed to create value for the owners of the company, in line to maximize firm value (Utama, 1997). EVA focuses on managerial effectiveness in a given year by estimating the economic profit obtained from the company's business activities in one year. EVA measures the extent to which the company has increasing value for shareholders (shareholder value). Therefore, if the manager focuses on the EVA, this can ensure

* Corresponding author: riodhani@unsoed.ac.id

consistent operating managers in maximizing shareholder wealth. Positive EVA indicates that the company succeeded in creating shareholder value. Shareholder value is measured in two ways, namely Market Value Added (MVA) and the Created Shareholder Value (CSV). MVA was introduced to overcome weaknesses or criticisms of EVA. For example, EVA is not counting the opportunity to grow attached to a company's investment decisions (Ramezani, Soenen and Jung, 2002).

There is a relationship between EVA and shareholder value. If viewed the relationship between EVA and MVA, it can be said that the company has a history of negative EVA, will have the possibility that the MVA value of the company will also be negative, and vice versa if the company has a history of positive EVA, will have the possibility that the MVA value of the company will also be positive.

Based on the above explanation, the researchers wanted to test the effect of EVA on shareholder value, where shareholder value is measured by the method of MVA. The methods are used to address and improve the methods of EVA as a measurement of value creation for shareholders. The purpose of this study was to obtain empirical evidence regarding the effect of EVA on shareholder value, through the two methods is the method of MVA. The results of the research will further clarify the relationship between EVA with shareholder value as measured through the method of MVA so it can be used as a benchmark in measuring corporate performance and value creation for shareholders.

2 Figures and tables

2.1 Population and Sample

The population in this study were the companies listed in Indonesia Stock Exchange (BEI). The study sample was non-financial companies listed in the population. The sampling method is purposive sampling, with sample criteria are as follows: Non-financial companies listed on the Stock Exchange during the last five years, starting from 2013 to 2017 and there is every year, The company that publishes the annual report ended December 31, during the period of observation, Financial statements denominated in rupiah, Having all the data used to calculate the variables that are the focus of this study.

2.2 Data Collection

The data used in this study are secondary data, ie data collected from a database of IDX's Annual Report on the Stock Exchange-listed companies available on Indonesian Capital Market Directory (ICMD)

2.3 Operational Definitions

2.3.1 Economic Value Added (EVA)

EVA is an independent variable in this study. EVA describes the company's residual income after deducting the overall cost of capital. In contrast to traditional accounting performance measures (such as ROE), EVA tries to measure the added value generated in a company, by reducing the cost of capital (cost of capital) incurred as a result of investments made. In other words, EVA measures the extent to which the company has increased shareholder value. EVA approach developed by the institute of management consultants from the United States named Stern Stewart Management Services in the mid-1990s. The operational definition of EVA is as follows:

$$EVA = NOPAT - Capital\ Costs \tag{1}$$

NOPAT = EBIT (1 - Tax)
Cost of Capital = Total Invested Capital x WACC
WACC = (cost of debt x proportion of debt) + (x proportion of the cost of share capital)
Cost of debt (kd *) = kd (1-Tax)
Cost of share capital (ke) = krf + (km-krf) β
Description:
EVA = Economic Value Added.
NOPAT = Net Operating Profit After Taxes.
EBIT = Earnings Before Interest & Taxes.
WACC = Weighted Average Cost of Capital.
krf =risk-free interest rate as measured by the interest rate of Bank Indonesia Certificates (SBI).
km = rate of return expectations in the stock market.
β = beta coefficient of stock that is used as an index of risk for the stock company.

2.3.2 Shareholder Value (SV)

Shareholder value is the creation of shareholder value. Shareholder value measured by the two approaches, namely the Market Value Added (MVA) and the Created Shareholder Value (CSV), as the dependent variable.

2.3.3 Company Size

Company size is a control variable in this study. The logarithm of the firm size measured by the total assets of the company. Firm size effect on firm value. The bigger the company, the greater the value of the company (Herawaty, 2008). Large companies tend to act fewer earnings management than smaller companies.

2.3.4 Corporate Leverage

Leverage the company is a control variable in this study. Leverage is total debt divided by total assets. According to the research and Machfoedz Siallagan (2006) in Rachmawati and Triatmoko (2007) stated that the leverage may reduce conflicts of interest between managers, owners, and bondholders. Companies that have substantial leverage may decrease firm value (Rachmawati and Triatmoko, 2007).

2.4 Data Analysis Technique and Hypothesis Testing

To test the hypothesis used regression analysis. Research hypotheses were tested using the equation:

$$MVAi,t = a0 + a1EVAi,t + a2LogASSTi,t + a3LEVi,t + \varepsilon i,t \tag{2}$$

Description:
VMAs,t = Market Value Added (MVA) firm I in year t.
CSVi,t = Created Shareholder Value (CSV) firm i in year t.
a0 = constant model
a1 = EVA relevant measurement values.
EVAi,t = Economic Value Added firm i in year t.
ASSTi,t = firm size measured by the logarithm of total assets at the firm i in year t, is the control variable.
LEVi,t = firm leverage as measured by dividing total debt by total assets at the firm i in year t, is the control variable.
εi,t = residual error

3 Results and Discussion

3.1 Data Collection

The population in this study are all companies listed on the Stock Exchange. The sample is non-financial companies listed on the Stock Exchange for which data are available to calculate the variables tested in this study. The study period was from 2008 to 2012. Companies listed on the Stock Exchange during the year as many as 300 companies. Of the 300 listed companies, excluded from the sample of 50 derived from financial companies (banks, credit companies other than banks, securities firms and insurance companies). Of the 250 non-financial companies that made these samples, 10 firms or incomplete data are expressed in dollars, so it should be excluded from the sample. Thus used as samples in this study were 240 companies over 5 years of observation with a total of 1200 observations. Briefly, sample selection can be seen in Table 1 below:

Table 1. Selection of Study Sample

Description	Company Amount / Year				
	2017	2016	2015	2014	2013
Companies listed on the Stock Exchange in 2013– 2017	300	300	300	300	300
Finance company	(50)	(50)	(50)	(50)	(50)
Companies for which data are incomplete or expressed in dollars	(10)	(10)	(10)	(10)	(10)
The total company that sampled	240	240	240	240	240
Total observations	1200				

3.2 Descriptive Statistics

Before testing the hypothesis it will be seen first descriptive statistics of each variable tested in this study. Descriptive statistics of variables that are the focus of this study is presented in table 2 below:

Table 2. Descriptive Statistics

Variable	N	Minimum	Maximum	Mean	Deviation Standard
EVA					
MVA	1200	1689.13^{12}	6644.10^{12}	5995.10^{9}-	2353.10^{11}
LOGASST	1200	1506.10^{12}	1149.10^{12}	6180.10^{6}-	8412.10^{10}
LEV	1200	5,027	13,315	10,482	1,869
	1200	7,389	5037,053	12,735	196,298

Table 2 above shows the mean value of EVA from the observed sample is 5995,109. EVA is the highest value while the lowest value is 6644.10 12-1689.1312 2353.1011 with a standard deviation. From these data shows that the lowest value of the company EVA is negative, indicating that the company has a total capital cost greater than the operating profit after tax, or in other words, companies can not create economic value-added.

The mean MVA of the sample being observed is -6180,106. MVA is the highest value while the lowest value is 1149.1012 -1506.1012 8412.1010 with a standard deviation. Mean values marked with a negative MVA indicate that on average the observed sample has a total share capital is greater than the market value of company stock or in other words, an average of the observed sample can not create value for shareholders.

3.3 Hypothesis Testing and Discussion

Hypothesis testing is done by regression analysis. This result test aims to determine the level of statistical significance of each independent variable and the dependent variable. The summary results of regression testing can be seen in Table 3. The hypothesis of the study examined the effect of EVA on shareholder value as measured using the MVA and CSV. The hypothesis proposed is as follows:

Hypothesis 1: Economic Value Added (EVA) affect the Market Value Added (MVA)

Table 3. Hypothesis Testing Results

Variable	Model 1
Dependent Variable: Independent Variable:	LogMVA Coefficient (t stat)
C	6,749 (12,260)*
LogEVA	0,538 (12,419)*
LogASST	0,085 (4,456)*
LogLEV	-0,302 (-2,825)*

R^2	0.385
Adjusted R^2	0,381
F	71,615
Sig.	0,000

Note: * significant 1%, ** significant 5%

Table 3 shows the regression results to test the research hypothesis. Value of F test on a model for the dependent variable of 71.615 and 27.336 MVA for the dependent variable CSV with a p-value of 0.000. This shows the relationship between the independent variable regression with the dependent variable, so it can be stated that there is a relationship between EVA regression with shareholder value as measured by MVA and CSV. R2 values conformity (Adjusted R2) of 38.1%. This indicates that the shareholder value variable is measured by EVA explained by the independent variables of 53.8% while the rest is explained by other causes outside the model.

EVA's influence on shareholder value is statistically significant at alpha 1%. This is demonstrated through the value t and t 12.419 0.000 7.330 with significance. Coefficient EVA relationship with shareholder value is positive 0.538 and 0.401 for the MVA to CSV, meaning that every 1% increase in EVA. This suggests that the higher EVA is the higher shareholder value.

The above hypothesis testing indicates that the first hypothesis is supported by second hypothesis statistics. Where the higher value of EVA, also mean the higher shareholder value. The results of this study are consistent with Stern (2004) which states that EVA is a positive effect on MVA. Besides, this study also supports the findings Ghanbari and More (2007) which states that there is a positive influence between the EVA to the MVA, the results of their study supports the results of the research Stern Stewart Management Services as an institution that developed the measurement of EVA and MVA. These results indicate that the performance of either produced by the company will be captured by the market, so the market responded positively to the company's performance. This will make investors increasingly interested and believe in the company, so the company's share price rises and an increase in market value (market value) company, which may ultimately create value for shareholders, so there is a positive relationship between EVA with shareholder value.

Firm size as control variables measured through the logarithm of total assets has a significant and positive impact on shareholder value in the form of the regression model. This shows that the larger the company the greater shareholder value. These findings support research Herawaty (2008) which states that the larger the company, the greater the value of the company. While the company leverage as control variables in regression models is only a significant effect on shareholder value as measured by the method of MVA. The leverage coefficient shows a negative sign which means the greater the company's debts, the smaller shareholder value. These findings support the results of research Triatmoko and Rachmawati (2007) which states that companies that have substantial leverage may decrease firm value.

4 Conclusion and Limitation

4.1 Conclusion

The research aims to obtain empirical evidence of the Economic Value Added (EVA) affects shareholder value by the method of Market Value Added (MVA). Using sample in this study were non-financial companies listed on the Stock Exchange from 2013-2017. The results can be summarized as follows: There is evidence to support that the Economic Value Added (EVA) positive effect on shareholder value or in other words, the higher EVA is the higher shareholder value and leverage had a negative effect on shareholder value as measured using the method of MVA. This means that the greater the smaller the debt the company shareholder value.

4.2 Limitation

The research only controls the effects of firm size and debt. Influence the type of industry that also affects shareholder value has not been included in the model, so further research is expected to enter the variable type of industry and other variables that affect shareholder value.

4.3 Research Implications and Suggestions

For investors, the expected results of this research provide consideration in determining the estimated shareholder value created by the company. In addition, Indonesia also expected investors can know and use the calculation of shareholder value in addition to using the MVA method can also be calculated using the CSV.

References

1. V. S. M. Ali M Ghanbari, IUP J. Account. (2007)
2. L. Booth, in *Present. Fed. Press "Creating Sharehold. Value" Conf.* (1998)
3. E. F. Brigham and M. C. Ehrhardt., *Financial Management: Theory and Practice*, 11th ed. (Thomson, South Western, 2005)
4. S. Chen and J. Dodd, J. Manag. Issues (2001)
5. P. Fernández, SSRN Electron. J. (2005)
6. P. Fernández, SSRN Electron. J. (2005)
7. I. Ghozali, *Aplikasi Analisis Multivariate Dengan Program SPSS* (Badan Penerbit Universitas Diponegoro, Semarang, 2009)
8. D. N. Gujarati, Basic Econometrics, 4th ed. (McGraw-Hill Companies, New York, 2005)
9. D. Irala and R. Reddy, PES Bus. Rev. (2006)
10. J. K. Kramer and J. R. Peters, J. Appl. Financ. **11**, 41 (2001)
11. D. Kyriazis and C. Anastassis, Eur. Financ. Manag. (2007)
12. R. D. Laksana, H. Hersugondo, S. Wahyudi, and H. Muharam, J. Appl. Econ. Sci. (2017)
13. A. RACHMAWATI and H. TRIATMOKO, Simp. Nas. Akunt. X Makassar (2007)
14. D. V. Ramana, SSRN Electron. J. (2011)
15. C. A. Ramezani, L. Soenen, and A. Jung, Financ. Anal. J. (2002)
16. S. Riceman, S. Cahan, and M. Lal, Eur. Account. Rev. (2002)

17. J. Stern, J. Appl. Corp. Financ. **16**, 91 (2004)

Deming's Quality Management Practices by Small Businesses in Rural Areas

Wiwiek Rabiatul Adawiyah[1*], *Bambang* Agus Pramuka[2], and *Zahrotush* Sholikhah [3]

[12] Universitas Jenderal Soedirman, Indonesia
[3] Universitas Negeri Yogyakarta, Indonesia

Abstract. Deming's Philosophy is a famous thought among Quality Management scholars and practitioner. The theory consists of 14 principles providing guidelines for appropriate firms' quality behavior and practices. Deming proposes that an organization shall experience high-quality service by lowering non-value added activities, such as scrap, rework, and conflicts, while increasing customers' loyalty. There is an overwhelming allegation on the apparent effect of Deming's theory on quality management theory and practices worldwide, mostly in larger organizations of developed countries. Few studies have discussed the relevance of Deming theories in small business management practices in developing countries. Moreover, there is a stigma among small business practitioners that all western originated management theories are less applicable in their context. At the same time, the currently available literature on Deming management methods is mostly a description of the 14 points. This study aims at elaborating the small businesses owners'/ managers' understanding of Deming's fourteen points. The rising numbers of Small Medium Enterprises in Indonesia is a positive sign of local economic development in the Industrial revolution 4.0. To remain competitive, SMEs are forced to be a quality focus and consistently innovate. The sample of the study was 335 SMEs' owners/ managers in Banyumas Central Java, Indonesia. Authors used descriptive statistics to analyze the data — results based on descriptive analysis. The result of the study confirmed the practics of Deming's theory among SMEs in Indonesia. Majority of respondents had a good understanding of the theory and felt the benefits of the theory while adopting it.

1 INTRODUCTION

Continues improvement is being given more significant consideration in portion by the inherent nature of global competition, which is driven by advances in consumer value and system (Adi, Wihuda, & Adawiyah, 2017; Hari Adi & Adawiyah, 2018). Therefore, small firms should consistently improve the quality of products and services offered. Continuous improvement connects higher value to lower expenses and more significant market share (meaning more fulfilled and faithful clients) and offers the rationale for ongoing

* Corresponding author: wiwiek.adawiyah@unsoed.ac.id

performance changes to organizations (Adawiyah, Pramuka, Najmudin, & Jati, 2015; Anderson, Rungtusanatham, & Schroeder, 1994). Consistently, continuous improvement implies higher performance with less variety, resulting from process management methods that take about incremental changes and innovations in procedures, products, and facilities (Anderson et al., 1994). Some studies over the years demonstrate that implementing quality management enables companies to improve their product or service quality systematically and sustainably. (see for example Adawiyah, Shariff, Saud, and Mokhtar 2011, Ahire, Landeros and Golhar 1995; Easton and Jarrell, 1998, Hendricks and Singhal 1996; Powell 1995; Yasir, Pervaiz, and Saifur, 2014) . The organization can adjust to the evolving setting in which it is currently active and maintain the dynamic edge (Huang and Lai, 2014, Kathleen, McFadden, Lee, Gowen III & Sharp (2014). Research results indeed promote the view that proper quality implementation is associated with creative procedures and goods (Kim, Kumar and Kumar 2012, Kathleen et al., 2014).

Although the Deming leadership technique has an impact on leadership practices, worldwide, empirical study assistance for its efficacy except for total proof is little (Anderson et al., 1994). This is partly because of the concept of depicting, explaining, and anticipating the effect of Deming leadership. The technique that was not submitted would be the direct advancement of the empirical investigator. In reality, the academic focus on Deming leadership was remarkably limited (Gartner & Naughton, 1988); this has still been comparatively unexplored for its position in formalizing and promoting the governance hypothesis.

The quality management frameworks can broadly be classified into three groups: 1) formal and quasi standards such as ISO 9000; 2) academic-research-based empirically validated models; and 3) road maps proposed by prominent gurus such as W. Edwards Deming, Joseph Juran, and Philip Crosby (Singh, Dean, and Chuong, 2013) . Management in many companies in the US and around the world embraces the Deming method (Hodgson, 1987). Its popularity is of various case studies that attributed the influence of the Deming method of managing organization (Anderson, Rungtusanatham, & Schroeder, 1994; Douglas & Fredendall, 2004; Khan, 2010).

This paper provides a model of system dynamics to demonstrate some of the elements of the Profound Knowledge System of Deming. It would act as a pedagogical instrument as a helpful guide to Deming's concepts for those fresh ones. Deming a knowledge system would direct people to new leadership from the 'Tyranny' of contemporary leadership (Deming, 1986). The scheme consists of four components (appreciation for systems, knowledge about variation, the theory of knowledge, and psychology) and emphasizes interconnection components. This study also presents how SMEs apply quality management based on W. Edwards Deming theory.

2 Literature Review

2.1 Deming's System of Profound Knowledge

Deming defined by the System of Profound Knowledge that gives insight into the transformation method of organizations. System of Profound Knowledge 's main topic is to undergo a conversion of the prevalent management style. The system consists of four interdependent components. The first component is an assessment of the scheme and the systems. A scheme is an organization that works together to attain organizational objectives. Managers must understand the links between all subsystems and individuals working in order to operate every scheme.

The comparison of the variance hypothesis is the second component of Deming's' theory. There are many causes of variability in the manufacturing method. It is not possible to understand the complicated relationship between materials, tools, machines, operators, and atmosphere differences between isolated bases. However, there is a statistical examination of the mixed impact of all different causes. Deming's second aspect of Profound Knowledge, "understanding of the variety," relates to not only recognizing and operating with variety. The checked performance is, according to Shewhart (1931), variable value and unidentified and fortunate triggers, as a result. Multiple schemes and complicated communication between these themes often affect performance.

The following aspect of Deming's System of Profound Knowledge is information theory, a philosophy that deals with the premises, essence, and extent of information. Deming is stressed that it can not create a theory. Managers are responsible for learning hypothesis and applying it. Psychology is the last component of the System of Profound Knowledge. Psychology enables executives to comprehend individuals. Managers must acknowledge and use psychologists to guarantee the beneficial inherent characteristics of each individual. The 14 leadership points of Deming's System of Profound Knowledge follows.

2.2 Deming's Fourteen Points

Place the figure as close as possible after the point where it is first referenced in the text. If there is a large number of figures and tables, it might be necessary to place some before their text citation.

Table 1. The Deming Management Method

Point 1	Create constancy purpose or toward the improvement of product and service to be competitive and to stay in business, and to provide jobs.
Point 2	Adopt the new philosophy
Point 3	Cease depends on mass inspection to improve quality. Eliminate the need for inspection on a mass base by building quality into the product in the first place.
Point 4	End the practice of awarding business based on price tags alone. Instead, minimize total cost. Move toward a single supplier for any one item, a long-term relationship of loyalty and trust.
Point 5	Improve constantly and forever the system of production and service, to improve quality and productivity, and thus consistently reduce costs.
Point 6	Institute training on the job.
Point 7	Institute leadership.
Point 8	Drive out fear so that everyone may work effectively for the company.
Point 9	Break down barriers between departments. People in research, design, sales, and production must work as a team, to address problems of production and use that may be encountered with the product or service.
Point 10	Eliminate slogans, exhortations, and targets for the workforce asking for zero defects and new levels of productivity. Such relationships only create adversarial relationships, as the bulk of the causes of low quality and low productivity belong to the system and thus lie beyond the power of the workforce.
Point 11	a. Eliminate work standards (quotas) on the factory floor. Substitute leadership.

	b. Eliminate management by objective. Eliminate management by numbers, numeric goals. Substitute leadership.
Point 12	a. Remove barriers right to pride on workmanship. The responsibility of supervisors must be changed from sheer numbers to quality. b. Remove barriers that rob people in management and their right to pride of workmanship. This means, among other things, the abolishment of the annual or merit rating and management by objective.
Point 13	Institute a vigorous program of education and self-improvement.
Point 14	Put everybody in the company to work to accomplish the transformation. The transformation is everybody's job.

Source: WE Deming, 1986, Cambridge: Massachusetts Institute of Technology, Center for Advanced Engineering Study, pp22-24

2.3 SMEs in Indonesia

Small and Medium Enterprises in Indonesia began to be regulated based on the Presidential Decree Number: 96 of 1993. The Decree implies that the Ministry of Cooperatives in Indonesia is responsible for fostering small entrepreneurs. Small business is considered as an influential founder of the nation's economy due to their agility. Beyond that, the mushrooming of small businesses in Indonesia help reducing poverty through job creation. The total number of registered small businesses in Indonesia have reached a figure of 4.076 billion. The development of SMEs in Indonesia is also evidenced by the increasing amount of loans channeled to small businesses.

3 Methodology

The selected samples using random sampling technique and data were collected from 335 SMEs in Indonesia. Research data includes primary data and secondary data. Primary data were obtained from an object of research to make use of n a list of questions (questionnaire), while the data secondary obtained from literature, literature books, magazines and reports related research problems in this study. The analytical method used in this study is descriptive statistics to describe the extent of understanding of the owner or small business manager of Deming's quality management theory.

4 Findings

4.1 Profile of Respondents

From the distribution of questionnaires obtained data regarding the profile of respondents of MSME owners. Of the 335 MSMEs, food and beverage businesses dominate more than half, namely 207 or 62 %. Followed by photocopy services of 14% or 48 MSMEs, and the same amount in the type of salon and fashion business by 12%.

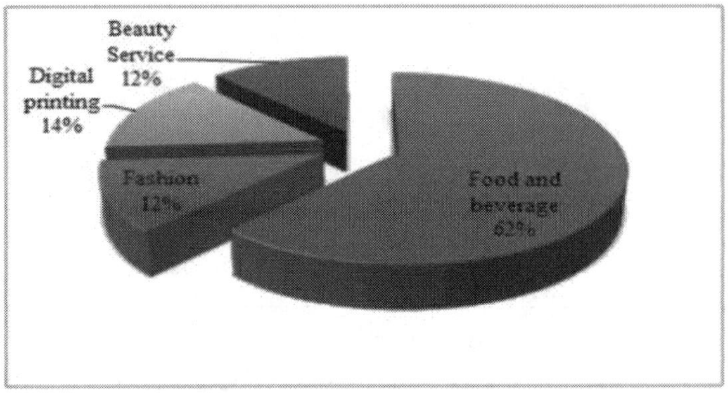

Fig. 1. Characteristics of respondents based on the type of business

While the number of respondents based on business age range <1 year with a percentage of 5 % or as many as 16 respondents, followed by a range of business ages 1-5 years (156 %), as many as 156 respondents, business age range 6-10 years (16%) as much 52 respondents, age range effort 11-15 years old (10 %) of 35 respondents, and the age of the business more than 15 years (23 %) of 76 respondents , as displayed in figure 2.

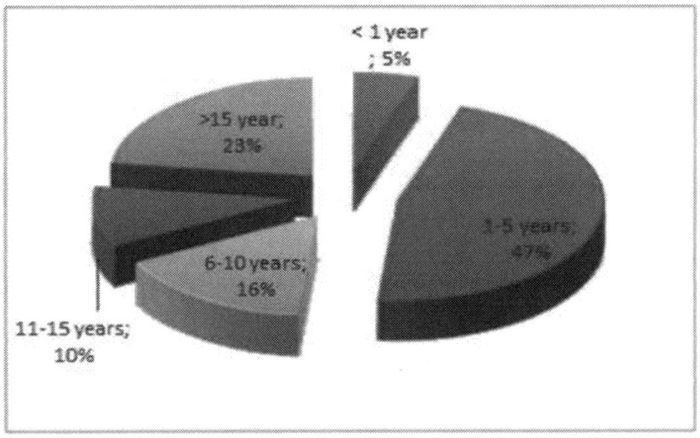

Fig. 2. Characteristics of respondents based on business age

Based on the number of employees, the highest number of employees in UMKM respondents was <10 employees, namely 231 respondents (69 %). The number of employees 10-25 people occupies the second largest number with 81 respondents (24%), followed by the number of employees 26-50 people (6 %), and finally the number of employees> 50 people only 3 people (1 %). For more details, can be seen in Figure 3 below.

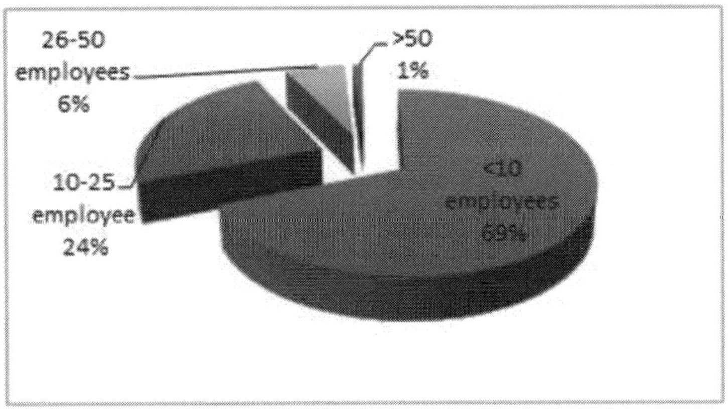

Fig. 3. Characteristics of respondents based on the number of employees

Based on monthly turnover, the most massive turnover of MSME respondents is> 100 million, which is 36 respondents (11 %). The total turnover of 51-50 million was 37 respondents (11%), followed by the turnover of 10-50 million (49 %) or 164 respondents, and the last was the turnover of <10 million there were 98 respondents (29 %). The following is a more precise explanation in Figure 4.

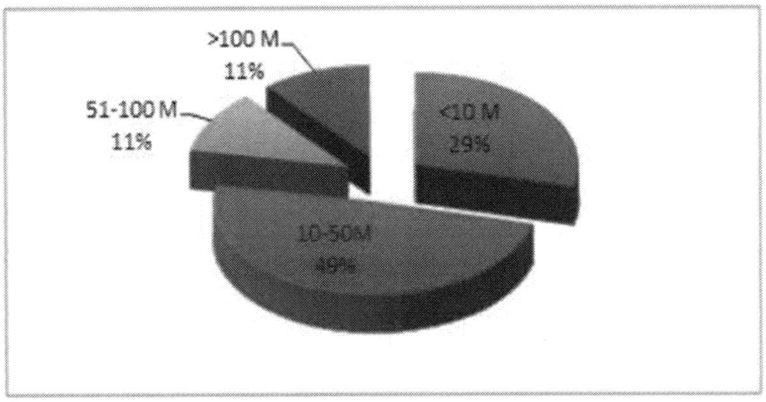

Fig. 4. Characteristics of respondents based on turnover per month

4.2 Descriptive Analysis

Respondents were asked on their daily quality management practice based the Deming's system of profound knowledge. Table 2 presents the results of the analysis using descriptive statistics.

Table 2. Descriptive Statistics

Points	N	Min	Max	Sum	Mean	Std. Dev	Variance	Std. Error
1	335	1	10	2389.25	7.13	1.63	2.65	0.13324
2	335	3.67	10	2700.67	8.06	1.43	2.04	0.13324
3	335	1	10	2258,33	6.74	1.87	3.5	0.13324
4	335	2.25	10	2464.75	7.36	1.4	1.97	0.13324
5	335	3	10	2622,33	7.83	1.37	1.87	0.13324

6	335	1	10	2308.75	6.89	1.86	3.45	0.13324
7	335	1	10	2366	7.06	2.1	4.39	0.13324
8	335	3.5	10	2448.75	7.31	1.36	1.85	0.13324
9	335	1	10	2232.67	6.66	1.85	3.42	0.13324
10	335	2	10	2618,33	7.82	1.36	1.85	0.13324
11	335	3.67	10	2608	7.78	1.37	1.89	0.13324
12	335	3.8	10	2445	7.3	1.28	1.63	0.13324
13	335	1	10	2329.5	6.95	1.78	3.17	0.13324
14	335	3	10	2499	7.46	1.35	1.82	0.13324

Based on Table 2. It shows that from 335 respondents the lowest average value in filling out the questionnaire is 1 and the maximum average value is 10. Then the average value of each attribute is ranged from 6.66 on point number six, which is the attribute "institutionalizing training" to 8.06 on point number 2, namely the attribute "throw away fear." While the standard error value for each attribute is 0.13324, this shows that the level of understanding of small business managers is in the excellent category. Although pedagogically managers do not recognize Deming's big names and their philosophical theories very well, at the practical level, small managers have implemented the theory as a daily practice in developing quality business management. The result of the study can be depicted in figure 5.

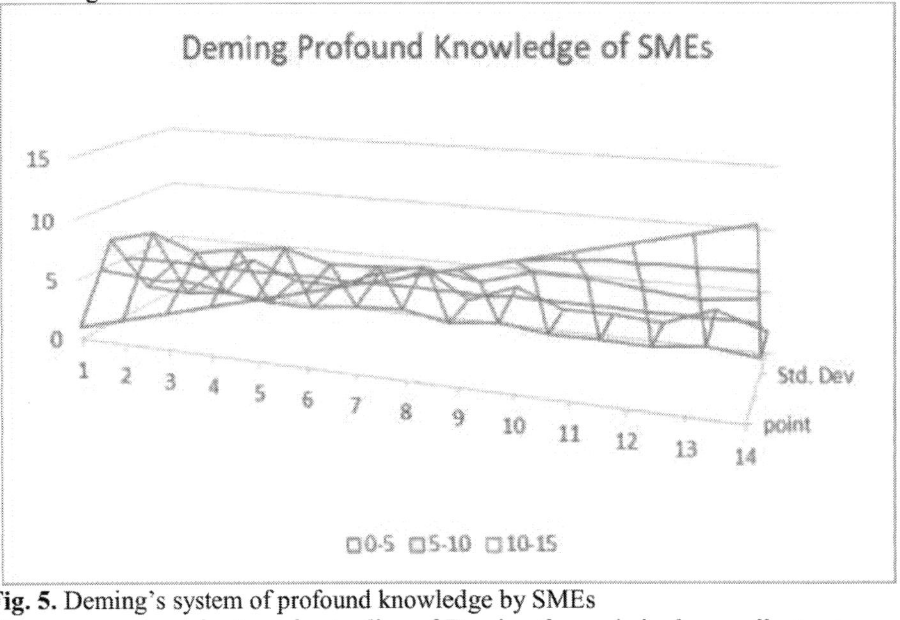

Fig. 5. Deming's system of profound knowledge by SMEs

The average respondent's understanding of Deming theory is in the excellent category. The top three points are the second point, "Adopt the new philosophy of quality", the fifth point is "Improve constantly and forever the system of planning, production, and service, to improve quality and productivity and constantly decrease costs "and tenth point," Eliminate slogans, exhortations, and targets for the workforce ". While three points with the lowest score are: "Cease dependence on inspection to achieve quality. Instead of require statistical evidence that quality is bunt in. ", " Institute modern methods of training on the job "and " Break down barriers between departments."

5 Conclusions

Overall, small business managers are practicing Deming's system of profound knowledge into their business operation. Managers adopt the new philosophy of quality to remain competitive in the market. Likewise, there is strong belief among the managers on the importance of continuous effort for improvement. They improve constantly and forever the system of planning, production, and service, to achieve high quality and productivity and continuously decrease costs. Small business managers also eliminate slogans, exhortations, and targets for the workforce. Managers of small business as a guideline when adopting quality management system can use the result of the study. The theory is easy to implement and have a good impact on organization performance.

References

1. W. R. Adawiyah, B. A. Pramuka, Najmudin, and D. P. Jati, Int. Bus. Manag. (2015)
2. W. R. Adawiyah, M. N. M. Shariff, M. B. Saud, and S. S. M. Mokhtar, Int. J. Bus. Soc. Sci. (2011)
3. P. H. Adi, F. Wihuda, and W. R. Adawiyah, Market-Trziste (2017)
4. S. L. AHIRE, R. LANDEROS, and D. Y. GOLHAR, Prod. Oper. Manag. (1995)
5. J. C. Anderson, M. Rungtusanatham, and R. G. Schroeder, Acad. Manag. Rev. (1994)
6. M. Asif Khan, Int. J. Bus. Manag. (2010)
7. W. E. Deming, Massachusetts Inst. Technol. Cent. Adv. Eng. Study (1986)
8. T. J. Douglas and L. D. Fredendall, Decis. Sci. (2004)
9. G. S. Easton and S. L. Jarrell, J. Bus. (1998)
10. W. B. Gartner and M. J. Naughton, (1988)
11. P. Hari Adi and W. R. Adawiyah, J. Islam. Mark. (2018)
12. K. B. Hendricks and V. R. Singhal, Manage. Sci. (1996)
13. Y. Latif, IOSR J. Bus. Manag. (2014)
14. K. L. McFadden, J. Y. Lee, C. R. Gowen, and B. M. Sharp, Qual. Manag. J. (2014)
15. T. C. Powell, Strateg. Manag. J. (1995)
16. M. Sadiq Sohail and T. B. Hoong, Benchmarking (2003)
17. W. A. Shewhart, Bell Syst. Tech. J. (1930)
18. P. J. Singh, C. M. W. Dean, and S. Chee-Chuong, Qual. Manag. J. (2013)
19. F. Talib, Talib, F.(2013),"An Overv. Total Qual. Manag. Underst. Fundam. Serv. Organ. Int. J. Adv. Qual. Manag. **1**, 1 (2013)

Productive Protection of Intangible Cultural Heritage Based on the Theory of Involvement: A Case Study on the Farmers' Paintings of Xin-Ji County, Hebei Province, China

Kong Xuhong [1*], *Hong* Jingjing [2]

[1,2] Management College, Hebei University, Baoding, Hebei Province, PRC, 071002

Abstract. The productive protection of intangible cultural heritages, always in the form of tourism development under the present context, is put forward by Chinese scholars, which are beneficial to both the protection of the heritage and the economy development of the locals. While not all intangible cultural heritages can be understood and accepted by tourists due to the reasons that the living circumstances and contexts of these heritages are changing and disappearing that it's hard for tourists to understand, neither do they desire to pay for it. Therefore, how to make tourists even including some craftsmen understand and accept the heritage means a lot to the protection and inheritance of these heritages. The paper argues that the Involvement Theory can be referred to analyze settle the problems. A case study of the farmers' painting in Xinji County, Hebei Province was carried out as the example, which is one of the most representative intangible cultural heritage of folk art in Hebei Province, China, with a long history of development, rich cultural connotation and high artistic value. A field investigation and deep-interview was carried out to gather the information of its status quo, problems of its inheriting and developing were analyzed, the paper found that with the development of the times and society, farmers' painting is losing its survival environment, the income of farmers' painting is not proportional to their putting-in and cost, the value of farmers painting can not be reflected, and the productive protection is seriously hindered. Therefore, based on the perspective of involvement theory, this study analyzed the bottleneck of productive protection of Xinji Farmers' Painting, suggested how to stimulate the involvement of tourists into the understanding and producing and creation of the paintings in order to promote the inheritance of the heritage.

Key words: Xinji Farmers' Painting; involvement Theory; Productive Protection; Intangible Cultural Heritage

1 Introduction

Intangible cultural heritage is our spiritual home left by our ancestors, playing more and more important roles in cultural recognition, economic development and rural revitalization. Tapping the intangible cultural resources is becoming a key way to

* Author: Kong Xuhong(1973-), Ph.D and Professor, specialized in tourist resource and rural development. E-mail: kongxuhong@126.com mobile:+8613831268236

reestablish local economy and develop local industry. The productive protection theory and practice is initiated by Chinese scholars. Due to its obvious value in tourism and economic development, the commercial utilization of intangible cultural heritage, usually in the form of tourism development, has been paid full attention these years. In order to avoid excessive commercialization and to differentiate it from the protection under the governmental financial sponsor and solve the dilemma of government protection, Wang (2006) initially put forward this concept in his book, introduced the idea of productive protection of intangible cultural heritage and pointed out that productive protection is one of the basic ways and principles of intangible cultural heritage protection.

While the productive protection is welcomed and practiced by craftsmen, not all intangible cultural heritage can be understood and accepted by tourists due to the reasons that the circumstances and contexts have changed a lot, it is hard for tourists to understand the spiritual contents of the heritage and why it is should be in that way, they are reluctant to pay for it. Therefore, how to motivate craftsmen manifest the art value, regional or ethnic cultural connotation and times background, make tourists understand and accept the heritage means a lot to the protection and inheritance of these heritages. The paper suggests that the involvement theory can be referential to the analysis and settlement of these problems, and a case study was carried out to analyze the dilemma the intangible cultural heritage confronting and how involvement theory can be used under this circumstance.

2 Research Review

2.1 Relevant Research on Productive Protection

The concept of productive protection of intangible cultural heritage was first put forward by Wang Wenzhang in 2006 in his book, where he argues that productive protection is one of the basic ways and principles of intangible cultural heritage protection. This was followed by the definition given in the Guidance on Strengthening the Productive Protection of Intangible Cultural Heritage which was formally formulated and issued by the Ministry of Culture of China in February 2012, therein said that “productive protection of intangible cultural heritage refers to the maintenance of the authenticity, integrity and inheritance of intangible cultural heritage in the course of practice of a productive nature, on the premise of effectively inheriting the skills of intangible cultural heritage, with the help of production, circulation, sales and other means, the intangible cultural heritage and its resources are transformed into the protection of cultural products.” The protection consciousness and cultural consciousness of the inheritors can be effectively enhanced and the economical benefits can be created to the relative stakeholders by the productive protection and conversion of intangible cultural heritage resources into cultural products (Liu Xiaochun et al., 2016). By taking into account economic benefits and cultural spirit, inherits and develops, productive protection has become a hot topic in the field of intangible cultural heritage study, achieved good effects and influence in the field of traditional skills, traditional fine arts and traditional medicine processing.

While more people recognized productive protection as a kind of a new approach to the protection of intangible cultural heritage, the implementation of productive protection is not as easy as their imagination, and the first is the question of the way of intangible cultural heritage innovation. With the fast pace of modernization and the urbanization, many intangible cultural heritage has been difficult to integrate into the life of the present generation, and how to make the modern consumers accept and understand the intangible heritage is a problem that needs to be solved urgently. There are a lot innovation practice in the field, such as the “ 1 + 3 + N” endogenous dynamic mechanism in which,“1” refers to " a set of inheriting mechanism, "3" refers to three driving systems, "N" means several inheriting carriers, this endogenous dynamic mechanism aims to stimulate the motive force of intangible cultural heritage to achieve the transmission of the external blood transfusion type into the power of blood generating type. The establishment and the good operation of the three driving systems (the conscious system, the motivation system and the social operation system) are the decisive force for the effective operation of the dynamic mechanism. It is directly applied to the intangible cultural heritage itself and the several inheritance carriers with a key and decisive function of promoting the productive protection of the intangible cultural heritage. (Chen J.X.2016). Some productive protection tries to build sacred consumption which means that only with "Gods, ancestors and ghost sacred consumption" the strong vitality and identity can be assured. In the process of protecting the traditional art, it is undoubtedly one of the effective ways to protect the traditional technology if it is introduced into the field of folk custom, belief and ceremony(Wen,2017). Liu(2012) suggested the studio inheritance mode of taking the quality route, the large-scale production imitating commercial development and the development of related derivatives , etc.

Secondly, when more people are generating the desire to buy, the question that needs to be further addressed is what channels should these heritages be sold through, including taking advantages of some new customs and festivals such as temple fairs, Valentine's Day, Christmas,etc.(Liu,2014) or selling through the display and exhibition in the meanwhile of producing in workshop (Yang,2016). A putting-out system can also be introduced, which is a kind of hand-made production mode the rural craftsmen and the buyers come to terms and cooperation in the aspects of capital, raw material supply and finished product sales, characterized by the informal contract and the dispersed production (Xu,2016).

Another problem productive protection method have to face lies in the authenticity of the intangible cultural heritage. The cultural connotation including the function, the generating background, the moral of the intangible cultural heritage should be paid attention to and protected and inherited in addition to the heritage itself and the process of production. How to make tourists understand, accept and approve the cultural contents is a significant problem concerning to the inheritance and widespread of the heritage. Based on the theory of involvement, the study tries to make some tentative analyse on the problems how to make tourists more involved and enhance their identity in the cultural heritage.

2.2 Involvement Theory

The concept of "involvement" was first put forward by Krugman (1965). He defined involvement as the connecting experience, connection, or the number of links built or recognized by audience between his own life and the stimulus in unit interval. From this definition, the substance of the involvement is a reflection of strong or weak relationship between the subject and the object. Since Krugman, a number of scholars have made a great deal of research on the involvement and have developed the involved concepts. Palanisamy and Wong(2003) defined involvement as the extent to which an stimulus or task is associated with the existing needs and ideals of the consumer. Zaichkowsky(1994) believes that involvement is the extent to which an individual perceives an advertisement or product associated with its inherent needs, life ideals and interests, while Laczniak,Muelhing and Grossbart(1989) pointed out that involvement is a state of individual motivation induced by a particular advertisement or situation.

In this study, involvement is understood as an unconscious experience, mainly referring to whether tourists feel the correlation between themselves and one of the factors of the intangible cultural heritage in need of productive protection. This is consistent with the fact that some traditional intangible cultural heritage can not be appreciated by tourists. Most of the research on involvement theory is applied in the field of marketing and pedagogy. Few scholars pay attention to the theory of involvement in the field of cultural heritage research, and have not yet discussed the productive protection of intangible cultural heritage from the perspective of involvement theory.

3 Case Study on Farmers' Painting

3.1 Profile of the Farmers' Painting of Xinji

3.1.1 Historical Review of the Xinji Farmers' Painting

Farmers' painting of Xinji County is one of the most representative intangible cultural heritage of folk art in Hebei Province, China, which is a reflection of farmers' life in different times with a very strong times color. During its long history of development, it has formed its own rich cultural connotation and high artistic value. As early as the Ming and Qing Dynasties, people here have folk traditions of painting, embroidery, cutting, weaving and so on. In the 1950s, the People's Art Publishing House had published two picture books about wall pictures of the county. In 1959, farmers' painting was exhibited in the Leipzig International Expo, Germany. In the 1960s and 1970s, the development of farmers paintings tended to be elegant and detailed, paying attention to shaping. A selection of amateur paintings by the masses was published.

During the Great Leap Forward Movement in 1958, Xinji people used cartoons and slogans to arouse the determination of the broad masses of people to actively fight against the enemy. Most of the farmers paintings at that time were based on the party's principles and policies, and also reflected a kind of revolutionary feelings. Most of the farmers' paintings in this stage are displayed in the form of wall paintings. Although the painting style is simple, most of them show the daily life of farmers, such as sowing, spring ploughing, fattening, harvesting and so on, they are vivid and powerful, which represents the farmers' yearning for socialist construction.

During the political movement of the Cultural Revolution, an extremely sensitive period, which was also a catastrophe for the art world, most of the art works at that time followed the party's principles and policies. Looking back at the Xinji farmers' paintings during the Great Revolution, it is not difficult to find that many of them are works with Mao Zedong as the theme. Although this period limited the expression of farmers painting works in the form and content too single, the painting level and color collocation have been significantly improved.

After the two turbulent periods of "Great Leap forward" and "Cultural Revolution", Xinji farmers' painting finally ushered in a period of free and open development. Farmers' painting in this period does not need to focus on political factors, but freely play their own imagination to creation. It was during this time under the leadership of Geng Zhanxiong that Xinji farmers' painting reached a new peak in the manifestation of the successfully held "Xinji Modern Folk painting Exhibition" at the National Art Museum of China on February 14, 1987, and the establishment of Wangxia Village Women Painting Society, which attracted the attention of the world because of its strong artistic flavor and strong folk art color. French embassy officials, British scholars, Mexican embassy officials in China came to Wang Xia Village to pay an exclusive visit to the female farmers' painting society and gave a high praise.

3.1.2 Artistic characteristics of Xinji Farmers' Painting

Hebei Xinji farmers' painting is obviously different from other painting types in terms of painting content and painting method, which embodies the unique cultural characteristics. The inspiration of farmers' painting comes from the experience of farmers' daily production and life, covering a very rich theme, ingenious conception and a strong local cultural flavor, which has strong characteristics of nationality, folklore, originality and creativity, expresses the customs of rural areas and farmers. Xinji farmers' painting is rich in subject matter, ingenious in conception, unfettered, exquisite and simple in painting, and beautiful in color. With the simple lines, farmer's experience and vision of rural life are vividly and incisively expressed, from big life events such as spring ploughing and autumn harvest, weddings and funerals, to as small as a single brick and tile, trees and bushes, all exudes a strong local flavor and regional amorous feelings. In addition, it also absorbs the quintessence of paper-cut, printing and dyeing, opera and other arts, and integrates into the painting, forming a unique style of childish simplicity that suits both refined and popular tastes.

3.2 The Problems of Productive Protection of Xinji Farmers' Painting

Xinji farmers' painting was listed as an intangible cultural heritage representative of Hebei Province in 2006. Productive protection was adopted and a creation base was established. Farmers painting are becoming well-known and popular in the market.

While there are some problems existed in the productive protection of farmers' painting. Firstly, there is an imbalance of development between villages. In 2017, Wangxia Village was listed as a characteristic town featured painting creation which got a 6million yuan investment, there are almost a hundred farmers painters between 30-50 years old. While in other villages, some painters left villages to work or do businesses in cities because the lack of sale, the left painters are aging seriously, the farmers' painting is losing successor.

Secondly, the painting is losing its creation roots. Farmers' painting is generally based on the productive labor of farmers or the simple scenery of rural areas, and the traditional folk villages are the creative environment of most farmer paintings. With the acceleration of urbanization in rural areas, the traditional farming life has undergone earth-shaking changes, the traditional folk villages have been destroyed, the artistic expression of traditional farming life and its memory and the source of materials for farmers painting is becoming deficient.

Thirdly, the painter is changing. The creator of farmers' painting is supposed to be a farmer, living the rural life of farming when he is busy and painting when he is idle. Painting is an amateur activity. As the myth of farmer painters getting rich became popular in painting villages, and farmers' painting has been used in specialty and souvenir producing, the painting derivatives are not only painted on paper, but also printed on umbrellas, cups, fans. It not only brought high reputation at home and abroad, but also real benefits for the local farmers. Now the painting has been accepted in educational system, more painters are becoming professional, some villages are changing to painting villages during the process of productive protection. The painters evolve to two different ends, one is the high-end pursuing of farmers' painting masters, the other is the low-end mass production and batch copy of the unnamed farmer painting workers.

Fourthly, the market is changing. The prosperity of farmers painting has its background, when there is not enough artist forms and entertainment. The farmers painting meets the aesthetic demands of market. With the continuous development of the times, the cultural life of the Chinese people is no longer scarce, and a variety of art forms emerged endlessly. Problems such as the old school of painting works are not in line with modern aesthetics, the form of expression is old-fashioned and can not meet the needs of modern individualization, the content of expression is solidified and different from the contemporary thought, and the historical value, cultural value and artistic value behind it are not well known to the public reduced the farmers painting were largely shrugged off by the market. As a result, a large number of old farmer painters put down their brushes. Now only Geng Zhanxiong and Wang Xiaojun as well as Cao Zhihua, Jia Yan and Geng Zhanxiong's daughter, Geng Wenxiao working in Xinji Cultural Museum are the only old artists who are still insisting on farmers' painting creation.

Other problems such as high enterprise tax burden, the lack of funds are also interfering with the protection and development of farmers' painting. The most important the problem is how to understand farmer's painting and how to make it accepted by market, whereby can be analyzed and solved by the involvement theory.

3.3 Productive Protection Countermeasures of Xinji Farmers' Painting Based on Involvement Theory

First of all, it is important to note that, while promoting its inheritance and development in the form of sales in the production protection, to meet the needs of modern people can not be avoided. But one main line must be stick to, that is, to always follow the core skill and core culture of the farmers' painting. As shown in Fig.1, both farmers and tourists need to be involved in the producing of the painting, while the experience type involvement, the exposure type involvement should be strengthened together with the functional involvement, the ideological involvement and the emotional involvement to enhance the relevance between tourists and the painting, and ultimately the purpose of the productive protection of the intangible cultural heritage can be realized.

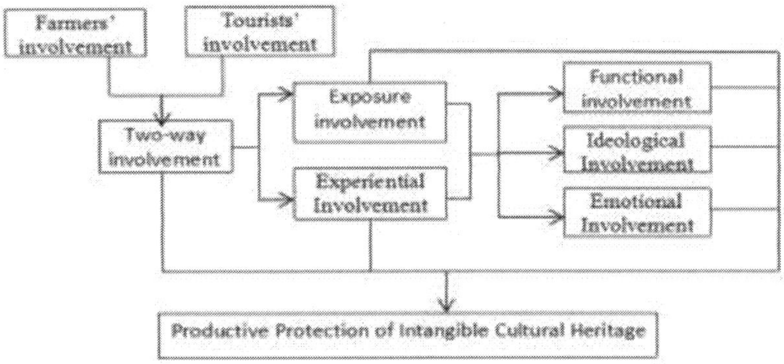

Fig.1: Involvement Analysis on the Productive Protection of In tangible Cultural Heritage

3.3.1 Two-way Involvement

Two-way involvement can be said to be the involvement of farmer painters and tourists, it is through the involvement of painters to promote the involvement of tourists, to realize the relationship with tourists from the innovation of artistic characteristics, which means the addition of popular art and innovative design in line with modern popular aesthetics on the basis of maintaining the core skills and core culture of Xinji farmers painting in order to promote the involvement of tourists.

3.3.2 Experiential Involvement

Because the involvement is the expression of the connection intensity between the subject and object, and as the object, Xinji farmers' painting is presented by the hand of the farmers, so it involves the problem of how to strengthen the association between the farmer and the tourists to maximize the bidirectional involvement. Experiential involvement is the key point of this problem, which combines the bidirectional involvement to a certain extent. And this kind of experiential involvement must rely on the rural and farmers field to achieve the effect.

Experiential involvement means that tourists can participate in the production of farmers painting in the field of rural and farmers' land, which can be led by traditional farmer painters, so that they can deepen their understanding of farmer paintings and their relevance to themselves in the process of creating farmer paintings.

3.3.3 Exposure Type Involvement

The so-called exposure involvement is based on the "exposure effect", that is, when tourists see or come into contact with something frequently, they will increase their preference for it. This is also an important basis to solve the problem of Xinji farmers' painting dissemination. With the development of new media technology, some farmer painting works with rich artistic characteristics can be collected and distributed to network media platforms by means of pictures, videos and so on to achieve certain exposure and enhance people's understanding of them, but the exposure frequency should not be too high, otherwise it will form a "hyperpolar effect" that too strong stimulation will cause boredom and resentment.

3.3.4 Functional involvement

Functional involvement can be achieved by linking the daily lives of modern people and the creation of derivatives. First of all, it should be associated with the daily life of modern people, such as a collection of farmer paintings, some can be displayed in the study, which can express some of the moral of studying hard and reading more; some can be put in the restaurant to express the moral meaning of valuing food; some can be put in the living room to express the moral of family and beauty, and share the moral of harmony. Secondly, the creation of derivatives, Xinji farmers' painting is mostly manifested in the form of paper painting, mural painting at present, so it can be used on other carriers combined with other objects to create popular derivatives.

3.3.5 Ideological Involvement

The involvement of ideological thought is the unity of the idea with the modern people's preference of humor and the connotation. Cultural elements and symbols catering to modern interests such as some popular slogans with certain philosophy or or some popular and profound lines in some popular old films can be added to the traditional meaning of the farmers' painting, such as piety, diligence and other positive emotions. Such personalized popular statement is the expression of the current generation, which will bring closer to the distance between the painting works and tourists.

3.3.6 Emotional Involvement

Emotional involvement refers to making people realize the significance of Xinji farmer painting emotionally, which represents the inheritance of human wisdom. Because most of the contents of farmer paintings express positive feelings, they can be combined with the spiritual civilization construction put forward by Chinese government today, such as displaying and publicizing the core values of socialism in the form of farmer paintings. Technically, we can make full use of the environment of the Internet, spread the culture of farmer painting, let people know the farmer painting and its profound historical and cultural connotation, from the emotional recognition.

4 Conclusion

At present, although people's living environment and way of life have changed, but we still need traditional culture. However, with the development of the times, most of the intangible cultural heritage such as Xinji farmer painting is facing the situation that the value can not be reflected and can not be sustained. It can not be separated from the fact that tourists can not appreciate some traditional intangible cultural heritage. Therefore, it is necessary to explore productive protection from the perspective of involvement theory, which provides a path for the subsequent protection of intangible cultural heritage. The humanistic value of farmer painting can be easier to be understood and appreciated by more and more people, so that the culture of farmer painting can be sustainable development based on the six strategies proposed in this study.

5 Acknowledgement

Thanks are due to the farmer painters of Xinji County interviewed in the study and scholars who made contributions to the study of productive protection and involvement. The study is supported by the Provincial Key Program of Cultural and Artistic Science Planning Project of Hebei Province(Grant No.HB18-ZD011).

6 References

Chen Junxiu. (2016).Research on the productive protection and utilization model of intangible cultural heritage.Journal of Guangxi National University(Philosophy and Social Sciences,38(4):64-71.(in Chinese)

Krugman.HE.(1965).The impact of television advertising: Learning without involvement. Public Opinion Quarterly,29:349-356.

Laczniak.RN,Muelhing.DD,Grossbart.S.(1989).Manipulating message involvement in advertising research. Journa lof Advertising, 18(2):28-38.

Liu Qiong.(2012).A probe into the productive protection of handwork intangible cultural heritage : A case sudy of Beijing silk figurine. Cultural Heritage,(4):62-68.(in Chinese)

Liu Qin(2014).The reconstruction of folk custom and the productive protection of Yang Liuqing woodblock new year painting.Art Observation,(10):22-23.(in Chinese)

Ministry of Culture.(2012,February 27).Guidance on strengthening the productive protection of intangible cultural heritage.Chinese Culture Daily,1.

Palamisamy.R,Wong.SA.(2003).Impact of online consumer characteristics on web-based banner advertising effectiveness.Global Journal ofFlexible System Manageent,4(1/2):15-25.

Wang Wenzhang. (2006). Introduction to intangible cultural heritage. Beijing: Culture Press.

Wen Xiaoxing.(2017). "Sacred Consumption" and Non-productive Protection of intangible cultural heritage of traditional craftsmanship-Case study of "Hengjiang Zhongzhi" production of Shicheng County of Jiangxi Province. Journal of Northwest Minzu University(Philosophy and Social Science,(4):7-12.(in Chinese)

Xu Jianghong.(2016).Productive protection of intangible cultural heritage under the path of putting-out purchase: A case study of Miao Batik in Danzhai, Guizhou Province. Guizhou Social Science,(06):88-93.(in Chinese)

Yang Yan(2016).Game and interaction between productive protection and traditional cultural exhibition-Taking the development of Mianzhu New year Painting before and after the earthquake as an example. Sichuan Opera,(04):90-93.(in Chinese)

Zaichkowsky.JI.(1994).The personal involvement inventory: Reduction,revision,and application to advertising. Journal of Advertising,23(4):59-70.

Weakened Patient Loyalty Model at Beauty Clinics:Based on Variety Seeking Behavior, Dissatisfaction, Negative WOM and Brand Switching

Herni Justiana Astuti [1*] and *Suryo* Budi Santoso [1]

[1] Universitas Muhammadiyah Purwokerto

Abstract. The purpose of this study was to design the Weakened Loyalty model through Variety Seeking Behavior, Dissatisfaction, Negative WOM and Brand Switching. The sample in this study was a patient of a beauty clinic who did treatment for one year or more and had moved from another clinic. The sample used was 173 respondents. Data analysis using Structural Equation Modeling with PLS approach. The data meets convergent validity and composite reliability. The results of the analysis prove that negative WOM, variety seeking behavior and dissatisfaction are factors that have the potential to weaken loyalty, because these three variables have a significant direct effect on brand switching. However, it does not directly affect patients to really weaken loyalty. Dissatisfaction can mediate relationship between Negative WOM and Brand Switching.

Keywords: Patient Loyalty, Variety Seeking Behavior, Dissatisfaction, Negative WOM and Brand Switching

1 Introduction

Competition in the business world in the modern era is undergoing many changes. Companies in any form look for the right strategy to reach the best position in the minds of consumers. One example is the competition in the beauty clinics in Indonesia. The development of a fast-growing beauty trend also affects the tastes of beauty products. Beauty that is identified with women, makes them try to look attractive not only fulfills the need for a beautiful face, but also attractive body and skin care. Nowadays there are more and more places of beauty clinics with various choices that suit customer needs and even dare to give promises about the desired beauty criteria.

Customer loyalty is absolutely needed by a company to be able to survive and be able to compete with other companies. To build customer loyalty, there is a need for companies to continue to provide the best quality in each product or service they have. Satisfied and loyal customers are opportunities to get new customers. Maintaining all existing customers will generally be more profitable than customer changes because the cost of attracting new customers can be five times the cost of retaining an existing customer (Kotler and Keller, 2009).

In today's increasingly competitive world, the word "loyal customers" is very rare. Due to the fact in the field there is no evidence that shows that there are consumers who are truly loyal to one particular brand. The average person must have consumed different brands in one category. One reason is post-purchase dissatisfaction. The greater customer

* Corresponding Author: +6289675955925
Email address: herni99@gmail.com

dissatisfaction, the more likely they are to switch providers, complain, and discuss their dissatisfaction with others (Zeelendberg and Pieters, 2004). Halstead (2002) supports the findings of previous researchers, that dissatisfied consumers significantly involved more word-of-mouth behavior than satisfied consumers, and would tell more people about their unsatisfactory experiences. Baird (2000) added that dissatisfied customers will tell 13 other people. Suharseno et al. (2013) explained that brand shifts carried out by consumers were caused by variations in search. The variety seeking behavior can be caused by boredom in product attributes. Consumers will be very easily affected when faced with a variety of brands that provide the desired product, causing consumers to be loyal to one brand. Based on the findings of Astuti and Nagase (2016), loyalty is prone to erosion when a patient decides to try treatment elsewhere receive negative information about service performance. These events can result in reduction or elimination of loyalty, and patients can switch to other service providers.

Based on the description above, researchers consider it important research to design models that cause customer loyalty to be weak. Is the decision to switch brands, dissatisfaction, variation seeking behavior, and talk from mouth to mouth causing loyalty to be weak? The results of the study can be a reference for beauty clinics to anticipate the causes of erosion of consumer loyalty. The purpose of this study is to design a model that weakens loyalty in terms of brand switching decisions, dissatisfaction, variation seeking behavior, and word of mouth conversation.

2 Literatures Review and Hypothesis

2.1 Patient Loyalty

Basically loyalty is the behavior of consumers to continue to use products or services and are based on attitudes towards products or services. The difference between loyal use and habits relates to the dynamics underlying the selection of a particular product or service. A loyal buyer, at some level, is involved in a relationship, while habitual buyers are indifferently involved in routine behavior (Knox, 1998). Dick and Basu (1994) treat the concept of customer loyalty as a relationship between one's attitude towards an entity (brand, service, shop, and vendor) and someone's patronage behavior. Customer loyalty is also determined by consumer characteristics. For example, some people don't like uncertainty and are very loyal to the first product they use. Others are more "adventurous" and want to try new products even though they like or are satisfied with the previous product.

2.2 Brand Switching

Brand switching is a condition in which a consumer moves his loyalty and switches to a competitor's brand and stops using his old product. Brand switching behavior is a complex phenomenon carried out by consumers for certain reasons. The vulnerability of consumers to continue consuming one brand can occur because of post-consumption dissatisfaction (Thawil, 2014), the existence of behavior seeking variation due to boredom or wanting to try other products (Arundina, 2016) or incessant negative talk by word of mouth (Kartajaya and Ridwansyah, 2012).

From the description above, the hypothesis can be drawn:

H1: Brand Switching has a significant negative effect on loyalty

2.3 Dissatisfaction

Consumer dissatisfaction occurs when the performance of a product is not in accordance with the perceptions and expectations of consumers (Kotler and Keller, 2009: 172). Consumers who experience service failure will react in various ways. According to Zeithaml et al (2012) that consumers have two ways to react; active and passive. Consumers who are passive do not react to complain, maybe because they do not know how to complain, or because they do not believe that the complaint submitted will be responded positively. Those who will actively react otherwise, he makes a complaint directly to the service provider, will disseminate his dissatisfaction to others for various reasons, he even uses a third party. All reactions, whether passive or active leads to the same goal: they leave or switch to another provider, or remain loyal to the product. Wangenheim and Bayon (2004) found that 68.9% of respondents were predicted to remain loyal or switch. One consideration is satisfaction. Dissatisfied consumers will switch brands (Yunita and Rosa, 2016; Andriani and Untarini, 2015; Astuti and Nagase, 2014), while satisfied consumers will remain loyal (Astuti and Nagase, 2014; Astuti and Nagase, 2016).
From the description above, the hypothesis can be drawn:
H2: Dissatisfaction has a significant positive effect on brands switching
H3: Dissatisfaction has significant negative affect on loyalty

2.4 Variety Seeking Behavior

Variation seeking behavior is the behavior of consumers who try to find brand diversity outside their habits because the level of involvement of some products is low. This behavior often occurs in some products with a low level of involvement. According to Assael in Waluyo (2003), said this behavior tends to occur at the time of the purchase of a product that raises minimal risks borne by consumers and when consumers lack commitment to certain brands. Variety seeking behavior arises when consumers feel fulfilled or are bored with the characteristics of products consumed before. The variety seeking behavior can be caused by boredom in the product. Consumers will be very easily affected if faced with a variety of brands that provide the desired product so that consumers are not loyal to one brand only.
From the description above, the hypothesis can be drawn:
H4: Variety seeking behavior has a significant positive effect on brand switching
H5: Variety seeking behavior has a significant negative effect on loyalty

2.5 Negative Word-of-mouth

Word of mouth is defined by Lau and Ng (2001) as verbal, person-to-person communication between what non-commercial communicators and recipients feel, about brands, products, or services offered for sale. Negative Word of Mouth is the act of giving consumers negative information to other consumers from someone to another (interpersonal) non-commercial both brands, products and services (Hasan, 2010). Previous research Harrison-Walker (2001) in Padmalia (2018) shows that every satisfied customer will tell 3-5 other people about his experience, while those who are dissatisfied will tell 10-11 people. This shows that consumers often tell dissatisfaction with products rather than satisfaction. Therefore, the company needs to pay attention to the negative transmittal about the company which will ultimately influence product purchasing decisions and brand switching (Kartajaya and Ridwansyah, 2012).
From the description above, the hypothesis can be drawn:
H6: Negative Word of mouth have a significant positive effect on brands switching

H7: Negative Word of mouth have a significant positive effect on dissatisfaction
H8: Negative Word of mouth have a significant negative effect on loyalty

2.6 Brand switching and Dissatisfaction as intervening variables that weakens loyalty

Brand switching is the consumer's decision to switch and use other product brands. Consumer decisions can be due to consumer behavior that is easily bored, and wants to try new products (Arundina, 2016). Brand switching can also be due to post-purchase consumer dissatisfaction (Yunita and Rosa, 2016). In addition, consumers decide to switch brands can be caused by negative word of mouth. The antecedents of switching brands cause consumers to be no longer loyal.
From the description above, the hypothesis can be drawn:
H9: Brand switching mediates the relationship between negative word-of mouth, variety seeking and dissatisfaction to loyalty
H10: Dissatisfaction mediates the relationship between negative word-of mouth to brand switching and loyalty

3 Research Method

3.1 Data Analysis

Data were analyzed using structural equation modeling with a partial least square approach.

3.2 Measures

PLS-SEM analysis usually consists of two sub-models, namely the measurement model or often called the outer model and the structural model or often called the inner model. Evaluation of models using SmartPLS 3.0 can be done by assessing the results of measurement models, namely through confirmatory factor analysis (CFA) by testing the validity and reliability of latent constructs (see table 1). Then proceed with the evaluation of structural models and significant testing to test the influence between constructs or variables.

Table 1. Rule of Thumb Evaluation of the Measurement Model (Gozhali and Latan, 2015)

Validity and Reliability	Parameter	Rule of Thumb
Convergent Validity	Loading Factor	1.> 0.70 for Confirmatory Research 2. 0.50 - 0.60 is still acceptable for the initial stages of development
	Average Variance Extracted (AVE)	> 0.50 for Confirmatory and Exploratory Research
Reliability	Composite Reliability	1.> 0.70 for Confirmatory Research 2.> 0.60 is still acceptable for the initial stages of development

Furthermore, after the outer model meet the criterias, the inner model was tested.

4 Findings

The response rate for the questionnaire was 96.1%. The total number of respondents was 173, they were patients at a beauty clinic who had changed services and were currently taking care for 1 year in the current beauty clinic.

4.1 Outer Model

A measurement model can proceed to the inner model stage if it meets valid and reliable criteria. It can be seen that table 2 describes a measurement model. All indicators are acceptable because loading factors are greater than 0.5. Likewise with AVEs, all variables meet the criteria for Convergent Validity which is greater than 0.5. So all indicators and variables are valid. The model is said to be reliable if the value of Composite Reliability is more than 0.6. If seen in table 2, all variables have met reliable criteria.

Table 2. Variables, Indicators, Loading Factor, AVE and Composite Reliability

Variables	Indicators	Loading Factors[a]	AVEs[b]	Composite Reliability[c]
Loyalty	Firmness (L1)	0.912		
(Astuti dan Nagase, 2014)	Defense (L2)	0.745	0.699	0.874
	Repeat purchases (L3)	0.842		
Brand Switching	Dissatisfaction (BS1)	0.846		
(Astuti dan Nagase, 2014)	Variation seeking (BS2)	0.785	0.590	0.811
	Negative WOM (BS3)	0.663		
Variaety Seeking Behavior	Differences between brands (VSB1)	0.826		
(Priansa, 2016)	Bored (VSB2)	0.847	0.553	0.781
	Just want to try (VSB3)	0.510		
Negative-WOM	Mutual dialogue.(WOM1)	0.639		
(Hasan, 2010)	Pass along effect (WOM2)	0.863	0.597	0.853
	Knowledge diffusion (WOM3)	0.876		
	Cause and effect (WOM4)	0.684		
Dissatisfaction	Bad treatment experience (DS1)	0.809		
	Disappointed (DS2)	0.841	0.684	0.866
	Not recommend to others (DS3)	0.830		

AVE: Average Varian Extract
[a]Acceptable value of Loading Factor is greater than 0.5
[b]Acceptable value of AVE is greater than 0.5
[c]Acceptable value of Composite Reliability is greater than 0.6

4.2 Inner Model

Evaluation of structural models or inner models aims to predict relationships between latent variables. The inner model is evaluated by looking at the magnitude of the percentage variance described, namely by looking at the R-Square value for the endogenous latent construct. Based on Figure 1, it can be seen that Negative WOM can explain the relationship with Dissatisfaction of 24.7%. Whereas Negative WOM, Variety Seeking Behavior and Dissatisfaction can explain the relationship with Brand Switching at 47.1%. Then, Loyalty is explained by all latent variables of 11.7%.

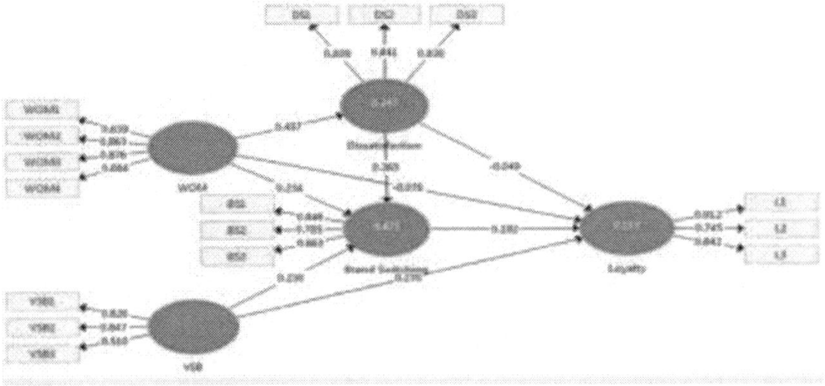

Figure 1. Structural Equation Modeling of Loyalty in Beauty Clinics

Based on the literature review that builds the hypothesis, 10 main hypotheses are obtained. Table 3 explains the results of the relationship, path coefficient, and P Value. Of the 10 hypotheses, there are 4 hypotheses received from the relationship of direct influence and 1 hypothesis received from the relationship of indirect influence. The relationship of variety seeking behavior with loyalty has a significant effect, but the direction of the relationship is positive. While the hypothesis is a significant negative effect, therefore the hypothesis is Potentially Weakens Loyalty

Table 3. Relationships, Path Coefficients, P Values, Result

Relationships	Path Coefficients	P values	Results
Brand Switching ➔ Loyalty (H1)	0.192	0.086	Rejected
Dissatisfaction ➔ Brand Switching (H2)	0.365	0.000	**Acceptable**
Dissatisfaction ➔ Loyalty (H3)	-0.049	0.628	Rejected
Variety Seeking Behavior ➔ Brand Switching (H4)	0.230	0.002	**Acceptable**
Variety Seeking Behavior ➔ Loyalty (H5)	0.270	0.007	Rejected
Negative WOM ➔ Brand Switching (H6)	0.234	0.002	**Acceptable**
Negative WOM ➔ Dissatisfaction (H7)	0.497	0.000	**Acceptable**
Negative WOM ➔ Loyalty (H8)	-0.076	0.384	Rejected
Negative WOM ➔ Brand Switching ➔ Loyalty (H9a)	0.045	0.119	Rejected
Variety seeking ➔ Brand Switching ➔ Loyalty (H9b)	0.044	0.187	Rejected
Dissatisfaction ➔ Brand Switching ➔ Loyalty (H9c)	0.070	0.110	Rejected
Negative WOM ➔ Dissatisfaction ➔ brand switching (H10a)	0.181	0.000	**Acceptable**
Negative WOM ➔ Dissatisfaction ➔ Loyalty (H10b)	-0.025	0.643	Rejected

5 Discussion

Based on the theoretical building described earlier in this paper, factors that can weaken loyalty are brand switching, dissatisfaction, negative WOM and variety seeking behavior. However, some hypothesis test results are different.

5.1 Potentially Weakens Loyalty

A patient has switched brands from a previous clinic because of their behavior, which is easily bored and might like to try new products, in fact makes them loyal to the beauty clinic they are currently using in the past year. A patient who experiences post-purchase dissatisfaction will weaken loyalty, but this relationship does not significantly weaken loyalty. This relationship also occurs due to negative WOM towards loyalty. Beauty clinic patients who have the potential to switch brands because of getting negative information, due to their adventurous behavior and dissatisfaction, apparently do not significantly affect loyalty.

In theory, these three variables make loyalty weak and have the potential to switch products or services. This has indeed happened to beauty clinics before, they have switched brands to beauty clinics now and for the past year or more they are still doing maintenance and buying products at the current beauty clinic. Beauty clinic service providers must be aware of their potential to move.

5.2 Really Weakens Loyalty

Loyalty is weak when the desire for brand switching arises. The desire to move can be seen from the incessant negative information received by patients from word of mouth, patients who are easily bored and have the desire to try other products, or unpleasant experiences. These three variables directly have a significant effect on switching brands, even negative information received causes patients to begin to feel dissatisfied and to compare actual expectations with the performance of beauty clinics.

6 Conclusions and Managerial Implication

6.1 Conclusions

A service company such as a beauty clinic continues to strive for its strategy to have loyal customers. Negative WOM, variety seeking behavior and dissatisfaction are factors that have the potential to weaken loyalty, because these three variables have a significant direct effect on brand switching. However, it does not directly affect patients to really weaken loyalty.

6.2 Managerial Implication

Respondents of this study were those who moved from the previous beauty clinic and now they are still at least 1 year of care at the current beauty clinic. The factors that cause loyalty are weak and the potential for brand switching must be the concern of beauty clinic managers. Customers feel comfortable and suitable for doing treatment at the current beauty clinic, but when they get unpleasant experiences they have the potential to move. Therefore beauty clinics must strive so that when they connect with the clinic, they must be satisfied. Managers must also anticipate dissatisfied consumers because they have the potential to share their negative experiences with potential customers. Likewise, to deal with consumers who get bored quickly, managers also try to innovate attractive care packages so as to minimize customers to switch brands.

References

Arundina, Debora Ocvia., Sriwidodo, Untung., and Wibowo, Edi. (2016) Influence Analysis of Prices, Product Quality, Variety Seeking Behavior And Product Competitor Toward Brand Switching Decision in Charm Women's Consumers. Jurnal Ekonomi dan Kewirausahaan Vol. 16 (Special edition): 132 -142 (In Indonesian)

Astuti, HJ dan Nagase. K (2016). A framework for conceptualizing patient loyalty to healthcare organizations. Health Services Management Research.Vol. 29(3) 70–78.
Dick AS and Basu K. Customer loyalty: Toward an integrated conceptual framework. J Acad Market Sci 1994; 22(2): 99–113

Gozhal
i, Imam., dan Latan, Hengky. (2015). PARTIAL LEAST SQUARE Konsep, Teknik dan Aplikasi Menggunakan Program SmartPLS 3.0. Semarang: Undip (In Indonesian)

Halstead D. Negative word of mouth: Substitute for or supplement to consumer complaints? J Consum Satisf DissatisfComplain Behav 2002; 15: 1–12

Hasan, Ali (2010). Word of Mouth Marketing. Medpress, Yogyakarta (In Indonesian)

Kartajaya, H., & Ridwansyah, A. (2012). Service with Character, Jakarta: PT Gramedia Pustaka Utama (in Indonesian)

Knox, Simon. (1998). Loyalty-based segmentation and the customer development process. European Management Journal, 16(6), 729-737

Kotler, Philip dan Lane, Keller, Kevin. 2009. Marketing Management. Penerbit Erlangga (In Indonesian)

Lau GTand Ng S. Individual and situational factors influencing negative word-of-mouth behaviour. Can J Admin Sci 2001; 18(3): 163–178

Padmalia, M. (2018). Determinants of Decisions in Purchasing Green Label Products Based on Green Products and WOM. Jurnal Manajemen Maranatha, 17(2), 91-102 (In Indonesian)

Suharseno, Teguh, Riskin H. dan Dyah Ayu L.D. 2013. Effect of Consumer Dissatisfaction and Characteristics of Product Categories on Brand Switching Decision with Variety Seeking As Moderating Variables .Jurnal Buletin Studi Ekonomi. 18(2):176-182 (In Indonesian)

Thawil, Ayu N. 2014. Effects of Post-Consumption Dissatisfaction and Variety Seeking Behavior on Brand Switching GSM Mobile Phones from Blackberry to Samsung Brands.Jurnal Riset Bisnis dan Manajemen.2 (4):73-89 (In Indonesian)

Waluyo, P., Pamungkas, A., 2003, Analysis of Consumer Switching Brand Behavior in Purchasing Mobile Products in Semarang, Jurnal Bisnis dan Ekonomi: Semarang (In Indonesian)

Wangenheim FV and Bayo´n T. 2004.The effect of word of mouth on services switching: measurement and moderating variables. Eur J Market ; 38(9/10): 173–1185

Yunita, Dessy and Rosa, Aslamia (2016). Influence Analysis of Variety Seeking, Dissatisfaction and Product Unavailability to Brand Switching. Jurnal Manajemen dan Bisnis Sriwijaya. Vol 14. No. 4: 539-558 (In Indonesian)

Zeelenberg M and Pieters R. Beyond valence in customer dissatisfaction: A review and new findings on behavioral responses to regret and disappointment in failed services. J Bus Res 2004; 57(4): 445–455

Zeithaml V, Bitner MJ and Gremler D. 2012. Services marketing, 6th ed. New York: McGraw-Hill Higher Education

Network governance in value chain management of traditional fisheries industry

Nur Choirul Afif[1*], and *Lusi* Suwandari[1]

[1]Faculty of Economic and Business, Jenderal Soedirman University, Indonesia

Abstract. Indonesia is one of the countries having the longest coastlines in the world. However, the productivity in fishing industry less contributes to the Gross Domestic Product (GDP). Besides, fishermen along the coastlines in Indonesia have a relatively very low standard of living. This study aimed to determine the role of network governance in the value chain of traditional fisheries industry on the coastal region of Cilacap, Central Java.This research was an exploratory study with the objective of exploring the concept of network governance in the traditional fisheries along the offshore Nusa Kambangan coast, Cilacap, Central Java. The sampling method used in this study was non-probability sampling with purposive sampling technique. The data collection techniques were observation, interviews, questionnaires, and documentation.The results of this study were expected to be able to get the actual value chain of wild fisheries off the coast of Nusa Kambangan, so that they can provide suggestions on the network governance model in the value chain of the traditional fisheries industry.

1 Introduction

Indonesia's economic growth continued to increase in the decade of 2006 to 2016. The economic growth was supported by the development of new economic zones. We saw that this activity was expected to be able to increase new economic growth areas. The central government tried to make a package of economic policies to increase economic growth in the new region so that the distribution of wealth was not only spread across the island of Java. The Nawa Cita (nine development goals) program created by the central government was expected to be able to build new economic zones in areas of 3 T, Terdalam (Inmost), Terluar (Outermost), and Terpencil (Remotest). The development of the areas was expected to be able to reduce income inequality between the people with low and high income. The efforts of the central government in resolving various economic problems of the community need to be appreciated. The appreciation was carried out with the support of various layers of society, both businessmen and local governments.

The central government is trying to build road, port, airport, and other infrastructures. However, was the infrastructure development in line with the improvement in the living standards of the people in the areas? The question is a fundamental thing that must be

[*] Corresponding author: author@email.org

answered by stakeholders, considering the purpose of development is to improve the welfare of the people, especially traditional fishermen. The program was considered futile when traditional fishermen along the coasts in Indonesia couldn't use these infrastructures. The program will not be enjoyed by traditional fishermen if the productivity in obtaining wild fisheries does not have a competitive advantage compared to other countries in the ASEAN region. The enactment of the ASEAN Economic Community (AEC) at the end of 2015 marked a reduction of government protection of Indonesian businessmen, including traditional fishermen. The central government also can no longer protect traditional fishermen from the entry of foreign companies in the fisheries sector, where they have the resources and technology that are far superior to the resources and technology possessed by local fishermen along the Indonesian coastlines. The concern is that we import fish, especially from countries in the ASEAN region, which have opened a marine fishing industry since the introduction of the AEC in early 2016.

The construction of such infrastructures must be able to sustain and enhance the competitive advantage of traditional fishermen in Indonesia. If the productivity of traditional fishermen is still relatively lower than that of the fishermen of other ASEAN countries, it means the infrastructures are mostly used by foreign multinational companies to enter and take market share in the country. Therefore, the role of various stakeholders is needed so that traditional fishermen have a competitive advantage in order to excel from foreign products. Competitive advantage can be realized with a holistic understanding of the value chain of the products. In this context, it is needed the value chain of fisheries industry. However, the value chain could not be fulfilled by fishermen, considering they have a variety of resource limitations. The most possible thing is to implement network governance in the management of the fisheries industry.

As it has been known, the purpose of the economic policy package is to improve people's welfare, which in this case is through the construction of fishing port infrastructure to reduce logistics costs. With the reduction in logistics costs, the prices of various fisheries commodities as industrial raw materials can be reduced. The decrease in the prices of the raw materials is expected to increase the competitive advantage of fisheries products. However, if the scale of production from traditional fishermen is low, the development of various supporting infrastructure is considered to have less impact. It means that if the orientation of the local or regional entrepreneurs has not been oriented to the national market or even to export, then the goal of infrastructure development has not been achieved. Therefore, the main issue in wild fisheries products is how to increase the productivity and competitive advantage of traditional fishermen in Indonesia.

The production capacity of fish caught by traditional fishermen must be increased so that the product expansion to other regions or countries can enter economies of scale. In this case, traditional fishermen at the local level cannot compete against medium and large-scale companies. Large-scale companies have the supports of superior and specialized human resources, technology, and company infrastructure that is far better than traditional fishermen. In such situations, traditional fishermen cannot compete with large-scale companies from within and outside the country. Moreover, traditional fishermen are currently faced with a fundamental problem, i.e., that the majority of traditional fishermen are still not incorporated. These conditions make traditional fishermen not have a competitive advantage to face large-scale companies. The competitive advantage of a business organization is determined by the extent of its ability to manage its value chain. The higher the ability to manage the value chain, the higher the competitive advantage of a company. Meanwhile, to develop organizational value chain management in an integrated manner, various resources are needed, i.e., raw materials, technology and human capital.

Peter Drucker stated, "Management is getting things through other people." Therefore, in principle the implementation of management science can only be carried out by the organization and cannot be implemented without an organization. If most traditional fishermen still do not apply organizational principles, i.e., vision, mission, and goals, impersonal relationships are more dominant than personal relationships, formal relationships are more dominant than non-formal relationships, and right man in the right place and are relatively permanent or long-term. Therefore, increasing the competitive advantage of traditional fishermen is done by converting from non-organizational to organizational, as can be seen in the picture below:

Fig. 1. The change of traditional fisheries management pattern.

This study aimed to obtain an actual value chain analysis model in the wild fisheries industry in Cilacap, Central Java. The reason for choosing the location was because Nusa Kambangan is one of the fish catching areas with relatively good fishing port infrastructure, so that it can be used as a role model for the development of network governance for fishermen in other regions in Indonesia. The results of this study were also expected to provide education to traditional fishermen, so that they are able to change from non-organizational to organizational management with the use of modern management science. It was hoped that by changing the management pattern towards the organization through network governance, it would be able to increase the productivity of wild fisheries, because of the application of knowledge of business organization management. The implementation of network governance was expected not only to increase fish catch productivity, but also to process the fish, to market them, and distribute them without ignoring the ecosystem.

Network governance is management based on the network. In the process, the governance of the network functions to build the value chain of an industry in a holistic and integrated manner. To build network governance in the context of traditional fisheries industry, it is required entrepreneurship-based leadership in the traditional fishing group. This study aimed to build a network governance model in the traditional wild fisheries value chain in Indonesia.

2 Theoritical

2.1 Wild fisheries production factors

To understand how to increase fisheries production, it is necessary to know various variables that form the basis for increasing fisheries production. In classical economic theory, the factors of production consist of natural resources, human capital, capital, and entrepreneurship. Thus, network governance aims to coordinate, integrate and synchronize these production factors in the value chain to improve competitive advantage.

This research focused on how to build leadership and entrepreneurship as the determining factors in developing network governance in the traditional fishing group. These three variables were to be explored to get dimensions and indicators to measure them. If the indicators are considered to have precision, then generalizations can be made for further research.

2.2 Entrepreneurship-based leadership in managing micro, medium and small enterprises (MSMEs)

2.2.1 Leadership

Leadership is the center for understanding entrepreneurial growth when the businesses develop with facilities of both individual and group efforts (Ensley et al., 2006). Entrepreneurial leadership is explained as the influence and regulator of the performance of a group of people involved and having entrepreneurship opportunity to achieve organizational goals (Gupta et al., 2004: 2420). Leadership is a group activity as well as the value of the team leader which has an influence on the company's strategy and results (Carpenter et al, 2004). Leaders in traditional fisheries / MSME groups have a high degree of flexibility in the organization and have a greater impact on the company's behavior and results in entrepreneurial leadership in small companies (e.g. Coglister and Brigham, 2004; Cope et al., 2011; Leitch et al., 2013; Ling et al., 2008). To understand the impact of leadership on MSMEs, this study used cognitive and motivational dimensions of the MSME leaders. Entrepreneurial cognition is related to people's mentality as the models used to make judgments or decisions involving the evaluation of opportunities, business creation, and growth. In entrepreneurship, knowledge has a positive influence on the number of markets (Gruber et al, 2008, 2012, 2013; Ucbasaran et al, 2009).

A MSME led by a team can improve their opportunity identification capabilities. The team consists of people with technical experience, marketing, as well as managerial and entrepreneurial experience to identify more opportunities. Intention is a motivating factor that influences behavior; it can be indicated from how much people's effort to do something when they experience difficulties. Entrepreneurs differ in terms of orientation and intention towards growth (Dutta and Thornhill, 2008). Furthermore, intention to grow (defined as aspiration and hope) predicts growth (Barringer et al., 2005; Delmar and Wiklund, 2008; Wiklund, Koryak et al., 1995, and Shepherd, 2003a).

Various studies show that entrepreneurial intentions depend on two antecedents: perceived desires and feasibility. The perceived desires are measured by individual attitudes to taking risks, taking autonomy decisions, work effort, and enjoyment of work (Chen et al., 1998; McGee et al, 2009). Feasibility is measured by entrepreneurial self-efficacy, that is, the strength of one's belief that he is able to successfully carry out various roles and tasks of entrepreneurship (Chen et al., 1998). Self-efficacy increases the focus, direction, perseverance, and intensity of action. Entrepreneurs with high self-efficacy believe that they have all the resources in themselves and in their hands to achieve their task goals (Chen et al., 1998).

According to Miftah (2007: 31), there are three main factors that influence the determination of entrepreneurs about which leadership behaviors to use to make decisions: entrepreneurs personality strength, strength in the subordinates, strength in leadership situations. The whole point of entrepreneurial leadership is that the entrepreneur can generate the best from each individual, team and organization. Remember that entrepreneurial leadership is: instilling the belief to think, behave and act in an entrepreneurial manner with a mindfulness of the real goals of the organization for profitable growth for all stakeholders involved.

2.2.2 Innovation

Drucker (1985), the father of management science, introduces innovation as a special tool for entrepreneurs that enables them to use change as an opportunity to offer services, and view innovation as an organizational capability that can be learned and applied. The most important thing from Drucker's definition is the organizational innovation capability that entrepreneurs can learn to turn threats into opportunities to gain profits. In the same context, Nohria & Gulati (1996) defined innovation as an acceptance of any policy, program, structure, market or product that is deemed appropriate by the manager. Innovation as the formation, acceptance, and implementation of new ideas, products, processes or services, and creative ideas in an organization (Amabile et al., 1996).

Differentiation Strategy includes the creation of a unique product. This can be said to be an effort to create innovation. The uniqueness of features or benefits providing superior value for customers can guarantee the success of the differentiation strategy. It means that innovation is also part of the company's differentiation strategy (Hull & Rothenberg, 2008: 781). Customers see products as unmatched products and no equals, so price elasticity tends to be reduced and customers tend to be loyal to the brand. Valuable innovation can avoid competition. However, this strategy needs additional costs related to the cost of creating different product features. Thus, it is needed a premium pricing strategy in an innovation.

In a study by Marques & Ferreira (2009), innovation consists of product innovation, process innovation, investment in R & D, and new distribution channels. Factors that influence Firm's Innovative Capacity are: (1) Size of the company: the bigger the company the greater the ability to innovate (2) Phase of life-cycle of the company: the company increasingly loses the ability to innovate along with each phase of the life-cycle that it passes (3) Quality of entrepreneurship: the higher the quality of entrepreneurship of a company, the higher the ability to innovate (4) Partnership and cooperation: a company partnering and working with other companies has higher ability to innovate.

The study presented by Marques & Ferreira (2009) is intended for companies in general. In its development, the dimensions of innovation can be simplified into four types of innovations (Hassan, Shaukat, Nawaz, & Naz, 2013): (1) Product innovation (2) Process innovation (3) Marketing innovation, and (4) Organizational innovation.

Product innovation includes offering new products and services to the market as well as major improvements to the functions or characteristics of the users of existing goods and services. Process innovation includes major changes to methods, equipment and / or software. One example is a new production method. Marketing innovation seeks to respond to customer needs better in order to increase company sales. Among the examples of this innovation are opening new markets and relocating company products on the market. Organizational innovation can be defined as an implementation of new organizational methods in commercial practices, workplace organizations, or corporate external relations. This study also does not focus on one type of company, but all types of companies.

Leadership and Innovation are related to the human resource performance of an organization. Nathaniel and Handy (2010: 3) stated that a leader at all times plays a role to direct, guide, supervise, and carry out several other managerial functions. Besides, the leader also shows a leadership style that will be applied in accordance with the conditions, situation and needs of the organization to increase innovation.

2.3 Network governance in the value chain of fisheries industry

The concept of the value chain was first coined by Porter in 1985 (Conney, 2011). The concept of social-ecological interaction adapts the value chain concept of Porters because it is considered to have similarities. The relations between systems are regulated in a governance system. The idea of value chain was developed to provide institutional value chain analysis, so that management principles can be applied. According to Kaplinski and Moris (2000), the idea of a value chain is expanded to bring more elements of institutional analysis (Kaplinski and Morris, 2000). At present, value chain is also applied to communities and politics-power (Keane, 2008).

Various references discuss value chain and governance, but it is still needed further explanation regarding the relationship between parts of value chain. In the context of fisheries, the value chain is related to health, welfare and the value chain besides being as commercial commodities (McConney, 2011). Thus, the main issues relating to value chain in fisheries industry can be explained as in Figure 2 below.

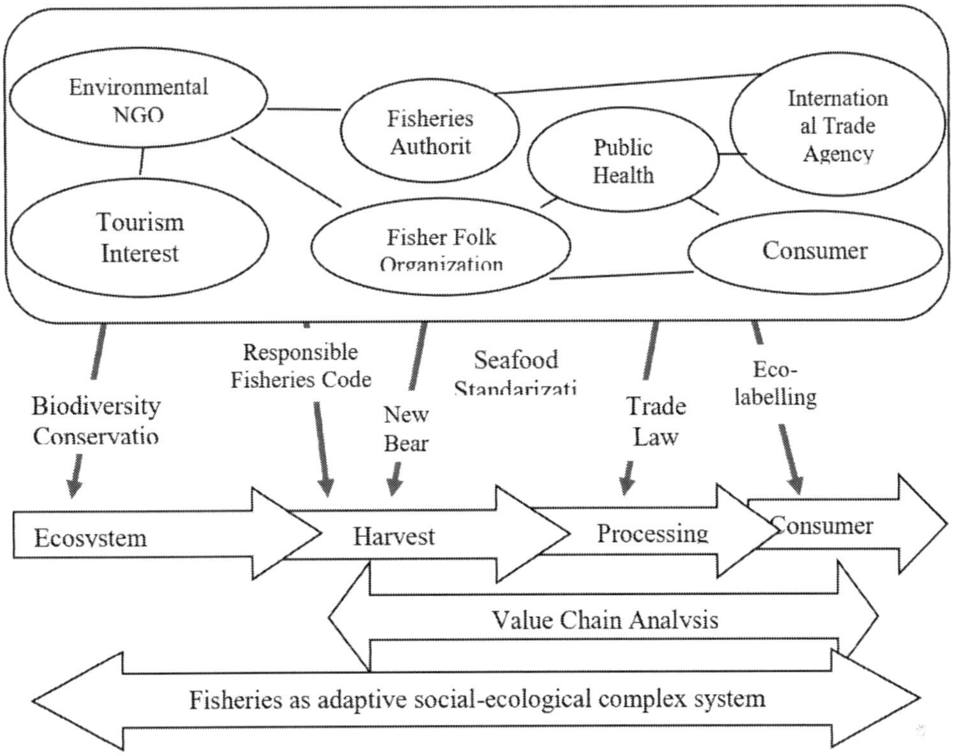

Source: McConney (2011:2)

Fig. 2. Network Governance in Fisheries Industry.

As can be seen in Figure 2, according to McConney (2011), the value chain analysis perspective should be based on good ecosystem health. Fish catches, marketing, and distribution also influence the health of the ecosystem. This is done through technology of harvesting to disposal of fish waste on land. If there are fish species that have characteristics of moving and spreading more widely, then the involvement of regional and international communities is much needed in fisheries management based on the network. Stakeholders are also involved in policy making in the fisheries sector.

According to McConney (2011), the network perspective can help stakeholders to develop fisheries value chains that contribute to sustainable social and economic development. A governance with the network perspective helps in thinking more broadly about how trade activities in one region have an impact in other regions. Regions refer not only to places, but also to legal institutional arrangements and virtual communities that reach the entire world. The network perspective places the fisheries value chain in a social-ecological context. This can help analyze lateral and vertical relationships in building development-oriented network governance.

As explained earlier, the network perspective will reveal how value chain of an industry is related to other institutions / organizations. According to McConney (2011), in the future, the linkages of networks at local and international levels and the possibility of changes in anthropogenic climate, ecosystems, and international trade patterns are mediated through value chain analysis. In the construction of physical infrastructure for landings and handling of living fish, the rising sea level and coastal inundation models should be considered. Besides, people are given an option to buy fish in the market if there is a shortage of supply. This network provides small-scale fisheries authorities to be able to deal with international / multinational companies, so that the welfare of fishing households is spread throughout the coastline.

Various studies have investigated the networks and governance of fish marketing in Grenville, Canada, where the value chain is adaptive and is very dependent on the social network of the fishermen. The condition can facilitate the marketing of fish with the role of cooperative. The function of network governance completely helps policy makers explain how to move fish catches from ships to consumers (McConney, 2011).

Like any fisheries industry in general that gets a variety of fish catches, commercial fisheries not only run the chains of international marketing, but also marine conservation areas for tourism purposes. Both perspectives must be holistic and integrated considerations. Considering only one of these perspectives can result in a loss to one party. For example, if marine conservation applies a prohibition zone for fish catches and results in reduced fishermen's fish catches, it will have an impact on the reduced supply of fish for local and foreign needs. The reduced availability of fish in the country will encourage importation of fish.

According to McConney, managers must be aware of the various inter-sectoral links that affect value chains. This encourages the implementation of management to be based on the ecosystem as part of the social-ecological network. McConney (2012) stated that groups of fishermen play a collective role in maintaining fisheries value chains to avoid imports. Based on various literature about network-based governance, traditional fishermen groups need to adapt the concept of network governance in accordance with natural resources, human capital, capital and the entrepreneurial spirit. The applied network governance can mobilize various resources in groups of fishermen, both tangible and intangible resources. This is expected to provide economic benefits to traditional fishermen groups by not ignoring other environmental impacts.

3 Methodology

3.1 Research type

This study was an exploratory research. It aimed to conceptualize the network governance in the value chain of the wild fisheries industry off Nusa Kambangan coast. The subjects in this study were fishermen living along the coast of Nusa Kambangan, Cilacap, Central Java. Data collection was done in the following ways:

3.1.1 Primary data

Primary data were obtained in the field directly from the sources. The data were obtained through observation and interviews with key-informants, i.e., the chairperson of a fishing group and the chairperson of a fish auction site.

3.1.2 Secondary data

Secondary data were obtained from literature study in the form of literature, written sources, and documents having relevance to this research, including documents from the

Department of Industry and Trade of Cilacap Regency, and other agencies related to this research. The secondary data included the general state of the study site.

3.2 Data collection techniques

3.2.1 Observation

This was collecting data by holding observations directly to the object or location of research to get a clear picture of the object under study (Nasir, 2003).

3.2.2 Interview

The interview technique used was depth interview, i.e., conducting in-depth interviews with traditional fishermen living along Nusakambangan coast, Cilacap, Central Java, regarding their opinions on the role of the leader of the fishermen group in Cilacap, Central Java. Interviews were conducted with key-informants, namely people who were considered to know aspects of the response. The following are the key informants interviewed in this study: Sumadi, Chairperson of the Fisherman Group and Chair of the Fish Auction Site of Menganti Kisik, Cilacap Regency ; Salimun, Supervisory Board of Fishermen Group and Fish Auction Site of Menganti Kisik, Cilacap Regency and Abnu, a farmer of cultivation of shrimp ponds in the area of the Fish Auction Site of Menganti Kisik, Cilacap Regency.

3.2.3 Questionnaires

The data collection method in this study also used questionnaires that had been systematically designed with closed and open questions.

3.3 Population and samples

The population in this study was fishermen living along Nusakambangan coastal region in Cilacap Regency, Central Java. Determination of the samples in this study used purposive sampling method, i.e., sampling with certain criteria. The criteria set out in determining the samples in this study were that they were traditional fishermen living in Nusakambangan coastal region, given that the traditional fishermen had knowledge of various phenomena that they had to face in increasing fisheries productivity. By using the purposive sampling technique, 30 traditional fishermen were taken as the research samples.

3.4 Data analysis method

Data analysis was carried out in a double way, i.e., interpretive qualitative analysis and quantitative analysis. Interpretative qualitative analysis was used to identify the respondents' characteristics and their responses to questions on the questionnaires. Quantitative analysis using descriptive analysis aimed to describe phenomena in the object being studied.

4 Results and discussions

4.1 Activities of fish auction sites in cilacap regency

Fish auction sites have an important role in the economic activities of coastal communities, including on the coast of Cilacap, Central Java. At this fish auction site, traditional fishermen sell the fish they have obtained to buyers. Therefore, fish auction sites have a significant role for the fishermen. In the coastal area of Cilacap, there are at least 9 (nine) fish auction sites:
Fish Auction Site (FAS) of Jetis
FAS of Menganti
FAS of Lengkong
FAS of Kemiren
FAS of Tegal Kamulya
FAS of PPC
FAS of Pandanaran
FAS of Sidakaya
FAS of Bonbaru

Each FAS has an average of 100 fishermen with 50 ships/boats. Each FAS has a coordinator/chairman who oversees 100 to 200 active fishermen in the Cilacap Regency. The FAS serves to collect fish catches and fish landing centers on the coast of Cilacap, Central Java.

4.1.1 Fishing activities

Wild fisheries activities in Cilacap consist of marine fisheries and fishing in public waters. Marine fisheries are a series of activities to find and take fish in sea waters with boundaries ranging from the coastline to the waters of the sea. Marine fisheries are carried out using fishing boats or boats equipped with certain types of fishing gear. Fishing boats or boats use propulsion engines. The size of the propulsion machine varies from small to large sizes adjusted to the size of the ship.

Fishing activities using fishing ships or boats are carried out from the coast to the open sea. There are no longer any ships or boats that are not machine-driven. The equipment used includes gillnets, bag nets, Danish seines, purse seines, shrimp nets, squid nets, crab nets, bubu traps, sero traps, and fishing rods. By using the equipment, the results obtained include fish, shrimp, squid, crab, crab, sea cucumber, and shellfish.

4.1.2 Range of fishing area

Limitation in reaching fishing areas for traditional fishermen is because most of the boats they have are small, so that they are unable to reach a wider area. They only sail around the offshore of Cilacap. Given the small capacity of the fuel that can be accommodated by the small boats, they try to reach only nearby areas. In addition to the limited range of fishing areas, there is an electric steam power plant (Pembangkit Listrik Tenaga Uap / PLTU) of Cilacap in one of the fishing areas. It results in increasingly limited space for fishermen to find fish.

4.1.3 Implementation of network governance in the ecosystem

Normally, the management of fisheries system cannot be separated from three dimensions inseparable from one another: (1) the dimension of fisheries resources and their ecosystems; (2) the dimension of fisheries resource utilization for socio-economic interests of the community; and (3) the dimension of fisheries policy itself (Charles, 2001).

4.2 The potentials of network governance implementation in fishing activities off the coast of cilacap, central java

Network governance is an effort to involve all stake-holders to overcome various problems involving external parties outside the organization to solve problems. This approach is considered as a solution to solve problems holistically and not partially. Based on the research results, there are several institutional potentials that can be done to create network governance in ecosystem management, fishing, processing and packaging, and marketing activities.

4.2.1 Ecosystem

Ecosystem damage on the coast of Cilacap, Central Java, includes coastal abrasion. The magnitude of the waves of the Indian Ocean resulted in the damage to the southern coast of Cilacap. Although in some areas breakwaters have been installed, but they have not been able to withstand the increasing rate of abrasion, particularly on the eastern coast of the Turtle Bay (Teluk Penyu).

Several institutions, both business (company) and government organizations, including Jenderal Soedirman University, Marine Affairs and Fisheries Office of Cilacap Regency, PT. Pertamina, and PT. Holcim, have tried to resolve these issues. The obstacles are not only technical, but also social. The community feels no interest in maintaining the coastal ecosystem of Cilacap, Central Java.

Jenderal Soedirman University and PT. Holcim collaborated in planting 3500 pine trees in 2015 in a part of the coastal area of Cilacap, Central Java, but one month after the tree had been planted, the community damaged the conservation area for the sake of making ponds. Diponegoro Regional Military Command (Komando Distrik Militer / KODAM) as the owner of the land rents out to the surrounding community, while the surrounding communities who rent the coastal area feel they have no economic advantage if the land rented is only used as marine pine forest. Therefore, they did land-use conversion into manufacture of shrimp farms, boat parking and other uses.

Based on the description above, various parties involved in network governance in solving the ecosystem problems are as follows: (1) Diponegoro Regional Military Command of Indonesian Armed Forces (KODAM TNI) as the landowner (2) University as the party conducting the study using a scientific and objective perspective (3) Marine Affairs and Fisheries Office of Cilacap Regency (4) Environment and Forestry Office of Central Java Province (5) Companies (state-owned enterprises and private companies).

The involvement of these institutions must be under the coordination of the Authority/Regional Management Agency of the South Coast of Cilacap Regency, Central Java.

4.2.2 Fishing

The people of the coast of Cilacap, Central Java, make fishing the main activity to earn income. However, the number of fish catches is still relatively low. The fish catches are only able to fulfill their daily needs. There are at least 3 (three) issues relating to the fishing activities of the fishermen: (1)unpredictable fish catches (2) limited boat capacity (3)The fishermen's position only as the price takers, due to the small number of middlemen who control all the Fish Auction Sites (FAS) on the coast of Cilacap.

Unpredictable catches due to weather and climate are conditions that cannot be controlled. Based on the Chairperson of the Fishermen and FAS Group, Cilacap, the fishermen need for fuel for one day's operation to go to sea is 20 liters or Rp140,000 to Rp150,000. The highest fish catches in harvest seasons are 100 kilograms, while the catches in normal seasons are 20 kilograms. In famine seasons, fishermen are sometimes unable to meet operational needs, especially fuel. Therefore, efforts need to be made to create a system of Social Security Needs so that it can be ensured that the condition of fishermen can still meet their needs when they do not get fish catches.

Limited boat capacity is also a very relevant issue to solve. One fisherman's boat usually carries 4 crew members, while the boat is 10 square meters with the height of 60 cm. With the relatively small capacity, it cannot carry large quantities of fish catches in harvest seasons. Thus, this problem needs to be solved, i.e., through the increase in the boat capacity.

The issue of selling prices of fish also needs to be solved. During famine seasons, the fish caught by fishermen are relatively low in number, even in certain situations they cannot cover operational costs at sea, especially fuel. During harvest seasons, fishermen also do not get much benefit from the fish catches due to the falling fish price. The fishermen also cannot take the initiative to do fish cooling and drying because the investment needed to buy the equipment is relatively high for them and the capacities of fish cooling machine and fish drying machine cannot accommodate fish catches. This condition is used by middlemen in pricing the fish production. As a result, the fishermen do not experience a significant increase in income when the production is abundant.

Various issues regarding fishing in the sea need to be solved by using a network governance approach. In the implementation of network governance, the parties that need to be involved in managing fishing activities are as follows: (1) fishermen groups (2) fish auction sites (3) fishermen's cooperatives (4) marine affairs and fisheries office

4.2.3 Processing and Packaging

The number of fish catches that cannot be predicted, both due to weather factors and the availability of fish in the sea need to be anticipated by the fishermen. This is to prevent the fishermen from accepting the price set by the middlemen. When the production of fish catches decreases, the fishermen can get a high price, but when there are abundant fish catches obtained, the fishermen need to keep the price from falling dramatically. This can be done by processing and packing fish catches to last a long time, either with freezers or drying machines (ovens). However, it is not possible for them individually or in groups to meet the needs of the fish processing and packaging machines. They have limited funds to buy the equipment. It is large-scale companies having sufficient funding sources that have the possibility to purchase the equipment.

In this case, the establishment of a state-owned enterprise (Badan Usaha Milik Negara / BUMN) or a regional-owned enterprise (Badan Usaha Milik Daerah / BUMD) in the field of fish processing and packaging should be considered in order to keep the price of fish caught not falling and harming fishermen. Processing and packaging of fish caught by fishermen have several objectives, including: increasing the durability of these fishery products so that when there is abundant production of fish, they can be stored for sale when the famine season arrives, expanding the distribution range, and creating hygienic fish production. To achieve the quality of these products, traditional fishermen have limited resources. Thus, the implementation of network governance in processing and packaging needs to be done. The institutions to be involved are as follows: (1) Indonesian National Agency of Drug and Food Control (Badan Pengawas Obat dan Makanan / BPOM) (2) Fish processing companies (BUMN/BUMD/private companies) (3) Center for Appropriate Technology (Balai Besar Teknologi Tepat Guna / B2TTG) of Indonesian Institute of Sciences (Lembaga Ilmu Pengetahuan Indonesia / LIPI)

The involvement of these institutions is expected to improve the ability of traditional fishermen to produce quality fish, reach wider consumers, and be able to meet national and international needs. Having good quality is a requirement for any product to be accepted by the market. Therefore, the processing and packaging of fish products need to be carried out in accordance with health and safety standards.

4.2.4 Marketing

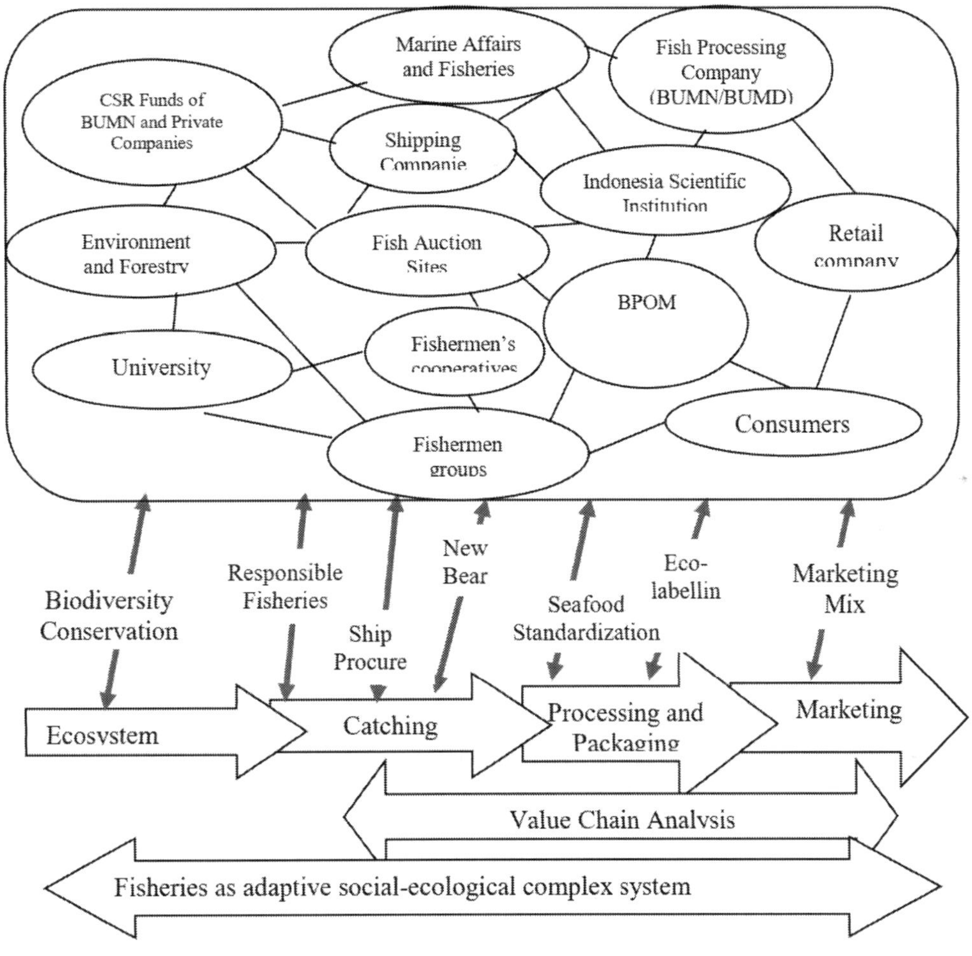

Fig. 3. Network governance model of traditional wild fisheries in indonesia.

The marketing function is done by creating, communicating, and delivering fishery products to consumers. It is hoped that by optimizing the marketing function of these fishery products, traditional fishermen can have loyal customers and obtain maximum profits. The obstacle faced by fishermen is in meeting market needs outside the region. So far, fishermen have relied on selling to middlemen. The average number of middlemen in each Fish Auction Site is 5 to 7 people. They control the price of fish and reach consumers outside the region. In this case, traditional fishermen need to be given an understanding of the environment of fish marketing, so that the dependence on middlemen can decrease. Thus, network governance in the activities of marketing fish products involves at least the parties as follow: (1) Indonesian Consumers Protection Foundation (Yayasan Lembaga Konsumen Indonesia / YLKI) and (2)Retail entrepreneurs (supermarkets) who are willing to accommodate fresh fish catches directly from fishermen.

Management of wild fisheries value chains from ecosystems, fishing, processing and packaging to marketing activities must be carried out in an integrated manner. Network governance seeks to integrate each of these actors as a network that has mutual connection between them. It is hoped to increase the productivity of fish catches while maintaining the marine ecosystem as can be seen in Figure 3.

5 Conclusions

5.1 Conclusions

The fishermen groups apply entrepreneurship-based leadership. It can be seen from the leaders of fishermen groups carrying out business activities to generate added value. In terms of creativity, the leaders of the fishermen groups have created new ideas for the development and welfare of fishermen in the coastal area of Cilacap Regency. The innovations made by the leaders of the fishermen group can be seen in the efforts to collaborate with multinational companies to invest in the fishing area, where the investment is expected to absorb large numbers of workers. However, this leadership has not been supported by managerial competence, so that management principles have not been applied in the fishermen groups.

Based on the results and discussions, there are several institutional potentials that can be realized to create network governance in the management of ecosystem, fishing, processing and packaging, and marketing activities. In managing ecosystems, the parties involved in implementing network governance on coastal management are Environment and Forestry Office, University, and CSR of private companies / BUMN / BUMD. Fishing activities are carried out by involving Marine Affaris and Fisheries Office, Shipping companies, Fish Auction Sites, Fishermen Groups, and Fishermen's cooperatives. Processing and packaging activities are carried out by involving fish processing companies, Center for Appropriate Technology of Indonesian Institute of Sciences (B2TTG-LIPI), and Agency of Drug and Food Control (BPOM), and marketing activities are carried out by involving the Indonesian Consumers Protection Foundation (YLKI) and retail companies (department stores) to distribute fresh fish to end consumers.

5.2 Recommendations

Menganti fishermen group implements entrepreneurial leadership pattern but do not yet have managerial competencies in business management. Thus, they are required to have managerial competencies in managing the group.

The management of wild fisheries value chain from ecosystem, fishing, processing and packaging, to marketing activities must be carried out in an integrated manner. Network governance seeks to integrate each of these factors as a network where there are mutual relationships among its parts, i.e., the activities in managing the ecosystem, fishing, processing and packaging, and marketing. The involvement of various actors in implementing network governance is not possible if it is established by the groups of fishermen themselves, so a government policy is needed to manage the coast in an integrated manner. The management authority of the coastal region of Cilacap needs to be established, so that network governance in the management of the Cilacap coast can be realized.

References

1. T. M. Amabile, Harvard Bus. Sch. Backgr. Note 396 (1996)
2. M. Atalay, N. Anafarta, and F. Sarvan, Procedia - Soc. Behav. Sci. **75**, 226 (2013)
3. C. H. Chang, J. Bus. Ethics **104**, 361 (2011)
4. P. F. Drucker and J. L. Noel, J. Contin. High. Educ. **34**, 22 (1986)
5. M. U. Hassan, S. Shaukat, M. S. Nawaz, and S. Naz, Pakistan J. Commer. Soc. Sci. **7**, 243 (2013)
6. C. E. Hull and S. Rothenberg, Strateg. Manag. J. **29**, 781 (2008)
7. Jogiyanto, Metodologi penelitian (BPFE UGM, Yogyakarta, 2010)
8. L. C. Leonidou, D. Palihawadana, and M. Theodosiou, J. Int. Mark. **19**, 1 (2011)
9. C. S. Marques and J. Ferreira, J. Technol. Manag. Innov. **4**, 53 (2009)
10. M. Mohd Rosli and S. Sidek, *Innovation and firm performance: evidance from malaysian small dan medium enterprises,* in Entrepreneurship Visison 2020: Innovation, Development Sustainability and Economic Growth, (2013)
11. N. Nohria and R. Gulati, Acad. Manag. J. **39**, 1245 (1996)
12. M. . Porter, *Keunggulan* bersaing: menciptakan dan mempertahankan kinerja unggul (Airlangga, Jakarta, 1994)
13. M. . Porter, *Strategi* bersaing: teknik menganalis industri dan pesaing (Airlangga, Jakarta, 1992)
14. M. E. Porter, *The Competitive Advantage of Nations* (1990)
15. S. X. Zeng, X. M. Xie, and C. M. Tam, Technovation **30**, 181 (2010)

REMUNERATION AND PERFORMANCE

Iriani Ismail

Universitas Trunojoyo, Madura, Indonesia

Abstract. The aim of this study is to explain the role of remuneration in improving the performance of library employees. This study uses the qualitative method which the population is all employees numbering 17 people. Based on interactive analysis methods and ethnographic techniques, its took In-depth interview and observations. Using this analysis, the result showed that the remuneration has an important role in improving employees performance. Some internal and external factors influence it so that optimal performance is obtained. Generally, employees expect that the remuneration has been high to be accepted like as expected. Most employees demands such remuneration to meet their daily needs. External factors also has a strong role so the employees demand high, but not balanced with high performance.

1 Inroduction

Performance is something that is often discussed to determine or provide direction for the development of the next employee. This human resource becomes a center that must be considered and addressed its performance. Human resources as employees are an important component in the delivery of services to the community. Employees will be assured of their welfare if all of the employee's needs can be fulfilled, including economic needs which in this case are realized in the system of providing performance benefits (remuneration). Therefore, the needs to be a change in the reward system that is only based on position and education, without considering the performance that has been done by employees for the organization (Siagian, 2014: 262)

Remuneration is a part of reward received employees as a result of their task in the organization, including gifts, awards or promotions. Performance cannot be achieved optimally if remuneration is not given proportionally. In addition, the individual's attachment to work is the key to the success and profitability of the organization. The use of remuneration in return for educators in the university environment as well as being a form of independence is also the responsibility of the leadership in improving job satisfaction and work motivation. Basically, the existence of these benefits has fulfilled the expectations of the teaching staff at work. Rivai, et al (2014) explained that someone can act something (in an effort to achieve goals and fulfill responsibilities) that tends to be due to the expectation of the results to be obtained. The statement explains that the nature of the employee works (doing something) because there is hope to get a reply that can be in the form of remuneration. So that when the needs and expectations of employees are fulfilled, work satisfaction can be achieved which in turn will affect work motivation. The institution strives to improve job satisfaction and work motivation, not least because basic needs are to get quality work from its employees. The quality of work, enthusiasm for work and morale can be realized when the employees were satisfied and work motivation can be obtained. The function of remuneration is clearly a stimulant for job satisfaction and work motivation, in general by referring to the theory of job satisfaction and previous research which examined how much influence the reward with job satisfaction and work motivation with the acquisition of positive influential results. The implementation of remuneration in the university environment is considered new, so the research is intended to examine the effect of remuneration on job satisfaction and work motivation in the university environment, especially in library unit.

Companies really need employees who have high employee engagement because if the company has high employee engagement, it will be able to provide higher returns compared to companies whose employees do not have high employee engagement with their work. Employees who have a high sense of employee engagement, will work seriously so that the resulting work can contribute well to the progress of the company.

This study was conducted at the Library unit (department). The performance of employees in this institution is quite good because it has been able to achieve performance as planned by the University. The library unit is got a new ISO and accredited. ISO 2015 is the latest ISO, already owned. Performance appraisal in that section is held once a year, this is intended to determine the increase or decrease in the performance of its employees, which will later become an evaluation for the library to improve employees performance. Employees do their works referring to the SOP (Standard Operating Procedure) that has been set.

From the results of interviews with several employees, the remuneration that the employees felt it was good enough and fair and decent. This can be proven by a system of remuneration based on performance appraisal carried out by library management. Provision of remuneration is given objectively in accordance with the workload and performance that has been generated by these employees. The amount of remuneration at this institution is based on the grade and or position. In the certain contexts related to discipline, the grade is ignored. This makes employees more careful so that can reach the expected and set value. Non-financial remuneration given to employees is in the form of a promotion or a grade. Employees have a high sense of employer engagement, this can be evidenced by direct observation in the field when working, where the employees look earnest and passionate even though they are not supervised by the leadership, but they remain fully concentrated and serious doing their jobs.

2 Literature Review

Remuneration

Sofa (2008) found that remuneration is a reward, or remuneration is given to workers or employees as a result of the achievements that have been given in order to achieve organizational goals.

Organizations in giving responsibility to employees to generate contributions in achieving goals must be balanced with reward. Providing rewards is the obligation of the organization to provide services to employees who have carried out their duties or responsibilities. Giving rewards in practice uses 3 determining indicators, namely giving based on performance (Pay for performance), giving based on position or position (Pay for position) and giving based on individuals (pay for person). Current pay for performance that starts frequently and is generally used to determine the amount of compensation, where the approach is often referred to as the remuneration system. Remuneration System has a remunerate base word which means paying or can also be called payroll (Echols and

Shadily in Roberia, 2009). In practice, remuneration is given to employees on the basis of an employment relationship, where when a job has been completed, remuneration can be received. In general, there are many opinions that judge that remuneration has similarities with compensation. But what needs to be understood is that in remuneration there is no acknowledgment of non-financial returns so that there is a difference with the concept of compensation for that (Roberia, 2009). Remuneration has become an indispensable instrument from an organization both private and public, especially for public organizations that have used a lot of remuneration systems. In the context of the company, remuneration is interpreted as an act of repayment of services or compensation received by employees / workers from employees for the achievements given by workers in order to realize company goals (Pora, 2011: 3).

Mahmudi (2007: 167) said that the main purpose of giving remuneration is as follow:

a. To attract people who are competent, qualified, and characterized by joining the organization.

b. To maintain people who have excellence, competence, quality and good character who have joined the organization so as not to leave the organization.

c. To keep the people in the organization willing to work because inadequate remuneration allows employees to carry out a strike which actually is very detrimental to the organization.

d. To motivate employees to achieve the best performance.

There are five remuneration indicators according to Santoso (2012) including the following:

1. Merit system, namely the determination of employee income based on position.
2. Fair, in the sense that the position with the workload and responsibility of work with the same weight is paid equally and work that demands higher knowledge, skills and responsibilities is paid more.
3. Eligible, that is, it can fulfill the needs of a decent life (not minimum).
4. Competitive, where the salary of civil servants is equivalent to the salary of employees with the same qualifications in the private sector, in order to avoid brain drain.
5. Transparent, in the sense that employees only get official salaries and benefits.

Performance

Performance is the result of work that can be achieved by a person or group of people in the organization, in accordance with their respective authorities and responsibilities, in order to achieve the objectives of the organization in question, not illegal and in accordance with morals and ethics (Sedarmayanti, 2011: 239) .

Moeheriono (2012: 95) said that performance is a description of the level of achievement of an activity program or policy in realizing the organization's goals, objectives, vision and mission as outlined through the organization's strategic planning.

Mangkunegara (2009: 68) said that the factors that influence employee performance are as follows:

a. Ability Factor

Psychologically, employee's ability consists of potential ability (IQ) and reality ability (knowledge + skill). Employees need to be placed in jobs that match their expertise.

b. Motivation Factor

Motivation is formed from the attitude (attitude) of an employee in the face of work situations. Motivation is a condition that moves employees who are directed to achieve organizational goals.

In this study the indicators used in evaluating employee performance include:

1. Individual Personality Ability, which consists of; integrity, enthusiasm, creativity, organizational commitment, guiding, taking strategic decisions, service orientation, and leadership
2. Job Results Targets consisting of; output of work, level of discipline, communication and performance alignment.

Relationship of Remuneration to Employees Performance
Jumaslini (2011: 110), stated that the performance itself cannot be achieved optimally if the remuneration is not given proportionally. Remuneration is basically a tool to realize the vision and mission of the company/organization with the aim of attracting competent and experienced employees, maintaining qualified employees and motivating employees to work effectively. Its existence in an organization cannot be ignored simply because it will be directly related to the achievement of goals (Fitria et al, 2014: 4).
With the provision of remuneration in accordance with what is expected, it will be able to encourage employees performance improvement, so that they are motivated to always try to be the best, this will benefit employees and the company. Effect of remuneration in improving employee performance by increasing employee welfare, with special attention to employee welfare through remuneration.

Framework of thinking

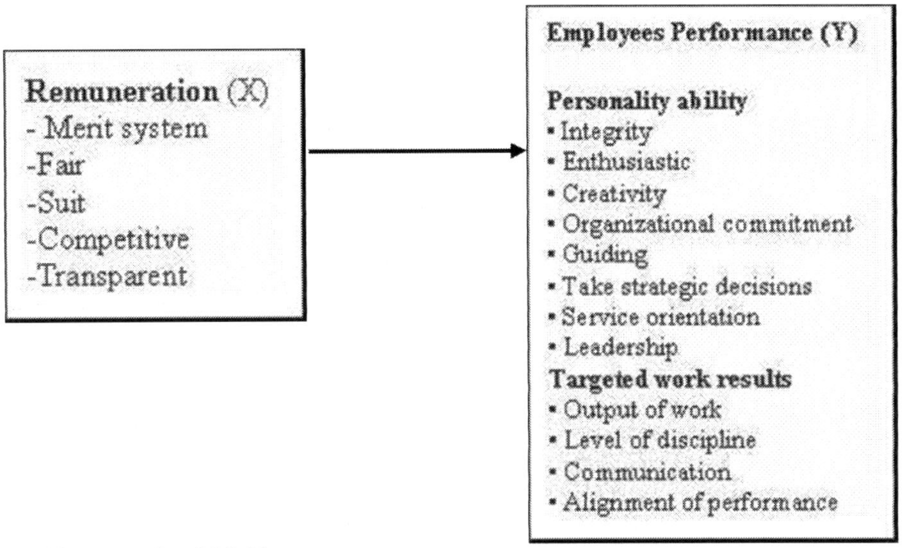

Fig. 1. Framework of thinking
Source: Retnoningtyas (2014)

3 Figures and tables

Object of research
The object of this study is Library Unit in Trunojoyo Madura University with 17 employees. The type of this research is qualitative research that examines participant perspectives with strategies that are interactive, flexible and more descriptive in nature. Qualitative research using qualitative descriptive research methods. Qualitative research has several strategies, Sukmadinata (2011: 61-66) said that the strategy in qualitative research is interactive and non-interactive. The interactive qualitative method is an in-depth study that uses the technique of collecting data directly from people in their natural

environment. There are 6 (six) kinds of interactive qualitative methods, including: ethnographic studies, historical studies, phenomenology studies, case studies, basic theories, and critical studies. Non-interactive qualitative methods are analytical thinking, conducting studies based on document analysis. The data source in this study is documents. Based on the explanation above, as for this study using an interactive qualitative strategy, with the design of phenomenological study research that is, looking for meaning from experience in life, with the aim of phenomenological research is to find meaning from the basic things of life experience. The study was conducted by in-depth interviews with participants or resource persons / informants. Research informants are subjects who understand information as actors or other people who knows about the research conducted. Informants (informants) of this study amounted to 5 people who have information (data) that is a lot about the object being studied, to be asked for information about the object of the research. The informants in this study came from direct interviews which were referred to resource persons.

Using a purposive sampling techniques, namely the technique of sampling data sources based on certain considerations (Sugiyono, 2016: 218). These considerations are packaged in the characteristics of the informants that have been described in the research method. Based on the characteristics of the informant, the library staff used to be the source of this study

Research uses data collection techniques, namely: observation, interviews and documentation. The first data collection observation, the type of observation carried out by researchers was passive participation observation, explained in Sugiono (2016: 228), for this type of observation, researchers came at the activity site of the observed person but were not involved in the activity.

Researchers also use frank or disguised observation, Sugiono (2016: 228). Researchers in carrying out data collection stated frankly to the data source, that he was conducting research. The second data collection interview, the type of interview conducted by the researcher was in-depth interviews. Sugiono (2016: 223) explained that this category is a type of interview that is more free than structured interviews. Interviews were conducted with informants who had been selected by the researcher in accordance with the characteristics of the informants needed and the researchers also received approval from informants that they stated they were willing to be interviewed. Interviews were conducted with question and answer directly with interview guides set by the researcher. The third data collection is documentation, this documentation is used to produce important evidence, here the researcher took documentation when conducting interviews with informants in the form of photographs, in accordance with the agreement with the informants, in addition, photographs taken were interview notes, which were previously in the notebook then arranged by researchers in the form of transcripts of interviews presented in the results of the study in accordance with the interview guidelines used.

The steps used are editing and interpretation. This study uses data validity techniques with persistence observation techniques, checking through discussion, holding member checks, and triangulation. This study uses qualitative data analysis with the stages of reduction, presentation of data, and verification of data in order to answer the formulation of existing problems.

https://doi.org/10.1051/shsconf/20208601034

4 Equations and mathematics

Remuneration and Employee Performance
Based on the results of observations and in-depth interviews, it is known that remuneration plays a very high role in improving employee performance. This can be seen in the behavior shown by employees when carrying out the tasks assigned to them, especially obtaining additional tasks. Employee performance moves parallel to remuneration received for the majority of employees. Likewise with the absence or lack of remuneration, it will have the same impact on employees performance. That is, if the remuneration is good, the employees performance will also increase, and vice versa.
In the field, it was found that the majority of employees benefited greatly from remuneration. Some to fulfill daily life, a small part to fulfill their lifestyle, some to cover the costs owed, some to reserve. Of all existing employees (17 people), as many as 15 employees have shown improved performance. Some indications of this increase are through timely reporting, increased creativity and innovation in anticipating institutional work programs, readiness to carry out work or additional tasks, and increasing motivation for human resource development.
This study is supported by Imam's research (2016) entitled "The Effect of Leadership, Work Motivation and Remuneration on Employee Performance at the East Java Regional Directorate General of Tax Office I". The results or findings from this study indicate that remuneration has a positive and significant effect on employees performance. Thus, good remuneration will improve performance and vice versa.
The findings of this study are also supported by the results of research conducted by Juairiah, and Rosyidah Umpu Malwa (2016) with the title of the Relationship between Giving Remuneration to Employee Performance at Siti Khadijah Islamic Hospital, Palembang, said that remuneration has increased employees performance.
In observation and in-depth interviews with employees, improving employees performance is depend on stimuli both internal (self-motivated) and external motivation such as remuneration. In this case both direct and indirect remuneration is one of the drivers of work morale and employees work productivity, by encouraging and maximizing the human resources that at last enables optimal performance to be achieved.

CONCLUSIONS AND RECOMMENDATIONS
Conclusion

1. Remuneration has a significant role in improving employees performance.
2. Remuneration is the main objective of implementing additional employees duties.

Suggestion

1. It is better to pay remuneration to employees because it influences employee performance. Provision of remuneration is adjusted to the workload carried out, must be on target in accordance with the needs in the field of work, by providing socialization regarding the provision of remuneration so that there is no misperception between senior employees and new employees
2. Next study
   To optimize the results of this study, it is expected that the future reviewers if with the same problem can add other variables that are deemed relevant and deeper in discussion in order to obtain maximum results.

References

1. Bakker, A. B., & Sanz Vergel, A. I. 2012. Weekly Work Engagement and Flourishing The Role of Hindrance and Challenge Job Demands. Journal of Vocational Behavior, Vol.83: 397-409.
2. Calvin, Ojeleye Yinka. 2017, The Impact Of Remuneration On Employees' Performance (A Study Of Abdul Gusau Polytecnic, Talata-Mafara And State College Of Education Maru, Zamfara State), Arabian Journal of Business and Management Review (Nigerian Chapter) Vol. 4, No. 2, 2017
3. Mahmudi. 2007. Public Sector Performance Management. Yogyakarta: Publishing and Printing Unit of YKPN College of Management.
4. Mangkunegara, Anwar Prabu A.A. 2009. Human Resource Management. Bandung: PT. Teenager Rosdakarya.
5. Marciano, Paul L. 2010. Carrots and Sticks Don't Work Build a Culture of Employees Engagement with the Principles of Respect. Mexico: McGraw Hill.
6. Moeheriono. 2012. Competency Based Performance Measurement. Jakarta: Raja Grafindo Persada.
7. Robbins S.P & Judge. 2011. Organizational Behavior of the second Book. Jakarta: Salemba Empat.
8. Santoso, A., S. 2012. Development of Remuneration Model at Tugu Ibu Hospital in 2011. Hospital Administration Masters Thesis, Faculty of Public Health. University of Indonesia
9. Schaufeli, W.B., Bakker, A.B., & Salanova, M. 2006. The Measurement of Work Engagement with a Short Questionnaire A Cross-National Study. Educational and Psychologycal Measurment, Vol. 66: 701-716.
10. Sofa. 2008. Human Resource Management. Jakarta: Bumi Aksara.
11. Pratama, Wildan Avian and Arik Prasetya. 2017. Effect of Remuneration System on Job Satisfaction and Work Motivation in Higher Education, Journal of Business Administration (JAB), Vol. 46 No.1, May 2017.
12. Juairiah, and Rosyidah Umpu Malwa. 2016. The Relationship Between Giving Remuneration to Employee Performance at Siti Khadijah Palembang Islamic Hospital, Psychic Journal, Vol. 2 No. 2

Spending Behavior and Coping Strategies of The Low Income Group in Kedah, Malay

Siti Hadijah Che Mat [1*], *Wan Roshidah* Fadzim [1], *Mukaramah* Harun[1], and *Norzita* Jamil[1]

[1]School of Economic Finance and Banking
Universiti Utara Malaysia
Kedah, Malaysia

Abstract. This study was conducted to examine and understand the behavior of low income households (B40) which are defined as households with income below RM3800. The data was collected from face to face interview based on structured questionnaire to cover 124 head of household (respondents) randomly selected in the Kubang Pasu district. The results show that most respondents spend their income on food items. Household heads are mostly employed in the agriculture, forestry & fisheries sector (27%) and most of them have an income lower than RM2000. The respondents also stated that they do side jobs and have to be frugal to cope with the rising cost of living. Nearly all (92.7%) of household heads received government assistance, particularly BR1M and a majority found that the assistance received was at a moderate level of satisfaction. There are many suggestions given by the household to overcome the rising cost of living such as eliminating the Goods and Services Tax (GST), increasing the amount of government assistance and controlling the prices of goods and fuel.

1 Introduction

Cost of living is divided into two main categories, namely daily cost of living and seasonal cost of living (Shaharudin Idrus, 2016). Daily cost of living is the day-to-day costs incurred such as the cost of purchasing groceries, and monthly fixed costs such as hire purchase installment, housing loan installment, utilities, telephones and others. Meanwhile, seasonal cost of living is the cost incurred during certain seasons such as costs incurred during festive seasons, wedding ceremonies, or at the beginning of the school year. Such costs provide additional burden to low-income individuals (Shaharudin Idrus, 2016).

Generally, there are various factors that can cause rising cost of living. One of the factors is when the increase in income is not in line with the increase in the prices of goods (Nuradilla Noorazam, 2016). Changes in taste and technology also contribute to the rising cost of living. A good that was once considered a luxury item becomes a necessity with the

* hadijah@uum.edu.my

change of time and technology. We have to admit that certain goods have become a necessity (Zulkifly Omar, 2016). A research led by Prof. Datuk Redzuan Othman involving 4468 respondents throughout Malaysia found that 82 per cent of the respondents agreed that the GST introduced in April 2015 has resulted in rising cost of living and they rejected the implementation of the GST (Sinar Harian, 2017)

Although the average monthly household income in the urban increased by 5.8 per cent per annum from RM6,833 to RM7,671 during 2014 to 2016 and average monthly household income in rural areas increased by 6.5 per cent per annum from RM3,831 (2014) to RM4, 359 (2016) (Jabatan Perangkaan Malaysia 2017), but the government is still carrying out various initiatives to help the low-income group. For example, the government has implemented the minimum wage law, 1 Malaysia People's Aid (BR1M), eRezeki Program, Digital Ability Program and others. It is hoped that these assistance can help ease the burden and increase the income of the low income group.

In the context of poverty in Malaysia, poverty is defined by a financial perspective or income perspective (Norzita, Siti Hadijah, 2014). In Malaysia, poverty rate has significantly reduced from 52.4 per cent in 1970 to 17.1 per cent in 1990 and further reduced to 3.8 per cent in 2009. Now poverty rate is about 1.7 (Jabatan Perangkaan Malaysia, 2012). Although report has shown that poverty is declining and the well-being of the people in the country is improving, the issue of rising cost of living is still widely discussed and much effort has been made by the government to help the people who are believed to be burdened by today's rising cost of living. Hence, the objective of this article is to describe the spending behavior of low-income households and the methods used by these households to cope with the rising cost of living.

2 Research Methodology

The data used in this study were collected using a survey on 124 households from the lowest 40 percent income group in the income hierarchy (B40) in the Kubang Pasu District in Kedah. The data collection process was conducted in February and March 2017 through face-to-face interviews with the respondents. Respondents in this study is head of household but if the respondent are not at home and if the household can answer on behalf the respondence the interview can go on. Household here refers to a family consisting of parents and children (married or unmarried) who share a common kitchen and in which the income of any member is used to support the family as a whole.

The interviews were conducted using a questionnaire form. The questionnaire form has eight sections where each section in the questionnaire aims to collect information about the respondent such as on the demographic profile, spending patterns, asset ownership and others. The data were then analyzed descriptively and presented in tables to facilitate understanding.

2.1 Data and Study Sample

This study focused on the area around Mukim Temin and Mukim Keplu in the Kubang Pasu District. The technique of purposive sampling was used to select the respondents. Heads of households (HH) with a monthly income of RM3800 and below were included in the sample. The enumerators went from house to house in the villages located in the two sub-districts looking at the condition of the houses before approaching the households. The respondents were chosen based on whether the head of the household has a monthly income of less than RM3800. Interviews were then conducted with the respondents who fulfilled this criterion..

3 Findings

The following are the initial information obtained on the background of the heads of household. A total of 104 respondents or the head of household (83.9%) were male and the remaining 20 respondents (16.1 %) were women.

3.1 Demographic Information of Respondents

In terms of the age of the head of household, a majority were in the age group of between 31 to 70 years, representing 110 respondents (88 %). Most of them were married (76.6%), while 20.2 percent were widows or widowers, and 3.2 percent were divorced..

Table 1: Age of Head of Household

Age of HH	Frequency	Percentage
20-30	4	3.2 %
31-40	17	13.7%
41-50	32	25.8%
51-60	41	33.1%
61-70	20	16.1%
71>	10	8.1%
Total	124	100.0%

In terms of education level, very few respondents or 11 respondents (8.87%) have received higher education, or at the STPM level and above.

Table 2: Education Level of Head of Household

Education Level	Frequency	Percentage
No certificate	7	5.6%
UPSR or Std 5	8	6.5%
LCE/PMR/SRP	36	29.0%
SPM	62	50.0%
STPM/HSC	3	2.4%
Diploma	5	4.0%
Bachelor's Degree	1	0.8%
Master's Degree	2	1.6%
Total	124	100.0%

In addition to information on the age, gender, marital status and education level of the head of household, the study also shows the main types of occupation of the head of household. The study found that most of the HH worked in the agriculture, forestry or fishery sectors, or 33 respondents (26.6%). A majority of the residents in the Kubang Pasu District are involved in activities such as rubber tapping or paddy farming. The head of household were also involved in the services sector, involving 32 respondents (25.8%). They worked as teachers, business owners, tailors and others. About 12.1% of the total respondents worked in the transport and storage sector. A majority of respondents working in this sector worked as lorry drivers.

Table 3: Job Sector of Head of Household

Job Sector	Number of Household	Percentage
Agriculture, forestry & fishery	33	26.6%
Manufacturingl	6	4.8%
Electricity supply, gas and air conditioning	2	1.6%
Water supply, sewerage, waste management & restoration activity	1	0.8%
Construction	7	5.6%
Wholesale & retail trade	2	1.6%
Transportation & storage	15	12.1%
Dining outlets/restaurant	3	2.4%
Services	32	25.8%
Unemployed/retired	23	18.5%
Total	124	100.0%

3.2 Financial Assistance and Satisfaction Level towards Assistance

Almost all respondents (92.7%) received financial assistance either from the government such as BR1M (1 Malaysia People's Aid), Social Welfare Department, SOCSO, alms (zakat) remittences (from others such as those received from working children). The remaining 7.3 percent of the respondents did not receive any financial assistance. The reason the respondents did not receive financial assistance is because they have Suruhanjaya Syarikat Malaysia (SSM) certificates. Some of them did not apply for financial aid as they felt that the process of financial aid application is troublesome as they need to fill out forms. A number of respondents said they were embarrassed to apply for financial assistance provided by the government.

When asked about the level of satisfaction with the assistance received, out of the 115 respondents receiving financial assistance, 56.5 percent of them agreed that the amount of assistance received was at a moderate level of satisfaction. The meaning of moderate in this context is the respondents felt that the amount of assistance received is sometimes sufficient and sometimes inadequate to cover their expenditure. Another 23 respondents (20%) stated that the amount of assistance received is sufficient, 16 respondents (13.9%) felt that the amount is inadequate and 9 percent felt is is very inadequate.

The majority of respondents who said the amount of financial assistance is insufficient suggested that the value of BR1M or other financial assistance is increased. That amount is insufficient due to rising prices of goods. However, 2 respondents (1.7%) stated that the financial aid received is very adequate.

Table 4: Satisfaction Levels towards Assistance Receive

Satisfaction	Frequency	Percentage
Very inadequate	9	7.8%
Inadequate	16	13.9%
Moderately adequate	65	56.5%
Adequate	23	20.0%
Very adequate	2	1.7%
Total	115	100.0%

3.3 The relationship between Employment and Income of Head of Household

Table 5 shows the sector of employment and income earned by the respondents or head of household. Most respondents (26.61%) worked in the agriculture, forestry and fishery sector and most of them have a monthly income of below RM2000. About 25 percent of the respondents worked in the services sector, most of whom earned less than RM2000 per month. Overall, for the average respondents, no matter in what sector they worked in, the amount of monthly income received is low which is below RM2000.

Table 5: Job Sector and Income

Job Sector	Below 1000	1001-2000	2001-3000	Above 3000	Total
Agriculture, Forestry & Fisheries	19(15.32)	11(8.87%)	3(2.42%)	0(0%)	33(26.61%)
Manufacturing	3(2.42%)	2(1.61%)	10.81%)	0(0%)	6(4.84%)
Electricity supply,gas, steam & air conditioning	2(1.61%)	0 (0%)	0(0%)	0(0%)	2(1.61%)
Water supply, sewerage, waste management & restoration activity	1(0.81%)	0(0%)	0(0%)	0(0%)	1(0.81%)
Construction	5(4.03%)	2(1.61%	0(0%)	0(0%)	7(5.65%)
Wholesale	1(0.81%)	1(0.81%)	0(0%)	0(0%)	2(1.61%)
Transportation & Storage	8(6.45%)	4(3.23%)	2(1.61%)	1(0.81%)	15(12.10%)
Dining/outlets/restaurant	2(1.61%)	1(0.81%)	0(0%)	0(0%)	3(2.42%)
Services	18(14.52)	9(7.26%)	5(4.03%)	0(0%)	32(25.81%)
Unemployed/retired	15(12.10)	6(4.84%)	1(0.81%)	1(0.81%)	23(18.55%)
Total	74(59.68)	36(29.03)	12(9.68 %)	2(1.61%)	124(100%)

Table 6: Monthly Income of Household

Household Income (RM)	Frequency	Percentage
0<1000	44	35.5%
11001-2000	55	44.4%
22001-3000	18	14.5%
33001 >	7	5.6%
Total	124	100.0%

Table 6 shows the total household income comprising HH's main income, side income such as rental of houses, income from side jobs and government assistance such as BR1M, alms (zakat) and others. Household income also includes income earned by all members of the household including spouse, children, and grandchildren living under the same roof.

Based on the analysis, as shown in Table 6, 44.4 percent of the households have a total monthly income of between RM1001-RM2000. This represents 55 out of a total of 124 households. This is followed by 44 households (35.5%) that have a total montly income of below RM1000.

3.4 Spending Behavior

The study also analyzed the respondents' spending patterns. Four main items were selected to analyze the respondents' spending behavior patterns namely expenditures on food, utilities, transport and education. Even though information on other expenditure items were also asked from the respondents, most respondents were unable to provide any answer while some did not have any answer for the items. These expenditure items include insurance, medical expenses and housing costs.

Almost all respondents did not buy any personal insurance policies or education insurance for their children. The reason is they no longer have any children in school and they also did not know about the importance of such policies. Most respondents only went to the government's health clinics where no charge is imposed on consultation and treatment. It is convenient for them to go to the government's health clinics and 1 Malaysia clinics near their homes. A majority of the respondents also lived in their own house or village house which does not cost them any rental.

Table 7: Expenditure Behavior

Type of Expenditure	Frequency	Percentage
Food	104	83.9%
Utilities	4	3.2%
Transportation	14	11.3%
Education	2	1.6%
Total	124	100.0%

Table 7 shows the behavior of the respondents in terms of allocating their income for expenditures. From the total 124 respondents, 104 respondents allocated the bulk of their income on food. The main expenditures were on rice, fish, milk, chicken and cooking oil. Livestock products such as fish (was regularly eaten by respondent) and chicken are important foods are crucial nutritional content of the tribal diets and are thus considered important component in dietery. while 14 respondents (11.3%) spent most of their income for transportation, 3.2% spent most of their income for utility expenditures i.e. electricity bill, water bill and Astro bill (for those respondents who subscribed to ASTRO) and the remaining 1.6% allocated the bulk of their monthly income for educational expenses.

3.5 Methods of Coping with Rising Cost of Living

The respondents were also asked about what they do to cope with the rising cost of living and their suggestion to reduce their cost of living. These questions are open-ended questions where the respondents can freely express any thoughts and suggestions according to their own common practice.

However, 60 respondents (48.4%) did not provide any response in this section. Among those who did not provide any answer, it is most probable that they thought there was no way or method they could do to overcome the rising cost of living. Perhaps they were also not involved directly in spending. For example, couples whose children are no longer staying at home would not suffer much from rising prices. For some, their children even brought cooked food and cooking provisions when they visited their parents.

As shown in Table 8, there are six suggestions given by the respondents to cope with the rising cost of living. A total of 20 respondents (31.25%) stated that they would be thrifty to cope with the rising cost of living. They also compare prices before making any purchases. By changes in purchasing behaviour was found to be a predominant form of coping strategy: it can be regarded as an 'income enhancer' strategy, in that it involves controlling the purchase of unnecessary goods and thus increasing the amount of money available for purchasing food. A large portion of their income was spent mostly on food items, followed by some spending transport and utilities (see Table 7). They thrifty also on some spending like on clothes and other necessary items. Parent try to avoid to purchase of clothes for their self, most of the respondent said they just bought for their children only. In many instances younger children in the family were asked to wear used clothing handed down by older brothers and sisters, simply to avoid spending on new items. Whenever purchases were necessary there was a preference for buying less expensive clothes for all the children once a year normally during celebrate hari raya. They also take a traditional medicines not because they have no access to modern health services but they more belief in their traditional medication. They also go to government clinic or general hospital rether than private clinic or private hospital.

In addition, 21.88 percent of the respondents have side jobs. They felt that having more than one job could increase their income so that there is sufficient income to cover monthly expenses. (Siti Hadijah, Mukaramah, Zalina and Zakiyah, 2017) found that a single mothers feel that an easy way to get financial resources is by participating in business activities or entrepreneurial activities. The study found that the income from entrepreneurial activity earned by single mothers succeed in bringing them out of poverty line without the need to rely on non-entrepreneurial income and non-work income. The study indicates that by participating in entrepreneurial activities, the poverty rate is only 12 percent compared to 76 percent if they are not involved in entrepreneurial activities. In addition, the results also show that the income from entrepreneurial activities is successful in shortening the length of time to break out of poverty for the single mothers. Doing side job such as getting non farm income among farmers also can increase income for the farmers (Siti Hadijah, Nor Aznin and Ahmad Zafarullah, 2012, Siti Hadijah and Roslan (2010). Other respondents worked overtime (9.38%) while others only spend according to needs (15.63%).

Table 8: Methods to Handle Rising Cost of Living Problem

Action by House Hold	Frequency	Percentage
Being Thrifty	20	31.25%
Do Side Job	14	21.88%
Spend According to Needs	10	15.63%
Grow Own Food	7	10.94%
Work Overtime	6	9.38%
Reduce Utility Consumption	7	10.94%
Total	64	100.0%

Note: Only considers the first suggestion by each respondent who answered this

This study also asked for the suggestions or opinions of the HH about the methods or ways to reduce the burden of cost of living. The respondents were also asked about their main suggestion about what could be done to help reduce the rising cost of living. As shown in Table 9, the study found that 38 respondents (30.65%) suggested that the Goods and Services Tax (GST) is eliminated. This means that even though are living in the rural area and most of them are elderly, they are aware about current issues such as the GST being the cause of rising prices and rising cost of living.

This is followed by the suggestion to control the prices of goods and fuel. Controlling prices is important so that sellers cannot arbitrarily increase prices. Most respondents found that there is a huge difference in the price of goods at sundry shops and at hypermarkets. However, most of them did not have the ability to shop at these hypermarkets due to distance and transport factors. A total of 26 respondents (20.97%) felt that controlling prices can curb the rise in cost of living.

A number of respondents also proposed that government aids such as BR1M are increased. BR1M or 1Malaysia People's Aid program is devised by the Malaysian Government as the effort to ease the burden of lower income group in Malaysia. The program is devised to help the lower income group cushion the high cost of living in Malaysia (Shazida Jan, Norehan, Siti Hadijah, Abu Sufian, Jamal and Hussin, 2018). They believed that an increase in the amount of BR1M would be able to lessen the effect of rising cost of living. They also felt that the amount of BR1M is too small compared to their daily expenses. A case study of rural poor in Kedah by Kalthum, Rohana and Norehan (2016) concluded that the livelihoods strategies adopted by the households could be translated into opportunities for the poor households to improve their socio-economic status if the members receive proper provision of trainings and capitals.

Table 9: Suggestions to Cope with the Rising Cost of Living

Suggestions by Respondents	Frequency	Percentage
Eliminate GST	38	30.65%
Raise Rubber and Paddy Prices	7	5.65%
Control Goods and Fuel Prices	26	20.97%
Higher Government Aids	24	19.35%
Raise the Minimum Wage	10	8.06%
Control the Entry of Foreign Workers	8	6.46%
Increase Employment Opportunity	11	8.87%
Total	124	100.0%

Note: Only considers the first suggestion by each respondent

Besides that, the respondents believed that additional employment opportunities in the labor market facilitate the employment of young people, which then provides opportunities for the children to help with their family's finances. Study by Aznita, Nur Fatihah, Norehan, Selamah, Hussin and Sabri (2018) concluded that technology or R&D significantly determine job creation. Thus, R&D policy can contribute a positive effect on job creation in the Malaysian manufacturing sector. Moreover, 6.46 percent of the respondents felt that by controlling the influx of foreign immigrants would lead to increased domestic employment opportunities especially in the manufacturing sector. This would have a positive effect, as the locals would be more likely to secure a job. Malaysia might reduce its dependency on foreign worker by identifying the occupational interest of rural youth. A survey on rural young labour suggets several attraction factors that can possibly increase local youth participation in plantation sector includes improvement of working environment, job status, facility and benefit in plantation field (Norehan, Siti Aznor and Mohamad Amizi, 2016) .

4 Conclusion

Most respondents allocate a huge portion of their income for food expenditure. This is followed by transportation, utilities, and education. To cope with the rising cost of living, various ways have been practised such as doing side jobs to earn additional income.

The respondents also suggested several ways to reduce the cost of living such as eliminating the GST. They claimed that the implementation of the GST has led to the rise in the prices of goods in the market and this has caused their daily expenditure to rise, particularly for food. In addition, they also argued that if the amount of financial assistance from the government could be increased, this would reduce their burden. In conclusion, this study found that due to pressing cost of living problem, the lower income group would face difficulties and would have to bear a higher burden than ever before.

The results of the study indicate that the main cause of the B40's low income is low rubber prices faced by the respondents. Most of those interviewed were farmers and they claimed that rubber price is too low (RM 1.60/kg) at the time the data was collected (in September 2017) causing their income to be much lower compared to other farmers.

Some of them have had to reduce their daily expenditure to ensure that their money lasts for a month. In addition, they also need to be thrifty and work in more than one job to ensure sufficient monthly income. Increased food prices have caused a rise in their spending. The findings reveal that financial aid can help ease their burden, especially for the HH with many dependents. However, the satisfaction level associated with the assistance is only at a moderate level.

Acknowledgements

The authors would like to express their deepest gratitude to Universiti Utara Malaysia (UUM) for funding the study under the Centre of Excellent Grant research project (S/O Code: 13711 in which this study represents a part of the overall research project.

Integrated Mobile Payment Technology for High Speed Financial Literacy in Cooperatives

Bambang Agus Pramuka [1*], *Poppy* Dian Indira Kusuma [2], Sugiarto[2] and *Wiwiek* Rabiatul Adawiyah[2]

[1] School of Accounting Faculty of Economics and Business
[2] School of Management Faculty of Economics and Business
Universitas Jenderal Soedirman Indonesia
Email: *bpramuka@gmail.com

Abstract. Digital money receives tremendous attention globally, from buyers to sellers, as an alternative form of payment. The passions on this technology enormous and infectious, therefore attracting more followers from large merchant to smaller one including the cooperative. School cooperative serves as a center of economic with a wide range of activities, such as mini market, canteen, to serve the needs of students, teachers, and people in the surrounding area. One of the programs launched by the Central Bank of Indonesia is a financial inclusion to increase the financial literacy of people living in the remote area. In this case, Islamic conventional boarding school cooperative can become influencers in most of the surrounding community to attract people's interest to adopt the technology. The purpose of this inquiry is to find out factors influencing the adoption of mobile payment and user's intention and decision to adopt m-payment. Authors proposed an innovative research model based on the theory of technology acceptance model with the innovation characteristics of the diffusion of innovations, perceived security and intention to recommend the technology. The research model has empirically tested at 205 employees Islamic boarding schools across four regions. The independent variables under study include perceived ease of use, perceived usefulness, perceived credibility, subjective norms, and trust while the dependent variables are mobile payment adoption and intention to recommend the technology. The study uses structural equation modeling as a tool for analysis. The result indicated that perceived ease of use, perceived usefulness, and subjective norms have a significant influence on the adoption of mobile payment. The study also confirmed the relevance of users' intention to recommend mobile payment technology. Thus, future research needs to include social media campaigns and technology adoption studies. For researchers, the result of this study act as a basis for further refinement of personal technology acceptance model. For practitioners, understanding the key constructs is crucial to design, refine, and implement mobile payment services, applications, and products that achieve high consumer acceptance, value, and high rates of positive recommendations in social networks.

* Email: *bpramuka@gmail.com

1 Introduction

The financial inclusion phenomenon is currently the center of attention in some countries. Limited knowledge and experience to use financial services is a barrier for certain parties to get financial services. Financial inclusion has gained attention from academics and policymakers for several reasons such as it has a positive impact on the economic development of a country [1], enhancing world development [2-4] and sustainable development [5]. Financial inclusion affects both the macro and microeconomics of a nation. At the micro level individuals who do not have a bank account shall not receive benefits in the form of increased value of savings[2, 6], By possessing an account a women shall have better economic empowerment [7, 8], as well as entrepreneurial productive consumption and investment [9, 10].

Islamic traditional boarding schools provide educational services to the surrounding people regardless of their social status. Most schools are located in rural areas to provide an opportunity for people in rural areas to participate in the education process. Therefore, the schools have the potential to act as a distribution channel in providing access to the majority of people in rural areas. Previous research shows that "disadvantaged people" come from women, young and the poor [4, 8, 11]. One of the factors that drive individuals to use financial services is ease of access. Individuals use financial services to meet several basic needs including education, health, and shelter, starting and operating a business. Financial inclusion increases poverty alleviation and promotes economic growth [12, 13].

In the digital era access to financial services should be open to anyone. The digital era stimulates the growth of internet-based business activities [14]. The number of mobile phone users has increased sharply and massively throughout the world. Technology innovation has an indispensable role in expanding access of all parties to financial services, so there is no more financial exclusion. People who do not have a formal bank account can use cellular technology to conduct transactions such as paying a taxi, using go pay. High interest in the use of financial technology has emerged since 2004 [15, 16]. Even among the 20 percent of the poorest households, almost 7 out of 10 have cell phones [17].

This paper uses technology acceptance and innovation by individuals to explain the behavior of the use of information technology in the financial sector. What are the factors that influence the interest of Islamic boarding school cooperative managers to use digital financial services? The results of this study contributed to the development of the literature on digital financial inclusion in Indonesia. The Central Bank of Indonesia initiates to develop an ecosystem of non-cash payments by implementing e-money in Islamic boarding schools. However, data on the potential for digitizing financial services for Islamic boarding schools are not yet available.

2 Literature Review

2.1 Financial Inclusion at Individual Levels

Although previous research on financial inclusion is enormous, few discussed financial inclusion policies at the individual level. With financial inclusion strategy, the Indonesian government expects to narrow down the exclusion for certain circles such as women, young people and the poor. To date, the literature on financial inclusion across the country rely on the state-level proxy based on the data collected by the policy-making, in this case, the Central Bank [12, 18, 19]. Non-asymmetric information narrows the generalization of existing research because data from central and macro banks is difficult to translate at the micro level, for example how the government policies increase the financial access as individual differences in terms of income level, gender, location of residence, or age [2, 6].

Specific interventions can encourage the use of digital-based Instruments due to the perception of the benefits of the device [7]. Financial inclusions provide some benefit to the individual in the form of low-cost accounts, financial security, improve networking, as well as legal rights. Nonetheless, the regular benefits have not yet motivated the disadvantaged group (women and the poor) to use financial services. Some of the reported barriers include money shortage, and there are no immediate needs for bank accounts and lack of trust. Therefore, the government policies to promote inclusion can increase the likelihood that people perceive financial services are within their reach [2].

2.2 The Feature of E-Money

The results of research in the field of financial management related to the digital economy is the emergence of digital finance and financial inclusion in the pieces of literature [20]. Digital banking through financial technology has a positive impact on financial composition for developing countries and developed countries because the digital economy provides individual comfort. The marketing field is carried out by researchers [21] about online retailer's consumer perception of e-loyalty intentions reported that the higher the online retailer's consumers, the more likely consumers to make purchases online. The phenomenal growth in cellular technology has made it easy for consumers to transact. Consumers do not need to bother carrying cash because transaction payments can be made using a cellular device. Mobile-based payments can be categorized into three perspectives, namely foreign mobile commerce, mobile proximity payments, and peer-to-peer mobile [22].

Payment technology using mobile phones also has several limitations which can disrupt transactions such as ergonomics and usability of devices such as small screen size and small keyboard [23], unstable internet connection [24], complicated and tedious payment procedures [25], confidentiality during authentication [26]. Therefore, mobile payment users require more significant effort and high learning speed to use the m-payment application which sounds discouraging. To increase users' interest in using e-money, the payment must be simple, easy to use, and do not require much energy to execute it [27].

2.3 The Technology Acceptance Theory

The potential of mobile payment in the future is enormous. Small firms have started to use digital payments for several reasons such as consumer demands and peer pressure. The euphoria has spread over the Islamic boarding schools, where parents have demanded an easy way of paying their children's' tuition due to geographical constraints. Most of the students who study at the boarding schools come from the distance. Likewise, the community living around the schools have limited access to financial services which requires them to travel to the city for payment of bills. As a consequence, the use of digital payment by Islamic boarding schools does not just prosper the pupil but also the surrounding people.

The limited number of schools have adopted digital payment. Among the factors influencing the adoption, the decision is the perceived usability and perceived ease of use of the technology [28]. This theory explains some of the reasons for individuals take electronic mail systems include lower operational costs and the speed of information spread. Mobile payment is voluntary and is used by individuals for the benefit of transactions that are generally personal. Therefore, all adoption costs of m-payment are borne by individuals and adoption decisions are not solely based on practical reasons. To use an m-payment, a person must have sufficient knowledge and skills about [29].

2.4 The Research Model

Perception against Compatibility of a technology depends on the extent to which innovation can provide meaningful experience and ability to meet the needs of the user wholeness, as well as offering better values for living. Almost all banks in Indonesia have proposed a mobile payment to anticipate growing business transactions. With the emergence of virtual business during the disruptive era (such as online transportation and online shopping), consumers in Indonesia regardless of their backgrounds have used m-payment [14]. The financial technology does not only offered by financial institutions. Some of the firms providing financial technology include cellular providers (t-money), Telkom, Go-jek, Grab, and Tokopedia. M-payment is ubiquitous because it allows for faster, safer and reliable financial services.

Perception of ease of use means that technology tools are simple to use [30]. The first step to do before start using a tool is understanding the complexity of the technology itself. Therefore, perceived ease of use is an essential determinant of technology adoption. Moreover, users also consider some benefits obtained from adopting technology as an important influencer. The application of a mobile payment provides some gains in the form of minimum fees, convenient to use. When consumers feel that mobile payment is easy to learn and use, they will be more likely to adopt the innovation. Consumers first impression of a technology offered is an essential determinant of adoption decision. When consumers believe that they will succeed in utilizing the technology, then the choice is immediately executed. In other words, the ease of perception has a positive effect on the intention to use a technique which later is followed by adoption [31]. Past studies have also discussed on the antecedents of consumer decision to adopt mobile payments [32-36].

Perceive the benefits of a technology tool refers to the relative advantages of a technology that can improve individual performance [30]. Understanding of benefits can increase an individual's desire to adopt a system [15, 37, 38]. Consumers get greater benefits by using online transactions using mobile devices. If compared with traditional e-payment methods, consumers who use m-payments benefit from services that are easy and fast. A study at Malaysia concludes that consumer perception of high benefits will result in a higher intention to adopt m-payment[25]. Similarly, m-payment adoption is useful regarding time savings and portability [34].

The use of technology in transactions does not mean without risk. From hard perspectives, a technique should possess multiple levels of a security system to prevent hijacking. The phenomena of increasing cybercrime are discouraging and become a severe challenge to technology adoption. The development of hardware should be accompanied with a soft approach. To improve one's attitude towards technology we need to strengthen the power of the subjective norms. For example by providing education about religious standards and legal rules that are expected to control behavior[39, 40].

Perceptions of security risks are an essential factor that determines consumers' decisions on whether or not to adopt a technology [14]. The knowledge of protection against the risks of using cellular transactions is crucial. The risk of losing confidential data that causes financial losses needs to be well anticipated to motivate users to adopt m-payment. Perception of security in transactions has a positive effect on the intention to take m-payment. Trust starts from a belief that everyone is sincere in his or her words, attitudde and behavior. Hence the technology of m-payment offered to the public is carefully prepared to prevent data leakage. The level of trust in adopting m-payment assumes that everyone will not behave opportunistically by expecting the promised service [10]. It will be more difficult for sellers and buyers to build online trust because there is no interaction between the two parties while performing the transactions. The excellent system offers protection to users' data, assets during the online deals.

3 Methodology

This research is quantitative in nature. The research model was empirically tested at Islamic traditional boarding schools across four regions. The independent variables under study include perceived ease of use, perceived usefulness, perceived credibility, subjective norms, and trust while the dependent variables are mobile payment adoption and intention to recommend the technology. There were 27 items used in the questionnaire to measure those variables. The study uses structural equation modeling as a tool for analysis. A survey distributed to respondents using purposive sampling. Criteria for respondents are boarding school cooperative managers who handle finance and have worked at least one year. Before distribution, participants were given a brief explanation of the research objectives. Also, to avoid response bias in the survey, the respondent's name is filled with anonymous. The respondents got an explanation of the variables and terms used in the questionnaire. A total number of questionnaires distributed to respondents is 205 with a response rate of 100 percent, so that all are processed further for further analysis. The data collection period was one month.

4 Result and Discussion

4.1 Descriptive Statistic

The results of the analysis of factor demographic respondent can be seen in table 1.

Table 1 Demographic profile of respondents

Item		Frequency	Percent
Gender	Male	101	49.3
	Female	104	50. 7
Age	<20	84	41
	20-30	76	37.1
	31-40	21	10.2
	41-50	16	7.8
	> 5 0	8	3.9
Education Level	High school	118	57.6
	Diploma	24	11.7
	Bachelor's degree	57	27.8
	Post Graduate	6	2.9
Income	<Rp. 3,000,000	130	63.4
	> Rp.	75	36.6

3,000,000

There were the almost equal proportion of male and female respondents, with female dominance, participated in the survey. The average age of the respondents was between 20 to 30 years, indicating that they are still in their productive age. Majority of the respondents were high school graduate. There was 27.8 percent of university graduates, with bachelor degrees, working at Islamic boarding schools. Majority of respondents earn monthly income from below three million rupiahs. Table 2 presents the descriptive statistics of the variable under study.

Table 2 Descriptive statistics of variables

	N	Minimum	Maximum	Mean	Std. Deviation
Perception Benefits	205	3.00	5.00	4.2476	.49072
Perception ease of use	205	2.00	5.00	4,0951	.49269
Perception Credibility	205	2.50	5.00	3.8793	.48984
Subjective Norms	205	1.00	5.00	3.7207	.66679
Trust	205	2.30	5.00	3.9137	.58746
Attitude	205	3.00	5.00	4,1005	.52192
Interest	205	2.60	5.00	3,8907	.42767
Valid N (listwise)	205				

Table 2 presents the descriptive statistics of each variable. The average response was statistically viable concerning perceive benefits, ease of use, credibility, subjective norms, trust, attitude and interest to use e-money. Perceive benefits on the technology is the most critical factors of consumers' attitude towards e-money, followed by perceived ease of use. Trust is an essential determinant of intention to use e-money or mobile payment for the transaction. Perceive security of the operation is crucial as everyone wants to protect valuable assets.

4.2 Validity and Reliability Test

Table 3 Validity and reliability test of the instruments

Variabel	Loading (L)	Error (E)	L	E	L	E	L	E	L	E	L	E	L	E
X1_1	0.85	0.15												
X1_2	0.82	0.18												
X1_3	0.67	0.33												
X1_4	0.62	0.38												
X2_1			0.63	0.37										
X2_3			0.81	0.19										
X3_1					0.69	0.31								
X3_2					0.59	0.41								
X3_3					0.55	0.45								
X4_1							0.58	0.42						
X4_2							0.48	0.48						
X4_3							0.89	0.11						
X4_4							0.80	0.20						
X5_1									0.67	0.31				
X5_2									0.83	0.17				
X5_3									0.76	0.24				
Y1_1											0.65	0.35		
Y1_2											0.77	0.23		
Y1_3											0.67	0.33		
Y2_1													0.49	0.51
Y2_2													0.49	0.51
Sum of Std.Loadings	296		1.44		1.83		2.75		2.26		2.09		0.98	
Sum of Measur. Error		1.04		0.56		1.17		1.25		0.74		0.91		1.02
RELIABILITY	0.894		0.787		0.741		0.8581		0.873		0.828		0.485	

According to Hair (1995), an indicator is considered valid provided it has a loading factor above 0.40. Based on the data above it is known that the data is valid. The next step is to test the reliability of the items used to measure the variables.

$$Contruct\,\mathrm{Re}\,liability = \frac{(\sum Std\,Loading)^2}{(\sum StdLoading)^2 + \in .j}$$

The result of the reliability test revealed that two variables did not meet the reliability criteria. The variables were Y2_1 and Y2_2. The limit value used to assess an acceptable level of reliability is 0.7. However, this value is not a dead value. This means that if the value below 0.70, an item can still be accepted if accompanied by strong empirical reasons. Nunally and Bernstain (1994) in Ferdinand, 2005, stated that the reliability value between 0.50-0.60 was enough to justify the results of the study.

4.3 The Structural Equation

The next step in the analysis is to put the variable into the structural equation. Picture 1 reflects the model.

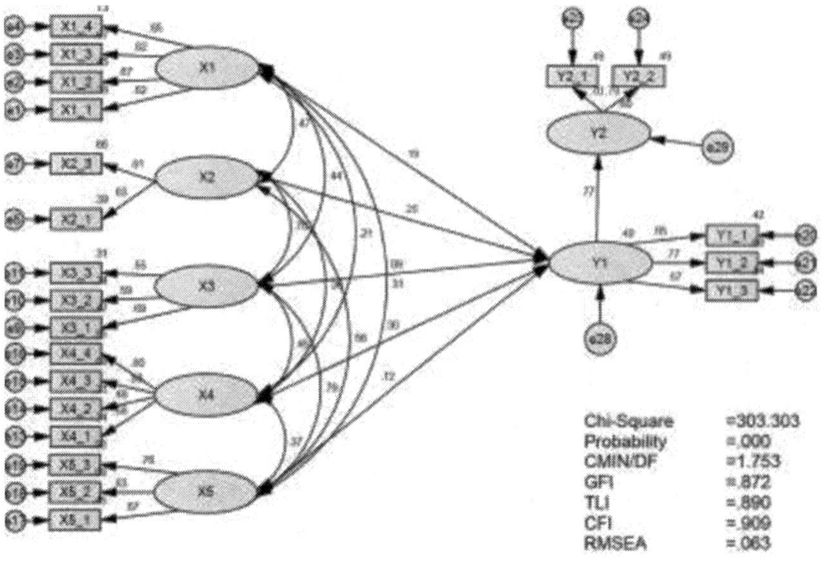

Picture 1. The structural equation modeling of the framework

The above output is the result of M-Payment testing using a mediating variable model with the following variables:

X1: Perceive benefits
X2: Perceive ease of use
X3: Perceive credibility
X4: Subjective norm
X5: Trust
Y1: Attitude
Y2: Interest

4.4 Direct Effect Analysis

Based on the output from testing the model using Amos shows that the immediate effect of perceived benefits on attitudes is 0.19; direct influence on perceived ease of attitude of 0.25; direct influence of credibility perception of 0.31; subjective norm's direct influence on attitudes of 0.30; the immediate effect of trust on attitude is 0.12; direct impact on attitudes towards interest of 0.77. The loading factor value shows a strong direct relationship between independent variables and mediating variables. The negative value on the loading factor shows the rejection of the influence of the variable, but based on the output above shows that the independent variable has a good effect, then the mediating variable has a good impact on the dependent variable.

4.5 Evaluate the Goodness of Fit Criteria

Based on the results of the analysis that we display on Amos Graphics or contained in the output table and the text output, we compare the values with the Cut of Value value for each criterion, so that the following results are obtained:

Table 4 Goodness of fit of the model

The goodness of Fit Index	*Research Model Results*	*Cut off Value*	*Model Evaluation*
X2-Chi-Square	303,303	Small expected <1000	Well
Significance Probability	0,000	≥ 0.05	Well
RMSEA	0.063	≤ 0.08	Well
GFI	0.872	≥ 0.90	Pretty good
CMIN / DF	1,753	00 2.00	Well
TLI	0.890	≥ 0.95	Pretty good
CFI	0.909	≥ 0.95	Pretty good

Based on the output above, it was found that (Ferdinand, 2005):

1. ^{2} - Chi-Square statistics, the smaller the value χ^2, the better the model, and the probability is accepted by the cut-off value of $p > 0.05$. The results of this study indicate that this research model is in the right category because it has a small chi-square value.
2. RMSEA (*The Root Mean Square Error of Approximation*), is an index used to compensate for *chi-square* in a large sample. RMSEA values that are small or equal to 0.08 are indices for accepting the model based on *the degree of freedom*. The RMSEA value of this research model is 0.063 which means that the model is good so that the model can be accepted.
3. GFI (*Goodness of fit Index*), is a *non-statistical* measure that has a range of values between 0 and 1. A high value in this index shows a "*better fit.*" The GFI value of this study is quite good because it approaches the minimum criteria required.
4. CMIN / DF (*The Minimum Sample Discrepancy Function Divided with the degree of Freedom*), is a chi-square X^2 statistic divided by the degree of freedom so that it is called X^2 relative. Relative X^2 value less than 2.0 or 3.0 is an indication of *an acceptable fit* between model and data. CMIN value of 1.753 indicates that this model is excellent and fit.
5. TLI (Tucker Lewis Index), an incremental index that compares a model that is tested against a *baseline* model, where the recommended value is ≥ 0.95. However, the most important thing is that the value close to 1 indicates *a perfect fit*.
6. CFI *(Comparative Fit Index), the* range of values is 0-1, which is getting closer to 1, indicating the highest level of fit.

Based on the results of the above analysis it can be concluded that the M-Payment testing model can be accepted. This is because the test output shows good results

The results of this study support the results of previous studies, which concluded that perceptions of benefits affect the interest in adopting e-money in transactions. This is consistent with several previous researchers [14]. Besides that, the perceived ease of operation also affects the interest in using e-money. The existence of government policies that began to lead to e-payment at individual facilities and the encouragement of using financial technology impacted consumers' decisions to use e-money. Perception of ease of use means that technology tools are easy to use and uncomplicated [30]. This supports previous research that the ease of positive influence on the intention to use a technology which later is followed by adoption[31]. Past studies indicate that customers' decision to adopt mobile payments have been used easily [32-36].

The results of this research study support previous research (such as [25, 34]) in Malaysia which concluded that the perception of high benefits would result in a higher intention to adopt m-payment. M-payment adoption is useful concerning timesavings and portability. Perceptions of technological security risks are important things that determine consumers' decisions on whether to adopt innovations [11]. Therefore, perceived security in the transaction has a positive effect restricted ap intention to take m-payment. Starting from the belief that everyone will be sincere in transacting so that there is no fraud, the level of trust in payment assumes that no one will behave opportunistically by expecting the promised service[10].

5 Conclusion

The study revealed that perceived ease of use, perceived usefulness, and subjective norms have a significant influence on the advertising of mobile payment. The study also confirms the relevance of users' intention to recommend mobile payment technology. Thus, future research needs to include social media cam campaigns and in future technology adoption studies. For researchers, the result is a basis for further refinement of personal technology acceptance model. For practitioners, understanding the key constructs is crucial to design, refining, and implementing mobile payment services, applications, and products that are high consumer acceptance, value, and high rates of positive recommendations in social networks.

References

[1] W. R. Adawiyah, Int. J. Appl. Bus. Econ. Res. (2015)

[2] W. R. Adawiyah and B. A. Pramuka, J. Manag. Dev. **36**, 877 (2017)

[3] W. R. Adawiyah, B. A. Pramuka, Najmudin, and D. P. Jati, Int. Bus. Manag. (2015)

[4] P. Hari Adi, F. Wihuda, and W. R. Adawiyah, Market-Tržište **29**, 39 (2017)

[5] F. Allen, A. Demirguc-Kunt, L. Klapper, and M. S. Martinez Peria, J. Financ. Intermediation **27**, 1 (2016)

[6] N. Ashraf, D. Karlan, and W. Yin, World Dev. **38**, 333 (2010)

[7] T. Beck, A. Demirguc-Kunt, and M. S. Martinez Peria, J. Financ. Econ. **85**, 234 (2007)

[8] M. BRUHN and I. LOVE, J. Finance **69**, 1347 (2014)

[9] S.-N. Cheong, H.-C. Ling, and P.-L. Teh, Expert Syst. Appl. **41**, 3561 (2014)

[10] F. D. Davis, R. P. Bagozzi, and P. R. Warshaw, Manage. Sci. **35**, 982 (1989)

[11] A. Demirgüç-Kunt, T. Beck, and P. Honohan, Financ. All? Policies Pitfalls Expand. Access (2008)

[12] A. Demirgüç-Kunt and L. Klapper, Brookings Pap. Econ. Act. **2013**, 279 (2013)

[13] P. Dupas, S. Green, A. Keats, and J. Robinson, *Challenges in Banking the Rural Poor: Evidence from Kenya's Western Province* (Cambridge, MA, 2012)

[14] P. Dupas and J. Robinson, Am. Econ. J. Appl. Econ. **5**, 163 (2013)

[15] V. Dutot, J. High Technol. Manag. Res. **26**, 45 (2015)

[16] Ferdinand, A. Structural Equation Modeling. Semarang: The Board of Directors of Diponegoro University (2005)

[17] *World Development Report 2016: Digital Dividends* (The World Bank, 2016)

[18] P. Hari Adi and W. R. Adawiyah, J. Islam. Mark. **9**, 841 (2018)

[19] P. Honohan, J. Bank. Financ. **32**, 2493 (2008)

[20] Istiqomah and W. R. Adawiyah, Int. J. Entrep. Small Bus. **34**, 330 (2018)

[21] J. Kendall, N. Mylenko, and A. Ponce, Policy Res. Work. Pap. (2010)

[22] Klapper, L., El-Zoghbi, M., & Hess, J. Achieving the sustainable development goals. The role of financial inclusion. (2016)

[23] L. Lu, H. Chang, and S. Yu, Internet Res. **23**, 47 (2013)

[24] G. Madlmayr, J. Langer, C. Kantner, and J. Scharinger, in *2008 Third Int. Conf. Availability, Reliab. Secur.* (IEEE, 2008)

[25] J. Ondrus and Y. Pigneur, in *Int. Conf. Manag. Mob. Bus. (ICMB 2007)* (IEEE, 2007)

[26] K.-B. Ooi and G. W.-H. Tan, Expert Syst. Appl. **59**, 33 (2016)

[27] J. Owens, Innov. Technol. Governance, Glob. **8**, 271 (2013)

[28] P. K. Ozili, Borsa Istanbul Rev. **18**, 329 (2018)

[29] S. Y. Park, Educ. Technol. Soc. (2009)

[30] M. Pinasti and W. R. Adawiyah, Int. J. Bus. Soc. **17**, (2017)

[31] N. K. Putri, W. R. Adawiyah, and B. A. Pramuka, DLSU Bus. Econ. Rev. (2017)

[32] R. Saadé and B. Bahli, Inf. Manag. **42**, 317 (2005)

[33] G. W.-H. Tan, K.-B. Ooi, S.-C. Chong, and T.-S. Hew, Telemat. Informatics **31**, 292 (2014)

[34] A.-C. Teo, G. W.-H. Tan, K.-B. Ooi, T.-S. Hew, and K.-T. Yew, Ind. Manag. Data Syst. **115**, 311 (2015)

[35] A. C. Teo, G. W. H. Tan, C. M. Cheah, K. B. Ooi, and K. T. Yew, Int. J. Mob. Commun. **10**, 578 (2012)

[36] H. S. Tsai and R. LaRose, Comput. Human Behav. **51**, 344 (2015)

[37] E. Valcourt, J.-M. Robert, and F. Beaulieu, in *WiMob'2005), IEEE Int. Conf. Wirel. Mob. Comput. Netw. Commun. 2005.* (IEEE, 2005)

[38] V. Venkatesh, Inf. Syst. Res. **11**, 342 (2000)

[39] Venkatesh, Morris, Davis, and Davis, MIS Q. **27**, 425 (2003)

[40] F. Wihuda, A. A. Kurniawan, A. I. Kusumah, and W. R. Adawiyah, Tourism (2017)

[41] A. Zins and L. Weill, Rev. Dev. Financ. **6**, 46 (2016)

CORRUPTION AND CULTURE: REVISITING THE CLAIM OF ITS RELATIONSHIP

Kiky Srirejeki[1*]

[1]Department of Accounting, Faculty of Economics and Business, Universitas Jenderal Soedirman, Indonesia

Abstract. The researcher has long discussed ideas on how culture related to corruption. The buzzword "corrupt culture" was introduced to describe a group of people who act corrupt behavior. Some researcher proclaimed that particular culture permits corrupt behaviors, such as gift-giving. However, to claim that gift-giving is a corrupt behavior, prudent consideration should be taken place. What is being meant as gift-giving in a certain area could be understood as simple bribes in other areas. This paper highlight the importance of understanding cultural relativism before judging the claim of a vile relationship between corruption and culture. Further, strategy to combat corruption is also advised.

1 INTRODUCTION

Transparency International released reports where Indonesia in the past several years has been consistently listed as one of the most corrupt countries in the globe. For the past ten years, Indonesia ranked 126th out of 180 countries in 2008, 118th in 2012, 90th in 2016 and 96th in 2017. In 2018, the country ranked 89 out of 180 countries in the Corruption Perception Index (CPI) by Transparency International, with a score of 38 points out of 100 (Transparency International, 2018). Though the level of corruption seems to be declining, corruption has been so endemic inside the country. Currently, corruption is not only on the central power in Jakarta, but all the way goes to the remote village areas.

A report from the Indonesia Corruption Watch (ICW) in 2018 revealed that village funds were one of the five sectors with the most corruption cases in 2017. Compared to other sectors, corruption in village funds was ranked first with 98 cases, far higher than the case corruption in the government with 55 cases, education with 53 cases, transportation with 52 cases and social community with 40 cases (ICW, 2018). Data from the Corruption Eradication Commission (Komisi Pemberantasan Korupsi-KPK) also show similar tendencies. According to the KPK report, in the first six months of 2017, a total of 459 community reports were received for alleged corruption of village funds (KPK, 2018).

The vast spread of corruption cause for concern by many parties. Immense corruption leads to severe condition to the home countries. One of the earliest research on the effect of corruption done by Rose-Ackerman (1978) in which she argued that corruption hinders countries' economic development, a view that received an agreement among various

* Corresponding author: kiky.srirejeki@unsoed.ac.id

scholars. Mauro (1995, 1997) and Wei (1997) began the cross-country analysis on the relationship between corruption and growth. Their works find that corruption reduces private investment and the growth of gross domestic product (GDP). Corruption is also a hindrance to economic growth in Asian countries (Thach, Duong & Oanh (2017) and significantly deters the inflow of foreign direct investment (FDI) to host countries (Zhao, et al., 2003).

On another note, the effect of corruption on micro-level has been studied by Kimuyu (2007) stated that corruption impairs a firm's growth and the propensity to export due to the significant spending on informal payments. Husted (2002) also argue that corruption also reduces the level of human capital and shatters confidence both in public and private organizations.

Many countries have been on a long road in fighting corruption. There are lengthy efforts within the governmental and non-governmental organization as well as international organizations to establish specific guidelines to prevent bribery and unethical practices in business (Getz and Volkema, 2001). Notwithstanding, exterminating corruption is not a flipping of a hand, because it is a multifaceted social phenomenon that penetrates many facets of societies.

Until several years ago, the traditional approach to study about corruption was related to economic and political factors, for example, like on how corruption impacts the economic growth, openness to trade, and the flow of foreign direct investment as well as the type of government and the degree of power to the press freedom. Yet the continuing corruption cases revealed appears quite obvious that the corruption in a certain country may depend by causes other than economic and political factors.

A number of researchers have shifted their focus to put culture as a determinant factor in corruption. They believe that individuals' behavior has a basis that rooted from family, education, social norm and environment that influence the way of their thinking and acting. The buzzword of "corrupt culture", which imply a predisposition for a group of people who behave in corrupt ways, then, emerges as an alternative excuse to the issue.

However, the notion of "culture" has often been used to elucidate or justify an act of corruption. Gift-giving, for instance, it is sometimes acclaimed as part of certain country's culture but it can be mistakenly understood as bribery or corruption. To what extent we should tolerate the cultural practice as the normal acceptable practice rather than corrupt act? On this paper, in addressing those concerns, the questions to be answered is: To what extent the word corrupt culture is legitimate? What strategies can be deployed to relieve the "corrupt culture" in our society?

The structure of the paper is following. First, it will give context on the concept of corruption. Second, it will review existing theories that explain the causes of corruption. Third, it will explore the influence of culture in which the culture of corruption is judged to be inherent in it. Finally, it will prescribe the possible strategies to alleviate the corrupt culture.

2 Theoritical Perspective on the Cause of Corruption

Research and literature on the causes of corruption are voluminous. The potential causes of corruption, according to Arifianto (2001) based on a theoretical perspective, could be classified into three categories: the mainstream economic theory, patrimonialism, and kleptocratic state theory. Each of these categories will be explained briefly in this section.

2.1 Mainstream Economic Theory

According to mainstream economic theory, corruption is more likely to happen in countries where the state plays a dominant role in the economy. This condition leads to the monopoly of the state to control the country's economic activities. In a country where government plays a significant role to rule economic activities, the government released too many regulations that regulate the country's economy, which hampered the participation of the private sector. Consequently, for private sectors to be able to participate in economic activities, they have to make payoffs to government officials so that they can operate their economic activities with little intervention from the state. In mainstream economic theory, a country with no system of accountability and transparency that discourage corrupt behavior also triggered corruption.

Empirical research has support this notion stated that the more competitive and open one country is, the less likely corruption would become an endemic problem within it (Ades & Di Tella, 1995). Majeed (2014) also confirms this finding stated that the combination of trade openness and high bureaucracy quality deter corruption. These research confirm that corruption correlates with a high level of government intervention in the economy.

2.2 Patrimonialism Theory

The prominent figure of the patrimonialism is Max Weber. Weber opined that there are three types of authority: traditional authority, in which authority is based on the hereditary claims based on tradition and rulers tend to have absolute power; charismatic authority, in which authority is based on the charisma of the leader and his or her ability to convince the public to follow his or her orders; lastly is rational/legal authority in which the leader obtains authority through legal and professional means (Weber, 1974).

In a country based on traditional authority, government officials are not recruited based on merit consideration, instead of on the family relations or personal closeness to the leader. The power that the leader has in this type of authority could maintain government officers' loyalty by playing to give reward or punishment. This system of authority called patrimonialism.

In the patrimonial system, there is no separation between the public and private property of the leader. The leader often uses public resources to secure their power. As a consequence, the citizen must pay enormous rents to the government officials to receive government services. For those who would like to become the government, officers also have to provide large sums of money to occupy or show loyalty to the leader. The leader who get and maintain their power using these traditional ways often ignore the "formal" laws that prohibit corrupt behavior.

2.3 Kleptocratic State Theory

In kleptocratic state theory, endemic corruption exists because the ruling regime has a sole interest, which is to maximize their own wealth. Anderski (1968) defined a kleptocrat as someone whose primary goal is personal enrichment and possesses the power to further this goal while holding public office. In a kleptocratic system, the governmental elites maximize their economic interest through their power and position. Laws and regulation (such as taxes, subsidies, and privatization) are made not because of its requisites but to maximize the avail of the elites.

3 Cultural Perspective on Corruption

To understand the link between corruption and culture is not an easy job. First and foremost, we must understand what the meaning of culture is. Citing from Larmour (2012), culture, according to Williams (1983) is "one of the two or three most complicated words in the English language". Culture can be defined into three concepts (Larmour, 2012). First, it refers to a process of intellectual, spiritual, and aesthetic development related to the idea of civilization. Second, it focuses on a particular way of life, and third, it narrowly used to artistic and intellectual endeavors, for example, like dance, music, and film.

In the context of this paper, we prone to refer culture as the particular way of life, because of its relevancy to what is so-called as "cultural relativism" in which the perspective of something that is regarded as corrupt in one culture might not be regarded as corrupt in another. Despite cultural relativism, where certain act, for example, gift-giving, might be permissible in particular culture but sometimes mistakenly as bribery or corruption, this section would suggest how we should revisit the link between corruption and culture.

Wijayanto (2013) describes why the practice of corruption, especially in Javanese culture is acceptable. On his paper, he noted that Javanese culture perceives power as the "no-question" of legitimacy, meaning that power derives from a single homogenous source, which is God. The idea that power does not raise a question on legitimacy brings the Javanese culture to its exposure to the abuse of power. People should just accept whatever the ruler (king) do. The King is allowed to accumulate the wealth for his own interest, although it should belong to the people. There is no separation on what is the public and private wealth. The example of this view can be seen in how Javanese culture see the prosperity of the King reflects the prosperity of society.

Another claim that is served as a rationale for the corrupt behavior in Javanese culture is related to how Javanese views about family. Javanese is educated to be somebody who should bring a good thing to the family. When someone already succeeds, he or she must not forget his or her family. In terms of corruption practice, it implies that if Javanese becomes a leader, he or she should put his or her family in the priority. For example, if there is a project, he or she should give it to the people who have the closest family relationship. Prioritizing family is a kind of moral obligation in the Javanese culture.

Another example of how corruption seems to be prevalent and accepted by society could also be found in African culture. Olusegun Obasanjo, former president of Nigeria once stated that in the African concept, gifts are considered as a token and not demanded. The value of the gifts is in the spirit rather than the material value, and it usually did in the open and never in secret. Therefore, people should distinguish between gifts and bribes.

In understanding the value of the gift, there are three obligations that underpin the needs to gift. First, the obligation to give, when the leaders are obliged to share their food with relatives or friends and so on. Second, the obligation to accept, which means that any refusal upon the gifts will offend. Third, the obligation to reciprocate, which means that when someone receives (receiver) one thing from his or her friend (giver), the receiver should give back with minimum the same value or even more to the giver (Mauss, 2002 as cited in Larmour, 2012).

In regards to the explanation on corruption and culture, to what extent does corruption and culture related? Does the lack of significance of formal organizations imply that corruption is ingrained in highly corrupted countries?

Culture is often fused with the general moral orientation of the people in a particular country. However, there is at least three problems with this tenet of corruption. First, the concept of cultural relativism implies that corruption should not be defined as a "one size that fits all". The definition of corruption should have a cultural context rather than culture's definition which imposed upon everyone else. One imaginary example is when an incumbent president competes with a non-governmental figure that competes in an election.

Long story short, the incumbent president won the elections because during his presidency he allocates a lot of developmental budget to the prosperity of the people. So, despite competitor's allegation against the use of the national budget as a means of campaign material, people want the incumbent president to lead them in the future years. Was the incumbent president the previous action is the inducement for the votes? Was this considered a corruption?

The second reason on why we should be careful in linking corruption and culture is due to the fact of lack of empirical support. Barr and Serra (2010) made an experimental analysis to seek the linkage between corruption and culture. They conducted thirteen experimental sessions, involving 195 participants to assessed whether the tendency to corrupt related to the reference of the level of corruption in the participants' home country. Their finding suggests that individuals should not be prejudged with reference to their country of origin.

Another survey to the respondents of the eighteen sub-Saharan African countries was asked their views on the corruption dilemma, for example like giving a job to someone from his or her family who does not have adequate qualifications. The result shows that between 60 percent to 76 percent of the 25,086 respondents consider the case of corruption dilemma presented as wrong and punishable (Rothstein, 2017).

Finally, we should be prudent to relate corruption and culture because designated the culture of a nation as corrupt would take us nowhere. The discourse of culture in understanding corruption should give us the benefit instead of the detriment. When we are pointing out that certain culture reserved corrupt behavior, it is not a good start to achieve abroad policy change. We should embrace the cultural values as a basis to prevent corruption. Javanese culture with its "family pride" value can be used to against corruption by accentuating the importance of preserving the good name of the family with achievements rather than merely the distribution of power. The pride of having legitimate achievement should be underscored. The cultural value of being prudent is also can be put to point out the account of future consequences, such as considering all alternative actions to the possible aftermath. In short, the value of culture should not be seen as individual elements but as a comprehensive way of thinking and way of life. Lastly, we should distinguish the difference between cultural values and cultural practice. Hofstede (1991) suggests that national cultural differences consist of difference in values (moral) and practices. An ethical government officer in the systematically corrupt country may morally disapprove of the unofficial payment, but it will be less sensible to him to be the only honest officer in a system where the corrupt behavior is presumed as a social practice.

4 Strategies to Alleviate the Corrupt Culture

With regards to the statement on how we should be prudent in linking culture and corruption, we should be optimist in a way to use the culture as the strategy to fight corruption. Research by Fisman and Miguel (2007) show that culture can play as key role to prevent despicable behavior.

Honesty is a value upheld highly in Indonesia, however corrupt behaviors that obviously contrast to this value are frequently happens. One of the sources of this action is suspected due to permissive culture towards corrupt acts. Promoting anti-corruption values needs to be done through various of channels. One of the predominant way to initiate cultural change is through education.

Strategy could be directed to integrate anti-corruption values into the academic curriculum in schools and universities. The government could empower the anti-corruption education networks to propagate the awareness of corruption, its impacts and detriment as well as to strengthen the social sanctions. Teachers and lecturers can put teaching materials which require students to undertake various activities aiming to build students' knowledge about concepts, principles, causes and negative effects of corruption as well as preventing and overcoming corruption.

Some countries have tested the effectiveness of integrating anti-corruption value through education channel. Nyamsogoro (2015) did a pilot research to test the effectiveness of integrating anti-corruption program in three business schools in Africa; Nigeria, South Africa and Tanzania through the development of Anti-Corruption Toolkit. Each of the business schools employ different strategy to pilot the toolkit; in Nigeria, the toolkit is integrated to the MBA program curriculum, while in South Africa the toolkit is tested through the business ethics course and in Tanzania it was introduced in regular postgraduate courses and in executive education program. He argue that integrating anti-corruption toolkit through education received positive feedback from the students.

Another lessons are also provided by another country like Hong Kong with its Independent Commission Against Corruption (ICAC). One of the responsibilities of ICAC is to be responsible for the education and mobilizing public support in the fights against corruption (Chan, 2000). The examples of their works are explaining corruption to the public in layman's language so that it can be understood easily by citizen, providing anti-corruption curriculum material and give direct teaching and teacher training for schools and universities. They also create a campaign through creative art to attract the youth through creating a popular cartoon character.

5 Conclusion

Despite massive strikes on the support of the connection between corruption and culture, we should revisit its relationship with prudent consideration. Culture does indeed useful to understand how people respond to certain acts that are judged as corrupt behavior. However, it does not automatically deny the ethical basis of judgment about corruption. We should distinguish between what is cultural values and cultural practices. We can use cultural values to understand how people decide to behave as well as use it as a counterattack strategy to combat corruption. At the end of the day, we all agree that combating corruption is the job to everyone. Education is believed to be one of the primary ways to tackle down corruption. Lessons from the best practice from other countries show that education plays an important part to create sustain and strong community that share the values of anti-corruption.

References

1. A. Ades and R. Di Tella, Appl. Econ. Discuss. Pap. OXFORD Inst. Econ. Stat. (1995)
2. S. Andreski, The African Predicament 92 (1968)
3. A. Arifianto, Brand. Univ. Http//Www. Acad. Edu/1337711/Corruption_in_Indonesia_Causes_History_Impact S_and_Possible_Cures (April 1, 2013) (2001)
4. A. Barr and D. Serra, J. Public Econ. (2010)
5. A. Budiman, A. Roan, and V. J. Callan, J. Bus. Ethics (2013)
6. J. D. Chandler and J. L. Graham, J. Bus. Ethics (2010)
7. R. H. Franke, G. Hofstede, and M. H. Bond, Strateg. Manag. J. (1991)
8. K. A. Getz and R. J. Volkema, Bus. Soc. (2001)

9. B. W. Husted, J. Bus. Ethics (2002)
10. Indonesian Corruption Watch, Www.Antikorupsi.Org (2018)
11. P. Kimuyu, Int. J. Soc. Econ. (2007)
12. P. Larmour, *Interpreting Corruption: Culture and Politics in the Pacific Islands* (2012)
13. M. T. Majeed, J. Econ. Integr. (2014)
14. P. Mauro, Q. J. Econ. (1995)
15. W. J. Mommsen, *The Age of Bureaucracy: Perspectives on the Political Sociology of Max Weber* (Blackwell, 1974)
16. B. Rothstein, World Econ. Forum (2017)
17. T. M. Smith, Indonesia (1971)
18. N. N. Thach, M. B. Duong, and T. T. K. Oanh, Imp. J. Interdiscip. Res. **7**, 791 (2017)
19. S. Wei, NBER Work. Pap. (1997)
20. W. WIJAYANTO, Polit. J. Ilmu Polit. **2**, 5 (n.d.)
21. J. H. Zhao, S. Kim, and J. Du, Manag. Int. Rev. (2003)

Function of Social Cohesion Element in Participatory Empowerment towards Landless Peasants

Dumasari Dumasari [1], *Budi* Dharmawan [2], *Imam* Santosa [3], *Wayan* Darmawan[3], *Dinda* Dewi Aisyah[3], and *Sulistyani* Budiningsih[3]

[1]Faculty of Agriculture, Purwokerto Muhammadiyah University, Purwokerto, Indonesia
[2] Faculty of Agriculture, Jenderal Soedirman University, Purwokerto, Indonesia
[3]Faculty of Social and Political Sciences, Jenderal Soedirman University, Purwokerto, Indonesia

Abstract. Abstract. Social cohesion is one of the essential elements in participatory empowerment of landless peasants. This study aims to examine the categorization function within elements of social cohesion in empowering landless peasants using a participatory manner. The research location was set intentionally in Baturaden District, Banyumas Regency and Purbalingga Wetan Subdistrict, Purbalingga Regency, Central Java Province, Indonesia. The research method is using an intrinsic case study approach. Based on the research result showed that the predominant function within elements of social cohesion at the individual level are enthusiasm, trust, quality and closeness of social relations, solidarity and social values, while some element of social cohesion at the community level has a function towards peasant participatory in the empowerment include motives, mentality, actions, behavior, awareness of self-identity and morality. Lastly, the quality of social relations and collaboration networks is a manifestation of social cohesion which has a dominant function at the institutional level. All elements of social cohesion function are reinforcing energy and driving the activeness of landless peasant in participatory empowerment.

1 Introduction

Land scarcity is one of the urgent and crucial issues in the dynamic agricultural development area which has a sustainable and environmentally friendly characteristic. Conversion of agricultural land into non-agriculture land is a dilemma because on the one hand it benefits investors with businesses outside of agriculture and on the other hand, it harms the peasants and agricultural enterprises [1]. The land conversion rate is difficult to control, so it threatens the peasants to become a landless peasant [2]. The number of landless peasants has increased steadily over time. Landless peasants are prone to the poverty traps because of their status, which shifts from peasant landowners to farm laborers. The scarcity of land due to conversion is closely related to poverty [3]. Land conversion also impedes sustainable agricultural development [4]. Landless peasants face various challenges, including the difficulties in getting the job opportunities and social security [5].

The right of farmworkers to arable land is not optimal because the landowners determine all decisions and actions. Farmworkers only work on a plot of the farm owned by someone else and are given daily or weekly wages according to the agreement. Some farmworkers are not blown but get the harvest share. It is not uncommon for peasants to leave work on-farm and switch to off-farm or non-farm. Transfer of work is not easy for peasants. The habit of adopting an agrarian culture made it difficult for farmworkers to accept the urban-industrial culture. Landless peasants who transfer their livelihoods are prone to experiencing cultural shock. Land scarcity tends to occur in lowland rice peasants who act as food producers. If it is allowed to continue, a series of problems with peasants tend to create complex problems. Agricultural land conversion is not only trapped the farmworkers into the poverty cycle but also encourages the urbanization and reduces national food production [6, 7, 8]. The Government of the Republic of Indonesia has mobilized landless peasants in various participatory empowerment activities. Empowerment theory has broad insight into individual and collective behavior related to active participation in changing and forming social-environmental contexts [9]. The focus of participatory empowerment activities lies in developing diversified creative, productive livelihoods. Livelihood diversity has a positive and significant impact on income. Diversification is a strategy for peasants who are proactive in choosing the types of jobs available such as progressive diversification [10]. Diversification of products on a farm reduces marginal effects along with the income distribution of peasants, including landless peasants.

Landless peasants can have an entrepreneurial spirit, especially in developing business on-farm, off-farm and sustainable non-farm as the producers of honey, animal fodder, processed fruit, souvenirs or handicrafts, wood fuel, processed fruit, and livestock products [11, 8, 12]. However, it is known that peasants in the rural areas are difficult to be actively mobilized in participatory empowerment.

One strategic solution to mobilizing landless peasants in participatory empowerment is through the use of elements of social cohesion. The limits of the concept of social cohesion are not only included the in participation but also broaden the emphasis of this behavior to include the idea of trust, connectedness, and community involvement, including peasants' persuasion [9]. Various elements of social cohesion of collectivity, solidarity, value, social norms, tolerance, caring, self-identity, networks of cooperation, reciprocity, social control have essential functions for the smooth participation of empowerment. Social cohesion is strong energy to awaken the sense of togetherness of landless peasants to decide to act actively in participatory empowerment.

Based on the essential values of social cohesion encourage the selection of research themes. A significant theme is the empowerment of agricultural communities based on local resources. Social cohesion as a collective product of landless peasant needs to be studied in-depth to reveal its contribution to the awareness, motivation, and ability of active peasants in participatory empowerment. Selected themes are relevant to developing the capacity of peasants to be able to be independent and produce appropriately in the social, economic, and environmental fields.

2 Material and Methods

The research design about an element function of social cohesion for landless peasants' empowerment in a participatory manner use intrinsic case study with qualitative and quantitative approach [13]. A qualitative approach is useful for completing the depth and detail of the data. A qualitative approach helps researchers to understand the subject's personal experience from the point of view concerned. A qualitative approach keeps researchers from being trapped in stereotypically rigid theoretical thinking. The quantitative approach is intended to reveal the categorization of elements of social cohesion. A qualitative approach is appropriate to examine the relevance of the functions of social cohesion to participatory empowerment. Researchers use social reality in the field to explain holistically and deeply about the diversity of functions of elements of social cohesion that determine participatory empowerment of peasants. Realization of speech, deeds, and actions of landless peasants are the data which has been observed and analyzed during the research.

The ethical approach and the emic approach are utilized in the formulation of the reconstruction result of the participatory empowerment concept. The ethical approach is directed towards meeting data sourced from the researcher's perspective. The emic approach used to uncover data from landless peasants.

Determination of the location of the study was carried out by purposive area in Baturaden Subdistrict, Banyumas Regency and Purbalingga Wetan Subdistrict, Purbalingga Regency, Central Java Province. Both of these sub-districts were chosen based on the consideration of the majority of the people patterned as livelihoods as landless peasants (farm workers). The average peasant income level is IDR 650,000 per month. Landless peasants follow the participatory empowerment but still slow because of weakened social cohesion. The problems in the two sub-districts were chosen because of the relevance to the research theme.

The subject of the population included all farmers who experienced poverty in the two study locations. Determination of respondents as a quantitative primary data source using purposive sampling technique, while informants as qualitative primary data sources were selected by the snowballing technique. The number of primary data sources is not strictly determined but tend to based on consideration of the needs and completeness of the data.

Data collection techniques used were in-depth interviews, participant observations, focus groups discussions, and documentation analysis. All qualitative data collected is processed through a sequence of steps: writing data, editing data, classifying similar data, reducing data, tracing data on themes, and linking themes and presenting data. Processing and analysis of qualitative data is done while in the field (ongoing analysis). Qualitative data analysis techniques use the Interactive Model [14]. Quantitative data were analyzed by descriptive statistics with scoring values, tabulation, frequency distribution, and percentage values.

3 Result and Discussion

Social cohesion in landless peasants has an essential function in increasing participation in participatory empowerment. The social cohesion inherent in the respondents is divided into three interrelated levels. These three levels are individuals, society, and institutions. Social cohesion at the community level most strongly contributes to the intention, enthusiasm, willingness, and ability of respondents to be active in participatory empowerment. The enticement of social cohesion is indeed strong at the community level to respondents so that they are active in managing business on-farm, off-farm and non-farm. The social reality is closely related to social values and norms that are based on collectivity and the identity of socio-economic status as landless peasants. The gemainschaft social bond connects the existence of respondents. Respondents tended to follow something that has been done by other community members in the majority (> 50 percent). Landless peasants have a lack of the courage to act in connection with empowerment activities. A technology for spacing setting with a legowo pattern is adopted if the majority (> 50 percent) of a large number of landless peasants have applied for morning farming in monoculture fields. The actions and decision-making of respondents are determined by agreement.

Social cohesion at the weak individual level encourages respondents to participate in various participatory empowerment activities actively. Respondent awareness is slow because independence is classified as low in determining decisions and actions. Social cohesion in the respondent seemed to fall asleep until he was often unaware of the importance of participatory empowerment — the dominant level of social cohesion influences decision making and specific actions for community social activities.

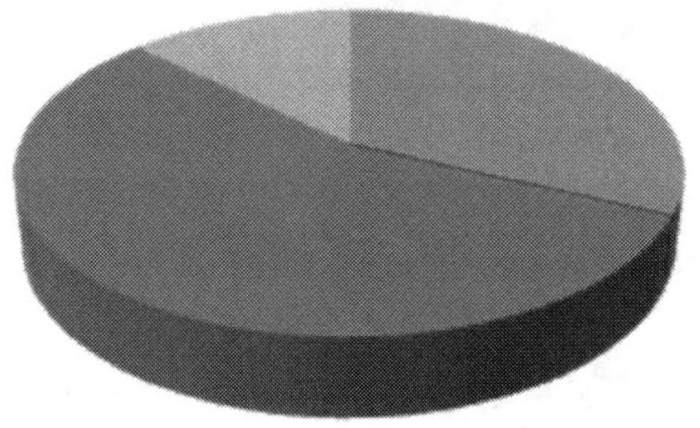

Fig 1. Contribution of the social cohesion function to landless peasant participatory empowerment

The ability of social cohesion at the institutional level is classified as weak in motivating respondents to be active in various participatory empowerment activities. Various institutional forms in the midst of the social structure of peasants who are less able to mobilize respondents' awareness of the strategic value of participatory empowerment. Innovations about the technology of processing corn crops into cornstarch delivered through the social gathering (arisan) of women peasants in the village took place slowly. Only a few (16 percent) of respondents participated in the trial. The institutional weaknesses that are less able to take advantage of social cohesion as a driver of landless peasants due to the far social distance and heterogeneous social levels. The condition of the function of social cohesion at the individual, community, and institutional level in contributing support to participatory empowerment in respondents was observed in Figure 1.

Social cohesion has an element that serves as an adhesive to the closeness of fellow respondents. Elements of social cohesion are at the level of individuals, communities, and institutions. The various elements of social cohesion at each level have similarities and differences. The condition of the elements of social cohesion before and after participatory empowerment shows different levels of functioning at each level. Respondents made adjustments to the elemental functions of social cohesion. Adjustments occur at each level during participatory empowerment.

The function in some elements of social cohesion within respondents is still weak especialy at the individual level. The weak condition of the function occurs before participatory empowerment takes place. Respondents were less eager to increase productivity and creativity. The respondent's attitude was surrendered and was willing to accept a limited situation due to the status of the charity. The respondents' mentality is weak, so they are reluctant to receive information about production technology, postharvest technology, and packaging techniques for participatory empowerment. Respondents rarely attend training sessions, counseling, group discussions, socialization, and demonstrations on ways. Trust weakened, and there was a suspicion of productive actions carried out by other craftsmen. The quality of social relations is terrible due to economic activities because there is a sense of competitiveness, jealousy, and suspicion. With some weakening elements of social cohesion, the function of the element of social control has also been reduced. Respondents carry out economic activities according to their abilities and are individualistic. The function of the element of strong social cohesion at the individual level is only in social activities in the community.

The function of elements of social cohesion at the individual level has increased and decreased after respondents actively participated in participatory empowerment. The most dominant spirit element has increased function. Respondents have the enthusiasm to develop productive businesses on-farm after post-harvest. Increased enthusiasm occurred after respondents learned the benefits of agricultural processed products that have added value. Trust among respondents who are active in participatory empowerment is increasing. A sense of competitiveness, jealousy, and suspicion are pressured until the quality of social relations improves among fellow respondents with other peasants. The function of respondent's solidarity element increases because there is a network of cooperation to help each other in providing information on technology, prices, and markets.

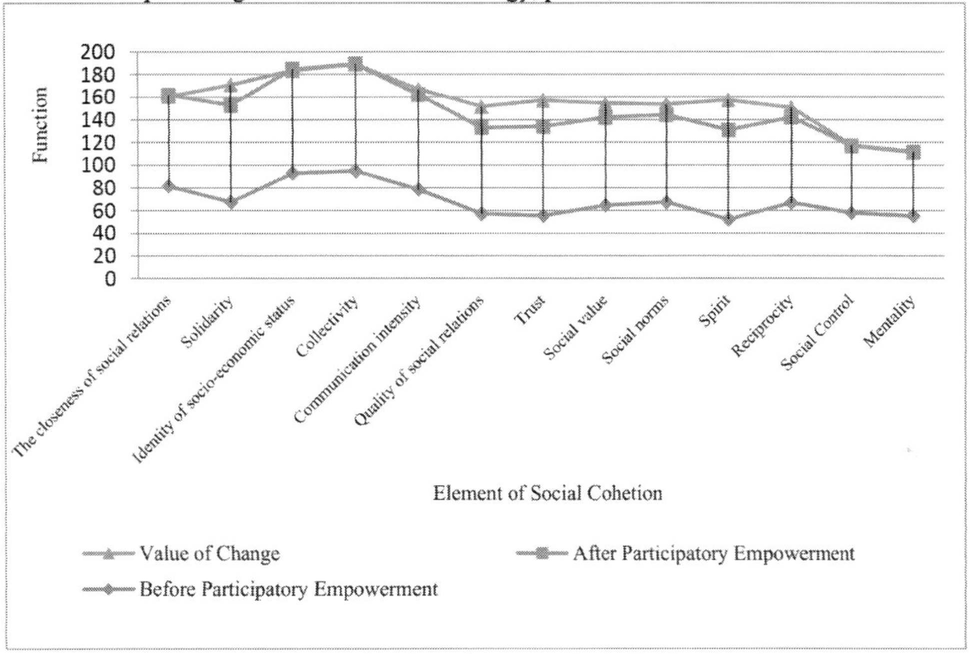

Fig 2. The function of elements of social cohesion at the individual level

Other elements that experience increased function are social norms and reciprocity. The element of social cohesion which has decreased function is the closeness of social relations and the identity of socio-economic status. Respondents claimed that after participating in participatory empowerment the forms of unproductive entertainment, social relations had been diminished. The identity function of the socio-economic status of landless peasants has eroded because the landless peasants have been able to continue farming with verticulture and aquaculture technology on a narrow yard. The function of individual-level social cohesion elements is shown in Figure 2.

Social cohesion at the community level has several elements that function important to the ability of respondents to actively participate in participatory empowerment. All elements of social cohesion have proper functions at the community level. The respondent's mentality element is not strong enough to bear the position of being squashed as a steady-state peasant when he has not participated in participatory empowerment. Motive function for productive and creative activities is low. Respondents only pursue temporary jobs that are as farm laborers, building laborers, small industrial workers, transport workers, and motorcycle taxi drivers, janitors, parking attendants, and security officers. Respondents only surrender to accepting any work to earn daily income. The function of self-action is limited only to subsistence economic activity. The function of the elements of social cohesion concerning community activities tends to be active even though respondents have not actively participated in participatory empowerment.

The elemental functions of mentality, motives, and actions of respondents experienced a significant increase after being active in participatory empowerment. The courage to start a new product business of on-farm, off-farm, and non-farm appears even slowly. Respondents manage farming vegetables and freshwater fish ponds with aquaponic technology on the home page. Some also started raising chickens and goats. Off-farm work is being developed, which is producing various foods and beverages sold around the village. Even so, the function of several elements of social cohesion has experienced a real setback after participatory empowerment. Some of these elements include adaptation, tolerance, and solidarity. The function of caring and empathy also experiences setbacks because respondents are busy managing productive businesses and feel increasingly accustomed to improved economic conditions. Therefore the respondent's empathy towards subsistence is reduced.

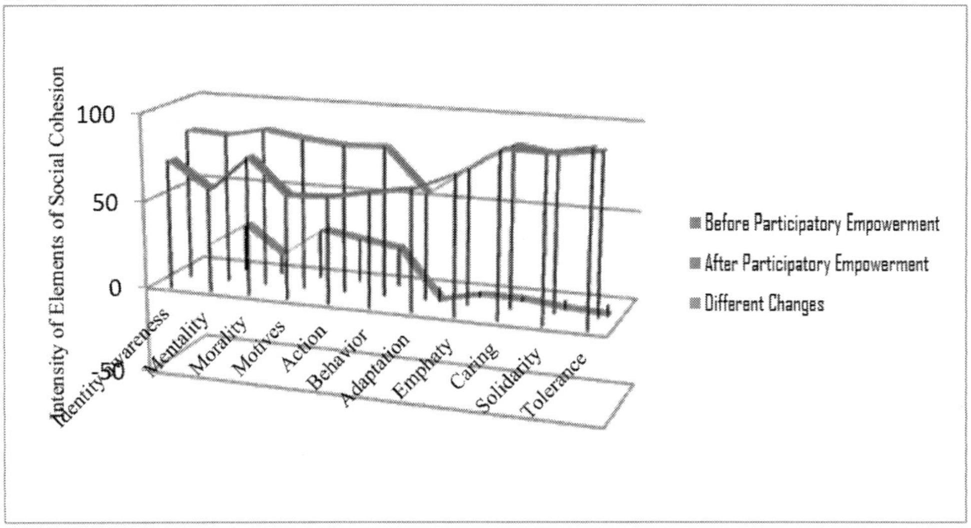

Fig 3. The function of social cohesion elements at the community level

Changes in the function element of social cohesion that took place at the community level show the attachment of the respondents in the collectivity of landless peasants is high. Empowerment carried out collectively, is easier to mobilize awareness, motivation, and ability as well as the willingness of active respondents, especially in developing diversified productive livelihood activities. The function of various elements of community-level social cohesion to participatory empowerment of smallholder peasants is detailed in Figure 3.

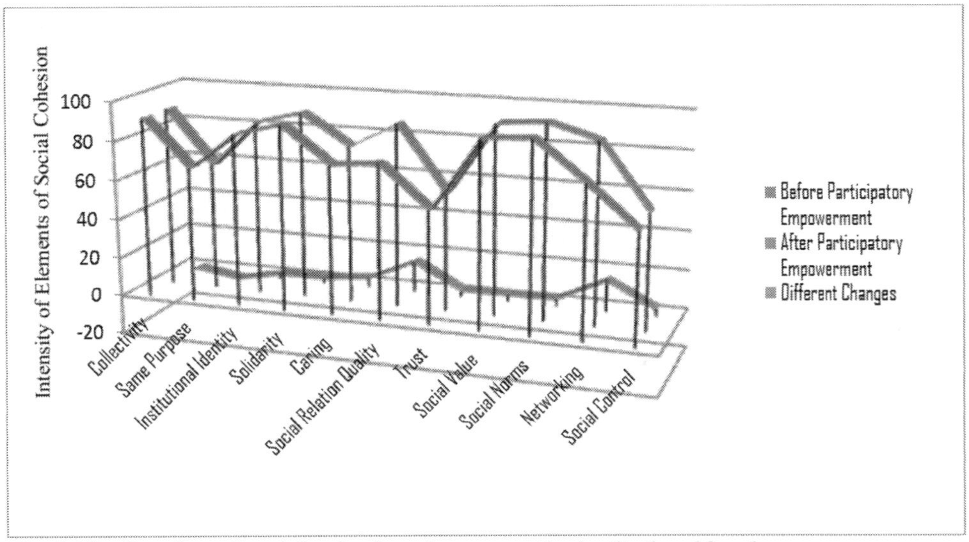

Fig 4. The function of social cohesion elements at the institutional level

Social cohesion at the institutional level has several elements that function to support participatory empowerment. The function of elements is more dominant in the interests of social activities such as helping, voluntary work, and cooperation. The function of social cohesion for the benefit of economic activity is still weak, especially before participatory empowerment. Elements that have a weak function, namely a network of economic cooperation, social control and trust. Social values and norms, solidarity and collectivity, function predominantly on the activity of respondents in participatory empowerment. The dominant function of the four elements of social cohesion is not very different between before and after the respondent follows the empowerment of active participation. All elements of social cohesion at the institutional level listed in Figure 4 only make a small difference.

The category of elements of social cohesion that has the dominant function of participatory empowerment at the individual level is a success, trust, quality of social relations, and social values. At the community level, the dominant element of social cohesion is participatory empowerment, namely motives, mentality, actions, and behavior. The element of social cohesion at the institutional level that has the dominant function of participatory empowerment is the network of cooperation and the quality of social relations. Strengthening the function of elements of urgent social cohesion is carried out in a priority starting from the community, individual, and institutional levels. Social cohesion cannot be separated from the success of participatory empowerment in landless peasants. The function of various elements of social cohesion is essential to strengthen and bring closer social relations between landless peasants, as stated by [15]. The strategic value of social cohesion cannot be ignored from every empowerment, including the landless peasants' community. Social cohesion is a vital construction, which is the heart of what is needed by humans in managing and developing various dimensions of life [16].

The results of the research also explain the details of their social cohesion, especially in their shared values, inequality, and quality of life. The function of social cohesion is essential as the axis that ignites the enthusiasm of the respondent and other landless peasants which active in participatory empowerment. Awareness of socio-economic identity drives the motivation of respondents to take decisions and actions to develop diversified productive livelihood patterns. Work as a farm laborer with perseverance as a creative handicraft or craftwork artisan, even as a breeder of freshwater fish in order to help them free from the problems of their family economic urgency. This social reality is in line with the results of research [17, 18, 19, 20].

Every element of potential social cohesion is utilized for the participatory empowerment of respondents as peasants in each research village. This social cohesion has strategic functions as a facilitator of entrepreneurial capabilities in agricultural development by maintaining a balance of economic, social, and environmental interests. The limitation on the ownership of fertile agricultural arable land causes some respondents to lose their motivation to work as peasants. Respondents' abilities regarding verticulture, hydroponic, aquaponic, and aeroponic technologies that are environmentally friendly and can be applied to yards with narrow land are still meager. Meanwhile, respondents experienced some severe problems when they switched to non-farm jobs because of their agrarian oriented behavior of work. The issue of job conversion from on-farm to non-farm could be overcome by utilizing elements of social cohesion at the individual, community and institutional level that directs peasants to have the ability to create a multiple livelihoods pattern at the same time while being in the on-farm, off-farm or even in a non-farm condition. Social, economic, and environmental cohesion values have an essential value for increasing their entrepreneurial spirit that supports the management of agricultural enterprises [21]. Enhancing social cohesion through a comprehensive policy package is crucial to sustainable poverty reduction [22]. Cohesive communities are more able to fight together with integrity based on social capital with a shared sense of belonging.

4 Result and Discussion

Social cohesion within landless peasants is distinguished according to both to individual, community, and institutional levels. The dominance of social cohesion at the community level is much stronger than the participatory empowerment of landless peasants in their individual and institutional levels. Some elements of social cohesion have been categorized at each level. The condition of the elements of social cohesion before and after participatory empowerment shows a different level of function. Landless peasants have also made adjustments to the elemental functions of social cohesion during their participatory empowerment.

The categorization of elements of social cohesion shows that at the level of individuals there is some dominant experience increase especially in enthusiasm, trust, quality, closeness of social relations, solidarity and social values, while the increase in the cohesion of social elements at the dominant community level is within their motive, mentality, action, behavior, awareness of self-identity and morality. There are also some slight increases in the element of social cohesion at the institutional level, namely the quality of social relations and collaboration networks. All the elements of social cohesion that have a dominant increase contribute significantly to the participatory empowerment of peasants within the community.

The use of elements in social cohesion needs to be considered while launching the participant empowerment of peasants. The management strategy for the function of elements of social cohesion at the individual, community, and institutional level requires an adaptive approach. Further research still needs to be done to examine the appropriate method model for the management of elements within social cohesion in order creates an ability of optimal function for the empowerment of peasants.

Acknowledgment

Thank you to the Directorate of Research and Community Service. Directorate General of Research and Development Strengthening. The Ministry of Research, Technology and Higher Education of the Republic of Indonesia has provided research grants to the research team. The research grant was given, including the category of centralization of basic research schemes in the first phase of the 2019 period.

References

[1] R. Harini, H.S. Yunus, Kasto, S. Hartono. Indonesian J. Geography. Vol **44** (2). (2012)

[2] D. Dumasari, B. Dharmawan, I. Santosa, W. Darmawan, P. Utami, D.D. Aisyah. J. Arts Humanit. (2019)

[3] Q. U. A. Memon, S. A. Wagan, D. Chunyu, X. Shuangxi, and L. Jingdong, J. Poverty **23**, 269 (2019)

[4] M. Rondhi, P. Pratiwi, V. Handini, A. Sunartomo, and S. Budiman, Land **7**, 148 (2018)

[5] J. Zhao. Thesis. University of Waterloo, China (2017)

[6] C. A. Francis, T. E. Hansen, A. A. Fox, P. J. Hesje, H. E. Nelson, A. E. Lawseth, and A. English, Int. J. Agric. Sustain. **10**, 8 (2012)

[7] T. H. T. Nguyen, V. T. Tran, Q. T. Bui, Q. H. Man, and T. de V. Walter, Land Use Policy **54**, 583 (2016)

[8] L. Han, H. Bao, and Y. Peng, Sustainability **9**, 1158 (2017)

[9] P. W. Speer, C. B. Jackson, and N. A. Peterson, Heal. Educ. Behav. **28**, 716 (2001)

[10] S. Martin, K. Lorenzen. World Development J. Vol **83**. (2016)

[11] M. Berhe, D. Hoag, G. Tesfay, T. Tadesse, S. Oniki, M. Kagatsume, and C. M. H. Keske, Pastoralism **7**, 12 (2017)

[12] D. Dumasari, W. Darmawan, A. Iqbal, B. Dharmawan, and I. Santosa, Int. J. Adv. Sci. Eng. Inf. Technol. **9**, 717 (2019)

[13] J. W. Creswell, *Qualitative Enquiry & Research Design, Choosing among Five Approaches* (2007)

[14] M.B. Miles, A.M. Mac Graw Hill Company, New York (1991)

[15] S. N. Mulunga and R. Yazdanifard, Glob. J. Manag. Bus. Res. (2014)

[16] X. Fonseca, S. Lukosch, F. Brazier. J. Social Science Research. Vol **32** (2). (2019)

[17] D. Pugersari, A. Syarief, and D. Larasati. J. Arts and Design. (2013)

[18] Y. Fu, in *Proc. 2015 Int. Conf. Arts, Des. Contemp. Educ.* (Atlantis Press, Paris, France, 2015)

[19] D. Dumasari, S. Budiningsih, W. Darmawan, and I. Santosa, J. Arts Humanit. **6**, 01 (2017)

[20] J. Fengfan, J. Arts Humanit. **6**, 12 (2017)

[21] E. Hudcová, T. Chovanec, and J. Moudrý, Eur. Countrys. **10**, 377 (2018)

[22] A. Kolev, (2015)

The Impact of Land Conflict on Darmakradenan Toward Social, Economic, and Politic Vulnerability in The Life of Farmers 'Community

W Handoko[1]

[1]Department of Political Science, Faculty of Social and Political Sciences, Universitas Jenderal Soedirman, Purwokerto

Email: waluyo.handoko@unsoed.ac.id

Abstract. This research is aimed to examine the impact of the Darmakradenan land conflict on the vulnerability of the social, economic and political life of the local farming community. This research is essential because the Darmakradenan conflict has been going on for around 20 years and all the conflict resolution efforts undertaken have not achieved results. This research is qualitative research that selects informants with a purposive method, including farmers, village heads and village officials, management of the farmer group *Stan Ampera*, plantation managers, community leaders and so on. The results of this research indicated that the two decades of Darmakradenan land conflict had an impact on the vulnerability of social, economic and political life, such as a lack of harmony between farmer groups, loss of livelihoods, tensions between farmer groups and village government, illegal acts, citizens easily ignited to act anarchist and so on. Such conditions have implications for the importance of considering social, economic and political vulnerabilities in developing efforts for real land conflict resolution.

Keywords: social, economic, political vulnerability, land conflict

1. Introduction

This study is aimed to examine the effects of land conflicts between farmers and the military which have lasted for 20 years and until now a resolution has not been reached. The intended impact is from a social, economic, and political perspective that has plagued farmers' lives for years. The land conflict is in Darmakradenan Village, Banyumas Regency, Central Java Province, Indonesia. The conflict is a conflict between peasants who were members of the Society of Peasants' Mandate (*Stan Ampera*) and *Diponegoro* Regional Military Command IV and *PT. Rumpun Sari Antan* (RSA). This study is expected to fill the gap between previous studies that examined land conflicts, including studies of important factors that can reduce escalation of conflict [1]; studies on land conflict resolution through traditional mechanisms in each community [2]; [3]; studies of the importance of the role of individuals and project managers in conflict resolution efforts [4]; and studies on conflict resolution efforts through traditional justice [5]. In the context of conflict resolution efforts in Indonesia, this study fills in research gaps that have not been reviewed by previous researchers who emphasize conflict resolution efforts based on customary institutions [6]; [7]; resolution efforts that emphasize customary law and Islamic rules [8]; study of the mechanisms of public consultation, negotiation, mediation and arbitration [9];[10]; and studies on conflict resolution efforts that are based on local wisdom [11].

The Darmakradenan land conflict began when the Dutch colonial period and continues to the present. At that time the colonial government issued the *Wet De Wall Agrariche Act* in 1870, where farmers were granted land certificates. The policy for granting land certificates was followed by the application of a high land tax of 30%. The implementation of this policy made farmers unable to pay taxes because it was not comparable to the results obtained from the land business. Facing this problem, the village head at that time suggested that the farmers rent out their land to the capital or foreign owners (Chinese citizens) to pay their tax debt to the colonials.

Entering the official state of Indonesia by changing several regimes, the land settlement has not yet been completed. Although various resolutions have been made starting from negotiations, mediation up to the court has been carried out, but between parties in conflict, namely the farmers and the TNI have not reached an agreement. It shows that there is a gap between theory and empirical (gap between theoretical and empiric). Also, the description of the chronology of land conflicts in Darmakradenan village shows that the land conflict has been very long and cannot be resolved, even though there are regulations that regulate land, namely the Basic Angrarian Law in 1960 (UUPA) in the Old Order and *TAP MPR* No. IX of 2001 and other regulations governing the land. Thus there is a gap between regulation and empirical facts (gap between regulation and empiric), where normatively regulations regarding land provide an opportunity to resolve the conflict resolution but at the practical level the solution is not finished.

The condition of such land conflicts is an important matter to examine how the impact felt by farmers who have been fighting for their rights for 20 years has undoubtedly consumed energy, time, power, and costs from farmers involved in the conflict. At present many of the lives of Darmakradenan farmers do not get real social, economic and political rights. In this case, the welfare of the Darmakradenan community is at a low level where 1,407 households or 52.62% are classified as underprivileged and prosperous families. The number of households included in this category consists of 273 households who experience land conflicts. Furthermore, regarding land ownership of the Darmakradenan village community is classified as low; this is because 47.71% of the people do not own land and farmers who have less than 0.5 Ha of land [12]. The description of the low level of farmer welfare and the lack of land ownership by farmers encourages the importance of studying the impact of land conflicts on the vulnerability of Darmakradenan farmers' wishes in a more comprehensive manner in the social, economic and political fields.

2. Materials and Methods

This study uses qualitative methods based on researchers' beliefs based on research experience and the nature of the problem [13]. Qualitative research methods have advantages concerning constructing social reality, cultural meanings, having a focus on interactive processes as well as events, and being flexible in determining research steps [14]. The selection of informants used purposive sampling and snowball sampling techniques [15]. The informants of this study consisted of founder *Stan Ampera*, chairman, and board of *Stan Ampera*, farmers, village head of Darmakradenan, head of PT RSA plantation, Darmakradenan village government, Banyumas Regency government, and so on. This study used data collection techniques with a combination of interviews, document studies, and observations [16]. Data processing and analysis started from developing analytical categories and coding data, as an analysis prerequisite [17]. The development of categories and data coding concerned to two things, namely 1) ensuring the analytical category was consistent with the research objectives so that the conclusions are valid; 2) balancing to keep too little and too much categorization. The analysis began with the process of clarifying the data to achieve consistency, followed by a theoretical abstraction on information and facts, which produced fundamental statements. By comparing research findings and deepening meaning, continued analysis of data would be obtained simultaneously throughout the research process.

https://doi.org/10.1051/shsconf/20208601039

3. Results and Discussion

3.1. The Land Conflict in Darmakradenan in Historical, Economic and Political Dynamics

Darmakradenan is a village in Ajibarang Subdistrict, Banyumas Regency, which is classified as beautiful with cocoa and rubber plantations supported by the fertile topography of hilly areas and valleys. However, this beauty is not matched by the welfare of the farmer's life. With a total village area of 1157.86 Ha, including 227.65 Ha of disputed land in it, the land beauty and the potential of its territory could not eradicate the poverty of the farmers. Prolonged and unresolved lands conflicts make farmers live in poverty. Official data from the village government shows that out of 3,038 households, the 1707 households (56%) are low-income families who do not have access to sufficient socio-economic resources. The Darmakradenan community is also a picture of a low-educated population of 5,198 people (57%) out of 9,050 inhabitants. [18].

The description of their helplessness in accessing social and economic resources can also be seen from a large number of unemployed people, which is as much as 15% and as farm laborers of 17% and 1,209 (13%) being farmers. The other 46% is divided into traders, entrepreneurs, civil servants, etc.[19] Another picture of poverty is the amount of not having a place to live. Families that do not have a home live with other families in one house. Even sometimes one house is occupied by 2-3 families. The well-being of the Darmakradenan community is thus included at a low level, where 1,407 households or 52.62 % belong to low-income families and prosperous families. Table 1 presents the complete level of welfare for the people of Darmakradenan Village.

Table 1. Population Data based on Amount and Welfare Level

No.	Categories	Total	Percentage
1	The number of men	4,550	49.93
2	The number of women	4,562	50.07
3	The number of heads of families	**2,674**	**100**
4	Pre-Prosperous Family	945	35.34
5	Prosperous Family 1	462	17.28
6	Prosperous Family 2	859	32.12
7	Prosperous Family 3	293	10.96
8	Prosperous Family 3 plus	115	4.3
	Total number	**9,112**	**100**

Source: The List of Darmakradenan village potency in 2015, has been processed

Farmers in Darmakradenan Village are poor farmers who expect land ownership status as a source of daily livelihood. They are a group of farmers who have tried all means to fight for the ownership status of 227.65 Ha of plantation land in Darmakradenan Village. The property is believed to be owned by farmers by a message given by their predecessors so that local farmers will continue to care for and develop plantation land. However, after almost twenty years the hopes of the Darmakradenan farmers to own the Darmakradenan plantation land has not come to them.

The origin of the Darmakradenan plantation land dispute is inseparable from the Dutch colonial policy that abolished cultivation in 1870, due to criticism from the European community on Forced Planting which had so much tormented on Indonesian community [20]. After the Forced Planting policy was abolished, the Dutch colonial issued the *Agrariche Wet De Wall Act* in 1870 which contained the establishment of a policy of granting land certificates to farmers in Indonesia. The policy for granting land certificates was followed by a high land tax assessment of 30 cents per 7000 m^2. Of course, the land tax policy maked farmers unable to pay taxes because it was not comparable to the results obtained from land management. In addition to the issue of land tax, the Law also regulated *erpacht* rights that apply to agricultural land and large plantations, agriculture and small plantations, as well as land for resting and yards.

The farmers 'objection to the policies of the Dutch colonial government also was felt by farmers in Darmakradenan Village. To overcome this difficulty, in 1892 the head of the village of Darmakradenan at that time, the Village Head of PR, agreed with the farmers to rent Darmakradenan plantation land to Dutch businessmen. The agreement was contained in the letter No. 5 *Verponding* agreement valid from July 15th, 1892 until July 15th, 1967, which covered an area of 230.10 Ha with *erpacht* status. As *erpacht* land, after July 15th, 1967, the land will be returned to the farmers. Since July 15th, 1892, the Darmakradenan plantation land was controlled by a Dutch businessman named Jan Abertus van Der Roeft, known as Mr. Maryon and planted with rubber trees. In its development, in 1940 the Dutch businessman transferred the management of plantation land to Chinese entrepreneurs. Since then, the rights to manage the Darmakradenan plantation land have been held successively by Chinese entrepreneurs [21].

But times were changing. The Dutch colonial occupation in Indonesia was not timeless. In 1945, the Indonesian people declared their independence. Along with the independence of the Republic of Indonesia, the Indonesian government through *PT Perkebunan Nusantara* (*PTPN*) and the Indonesian State Forest Company (*Perhutani*) began to nationalize companies previously controlled by the Dutch colonial government. Concerning nationalization, it was also applied to the Darmakradenan plantation. Along with the process of nationalization, when the government established the Agrarian Basic Law of 1960, the giving of the right of *erpacht* of Dutch businessmen was legally ended, and the Darmakradenan plantation land was under state control.

However, when the darkest political events in the Indonesian political history exploded, namely the rebellion of the Indonesian Communist Party (*PKI*) in 1965, plantation workers belonging to the Republic of Indonesia Plantation Labor Union (*Sarbupri*) became one of the organizations linked to the existence of the PKI. In addition to *Sarbupri*, the Indonesian Farmers' Organization (*BTI*) organization which consisted of farmers in rural areas was also associated with the PKI rebellion. During these times, many of the plantation workers who worked on the Darmakradenan plantation also became members of *Sarbupri*, so there was an assumption that they were also involved in supporting the PKI movement in 1965.

The *TNI* as a state army cleansed PKI supporters from the national level to various rural areas. This situation also affected the situation of workers working on the Darmakradenan plantation. Many workers fled because they were afraid of being accused of being involved in the PKI movement, including in this case Chinese businessmen who rented the plantation land. The Darmakradenan village government under the leadership of the village head at that time also did not dare to accept and manage plantation land. In the end, with reasons for security, the Darmakradenan plantation land was held by the *TNI* in the territory of Diponegoro Military District Command IV since 1965. What the *TNI* did at that time was an attempt to deal with the chaotic situation caused by the PKI rebellion, where associated with support for the PKI movement. This means that in this context we can say that the *TNI*'s control of the Darmakradenan plantation land was preceded by pure action to maintain peace in the form of political stability. In managing plantations, the *TNI* gave responsibility for plantation management to the *Rumpun Diponegoro* (*Yardip*) Foundation. Furthermore, from 1975 until now, *Yardip* leased Darmakradenan plantation land to *PT Rumpun Sari Antan* (RSA) [22].

That period was the initial period of the New Order's power that took power from the Old Order in 1966. New Order power grew to become a power that emphasized economic growth and political stability supported by a rigorous security approach. In the practice of New Order power, the penetration of power was felt from the national level to remote areas. Therefore, during the New Order regime, there was almost no effort by farmers who dared to fight for the ownership of the Darmakradenan plantation land. During that time, farmers relied on the limestone mining business, the results of which were very promising because the turnover of money in one month reached 1 billion Indonesian rupiahs. But when the monetary crisis occurred in 1998, the lime business fell; so many farmers who relied on the lime business went bankrupt. Such conditions have increasingly made farmers live in poverty, so farmers are beginning to desire to obtain ownership of plantation land managed by PT RSA. For this reason, the following sub-sections present in full the social, economic and political impacts of the life of the farmers of Darmakradenan.

3.2. The Impact of Land Conflict on Social, Economic and Political Life

The impact of land conflicts in the social life of farmers or the community can be seen from several perspectives, namely social, economic, and political consequences. The social change of Darmakradenan's farmers after entering 1999 began to appear when the escalation of conflict reached a new phase along with the on-going reform era in Indonesia. The flow of political change at the national level since the fall of the New Order government has spread to rural levels. The life of the Darmakradenan community, which was initially harmonious among fellow citizens, the village government and other government apparatus, began to appear friction. The disharmony of residents was triggered by the existence of a pro-contra against the efforts of some farmers to fight for land rights which at that time was still in the management of PT RSA. That condition led to the existence of two groups with differing views, for those who supported the presence of efforts to struggle the land viewed other groups as "enemies" or vice versa.

In addition to the friction between the community, the disharmony also obvious that *Stan Ampera* faced the village government. *Stan Ampera* is an organization formed by farmers as a forum for farmers' struggle in fighting for their rights. They saw that the village government always defended the *TNI*, which in this case was the party facing *Stan Ampera*. This was also felt by the Darmakradenan village officers where *Stan Ampera* considered that the village government was an extension of the TNI. For village government officers themselves, they have admitted that so far concerning land conflicts they have acted following the legal corridors. The legitimate HGU's holder was Diponegoro IV Regional Military Command, so the community saw that village government as part of the TNI. Thus, this land conflict has an impact on society's social change. This condition certainly agrees with Ferguson and John Millar [23], who see that social change can come from conflicting interests with each other.

Other social changes that arise after conflict were changes in people's behavior. A culture of violence and unlawfulness that has not been seen so far, after the conflict was apparent. Courage to commit violence and against the law was seen by carrying out anarchist acts in 1999 by conducting demonstrations by destroying company facilities. In 2014, residents also looted and felled the cocoa plantations owned by PT RSA which reflected selfish behavior and an attitude of arrogance.

Symptoms of violence in people who are experiencing a crisis due to critical change processes, both in the social, economic, political and cultural fields. The phenomenon that occurs in Darmakradenan where people have dared to commit violence and against the law is a situation of community crisis resulting from an important process of change in the economic field. Before the monetary crisis in 1998 the lives of farmers were supported by a very promising limestone business, but due to the financial crisis the limestone business became sluggish, and people lost their livelihoods. This economic situation results in a social change in society.

Regarding the economic perspective, this land conflict can be seen from economic indicators, of course. The indicators intended include the number of poor people in the area of conflict. People who joined *Stan Ampera* amounted to 325 families initially and now reach 600 families. The number was spread in 4 neighborhood units (RW) in Darmakradenan, namely *RW* 6, 9, 8, and 10. The data showed that the majority of the poor in the village were concentrated in the four RWs. Conditions in the field showed that in one house there were 2 or 3 households. This condition was not due to the existence of a culture of living together but due to the economic crushed where they cannot afford to own their own homes. This situation can be understood that during the 20 years of conflict, farmers with *Stan Ampera* in their struggle certainly needed citizen support in the form of material, energy, and mind. Besides, of course, for twenty years faithful farmers demanding the return of land rights were mostly those who only rely on access to the source of life from the land. While with unclear land status, of course, they cannot support their efforts to meet their needs. Thus the concentration of doing an economic enterprise cannot be done well so that economically the members of *Stan Ampera* did not experience an increase and even the opposite.

In political life, the condition of citizens who are not in harmony with the village government and economically unlucky certainly has an impact on their political life. One of the essences of politics is citizen participation in decision making concerning them and also related to development affairs in their area. Observing the conditions in the field of disharmony with the village government resulted in the participation of *Stan Ampera* members in many respects being low. Of course, this is very reasonable considering the relationship between them is not good so that on the one hand the citizens reluctant to take part in the activities of the village government and on the other hand the village government is also very rare to involve them in their activities.

4. Conclusion

The Darmakradenan land conflict shows a conflict in which parties face to face structurally lame in terms of political and economic power. The farmers are those who are economically poor and only hope for the existence of land as a source of livelihood. The *TNI* has a complete set of structures, strengths, and resources to maintain the authority to use plantation land which is indeed legally held by the *TNI*. However, the structural inequality did not significantly reduce the enthusiasm of farmers to fight for the ownership status of the Darmakradenan plantation land. Both parties have taken various efforts to achieve conflict resolution. However, the actual form of conflict resolution still cannot be achieved to date.

In fact, land conflicts have reduced the rights of farmers to manage land which is their source of income. This means that land conflicts have eroded the rights of farmers to enjoy a prosperous life. The struggle of farmers who have spent energy, time, and material that are widely used by them to fight for land rights also further reduces the rights of farmers to enjoy economic rights to live more properly, neglected farmers' social rights because they focus on the problem of achieving resolution conflict, as well as rights in the political field that are also pawned, and so on.

The study also identifies the impact of conflict from social, economic and political aspects. Socially, the lives of farmers experience a change in the social life order in which disharmony exists among the community members and with the village government. The change in violent behavior and courage to oppose the law is a result of a community crisis caused by an important process of change in the economic field. Loss of livelihoods is a major factor in emerging community crises. The problem of vulnerable economic life also makes prolonged land conflicts and increasingly causes the economic life of the community to not develop. Politically, the relationship between farmers and the village elite is also not harmonious so that farmers' participation in village government activities is too low.

The condition of farmers who are increasingly vulnerable and continue fighting for their rights makes their lives very urgent to be empowered. The hope is that the resolution of the Darmakradenan land conflict will restore the lives of farmers to regain their economic, social and political rights. A further study is needed regarding the model for resolving the Darmakradenan land conflict based on efforts to empower citizens along with the vulnerability of the economic, social and political life of the local farming community.

Acknowledgment

The author would like to thank the Indonesian Ministry of Research Technology and Higher Education through Dissertation Research Grant 2018.

References

[1] Ide, Tobias. 2015 "*Why do Conflict over Scarce Renewable Resources Turn Violent? A Qualitative Comparative Analysis,*" *Journal of Global Environment Change*, Vol. 33.

[2] Emanuel, Maria, and Ndimbwa, Tumpe. 2013 "*Traditional Mechanisms of Resolving Conflicts over Land Resource: A Case of Gorowa Community in Northern Tanzania,*" *International Journal of Academic Research in Business and Social Sciences,* November, Vol. 3, No. 11, pp.214-224.

[3] Chanya, Apipalakul, Prachaak, B, Ngang, TK *Conflict Management on Use of Watershed Resource, Journal of Procedia-Social and Behavioral Sciences, 136, 2014*, h. 481-485.
[4] Prieto-Remona, Thomas, Cobo-Benitab, Jose Ramon, Isabel Ortiz-Marcosa, Isabel. 2015. "*Conflict Resolution to Project Performance" Journal of Proceeding Social and Behavioral Sciences*, Vol. 194.
[5] Froje, John W. 2009. "*Traditional Justice and Reconciliation after Violent Conflict: Learning from African," African Journal on Conflict Resolution*, Vol. 9 No. 3. pp. 145 – 155.
[6] Sulastriono, 2014. "Resolusi Konflik Pengelolaan Sumber Daya Alam Berbasis Pranata Adat", *Jurnal Media Hukum*,Vol. 21 No. 2 December, p.213-224.
[7] Enggelina Iku, Maria, 2010. *Peran Lembaga Adat Mosalaki dalam Penyelesaian Konflik Hak Tanah Ulayat di Kelurahan Wolojito Kecamatan Wolojito Kabupaten Ende Provinsi Nusa Tenggara Timur*, Tesis, Universitas Gadjah Mada, Yogyakarta.
[8] Syam, Aliar, 2015. *Model Resolusi Sengketa Tanah Hak Ulayat dengan Asas Kepastian Hukum dalam Mewujudkan Kesejahteraan Masyarakat Adat di Kabupaten Kampar Propinsi Riau,* Disertasi Pascasarjana Unisba Bandung.
[9] Musdalifah, 2007. *Konflik Agraria dalam Relasi Antara Perusahan Perkebunan dengan Masyarakat (Kasus Konflik antara Petani dengan PT Lonsum di Kabupaten Bulukamba),* Disertasi, Universitas Hasanuddin.
[10] Alting, Husen, 2013. "Konflik Penguasaan Tanah di Maluku Utara: Rakyat Versus Penguasa dan Pengusaha." *Jurnal Dinamika Hukum*, Vol. 13, No. 2, pp. 266 – 282.
[11] Asmara, Galang, Arba, Maladi, Yadi. 2010 "Resolusi Konflik Pertanahan Berbasis Nilai-nilai Kearifan Lokal di Nusa Tenggara Barat", *Jurnal Mimbar Hukum,* Vol. 22. No.1.
[12] *Daftar Isian Potensi Desa Darmakradenan 2015*, Desa Darmakradenan: Pemerintah Desa Darmakradenan Kecamatan Ajibarang Kabupaten Banyumas.
[13] Strauss, Anslem, and Juliet, Corbin. 1990. *Basic of Qualitative Research: Grounded Theory Procedures and Techniques.* Beverly Hills, SA: Sage Publications.
[14] Newman, Lawrence, 1994. *Social Research Methods: Qualitative and Quantitative Approaches,* Boston: Allyn and Bacon.
[15] Harrison, Lisa. 2007. *Metodologi Penelitian Politik*, Jakarta: Kencana Prenada Media.
[16] Punch, Keith, 2006. *Developing Effective Research Proposal*, London: Sage Publication.
[17] Silbergh, David M, 2001. *Doing Dissertations in Politics: A Student Guide,* London: Routledge.
[18] *htpp://www.darmakradenan.desa.id* diakses 28 Februari 2018.
[19] *htpp://www.darmakradenan.desa.id* diakses 28 Februari 2018.
[20] Ricklefs, M.C., 2011. *Sejarah Indonesia Modern*. Yogyakarta: Gadjah Mada University Press.
[21] *Laporan Investigasi Kasus Tanah di Darmakradenan*, 1999. Desa Darmakradenan : Stan Ampera.
[22] *Kronologi Kasus Sengketa Tanah Darmakradenan*, 1999. Desa Darmakradenan : Stan Ampera.
[23] Hartanto, Sindung. 2012. *Spektrum Teori Sosial Dari Klasik Hingga Posmodern*. Jakarta: Ar-Ruzz Media.

M-Payment as Financial Inclusion Strategy in Islamic Boarding School Cooperatives

B APramuka[1,*], P D I Kusuma[1], Sugiarto[1]
[1]School of Accounting Faculty of Economics and Business Universitas Jenderal Soedirman Indonesia

*Corresponding author: bpramuka@gmail.com

Abstract. Financial inclusion refers to the use of financial services such as electronic money. Digital money receives tremendous attention globally, from buyers to sellers, as an alternative form of payment. The passions on this technology enormous and infectious, therefore attracting more followers from large merchant to smaller one including the cooperative. School cooperative serves as a center of economic with a wide range of activities, such as mini market, canteen, to serve the needs of students, teachers, and people in the surrounding area. One of the programs launched by the Central Bank of Indonesia is a financial inclusion to increase the financial literacy of people living in the remote area. In this case, Islamic conventional boarding school cooperative can become influencers in most of the surrounding community to attract people's interest to adopt the technology. The purpose of this inquiry is to find out factors influencing the adoption of mobile payment and user's intention and decision to adopt m-payment. Authors proposed an innovative research model based on the theory of technology acceptance model with the innovation characteristics of the diffusion of innovations, perceived security and intention to recommend the technology. The research model has empirically tested at 205 employees Islamic boarding schools across four regions. The independent variables under study include perceived ease of use, perceived usefulness, perceived credibility, subjective norms, and trust while the dependent variables are mobile payment adoption and intention to recommend the technology. The study uses structural equation modeling as a tool for analysis. The result indicated that perceived ease of use, perceived usefulness, and subjective norms have a significant influence on the adoption of mobile payment. The study also confirmed the relevance of users' intention to recommend mobile payment technology. Thus, future research needs to include social media campaigns and technology adoption studies. For researchers, the result of this study act as a basis for further refinement of personal technology acceptance model. For practitioners, understanding the key constructs is crucial to design, refine, and implement mobile payment services, applications, and products that achieve high consumer acceptance, value, and high rates of positive recommendations in social networks.

1. Introduction

The financial inclusion phenomenon is currently the center of attention in some countries. Limited knowledge and experience to use financial services is a barrier for certain parties to get financial services. Financial inclusion hasgained attention from academics and policymakers for several reasons such as it has a positive impact on theeconomic development of a country [11], enhancing world development [5,41] and sustainable development [22]. Financial inclusion affects both the macro and microeconomics of a nation. At the micro level individuals who do not have a bank account shall not receive benefits in the form ofincreased value of savings [4,5], By possessing an account a women shall have better economic empowerment [6,20] , as well as entrepreneurial productive consumption and investment [4] .

Islamic traditional boarding schools provide educational services to the surrounding people regardless of their social status. Most schools are located in rural areas to provide an opportunity for people in rural areas to participate in the education process. Therefore, the schools have the potential to act as a distribution channel in providing access to the majority of people in rural areas. Previous research shows that "disadvantaged people" come from women, young and the poor [20,41]. One of the factors that drive individuals to use financial services is ease of access. Individuals use financial services to meet several basic needs including education, health, and shelter, starting and operating a business. Financial inclusion increases poverty alleviation and promotes economic growth [7,8].

In the digital era access to financial services should be open to anyone. The digital era stimulates the growth of internet-based business activities. The number of mobile phone users has increased sharply and massively throughout the world. Technology innovation has an indispensable role in expanding access of all parties to financial services, so there is no more financial exclusion. People who do not have a formal bank account can use cellular technology to conduct transactions such as paying a taxi, using go pay. High interest in the use of financial technology has emerged since 2004 [27]. Even among the 20 percent of the poorest households, almost 7 out of 10 have cell phones [17].

This paper uses technology acceptance and innovation by individuals to explain the behavior of the use of information technology in the financial sector. What are the factors that influence the interest of Islamic boarding school cooperative managers to use digital financial services? The results of this study contributed to the development of the literature on digital financial inclusion in Indonesia. The Central Bank of Indonesia initiates to develop an ecosystem of non-cash payments by implementing e-money in Islamic boarding schools. However, data on the potential for digitizing financial services for Islamic boarding schools are not yet available.

2. Literature Review

2.1 Financial Inclusion at Individual Levels

Although previous research on financial inclusion is enormous, few discussed financial inclusion policies at the individual level. With financial inclusion strategy, the Indonesian government expects to narrow down the exclusion for certain circles such as women, young people and the poor. To date, the literature on financial inclusion across the country rely on the state-level proxy based on the data collected by the policy-making, in this case, the Central Bank [7,19,21, Ponce, & Mylenko, 2010). Non-asymmetric information narrows the generalization of existing research because data from central and macro banks is difficult to translate at the micro level, for example how the government policies increase the financial accessas individual differences in terms of income level, gender, location of residence, or age [5; Dupas et al., 2012).

Specific interventions can encourage the use of digital-based Instruments due to the perception of the benefits of the device [6]. Financial inclusions provide some benefit to the individual in the form of low-cost accounts, financial security, improve networking, as well as legal rights. Nonetheless, the regular benefits have not yet motivated the disadvantaged group (women and the poor) to use financial services. Some of the reported barriers include money shortage, and there are no immediate needs for bank accounts and lack of trust. Therefore, the government policies to promote inclusion can increase the likelihood that people perceive financial services are within their reach (Allen et al., 2016).

2.2The Feature of E-Money

The results of research in the field of financial management related to the digital economy is the emergence of digital finance and financial inclusion in the pieces of literature [28]. Digital banking through financial technology has a positive impact on financial composition for developing countries and developed countries because the digital economy provides individual comfort. The marketing field is carried out by researchers (Lu, Chang, & Yu, 2013) about online retailer's consumer perception of e-loyalty intentions reported that the higher the online retailer's consumers, the more likely consumers to make purchases online.The phenomenal growth in cellular technology has made it easy for consumers to transact. Consumers do not need to bother carrying cash because transaction payments can be made using a cellular device. Mobile-based payments can be categorized into three perspectives, namely foreign mobile commerce, mobile proximity payments, and peer-to-peer mobile [26].

Payment technology using mobile phones also has several limitations which can disrupt transactions such as ergonomics and usability of devices such as small screen size and small keyboard [25], unstable internet connection [37], complicated and tedious payment [33], confidentiality during authentication [9]. Therefore, mobile payment users require more significant effort and high learning speed to use the m-payment application which sounds discouraging. To increase users' interest in using e-money, the payment must be simple, easy to use, and do not require much energy to execute it [24].

2.3The Technology Acceptance Theory

The potential of mobile payment in the future is enormous. Small firms have started to use digital payments for several reasons such as consumer demands and peer pressure. The euphoria has spread over the Islamic boarding schools, where parents have demanded an easy way of paying their children's' tuition due to geographical constraints. Most of the students who study at the boarding schools come from the distance. Likewise, the community living around the schools have limited access to financial services which requires them to travel to the city for payment of bills. As a consequence, the use of digital payment by Islamic boarding schools does not just prosper the pupil but also the surrounding people.

The limited number of schools have adopted digital payment. Among the factors influencing the adoption, the decision is the perceived usability and perceived ease of use of the technology [10]. This theory explains some of the reasons for individuals take electronic mail systems include lower operational costs and the speed of information spread. Mobile payment is voluntary and is used by individuals for the benefit of transactions that are generally personal. Therefore, all adoption costs of m-payment are borne by individuals and adoption decisions are not solely based on practical reasons. To use an m-payment, a person must have sufficient knowledge and skills about [35].

2.4The Research Model

Perception against Compatibility of a technology depends on the extent to whichinnovation canprovide meaningful experience and ability to meet the needs of theuser wholeness, as well as offering better values for living. Almost all banks in Indonesia have proposed a mobile payment to anticipate growing business transactions. With the emergence of virtual business during the disruptive era (such as online transportation and online shopping), consumers in Indonesia regardless of their backgrounds have used m-payment [4]. The financial technology does not only offered by financial institutions. Some of the firms providing financial technology include cellular providers (t-money), Telkom, Go-jek, Grab, and Tokopedia. M-payment is ubiquitous because it allows for faster, safer and reliable financial services.

Perception of ease of use means that technology tools are simple to use [38]. The first step to do before start using a tool is understanding the complexity of the technology itself. Therefore, perceived ease of use is an essential determinant of technology adoption. Moreover, users also consider some benefits obtained from adopting technology as an important influencer. The application of a mobile payment provides some gains in the form of minimum fees, convenient to use. When consumers feel that mobile payment is easy to learn and use, they will be more likely to adopt the innovation. Consumers first impression of a technology offered is an essential determinant of adoption decision. When consumers believe that they will succeed in utilizing the technology, then the choice is immediately executed. In other words, the ease of perception has a positive effect on the intention to use a technique which later is followed by adoption [15]. Past studies have also discussed on the antecedents of consumer decision to adopt mobile payments [29,32,33,34,38]

Perceive the benefits of a technology tool refers to the relative advantages of a technology that can improve individual performance [38]. Understanding of benefits can increase an individual's desire to adopt a system [40]. Consumers get greater benefits by using online transactions using mobile devices. If compared with traditional e-payment methods, consumers who use m-payments benefit from services that are easy and fast. A study by Tan et al., (2014)at Malaysia concludes that consumer perception of high benefits will result in a higher intention to adopt m-payment. Similarly, Teo et al. [34] show that m-payment adoption is useful regarding time savings and portability.

The use of technology in transactions does not mean without risk. From hard perspectives, a technique should possess multiple levels of a security system to prevent hijacking. The phenomena of increasing cybercrime are discouraging and become a severe challenge to technology adoption. The development of hardware should be accompanied with a soft approach. To improve one's attitude towards technology we need to strengthen the power of the subjective norms. For example by providing education about religious standards and legal rules that are expected to control behavior (Adawiyah, Pramuka, Najmudin, and Jati, 2015; 31].

Perceptionsofsecurity risks are an essential factor that determines consumers' decisions on whether or not to adopt a technology [18]. The knowledge of protection against the risks of using cellular transactions is crucial. The risk of losing confidential data that causes financial losses needs to be well anticipated to motivate users to adopt m-payment. Perception of security in transactions has a positive effect on the intention to take m-payment.

https://doi.org/10.1051/shsconf/20208601040

Trust starts from a belief that everyone is sincere in his or her words, attitude and behavior. Hence the technology of m-payment offered to the public is carefully prepared to prevent data leakage. The level of trust in adopting m-payment assumes that everyone will not behave opportunistically by expecting the promised service (W. Adawiyah, 2015). It will be more difficult for sellers and buyers to build online trust because there is no interaction between the two parties while performing the transactions. The excellent system offers protection to users' data, assets during the online deals.

3. Material and Methods

This research is quantitative in nature. The research model was empirically tested at Islamic traditional boarding schools across four regions. The independent variables under study include perceived ease of use, perceived usefulness, perceived credibility, subjective norms, and trust while the dependent variables are mobile payment adoption and intention to recommend the technology. There were 27 items used in the questionnaire to measure those variables. The study uses structural equation modeling as a tool for analysis. A survey distributed to respondents using purposive sampling. Criteria for respondents are boarding school cooperative managers who handle finance and have worked at least one year. Before distribution, participants were given a brief explanation of the research objectives. Also, to avoid response bias in the survey, the respondent's name is filled with anonymous. The respondents got an explanation of the variables and terms used in the questionnaire. A total number of questionnaires distributed to respondents is 205 with a response rate of 100 percent, so that all are processed further for further analysis. The data collection period was one month.

4. Result And Discussion

4.1 Descriptive Statistic

The results of the analysis of factor demographic respondent can be seen in table 1.

Table 1 Demographic profile of respondents

Item		Frequency	Percent
Gender	Male	101	49.3
	Female	104	50. 7
Age	<20	84	41
	20-30	76	37.1
	31-40	21	10.2
	41-50	16	7.8
	> 5 0	8	3.9
Education Level	High school	118	57.6
	Diploma	24	11.7
	Bachelor's degree	57	27.8
	Post Graduate	6	2.9
Income	<Rp. 3,000,000	130	63.4
	> Rp. 3,000,000	75	36.6

There were the almost equal proportion of male and female respondents, with female dominance, participated in the survey. The average age of the respondents was between 20 to 30 years, indicating that they are still in their productive age. Majority of the respondents were high schoolgraduate. There was 27.8 percent of university graduates, with bachelor degrees, working at Islamic boarding schools. Majority of respondents earn monthly income from below three million rupiahs. Table 2 presents the descriptive statistics of the variable under study.

Table 2 Descriptive statistics of variables

	N	Minimum	Maximum	Mean	Std. Deviation
Perception Benefits	205	3.00	5.00	4.2476	.49072
Perception ease of use	205	2.00	5.00	4,0951	.49269
Perception Credibility	205	2.50	5.00	3.8793	.48984
Subjective Norms	205	1.00	5.00	3.7207	.66679
Trust	205	2.30	5.00	3.9137	.58746
Attitude	205	3.00	5.00	4,1005	.52192
Interest	205	2.60	5.00	3,8907	.42767
Valid N (listwise)	205				

Table 2 presents the descriptive statistics of each variable. The average response was statistically viable concerning perceive benefits, ease of use, credibility, subjective norms, trust, attitude and interest to use e-money. Perceive benefits on the technology is the most critical factors of consumers' attitude towards e-money, followed by perceived ease of use. Trust is an essential determinant of intention to use e-money or mobile payment for the transaction. Perceive security of the operation is crucial as everyone wants to protect valuable assets.

4.2 Validity and Reliability Test

Table 3 Validity and reliability test of the instruments

Variabel	*Loading (L)*	*Error (E)*	*L*	*E*	*L*	*E*	*L*	*E*	*L*	*E*	*L*	*E*	*L*	*E*
X1_1	0.85	0.15												
X1_2	0.82	0.18												
X1_3	0.67	0.33												
X1_4	0.62	0.38												
X2_1			0.63	0.37										
X2_3			0.81	0.19										
X3_1					0.69	0.31								
X3_2					0.59	0.41								
X3_3					0.55	0.45								
X4_1							0.58	0.42						
X4_2							0.48	0.48						
X4_3							0.89	0.11						
X4_4							0.80	0.20						
X5_1									0.67	0.31				
X5_2									0.83	0.17				
X5_3									0.76	0.24				
Y1_1											0.65	0.35		
Y1_2											0.77	0.23		
Y1_3											0.67	0.33		
Y2_1													0.49	0.51
Y2_2													0.49	0.51
Sum of Std.Loadings	**296**		**1.44**		**1.83**		**2.75**		**2.26**		**2.09**		**0.98**	
Sum of Measur. Error		**1.04**		**0.56**		**1.17**		**1.25**		**0.74**		**0.91**		**1.02**
RELIABILITY	**0.894**		**0.787**		**0.741**		**0.8581**		**0.873**		**0.828**		**0.485**	

According to Hair (1995), an indicator is considered valid provided it has a loading factor above 0.40. Based on the data above it is known that the data is valid. The next step is to test the reliability of the items used to measure the variables.

$$Contruct\,Reliability = \frac{(\sum Std\,Loading)^2}{(\sum StdLoading)^2 + \in .j}$$

The result of the reliability test revealed that two variables did not meet the reliability criteria. The variables were Y2_1 and Y2_2. The limit value used to assess an acceptable level of reliability is 0.7. However, this value is not a dead value. This means that if the value below 0.70, an item can still be accepted if accompanied by strong empirical reasons. Nunally and Bernstain (1994) in Ferdinand [16] , stated that the reliability value between 0.50-0.60 was enough to justify the results of the study.

4.3 The Structural Equation

The next step in the analysis is to put the variable into the structural equation. Picture 1 reflects the model.

https://doi.org/10.1051/shsconf/20208601040

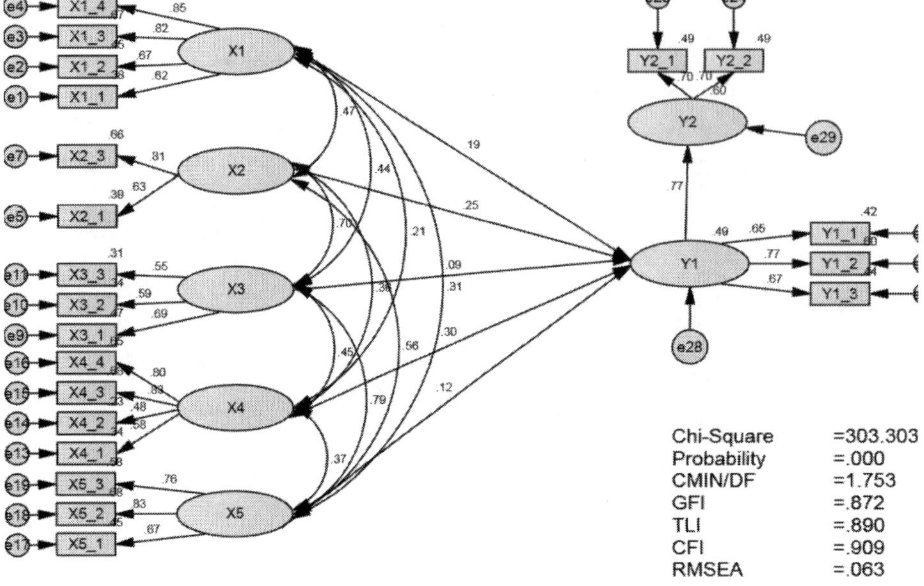

Picture 1 The structural equation modeling of the framework

The above output is the result of M-Payment testing using a mediating variable model with the following variables:

X1: Perceive benefits

X2: Perceive ease of use

X3: Perceive credibility

X4: Subjective norm

X5: Trust

Y1: Attitude

Y2: Interest

4.4 Direct Effect Analysis

Based on the output from testing the model using Amos shows that the immediate effect of perceived benefits on attitudes is 0.19; direct influence on perceived ease of attitude of 0.25; direct influence of credibility perception of 0.31; subjective norm's direct influence on attitudes of 0.30; the immediate effect of trust on attitude is 0.12; direct impact on attitudes towards interest of 0.77. The *loading factor* value shows a strong direct relationship between independent variables and mediating variables. The negative value on the *loading factor* shows the rejection of the influence of the variable, but based on the output above shows that the independent variable has a good effect, then the mediating variable has a good impact on the dependent variable.

4.5 Evaluate the Goodness of Fit Criteria

Based on the results of the analysis that we display on Amos Graphics or contained in the output table and the text output, we compare the values with the *Cut of Value value* for each criterion, so that the following results are obtained:

Table 4 Goodness of fit of the model

The goodness of Fit Index	*Research Model Results*	*Cut off Value*	*Model Evaluation*
X2-Chi-Square	303,303	Small expected <1000	Well
Significance Probability	0,000	≥ 0.05	Well
RMSEA	0.063	≤ 0.08	Well
GFI	0.872	≥ 0.90	Pretty good
CMIN / DF	1,753	00 2.00	Well
TLI	0.890	≥ 0.95	Pretty good
CFI	0.909	≥ 0.95	Pretty good

Based on the output above, it was found that [16]:

1. ^{2} - *Chi-Square statistics*, the smaller the value χ^2, the better the model, and the probability is accepted by the *cut-off value* of p> 0.05. The results of this study indicate that this research model is in the right category because it has a small chi-square value.
2. RMSEA (*The Root Mean Square Error of Approximation*), is an index used to compensate for *chi-square* in a large sample. RMSEA values that are small or equal to 0.08 are indices for accepting the model based on *the degree of freedom*. The RMSEA value of this research model is 0.063 which means that the model is good so that the model can be accepted.
3. GFI (*Goodness of fit Index*), is a *non-statistical* measure that has a range of values between 0 and 1. A high value in this index shows a " *better fit*." The GFI value of this study is quite good because it approaches the minimum criteria required.
4. CMIN / DF (*The Minimum Sample Discrepancy Function Divided with the degree of Freedom*), is a chi-square X^2 statistic divided by the degree of freedom so that it is called X^2 relative. Relative X^2 value less than 2.0 or 3.0 is an indication of *an acceptable fit* between model and data. CMIN value of 1.753 indicates that this model is excellent and fit.
5. TLI (TuckerLewisIndex), an incremental index that compares a model that is tested against a *baseline* model, where the recommended value is ≥ 0.95. However, the most important thing is that the value close to 1 indicates *a perfect fit*.
6. CFI *(Comparative Fit Index), the* range of values is 0-1, which is getting closer to 1, indicating the highest level of fit.

Based on the results of the above analysis it can be concluded that the M-Payment testing model can be accepted. This is because the test output shows good results

The results of this study support the results of previous studies, which concluded that perceptions of benefits affect the interest in adopting e-money in transactions. This is consistent with several previous researchers [4]. Besides that, the perceived ease of operation also affects the interest in using e-money. The existence of government policies that began to lead to e-payment at individual facilities and the encouragement of using financial technology impacted consumers' decisions to use e-money. Perception of ease of use means that technology tools are easy to use and uncomplicated [38]. This supports previous research that the ease of positive influence on the intention to use a technology which later is followed by adoption [15]. Past studies indicate that customers' decision to adopt mobile payments have been used easily [29; 32, 2005, 33,34,38]

The results of this research study support previous research [33] in Malaysia which concluded that the perception of high benefits would result in a higher intention to adopt m-payment. Similar results are also illustrated by [34] showing that m-payment adoption is useful concerning timesavings and portability. Perceptions of technological security risks are important things that determine consumers' decisions on whether to adopt innovations [18]. Therefore, perceived security in the transaction has a positive effect restricted ap intention to take m-payment. Starting from the belief that everyone will be sincere in transacting so that there is no fraud, the level of trust in payment assumes that no one will behave opportunistically by expecting the promised service (Adawiyah, 2015).

5. Conclusion

The study revealed that perceived ease of use, perceived usefulness, and subjective norms have a significant influence on the advertising of mobile payment. The study also confirms the relevance of users' intention to recommend mobile payment technology. Thus, future research needs to include social media cam campaigns and in future technology adoption studies. For researchers, the result is a basis for further refinement of personal technology acceptance model. For practitioners, understanding the key constructs is crucial to design, refining, and implementing mobile payment services, applications, and products that are high consumer acceptance, value, and high rates of positive recommendations in social networks.

Acknowledgement

Authors would like to thanks Universitas Jenderal Soedirman for funding this research

References

[1] Adawiyah, W. (2015). Impression management tactics, work-related behavior and perceived individual performance of sales clerks: a case of Indonesia. *International Journal of Applied Business and Economic Research, 13* (7), 5481-5501.

[2] Adawiyah, WR, & Pramuka, BA (2017). Scaling the notion of Islamic spirituality in the workplace. *Journal of Management Development, 36* (7), 877-898.

[3] Adawiyah, WR, Pramuka, BA, Najmudin, and Jati, DP (2015). Green Supply Chain Management and its Impact on Construction Small and Medium Enterprises Sector (SMEs) Performance: A Case of Indonesia. *International Business Management, 9* (6), 1010-1024.

[4] Adi, PH, Wihuda, F., & Adawiyah, WR (2017). The role of social media browsing intention for behavioral outcomes of young consumers. *Trziste = Market, 29* (1), 39.

[5] Allen, F., Demirguc-Kunt, A., Klapper, L., & Martinez Peria, MS (2016). The foundations of financial inclusion: Understanding ownership and use of formal accounts. *Journal of Financial Intermediation, 27*, 1-30. doi: https://doi.org/10.1016/j.jfi.2015.12.003

[6] Ashraf, N., Karlan, D., & Yin, W. (2010). Female Empowerment: Impact of commitment savings product in the Philippines. *World Development, 38* (3), 333-344.

[7] Beck, T., Demirguc-Kunt, A., & Martinez Peria, MS (2005). *Reaching out: Access to and use of banking services across countries*: The World Bank.

[8] Bruhn, M., & Love, I. (2014). The impact of improved access to finance: Evidence from Mexico. *The Journal of Finance, 69* (3), 1347-1376.

[9] Cheong, S.-N., Ling, H.-C., & Teh, P.-L. (2014). Secure encrypted steganography graphical password scheme for access control system near field communication. *Expert Systems with Applications, 41* (7), 3561-3568.

[10] Davis, FD, Bagozzi, RP, & Warshaw, PR (1989). User acceptance of computer technology: a comparison of two theoretical models. *Management Science, 35* (8), 982-1003.

[11] Demirgüç-Kunt, A., Honohan, P., & Beck, T. (2008). *Finance for all ?: Policies and Pitfalls in Expanding Access*: World Bank.

[12] Demirgüç-Kunt, A., & Klapper, L. (2013). Measuring financial inclusion: Explaining variation in use of financial services across and within countries. *Brookings Papers on Economic Activity, 2013* (1), 279-340.

[13] Dupas, P., Green, S., Keats, A., & Robinson, J. (2012). *Challenges in banking the rural poor: Evidence from Kenya's western province*. Retrieved from

[14] Dupas, P., & Robinson, J. (2013). Savings constraints and microenterprise development: Evidence from a field experiment in Kenya. *American Economic Journal: Applied Economics, 5* (1), 163-192.

[15] Dutot, V. (2015). Infected near field communication (NFC) factors adoption: An extended TAM approach. *The Journal of High Technology Management Research, 26* (1), 45-57.

[16] Ferdinand, A. 2005. *Structural Equation Modeling*. Semarang: The Board of Directors of Diponegoro University

[17] Group, WB (2016). *World Development Report 2016: Digital Dividends*: World Bank Publications.

[18] Hari Adi, P., & Adawiyah, WR (2018). The impact of religiosity, environmental marketing orientation, and practices on performance: A case of Muslim entrepreneurs in Indonesia. *Journal of Islamic Marketing*.

[19] Honohan, P. (2008). Cross-country variation in household access to financial services. *Journal of Banking & Finance, 32* (11), 2493-2500.

[20] Istiqomah, & Adawiyah, WR (2018). Development of rural entrepreneurship groups in Indonesia: benefits, problems, and challenges. *International Journal of Entrepreneurship and Small Business, 34* (3), 330-342.

[21] Kendall, J., Ponce, A., & Mylenko, N. (2010). *Measuring financial access around the world*: The World Bank.

[22] Klapper, L., El-Zoghbi, M., & Hess, J. (2016). Achieving the sustainable development goals. *The role of financial inclusion. Available online: http: // www . bro. Org. Accessed, 23* (5), 2016.

[23] Lu, L.-C., Chang, H.-H., & Yu, S.-T. (2013). Online shoppers 'perceptions of e-retailers' ethics, cultural orientation, and loyalty: an exploratory study in Taiwan. *Internet Research, 23* (1), 47-68.

[24] Madlmayr, G., Langer, J., Kantner, C., & Scharinger, J. (2008). *NFC devices: Security and privacy*. Paper presented at Availability, Reliability and Security, 2008. ARES 08. Third International Conference on.

[25] Ondrus, J., & Pigneur, Y. (2007). *An assessment of NFC for future mobile payment systems*. Paper presented at the Management of Mobile Business, 2007. ICMB 2007. International Conference on the.

[26] Ooi, K.-B., & Tan, GW-H. (2016). Mobile technology acceptance model: An investigation using mobile users to explore smartphone credit cards. *Expert Systems with Applications, 59*, 33-46.

[27] Owens, J. (2013). Offering Digital Financial Services to Promote Financial Inclusion: Lessons We Learned. *Innovations: Technology, Governance, Globalization, 8* (1-2), 271-282.

[28] Ozili, PK (2018). Impact of digital finance on financial inclusion and stability. *Borsa Istanbul Review*. doi: https://doi.org/10.1016/j.bir.2017.12.003

[29] Park, SY (2009). An analysis of the technology students 'acceptance model in understanding university' behavioral intention to use e-learning. *Educational technology & society, 12* (3), 150-162.

[30] Pinasti, M., & Adawiyah, WR (2016). CO-OPETITION TO PROMOTE GROWTH OF BATIK SMALL AND MEDIUM ENTERPRISES. *International Journal of Business & Society, 17* (3).

[31] Putri, NK, Adawiyah, WR, & Pramuka, BA (2017). Independence of Audit Ethical Decision Making Process: A Case of Indonesia. *DLSU Business & Economics Review, 26* (2), 1-1.

[32] Saadé, R., & Bahli, B. (2005). The impact of cognitive absorption on perceived usefulness and perceived ease of use in online learning: an extension of the technology acceptance model. *Information & Management, 42* (2), 317-327.

[33] Tan, GW-H., Ooi, K.-B., Chong, S.-C., & Hew, T.-S. (2014). NFC mobile credit card: the next frontier of mobile payment? *Telematics and Informatics, 31* (2), 292-307.

[34] Teo, A.-C., Tan, GW-H., Ooi, K.-B., Hew, T.-S., & Yew, K.-T. (2015). The effects of convenience and speed in m-payment. *Industrial Management & Data Systems, 115* (2), 311-331.

[35] Teo, AC, Tan, GWH, Cheah, CM, Ooi, KB, & Yew, KT (2012). Can the demographic and subjective norms influence the adoption of mobile banking? *International Journal of Mobile Communications, 10* (6), 578-597.

[36] Tsai, H.-y. S., & LaRose, R. (2015). Broadband Internet adoption and utilization in the inner city: A comparison of competing theories. *Computers in human behavior, 51*, 344-355.

[37] Valcourt, E., Robert, J.-M., & Beaulieu, F. (2005). *Investigating mobile payment: supporting technologies, methods, and use.* Paper presented at the Wireless and Mobile Computing, Networking and Communications, 2005. (WiMob'2005), IEEE International Conference on.

[38] Venkatesh, V. (2000). Determinants of perceived ease of use: Integrating control, intrinsic motivation, and emotion into the technology acceptance model. *Information systems research, 11* (4), 342-365.

[39] Venkatesh, V., Morris, MG, Davis, GB, & Davis, FD (2003). User acceptance of information technology: Toward a unified view. Quarterly *MIS*, 425-478.

[40] Wihuda, F., Kurniawan, AA, Kusumah, AI, & Adawiyah, WR (2017). Innovative behavior: A study from the hotel industry. *Turizam: međunarodni znanstveno-stručni časopis, 65* (3), 294-313.

[41] Zins, A., & Weill, L. (2016). The determinants of financial inclusion in Africa. *Review of Development Finance, 6* (1), 46-57. doi: 10.1016 / j.rdf.2016.05.001

Digital Literacy Movement by Former Migrant Workers in Wonosobo, Central Java

Wiwik Novianti, S. Bekti Istiyanto
Universitas Jenderal Soedirman
wiwik.novianti@unsoed.ac.id; wiwiknovianti27@gmail.com

The digital literacy movement is mostly centered in big cities even though internet users in rural areas are also very high. In Tracap Village, Wonosobo District, there is a community of former migrant workers who care about digital literacy. Living in a remote area does not mean that the people there are backward but it makes them more creative. This paper would like to describe the digital literacy movement carried out by former migrant workers in Wonosobo Regency, the goals to be achieved and their strategies in conducting digital literacy.

The research method used in this study is qualitative. Data collection is done by in-depth interviews, observation, and document analysis.

The study found that the experience of being a migrant worker who had difficulty communicating with families in the village was the main driver for former migrant workers in Wonosobo to share their knowledge about internet use with villagers. The goal is that communication between family members continues to exist even though one of his family is abroad.

The strategies carried out to provide knowledge and skills in using the internet started from subscribing to the internet which can be accessed free of charge by citizens to open internet classes for migrant workers' children. The internet which was originally only used as a medium of communication with families abroad developed into something productive, namely to help with school or work assignments, sell online and make games.

Keywords: literacy, digital, migrant workers, internet, Wonosobo

Introduction

From the survey report of the Indonesian Internet Service Providers Association (APJII) in 2016, in that year (2016) the number of internet users in Indonesia reached 132.7 million out of 256.2 million people in Indonesia. This means that 51.8 percent of Indonesia's population are internet users. Internet users are dominated by groups of workers / entrepreneurs as much as 62% or 82.2 million people followed by housewives as many as 22 million people or 16.6%. In the third rank, there were groups of students with a total of 10.3 million (7.8%) and student groups totaling 8.3 million (6.3%).

APJII also reports that social media is the most dominant content accessed. 97.4% or 129.2 million internet users access social media. The next most accessed content is entertainment as much as 96.8% or 128.4 million people. In the third rank, news content is the most accessed with a percentage that is not much different from entertainment content, which is 96.4% or 127.9 million. This was followed by educational content (93.8% or 124.4 million people), commercial content (93.1% or 123.5 million people) and public services (91.6% or 121.5 million people).

The large number of Internet users in Indonesia, as well as the high frequency of accessing information content and social media, does not guarantee that Indonesians are mature in using the Internet. Lots of internet abuse cases are found, ranging from addiction, violation of privacy and the number of hoaxes. This shows that the digital literacy of the Indonesian people is still low.

Various efforts have been made by a number of parties to improve the digital literacy of the community. But it still hasn't touched all levels of society. This is because the main drivers of digital literacy activities are still dominated by universities. Community groups or communities that are actors of digital literacy activities are still very limited (Kurnia & Astuti, 2017). Digital literacy activities are still centered in urban areas, even though internet usage is very widespread in rural areas.

In Wonosobo Regency, there is an interesting phenomenon of digital literacy activities in which the main actors of the activity are former migrant workers. After returning to Indonesia, the former migrant worker was called to provide education about internet use for residents around him. The Secretariat of the Indonesian Migrant Workers' Union in Wonosobo Tracap Village is a gathering place for citizens to be able to access the internet while increasing their digital literacy. The digital literacy movement which is usually driven by circles with higher education, in this remote village is actually activated by a group of former migrant workers. This made the writer interested in studying it, how the community of former migrant workers together with the surrounding community tried to improve their digital literacy skills.

Literacy means the ability to read and write. In accordance with the context, literacy as 'literacy' developed into various types of literacy, such as political literacy, financial literacy, legal literacy, including media literacy, information literacy and digital literacy ((Kurnia & Astuti, 2017).

According to Potter, experts initially defined media literacy as "the ability to access and process information from any form of transmission". The definition was then detailed into "... the ability to access, analyze, evaluate and create messages across a variety of contexts (Livingstone, 2003). Following the dynamism of the community and the rapid development of information and communication technology, media literacy alone is not enough to answer the complexity of existing problems so that the concept of digital literacy emerges.

The concept of digital literacy was presented by Paul Gilster for the first time in 1997 in a book called Digital Literacy. Gilster defines it simply as 'literacy in the digital age', or the ability to understand and use information through a variety of digital sources (Bowden in Lankshear & Knobel, 2008: 18). In its development, digital literacy or precisely digital literacies is defined as "practices of communicating, relating, thinking and" being 'associated with digital media "(Jones & Hafner, 2012: 13).

Digital literacy is very important to be developed not only because of concerns about the negative impact of media exposure, but also has several other important factors, namely First, the important role of information in the democratic process. Second, the important role of cultural participation and citizenship. Third, the development of popular culture makes children and teenagers more and more access to digital media (Koltay, 2011).

Research on the use of digital media has been carried out by Widyastuti, Nuswantoro and Sidhi in the Special Region of Yogyakarta. The research was then outlined in an article published in the 2016 ASPIKOM Journal entitled "Digital Literacy in Productive Business Women in Yogyakarta". One of the findings of the study was internet access to increase the sustainability of women's economy. In the research the writer will do is how the community of former migrant workers conduct digital literacy activities and together with the community around realizing an intelligent Indonesian society in using digital media.

Methods

This research was conducted using qualitative methods. In conducting research, data was collected through in-depth interviews, focus group discussions, observation and analysis of related documents such as mass media, archives and so on. This study involved 17 informants consisting of migrant worker activists, internet class teachers, and residents who benefited from this digital literacy activity.

Result and Discussion

Working overseas is still in great demand by the people of Indonesia. In the period January to June 2018, as many as 128,544 Indonesian migrant workers departed abroad.[1]

From 2016 to 2018, on average there are 20 thousands of workers departing each month. The number of Indonesian migrant workers abroad is around 6 million and gives Rp. 70 trillion to state finance.[2]

Indonesian migrant workers come from various provinces in Indonesia. The three provinces that contributed the most migrant workers were East Java, Central Java and West Java. Cilacap, Kendal, Brebes, Banyumas, Wonosobo and Grobogan are enclaves of migrant workers in Central Java..[3]

Wonosobo is one of the Regencies that sends the largest migrant workers in Central Java. Based on data in BNP2TKI, the number of migrant workers from year to year has decreased, meaning that many former migrant workers returned home. The return of migrant workers to the area of origin brings a positive impact on the equitable distribution of skills in using information and communication technology. Former migrant workers are relatively able to use information and communication technology to meet their needs. This capability is the main capital to be able to provide knowledge and skills to use information and communication technology to other villagers.

[1] http://www.bnp2tki.go.id/read/13374/BNP2TKI:-Penempatan-PMI-Januari-Sampai-Juni-2018-Sebanyak-128.544-Orang-.html last accessed November 1, 2018
[2] https://www.cnnindonesia.com/ekonomi/20180530183040-92-302325/pekerja-migran-indonesia-alirkan-devisa-negara-rp70-triliun last accessed October 31, 2018
[3] https://www.suaramerdeka.com/smcetak/baca/78642/pengiriman-tki-pati-nomor-tujuh-se-jateng last accessed November 1, 2018

Maizidah Salas, a former migrant worker belonging to the Indonesian Migrant Workers Union in Wonosobo Regency, has an initiative to subscribe to the internet at her home which also functions as the SBMI Wonosobo secretariat. The internet can be accessed by residents within a 20 meter radius for free.

Salas shared that his background provided free access because based on his experience as a migrant worker he felt how difficult it was to communicate with family in his hometown. When Salas became a migrant worker in Taiwan, he felt very homesick for his children, husband and parents. He could not contact his family because even though in Taiwan he could access the internet but his family in the village could not. Upon returning to their homeland, the experience as a migrant worker who had difficulty communicating with his family encouraged Salas to install a wifi network at his home. The wifi network in the SBI Wonosobo secretariat can be accessed free of charge for 24 hours by local residents.

It's not easy to introduce the internet to villagers. Salas admitted that at first there were many who did not like the free internet access for fear that children and teenagers would access pornographic content. Salas then tried to convince them that pornographic sites had been blocked so that they would not be accessible to children and teenagers there.

In order for citizens to be interested in using the internet facilities provided, it is often held watching together soccer or MotoGP live at the SBMI secretariat in Wonosobo. In addition, during the month of Ramadan, from 15:00 - 22:00 the wifi will be turned off so as not to interfere with the schedule of reciting and worshiping citizens. Salas said that at this time there were also many elderly people in his village who began learning to use the internet

> "In fact, now many old people who start using IT come with smartphone and ask how to use WhatsApp. Then I explained. I also said that I installed the Al-Quran application also from the internet. "[4]

The internet which was initially only used as a medium to establish communication with families outside the country was then also used as a medium of learning for citizens, both learning to recite and to fulfill school assignments.

[4] Maizidah Salas, July 21, 2018

 https://doi.org/10.1051/shsconf/20208601041

It didn't stop there, Salas and activists from SBMI Wonosobo then sought to make the internet more productive, such as developing online marketing of products produced by former migrant workers in Wonosobo. To realize this dream, Salas along with activists from SBMI Wonosobo made an internet class.

The internet class held by SBMI Wonosobo aims to enable children of migrant workers to be literate in IT and use the internet for something productive. "We really want to educate children of migrant workers to know the world of IT, can have the skills to process the internet properly, can make money".[5] By mastering IT skills, it is expected that migrant workers' children do not follow in the footsteps of their families who are migrant workers abroad.

The internet class was then opened at the end of August 2016 with around 12 children of migrant workers in Tracap Village. There are 2 teachers who provide material in the internet class. The material provided starts from zero, which is how the children can use computers, write to finally be able to manage the website. The ultimate goal of the internet class is to produce children of migrant workers who are able to manage the website because SBMI Wonosobo dreams of having a buying and selling website as an online media to market products produced by former migrant workers.

The implementation of the internet class is also inseparable from the support of migrant workers. Computers used by internet class students are obtained from donations of migrant workers in Macau. Funds used for internet class operations also use SBMI Wonosobo cash funds.

Unfortunately the internet class only lasted 5 months even though the children of trained migrant workers did not get the expertise they expected. Salas explained that the cessation of internet classes because teachers who train often do not come so that the children's enthusiasm becomes slack.

> "Now the problem is sometimes the children have left and the instructor comes late. Then the next week the children have gathered here and the ones who teach are not coming. Mostly, I can only give petrol money to the instructor, maybe that is not enough or what. But what was said was "sorry to be busy, sorry to be busy", so my friends just stuck there".[6]

---

[5] *Ibid*
[6] *Ibid*

https://doi.org/10.1051/shsconf/20208601041

When the FGD was conducted (September 10, 2018), researchers had asked Ahmad Khanafi as the teacher who gave internet training, Khanafi replied that at that time he was indeed busy. He also felt that he lacked the ability to provide internet class material because basically he did not have a scientific background about managing websites. He only taught himself. For this reason, Khanafi suggested that SBMI Wonosobo establish cooperation with more competent parties.

Even though the internet class is not running, free wifi is still available. When the FGD was conducted, there were 15 people who often access the wifi network. Most who access wifi are students and students who use wifi facilities to assist them in completing their school assignments. There are also PAUD teachers who use the wifi to find teaching materials for their students. Some even try to make games using the free wifi.

From the research conducted, it appears that SBMI Wonosobo and the children of migrant workers who have participated in internet classes still hope that the internet class which is currently stagnant can be continued. For that they need people who have internet expertise and want to share with them. They realize that remote village locations make people reluctant to come but the desire to create a website that contains online buying and selling continues to be the vision they will create.

Conclusion

Former migrant workers returning to their villages have great potential as actors of community change to reduce the mastery gap in information technology and communication between modern society and traditional communities. The digital literacy movement carried out by former migrant workers is motivated by their experience as migrant workers who are very difficult to establish communication with family and relatives in their hometowns. From the beginning only to meet communication needs, former migrant workers later realized that the internet could be productive, therefore efforts were made to improve their abilities and those around them to use the internet as a medium that can improve quality. their life.

References

APJII. 2016. Infografis Penetrasi dan Perilaku Pengguna Internet di Indonesia Survey 2016. Asosiasi Penyelenggara Jasa Internet Indonesia.

Jones, Rodney H. & Hafner, Christoph A. 2012. "*Understanding Digital Literacies: A practical introduction*". Oxford: Routledge.

Koltay, Tibor. (2011). "The media and the literacies: media literacy, information literacy, digital literacy". *Media, Culture, & Society*. 33(2). 211-221

Kurnia, N & Astuti, SA. 2017. 'Peta Gerakan Literasi Digital di Indonesia: Studi Tentang Pelaku, Ragam Kegiatan, Kelompok Sasaran dan Mitra yang dilakukan oleh JAPELIDI'. Paper dipresentasikan pada Konferensi Literasi Digital. Yogyakarta: Universitas Yogyakarta. 12 September.

Lankshear, Colin & Knobel, Michelle. (2008). "Digital Literacies: concepts, policies and practices". New York: Peter Lang.

Widyastuti, Nuswantoro & Sidhi. 2016. Literasi Digital pada Perempuan Pelaku Usaha Produktif di Daerah Istimewa Yogyakarta. *Jurnal ASPIKOM,* Vol.3 No.1, Juli 2016 p.1-15. jurnalaspikom.org/index.php/aspikom/article/view/95/91

Fostering Trust as a Catalyst for Sustainable Village-Owned State Enterprise

K Srirejeki[1], A Faturahman[1], S Supeno[1]

[1]Universitas Jenderal Soedirman

Email: kikysrirejeki@gmail.com

Abstract. While the obligatory mandate to establish Village-Owned State Enterprise (VOE) is final, the fact that there are still many villages do not have VOE yet, lead us to investigate what hamper its establishment. Using an exploratory case studies to several villages in Banyumas district, Central Java, we found that trust and reliance hold as key roles for the success of the establishment as well as the sustainability of VOE. We argue that by fostering trust and reliance among village government officials, rural communities and village-owned enterprise managers would give positive outcomes. Further, this study also shed light on ways to foster trust. The practical implication of our finding shows that for a village leader who wants to have a successful village owned enterprise needs to be able to trust and rely on Village-Owned State Enterprise managers.

1. Introduction

Data from World Development Indicators 2017, shows that half of the Indonesian are in rural areas. According to Central Bureau Statistics (Badan Pusat Statistik/BPS), 13,8 percent of them are poor, while data from World Bank 68.7 percent of them are classified as poor. In 2014, Indonesia through Village Law had attempted to address this issue. The aims are to advance general welfare, fostering community-based economy and create sociocultural resilience in 74,754 villages. Through this Law, the Indonesian government wants to empower village by strengthening the village position.

Indonesia before the Reformation Era put the village only as the development object. The village did not have power and authorities to manage its resources, and it became a habit because it was being practiced for decades. Currently, there is a shifting paradigm that place village as the subject or actor for its development. The village is hoping to be more independent in managing its resources and solving its problems. To be able to do that, the village is expected to be more independent economically. The village which does not rely on central government donors or subsidy.

Village-Owned State Enterprise is believed to become a very strategic tool to improve the village economy and solve rural problems. VOE is a legal village economic organization established and owned by the village and independently and professionally managed by the separated-management team. VOE's business model allows capital sharing between village government and private party or rural community.

VOE is the pillars of village economic and or social activity. VOE can play the role as the social institution and or commercial institution [18]. VOE in the context of social institution serves a purpose to contribute solutions towards rural social problems. While VOE as the commercial institution plays as the contributor to hold economic activities that can increase the village income (*Pendapatan Asli Desa/PaDes)*.

VOE can be a strategic tool to create an economically-independent village. The economic activities that have been carefully considered and then implemented through VOE can increase the *PaDes*. In that construct way of thinking, when *PaDes* can be increased trough VOE, this condition should encourage each village government to respond to the establishment of VOE positively. Not only logically correct, but VOE is also one of the mandatory and consequence of the implementation of the Village Law, it means VOE is also legally required. However, aside from its mandatory nature, recent data in 2017 shows that of 74,954 villages there is only 18,446 VOE. The data shows that the VOE establishment rate is only 24,6 percent. Based on this phenomenon gap, the research aim is to investigate what hampers the establishment of VOE. The significance of the research is twofold. First, there are not many research investigating what factors might hamper the establishment of VOE. Most research discussed the importance of VOE and how to improve it, but the main question regarding the phenomenon of why there are many villages do not have VOE yet has not been answered yet. Second, investigating factor that hampers the establishment of VOE is important to find the solution, in a hope, it will cue in how to initiate the establishment of VOE.

2. Research Method

The research was done using a qualitative approach, in particular, an exploratory case study research. This method allows us to conduct field research that investigates a current phenomenon within its real-life situation. The benefit of using this method is not only it focuses on the current situation, but also it gives very minimum access of researchers to manipulate the events. This method is also consistent with our analysis in that the research question is about asking *how* or *why* certain events occur.

The study focused at a micro and meso level analysis. Micro-level analysis means we analyzed based on single VOE while meso means the level of analysis is on the district level. The interviews with village government officials, VOEs managers, rural dignitaries, and local government authorities were primarily carried out using the semi-structured and in-depth interview. Questions list were open-ended, which allows interviewees to answer based on their situations. We conduct face-to-face conversations which lasted around one and a half hours for each interview. We recorded the conversations and taken the notes.

To improve the validity and reliability of our findings, we use triangulation technique. Patton (1999) stated that triangulation refers to the use of multiple methods and or data sources in qualitative research to develop a comprehensive understanding of the phenomenon [6]. We gathered empirical information through multiple sources of evidence, interviews as primary data, and as for secondary data, we gathered official documents, website, and press releases.

3. Result and Discussion

3.1. Trust Issue, Lack of Competence, and Structural Vagueness

Jamil and Askvik [10] argue that trust is a multidimensional concept and has various meanings and applications in the field of social science. However, among various concept of trust, one of the most widely cited definitions of trust is explained by Mayer et al. [15]. He defines trust as the willingness of a party to be vulnerable to the actions of another party based on the expectation that the other party will perform a particular action necessary to the trustier, irrespective of the ability to monitor or control the other party.

Jones and Shah [13] extended Mayer's model of trust by dividing the trust process into three concepts:

1. Trusting actions refers to behavior that demonstrates a reliance on others.
2. Trusting beliefs refers to perceived trustworthiness
3. Trusting intentions refers to the willingness to be vulnerable.

Based on the above model of trust, in the context of this study trust refers to involving belief and hope from village government officials to VOE's managers that they have the competence to manage VOE and act by the village interest. Trust becomes the salient issue that we found on our study. There are many VOE managers feel that village officials do not trust them completely to manage and to develop VOE. Statement from VOE Manager was:

> "We do really want to develop VOE, but we feel like VOE establishment was just because village officials do not want to violate the Law, because the Law stated that every village has to have

VOE. So, basically they (village officials) asked us to develop and manage VOE but our feet are tied, and we cannot move freely. So, we just got stuck."

The frustration keeps on going because on a planning level, the distrust is already in place. The distrust to VOE managers is not only by the Village Leader but also extended to other stakeholders, which seems to be initiated by the Village leader then extended to other stakeholders. One of the VOE managers stated that:

"In a village meeting, we (VOE management) had informed to the village officials and village representative assembly as well as prominent figures about our business idea. But because we have no source of fund, we have to consult every business idea to them, which is okay, because that's how the system works. But what sometimes saddens me is on so many occasions, we ended by forgetting our ideas and surrender. Because in the village meeting most of the time we only discuss the technical things, such as who will handle the projects, should the village's officials took part in the project or just be an observer, how much will it costs, etc. So, in the end, we just lost the moment, or the program just does not work."

Many research stated that there are many reasons for the (dis)trust. Those research argue that trust arises from partners' competence, which includes technical skills, experience, and reliability (as cited by Connelly et al., 2015). Hill & Lineback [9] argue that to build trust, competence is the key. They argue that competence should cover technical knowledge, operational knowledge, and political knowledge. Technical knowledge related to the understanding of how to perform work and basics management. Operational knowledge refers to the realization of technical knowledge, so it covers not what to do things but how to do things. While political knowledge covers the skill to get what is being wanted, understand how to justify specific action or request.

In the context of this study, the competence of the VOE's managers should cover their understanding of their job description, real action to do the job and persuasive skills to make the plan into realization. When in fact, the village government does not seem to trust the VOE manager indicating that there is no competence in place. It might be due to the fact that many VOE managers are not business professionals. They have no previous experience in managing the business. Their basic management skill, therefore, is questionable. Many of them are also not a prominent figure in the village.

This condition quite paradoxical, however, because, the VOE managers should be chosen not only by the village officials but also approved by the society. As being stated by the Law, the village meeting is an important forum where the VOE manager's selection process should be discussed and approved. When the chosen VOE managers do not seem to satisfy the village officials, it indicates there is something wrong with the process. This finding, lead us to dig more about the process of VOE managers' selection.

Regarding the VOE manager selection, we found that the VOE managers' lack of competencies cannot be separated from bungling election process. The election process does not base on competence. One of our interviewee, the Leader of VOE Forum, approved this, he stated that:

"The basis to choose VOE managers are not clear. It happens carelessly. Why careless? Because truthfully, many of village officials do not really understand what is VOE. They just know that every village should establish VOE was stated in the Law, so they just do it as a formative action to abort the obligation. So, because the process was like that, the VOE management was also formed carelessly".

The consequence of the careless VOE management formation impacting to the poor performance. Many of the VOE managers do not know what their duties and responsibilities are and what they supposed to do with the jobs. This lack of technical and operational competence lead to the distrust by the village officials.

Further, the condition is getting worse, because there are also many villages that do not have a solid village development master plan. Let alone the master plan, when the village officials were being asked what the village potential are, they were not sure how to answer that. They find it difficult to make a solid profile about the village, what is actually the potential village priority. When being asked what the village priority development is, one of our interviewees said:

> "Our village potentials are varied. We have plans to develop a tourism village because we have river flows that pass through our village. So, we have plans to make rafting tours. We also have many chicken farmer, so we want to help them in terms of capital sharing. In our village there are also many motorcycle repair shop, so we want to enlarge their work scale. We also have a plan to make a tent rental service, because if there is a marriage celebration or recitation celebration, our people can rent it from us (village government). Our village is also very well known for its traditional vegetable soup, called *lodeh*. It is the best in the district."

Many village officials were also stated similarly. It is difficult for them to choose their village development plan and it made them precarious in choosing the right type of business for their VOE's. If we would like to see this story from the bigger perspective, blaming VOE managers alone because of lack competence issue, is not fair. Village officials are also taking part in making this vicious circle do not meet its end. This is not a causality relationship where one cause is impacting other cause, but it is more like a reciprocal relationship. In other words, we cannot say that the lack of competence is causing the distrust or vice versa, but that the lack of competence and distrust are causing each other.

The solution toward this problem is nothing then improving the competence, especially VOE management team, as they are the one who will managed and face daily basic activities. In this notion, we argue that improving competence should also be specific for effectivity and efficiency. The competence improvement program should cover the management training program and business-related program. Management training program is needed because many of the VOE management team members are not professional, they should be given the training that covers managerial skill and leadership. In the program they learn how to make decision, how to establish solid teamwork, resolve conflict and how to lead a project. Business-related programs also needed as they are facing the technical problems related to the business development.

Aside from the problematic trust between the village leader and VOE management, the fact that both parties do not understand fully each rights and obligations make the situation even worse. The idea to put VOE as independent and professionally-managed economic institution seems to be the talk only. The reality is the village leaders do not want to lose sight on VOE activities not only caused by the excuse that VOE management needs directions but because the village leaders think that it is their job to interfere the upstream to downstream VOE activities.

> "I believe that in this initial stage, the VOE management needs guidance and directions, but I do almost everything. Say, the initiative process was come from me, I give them the business ideas, many business ideas. I went to other villages to do comparative studies. I want to learn from them who had proven to be successful in managing VOE. But when I got back to my villages, I share what I know to the VOE management them, hope that they will respond actively, but no, they just make wait passively, waiting for a commando".

We make clarification about that statement to the VOE management team, and they respond differently. According to them, what makes the VOE activities has not run yet, or just going nowhere, is because they believe that they have to wait for the directions or commando from the village leader.

https://doi.org/10.1051/shsconf/20208601042

"We are waiting for the command, we believe that it is the best for now. We are not going to take the risk by doing activities without approval. It is also something that we have been agreed together. Every business ideas should be approved by the village officials. They are the ones who know how much money we have, when the money will be allocated to us".

The structural vagueness within the VOE contribute to the late progress of the VOE establishment. This might due to the lack of understanding of rights and obligations among parties. We argue that village leaders assume that VOE is a project under their authority in which its management and activities are also under their control. In other hand, the VOE management think that they are just operational manager who do technical activities only. They are not aware that VOE management is expected to be the one who manage the entire activities of VOE professionally, while the village leader only serves as the advisor.

We cannot separate this phenomenon with paternalistic culture that might be contrary to the spirit of VOE autonomy. The village leader propensity to direct is reasonable. However, too many directions and restrictive actions has burdened VOE management in designing the VOE business plans, instead of assisting them toward smooth and real implementation.

3.2. Fostering Trust

Many types of research showed the importance of trust in an organization. Jiang, Henneberg, and Naudè [12] stated that trust has a positive influence on long-term orientation in business relationships. Therefore, they argue that trust is important not only in the early stage of business development but it is important to be sustainably managed by the management and the business partner. Trust is also very crucial for open innovation. Brockman, Khurana, and Zhong [5] argue that societal trust serves as an effective mechanism for improving relational governance among partners, thereby leading to better collaborative outcomes. Financial performance will result in higher managerial performance if it is mediated by interpersonal trust. Those researches put trust as an important factor in organization activities and to generate the desired output.

In the government context, especially democratic governance, trust is a prerequisite. Schlesinger and Heldman [16] argue that in order for the public administration to function smoothly and effectively, the government needs public trust. Lewicki and Tomlinson [14] support the notion that trust is an essential ingredient in the building of a competent state and helps to build good governance. Jennifer and Jennifer [11] report that public trust towards local government is higher than to the national/central government, though they do not neglect the importance of trust both in local and central government. The higher trust towards the local government compared to central government is due to the better understanding of political issues and that people feel most politically effective. Much of the actions done by the local government are in a reach of the people, they learn about the distant decisions of the local government leader from first-hand interactions so that people are more likely to interact with them.

The previous studies have been supported the argument of the importance of trust for effective organizations. In the context of this study, we also found that what might hamper the effectiveness and inefficiency of the VOE establishment is due to the lack of trust. The next important question, therefore is, how we foster trust among the village elements, more specifically between the village government officials and VOE management?

To answer that question, first, we must understand the nature of the relationship between village government official (leader) and VOE management? Do they are in the same position or the other is in the upper position towards another? Understanding the nature of the relationship is important to create an effective strategy.

According to the Village Law, the VOE management must be responsible to the Village Leader (*Kepala Desa/Kades)* regarding the VOE activities. It means that the relationship between those two parties is not equivalent. The village leader have a higher position, not only because the VOE management gives a report to them, but also in a sense that the village - leads by the village leader- have greater number of shares proportion compared to other parties. Does, then, the affiliation portrayed investor-management relationship? To the best of our understanding, the relationship between village leader and VOE management should not be treated under the principal-agent theory, where the agent serves the interest toward the agent. For some extent, it might be correct to conclude that type of relationship, that village government shares being put in VOE can be considered as an investment, does put the village as the investor and VOE management as the agent. However, the relationship is more complicated than that. The village leader is placed not only as an investor but as facilitator and advisor.

Financial and Development Supervisory Agency (*Badan Pengawas Keuangan dan Pembangunan/BPKP)* divide the structure of VOE into three parties; advisor, is served by village leader; operational management is the VOE management; and supervisor served by village consultative assembly (*Badan Permusyawaratan Desa)*. As advisor, village leader's job is to provide information and assistance. The village leader may assist VOE management by comunicating with them to understand the situation within VOE, acting as a sounding board, helping the VOE management to explore options to resolve certain situation and empowering the VOE management to decide which option/s that might be best to pursue. Based on this description, the dyadiac relationship between the village leader and VOE management is not only about money (capital) allocation that serves village leader as an investor but also about coordination-assistance between the advisor and the advisee.

To foster the trust in that type of relationship, study by Argento and Peda [1] might be a meaningful source. In the study, they stated that to foster trust, local government can use contracts either as a substitute or complement to trust. However, which can be either be substitutes or complements, is contingent upon the capacity of interacting individuals involved. Six [17] argue that trust and contract are complements, as they can coexist and reinforce each other. Contracts may also serve the various functions of coordination and as a sign of a credible commitment from both parties [20]. The notion of trust as the contract as the complement is also supported by Dekker and Broek [8]. In his study, he admits that trust complement formal control mechanism to a certain threshold. As a legal form, contracts can support trust development within organization [2]. Those arguments speak for contract as the complement of the trust.

Different opinion stated by Vosselman and Van der Meer-Kooistra [19], according to them, contract is a substitutes for trust. They argue that if there is trust between parties, then contract is not needed. However, the use of tight contract is advocated when low or no trust exist. In this context, when trust is in place between parties, then contract is an optional. But the problem with this argument is that there is no single benchmark to measure the trust. We tend to believe that contract complementing the trust, not substituting it. We believe that contract plays the role to foster the trust as it bind the parties to commit with specific roles and obligations.

Not only contractual agreement which seen as the tool to fostering the trust, a study by Becerra and Gupta [3] stressed the importance of communication frequency in the dyadic relationship between trustor and trustee. They argue that when the communication between trustor and trustee is frequent, their relative interest and specific linkage to the organization are now more apparent to both of them and therefore have a greater effect on how both trustor and trustee behave within their dyadic relationship. Argento and Peda [1] agree to this notion. In particular, they stated that interactions frequency among individuals within organization play a role in the dynamics of trust and therefore also impact the contractual needs. When the individuals have frequent interactions, within both official and informal settings, the contract might be loose, on a contrary when there is less interactions then the contract is tight used as a control tool.

We propose a framework for analyzing the relationship between trust and contract. Specifically, when contract can plays as a substitute or as a complement. Framework depicted in figure 1 suggest in-depth understanding of trust-contract as a fostering-trust mechanism can be achieved through examining how the interactions and the competence as variables between trust vis-à-vis contract and add plays as a complement or substitute.

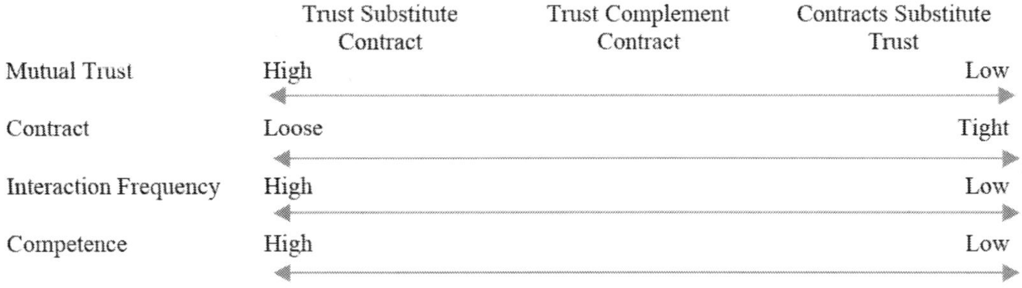

Figure 1.
Framework of the Relationship Between Trust and Contract

4. Conclusion

Our research show that one of the factors that hamper the establishment of VOE is the distrust from the village leader towards the VOE management. This distrust is because the village leader perceived the VOE management is lack of competence. We believe that improving the competence in terms of managerial skills, leadership and business-related skills is important and in terms of fostering trust, contracts can play as a complement toward trust. For a village with a low mutual trust between the village leader and VOE management, contract can place as the substitute. Binding related parties in a contractual agreement may one of a strategic tool to answer the lack of rights and obligations understanding between two parties. Contractual agreement may serve as a solution in which each duties, obligations even punishment and reward are stated. This mechanism, we believe will fill the vacuum of trust within the VOE.

Acknowledgment

The authors appreciatively thank to the BUMDes managers and village leaders in Banyumas region for their support, assistance and information during the research project. The authors also would like to gratefully thank to the Lembaga Penelitian dan Pengabdian Masyarakat (LPPM) Universitas Jenderal Soedirman for funding of this research project.

References

[1] Argento, D., & Peda, P. (2015). Interactions fostering trust and contract combinations in local public services provision. *International Journal of Public Sector Management, 28*(4/5), 335–351.

[2] Bachmann, R., & Inkpen, A. C. (2011). Understanding institutional-based trust building processes in inter-organizational relationships. *Organization Studies, 32*(2), 281–301.

[3] Becerra, M., & Gupta, A. K. (2003). Perceived trustworthiness within the organization: The moderating impact of communication frequency on trustor and trustee effects. *Organization Science, 14*(1), 32–44.

[4] Bone, H. (2017). The effects of financial and non-financial performances towards the managerial performances with interpersonal trust as a mediation variable. *International Journal of Law and Management, 59*(6), 1190–1202.

[5] Brockman, P., Khurana, I. K., & Zhong, R. I. (2018). Societal trust and open innovation. *Research Policy*.

[6] Carter, N., Bryant-Lukosius, D., DiCenso, A., Blythe, J., & Neville, A. J. (2014). The Use of

Triangulation in Qualitative Research. *Oncology Nursing Forum*, *41*(5), 545–547. https://doi.org/10.1188/14.ONF.545-547
[7] Connelly, B. L., Crook, T. R., Combs, J. G., Ketchen, D. J., & Aguinis, H. (2018). Competence- and Integrity-Based Trust in Interorganizational Relationships: Which Matters More? *Journal of Management*, *44*(3), 919–945. https://doi.org/10.1177/0149206315596813
[8] Dekker, P., & Broek, A. (2004). Civil society in longitudinal and comparative perspective: voluntary associations, political involvement, social trust and happiness in a dozen countries. In *Proceedings of the 6th International Conference of the International Society for Third-sector Research: Ryerson University, Toronto*.
[9] Hill, L., & Lineback, K. (2012). To build trust, competence is key. *Harvard Business Review*.
[10] Jamil, I., & Askvik, S. (2015). Citizens' Trust in Public and Political Institutions in Bangladesh and Nepal. In *Governance in South, Southeast, and East Asia* (pp. 157–173). Springer.
[11] Jennifer, F., & Jennifer, W. (2014). The roots of trust in local government in Western Europe. *International Political Science Review*.
[12] Jiang, Z., Henneberg, S. C., & Naudé, P. (2011). The importance of trust vis-à-vis reliance in business relationships: some international findings. *International Marketing Review*, *28*(4), 318–339.
[13] Jones, S. L., & Shah, P. P. (2016). Diagnosing the locus of trust: A temporal perspective for trustor, trustee, and dyadic influences on perceived trustworthiness. *Journal of Applied Psychology*, *101*(3), 392.
[14] Lewicki, R. J., & Tomlinson, E. C. (2003). Trust and trust building. *Beyond Intractability*, 114–139.
[15] Mayer, R. C., Davis, J. H., & Schoorman, F. D. (1995). An integrative model of organizational trust. *Academy of Management Review*, *20*(3), 709–734.
[16] Schlesinger, M., & Heldman, C. (2001). Gender gap or gender gaps? New perspectives on support for government action and policies. *Journal of Politics*, *63*(1), 59–92.
[17] Six, F. (2013). Trust in regulatory relations: How new insights from trust research improve regulation theory. *Public Management Review*, *15*(2), 163–185.
[18] Srirejeki, K. (2018). Empowering the role of village owned enterprises (BUMDes) for rural development: case of Indonesia. *Jurnal Akuntansi Manajemen Dan Ekonomi*, *20*(1), 5–10.
[19] Vosselman, E., & Van der Meer-Kooistra, J. (2009). Accounting for control and trust building in interfirm transactional relationships. *Accounting, Organizations and Society*, *34*(2), 267–283.
[20] Woolthuis, R. K., Hillebrand, B., & Nooteboom, B. (2005). Trust, contract and relationship development. *Organization Studies*, *26*(6), 813–840.

Management Capacity of Educational Office in The Provision of Basic Education Services (Studi in Banyumas Regency)

M Faozanudin[1], LS Sulistiani[1]
[1]Public Administration Department, FISIP Unsoed Purwokerto

Email: muslihfz@gmail.com; lilissfz@gmail.com

Abstract: This paper aims to analyse the local government management capacity in the implementation of education services in Banyumas Regency. The research is based on the perception of the principal against the management process undertaken by the education service District Banyumas, which is carried out by the survey method, with the respondents of the heads of State Primary Schools in the Banyumas Regency. The results of the study indicate that in general the implementation of basic education management at the education office is already underway, and still needs a lot of improvement. Perspective index value The principal of the Office of Education is only 66.96. Some of the problems that need to be improved include: first, there is still a lack of coordination in the process of preparing programs and activities in the education office, which concerns education services in education units. Second, the management decision-making process tends to be slow in the education office, although in some cases, the decision-making process has involved subordinates and stakeholders. Third, there is still a less of transparency in the budgeting process carried out by the education office, especially for the budget allocated for education units, and the fourth, the procedures for the procurement and maintenance of educational facilities and infrastructure are also still less transparent.

1. Introduction

Education, especially basic education, is one of the sectors that has been explicitly included as a government affair that has been decentralised to the district / city government. It is a concurrent affair, -shared responsibility of government affairs for all level of governments, i.e. the central government, provincial government and Regency/city government (Indonesian Act No. 23-2014). Decentralization of the education sector is intended so that the services of basic education affairs become better quality, and bring services closer to the community {15,18} Similarly, when decentralization can be applied well in the field of education, it will improve the quality of education services {8,24).

The management capacity of local government, in the other hand, is very crucial for implementation the duties and functions, especially for giving a good decentralised service to the community (25). It can be defined as a process, and the ability to achieve goals. As a process, management capacity is interpreted as the ability of an organisation to implement, develop, and control every task and programs to achieve its vision and mission and its objectives effectively{ 1, 2, 4, 12}.

In the public sphere, management capacity means an activity to process public policy, and the ability to get an act together, to institute structure, the routine, and coordinated efforts of talented people to convert a policy massage into a set of real achievement "{6}. Thus, the capacity of local government management can be elucidated as the ability to make maximum use of human and other resources (finance, infrastructure, and technology) by implementing the management principle function to carry out their duties and functions in accordance with the stipulated policies to achieve organisational goals effectively and meet the needs of stakeholders.

The implementation of basic education services by regency has actually been going on for a long time, but the quality of provision basic education service has not changed significantly yet. Education services in local level still faces many problems, ranging from the lack of resources, e.g., human resources (teachers and non-teachers) to operational budget as well as educational infrastructure. The fulfilment of teachers, as the main resource in the teaching and learning process until now has also not been fulfilled. Most of them actually filled with honorary teachers. Likewise with the problem of educational infrastructure, there are still some schools that do not meet the Service minimum standard (SPM), those are the lack of laboratory appliance, lack of library room for supporting learning process, etc.

Another problem related to basic education is a limited funding for school operations. School operational budget is now allocated to Budget aid for school operational named BOS, which are set limited and very rigid so the school. Therefore, sometimes difficult for using school budget when out-listed-affairs need to be funded.

Based on this fact, it is urgent to discuss the capacity of local government management in providing good basic education services.

This paper is aimed at describing the condition of the capacity management of Banyumas regency in providing educational service.

2. Material and Method

This research was conducted in Banyumas regency, especially the Banyumas education office as the party entrusted with the responsibility to carry out the functions of administering basic education services. Quantitative survey was applied. Quantitative data was collected by questionnaire, which based on respondent perception. The head of elementary school as object of the research give assessment for the process of management capacity conducted by the educational office. The number of them were 142 respondent who had chosen through proportional random sampling from total number of 804 of populations. The area of research widespread out of 15 districts in Banyumas regency. The collected data, then analysed by descriptive statistic, applying simple model i.e. mean, median and modus.

3. Result and Discussion

Banyumas regency has about 745 public elementary schools and 40 private elementary schools that employe about 7,355 teachers. These consist of 5,045 males and 2,310 females. From this number also consisted of 5,031 (68.4%) teachers with civil servant status, the remaining 2,327 (31.6%) were honorary status. With the large number of schools being managed, adequate management capacity is needed, so that basic education services and activities can run well

Based on descriptive analysis, the results showed that the management capacity index of the Banyumas Regency Government in the provision of educational services amounted to 66,96. (sufficient category). this category shows that the management capacity of the local government is very urgent for improvement. (the detailed result of the research is presented in table 1) management capacity index of local governments in the provision of education services is 66.96 indexes with sufficient catagories. The results of the complete analysis are presented in table 1 below.

Table 1. Distribution of the Management Capacity index

No	Dimension	Value	Indeks	Category
1	Planning and Program Process	482,72	68.00	Sufficient
2	Decision making process	581	66,8	Sufficient
3	Coordination and Communication	553	77,9	Good
4	Clarity of education infrastructure procurement procedures	416	58,6	Sufficient
5	Monitoring dan evaluation	519	73,2	Good
			66,96	Sufficient

Source: Primary Data Processed: 2016

From the five dimensions studied, it was seen that the highest index score was in the coordination and communication dimensions with the value of 77.9, while the lowest index score was the dimension of the process of procurement of educational facilities and infrastructure with a value of 55.8. In more detail, the discussion of the five dimensions is as follows

Firstly, the dimension of planning as a process that must be done at the beginning of each activity was only get an index score of 68.00, with enough categories. This value indicates that the capacity of the Banyumas District Education Office in the process of preparing the activity plan has not gone well. The results of interviews with respondents related to the activity planning process are known that many of them feel that they are often not directly involved in the activity planning process. This condition causes the range of planning made by the agency, on the aspirations and needs of developing basic education, which are not well-established. There are still many things that must be resolved related to education problems in Banyumas Regency, starting from the condition of infrastructure, learning systems, to the problem of teachers as educational resources

Secondly, coincide with the planning dimension, the deccsion making process was also in sufficient category, the index was just 66,8. These results indicate that, officials in the education office in Banyumas Regency in the decision making process were still considered to be not fast enough. The chain of bureaucracy is still considered to be the thing that causes the process to delay. The results of interviews with informants at officials at the education office stated that, the speedy decision-making process in handling a problem with the education office couldn't be separated from the instinctive condition of government bureaucracy that tends to be 'obedient' because of the hierarchy that must be passed. the construction of government bureaucracies that adopt the Weberian model, which implements hierarchical-authoritative organizational structure, with a high level of formalization, is indeed a problem in the speed of the decision-making process. The leader as a strategic apex is very vital, because the position as the person in charge of the entire organization, while other bureaucratic personnel who are in the lower layer are in a limited position for decision making {22}. With such conditions, it often causes the slow decisions made by officials.

Ideally, the decision-making process in democratic organization use participation model, especially if the problems decided are related to public interests, including asking subordinates to participate in the decision-making process. The involvement of subordinates or staff in an open retrieval process is the best approach.

Thirdly, the management capacity in coordination and communication conducted by the Banyumas District Education Office received a 'good' rating with an index value of 77.9. These results illustrate that coordination and communication functions have been carried out well. Coordination and communication in the implementation of work has a strategic function in the management process, especially to equalize perceptions of work. The same understanding of officials and staff about work to be done will facilitate the completion of tasks and work. Good coordination and communication will result in harmonization of work carried out in teamwork, so that work problems can be immediately resolved.

Completion of work in an organization is a job that requires a strong teamwork. Many jobs can only be completed in collaboration with many people. A leader will be very difficult if only rely on one person. Therefore, the function of coordination and communication both between personal and units within the organization has an influence on the quality of work results.

Fourthly, In the management of procurement of goods and services, there are 2 main things that need to be considered, namely the Standard operating procedure (SOP), and the availability of budget which prepared for the realization of procurement of educational facilities. The results of the study show that the index score for this dimension is assessed as only a category: "sufficient" with an index score of 58.6. For the first, it could be illustrated that the systems and procedures related to the procurement, maintenance of facilities and infrastructure conducted in the educational regency office could be "less transparent". This condition was be reinforced by the result of the study and interviews with respondents, which stated that the Education office was still considered to be less informative in providing educational facilities, both in the form of educational equipment, teaching aids and other school facilities. Many respondents felt that they had applied for educational facilities and infrastructure, but information about status from demand was often unclear. Conversely there are some respondents who state that the school does not apply for teaching aids and equipment, but obtains educational goods.

For the second, the limited budget provided by Banyumas local government for supporting operational educational causing the allocation of it to all primary schools could not be equally. Sometimes the respondents feel that they are struggling to ask for a budget for the provision of school supplies. Regularly, all of the level government,-central, province and regency government-, simultaneously provided budget according to their respective proportions. Unfortunately because of the limited budget owned by the Banyumas district government, not all primary schools get an allocation for the use of school facilities and infrastructure. Budget assistance originating from the local government, is more intended for school operational activities, with a size of around 5 to 10 million, depending on the condition of the school.

Based on the results of the analysis it can also be said that, when the financial capacity of the local government is still limited in financing educational activities, it has an impact on the lack of attention to educational facilities and infrastructure at the elementary level. Local governments rely more on financial assistance to repair facilities and infrastructure from the central government and the provincial government. Hadna {7} staled that, For regions that do not have sufficient natural resources sufficient to support financing, ultimately more burden and depend entirely on subsidies from the central government in the form of „General Allocation Funds (DAU).

This condition can also illustrate that the implementation of regional autonomy and decentralization in Indonesia has not yet been accompanied by regional independence,

 https://doi.org/10.1051/shsconf/20208601043

especially in the financial sector, so that public services which are delegated to the regions tend to be lack of optimal {10}. Likewise seen from the process of hand over of authority and responsibility, the Central Government still seems half-hearted in giving authority to the regions, namely the authority given to the regions is not proportional to the financial resources and other resources submitted to the regions, so the implementation of regional autonomy in particular in the field of education has not run optimally. Most of the potential financial resources are controlled by the Central Government, so that local governments are unable to explore more financial resources. Therefore, when the decentralization process continues as above, it is difficult for regions to be more independent in implementing decentralization. {23}.

Finally, the results of the analysis of the dimensions of monitoring and evaluation carried out by the regional education office on programs and activities in the elementary school were categorized as "good" with an index score of 73.2. These results indicate that the capacity of the education office in carrying out the function of evaluation and monitoring of activities carried out on activities still needs to be improved.

Monitoring and evaluation are important principles in the management process. Monitoring is an activity to monitor the implementation process of a program and activity in an organization, while evaluation is an activity to assess whether a program and activity has reached the desired goal of the organization. The evaluation process is carried out according to certain terms of an activity process. Evaluation and monitoring functions to maintain that planned programs and activities can achieve the targets as planned and when problems occur in the implementation of activities and programs, can be immediately above.

The results above indicate that, there is no synchronization between the function of monitoring and evaluation with follow-up activities. Feedback related to monitoring and evaluation of the implementation of programs and activities is very important to be done to solve the problem. As a result of these conditions, there are several problems that have not been completely resolved by the Regional Government, for example the lack of teachers. Many schools found that most of the teachers were still in honour status, with very low salaries.

Having been analysed as above, can be said that the management capacity for local government organizations is very important for the effectiveness of the implementation of policies and programs, this in line with the Solanke {21} and Pope *et al* {16} stated that "the capacity of an individual organization to deliver programs or services that are funded, and to the required standards, will have strong influence on their subsequent in meeting the outcomes sought by government ". And it is reinforced by Solanke (21) that organizational capacity is very important for the effectiveness of the implementation of policies and programs. it includes a broad range of attributes, including resource sufficiency and stability, professional skill, management competence .... " {25}

The capacity of local government organizations can show the level of independence of local government organizations themselves. The higher the capacity of local government management, the higher the level of independence of a region. This is in accordance with the objectives of regional autonomy, namely that the regions become independent (self standing) (10). To carry out the autonomy function, the region must have "the discretion to change in fiscal, functional and organizational activities without restraining from higher levels of government; and the capacity or means to achieve their policy and governance preferences {25}. Independence can be seen from the ability of the area in (1). Meet all the needs needed to manage the household, which means having adequate resources. Like an adequate budget, human resources are capable of carrying out the tasks of autonomy. (2) managing and managing their household affairs in an accountable and accountable manner to all parties (stakeholders from the local government). The extent of this responsibility is because the concept of the institution from the name of the state as a public organization is different from the concept of other institutions, not only from the scope of its activities, but also related to the authority relating to its citizens. Ideally the autonomous region must be able to explore the resources it has, as financial resources that are used to manage the financing of the administration of its regional government {11}.

4. Conclusion

Based on the results of the analysis in general, it can be concluded that the capacity of the Banyumas education office management in administering education services still needs improvement. This can be seen from a number of things, Firstly, because in the process of drafting program plans and activities in the education office the principals were not directly involved, so many people said that they did not understand the strategic plans made by the education office. The strategic plan made by the education office is deemed unable to reach all that is needed in the education sector in Banyumas Regency. Secondly, the management decision-making process at the education office is still considered slow, although in some cases, the decision-making process has involved subordinates and stake holders. Thirdly, judging from the organizing process, the regional government and the education office have done quite well, the function of coordination and monitoring has also been going well, although feedback on the results of monitoring is still considered slow. Fourthly, the procedure for the procurement and maintenance of educational facilities and infrastructure is still unclear and transparent. There is less clear information about the criteria and categories of conditions so that the facilities and infrastructure of the school can be obtained. The limited budget that is owned by the local government has to be an obstacle in the implementation of quality basic education services.

References

[1] Chaskin, Robert J. (2001). Building Community Capacity: A Definitional Framework and Case Studies from a Comprehensive Community Initiative. Urban Affairs Review 36(3): 291- 323.

[2] Connolly, Paul and Carol Lukas (2008) 'Component of Organization Capacity' dalam http://www.fieldstonealliance.org/client/articles/Article-Components_Org_Capacity.cfm, diunduh, 12 December 2011

[3] Creswell, John W, (2009). Research Design: Qualitative, Quantitative , and Mixed Methods Approaches, Third Edition, California: Sage Publication, Inc.

[4] Eisinger, Peter. (2002). Organizational Capacity and Organizational Effectiveness Among Street-Level Food Assistance Programs. Nonprofit and Voluntary Sector Quarterly, 2002 31(1): 115-130. Online http:// slg.sagepub.com/content/39/3/144

[5] Gazley, Beth and Christensen, Robert K. (2007). Salvaging Capacity's Usefulness: Analysis of Meaning and Measurement, Paper presented in 'Ninth Public Management Research Association Conference', October 25-27 October 2007. USA).

[6] Goggin, Malcom L, et.all. (1990). Implementation Theory and Practice: Toward Third Generation, Scott, Foiresman/Little, Brown Higher Education, Scott, Foresman and Company, USA

[7] Hadna, Agus Heruanto, (2010). Simbiosis Mutualisme Antara Birokrasi dan Politik di Daerah, dalam, Kumorotomo dan Widianingrum, 2010, *Reformasi Aparatur Negara Ditinjau Kembali*, Yogjakarta; Gava Media

[8] Hanson, E. Mark. (1997). Educational Decentralization Issues and Challenges, Working Paper No. 9: http://www.the dialogue.org. /publication files/ PREAC%209-English.PDF. (Unduh, 13 October 2012)

[9] Horton, Douglas, et. all. (2003). Evaluating capacity development: experiences from research and development organizations around the world. The Netherlands: International Service for National Agricultural Research (ISNAR); Canada: International Development Research Centre (IDRC), the Netherlands: ACP-EU Technical Centre for Agricultural and Rural Cooperation (CTA).

[10] Kartiwa, A dan Nugraha, (2012). Mengelola Kewenangan Pemerintahan, Bandung: LePSinDO

[11] Koswara, E. (2001) Otonomi Daerah Untuk Demokrasi dan Kemandiran Rakyat, Jakarta: Yayasan Pariba

[12] Letts, Christine, William P. Ryan, and Allen Grossman. (1999). High Performance Nonprofit Organizations. New York: John Wiley.

[13] Lusthaus, Charless, et,.all, (2002). Organizational Assessment. A framework for improving performance. Ottawa: Inter-American Development Bank, Washington DC and International Development Research Centre

[14] Morgan, Peter, (1997). The Design and Use of Capacity Development Indicators, Hull: Policy Branch, CIDA.

[15] Pollitt, Christopher. (2007) Decentralization: A Central Concept in Contemporary Public Management, dalam Ferlie Ewan, Lauranence Lynn nad Christopher Pollitt, 2007, The Oxford Handbook of Public Management, Oxford: Oxford University Press.

[16] Pope, Andrew, et. all.. 2012. *Capacity Development For Indigenous Service Delivery,* Canbera: Australian National Audit Office: Canbera: http://www.anao.gov.au, di unduh 12 Mei 2012.

[17] Rohdewohld, Reiner dan Manfred Poppe. (2005). Guidelines on Capacity Building in The Region : Module A: The Capacity Building Cycle From Capacity Building Needs Assessment (CBNA) Toward the Capacity Action Plan, GTZ-SfDM (Support for Decentralization Measures), Version 2.0, SfDM Report 2005-2

[18] Rosenbloom, David H., Robert Kravchuk and Richard M. Clerkin. (2009). Public Administration: Understanding Management, Politics, and Law in the Public Sector, 7th Edition, New York: McGraw-Hill International Edition.

[19] Rubbin, Irene S, (2010) .The Politics of Public Budgets, dalam Richard J Stillman II, 2010, Public Administration: Concepts and Cases, Ninth Edition, International Edition, Wadsworth Cengage Learning, USA

[20] Singarimbun, Masri, dan Sofyan Effendi, (1989) . Metode Penelitian Survey, Jakarta: LP3ES.

[21] Solanke O, Eziyi O Ibem, (2011). Assessing Organizational Capacity in Housing Provision: A survey of Public Housing Agencies in Ogun State, Nigeria, Journal of Social and Development Sciences Vol 2 No.6 pp 275-288.

[22] Wicaksono, Ari, (2010). Peluang Penerapan *Servant Leadership* dalam Birokrasi Publik, dalam Kumorotomo dan Widianingrum, 2010, *Reformasi Aparatur Negara Ditinjau Kembali*, Yogjakarta; Gava Media.

[23] Widaningrum, Ambar. (2010). Decentralisasi, Kapasitas Daerah dan Pengelolaan Jaringan alam Penyelenggaraan Pelayanan Publik, dalam Wahyudi Kumorotomo dan Ambar Widaningrum, (Editor), Reformasi Aparatur Negara Ditinjau kembali, Yogjakarta: Gava Media dan MKKP dan MAP, UGM

[24] Winkler, Donald R. and Boon-Ling Yeo, (2007). Identifying the Impact of Education Decentralization on the Quality of Education (Working Paper), Washington, D.C.: Educational Quality Improvement Program 2(EQUIP2), United States Agency for International Development (USAID), Academy for Educational Development (AED)

[25] Wolman, Harold; Robert McMAnmon, Michael Bell, and David Brunory, (2008). Comapring Local Government Autonomy, *Proceeding of The 101st annual Conference on Taxation:* National Tax Association, Washington, 2008,

[26] World Bank, (1998) Assessning Aid: What Work, What Doesn't, and Why, Oxford: Oxford University Press.

[27] http://www.fieldstonealliance.org/client/articles/Article-Components_Org_Capacity.cfm, downlaoded, 12 Desember 2011

Gender Mainstreaming Through Woman Political Participation at Representative Council in Banten Province

I A Indriyany[1], M D Hikmawan[1], M R Godjali[1]
[1]Sultan Ageng Tirtayasa University
Email: ikaarinia@fisip-untirta.ac.id

This research aims to observe gender mainstreaming by women representative council in Banten Province. To analyze the process of becoming, Gender mainstreaming Become Several important for reasons, first, the equity between the Human Development Index and the Gender Development Index, Banten Province indicates that human development is not always in line with gender-based development. Gender-based development is Often neglected because development is too focused on human development. Second, women are still stereotyped by gender Injustice. In the local context of Banten, the injustices experienced by women are legalized through social, cultural and religion institutions. Third, in the political realm, gender mainstreaming is not yet be maximized. This is because the affirmative policies for women are not in line with women awareness to be Involved in public affairs.

1. Introduction

Gender inequality in Indonesia is still a phenomenon that is encountered both in the lives of families, communities up state. Gender inequality as a result of gender inequality is reflected through the existence of a disparity in access and opportunities to acquire development resources between men and women in various fields that have an impact on participation and also controls each of the parties which resulted in the difference between the benefit of the results of development. Gender inequality seen through a data Gender Development Index (GDI), which has yet to reach 100%. Gender inequality in the field of development visible through a gap between the Human Development Index (HDI) with the IPG. HDI itself is an index that is used to view the progress of quality its trend of human life increases from year to year. This can be seen through the following table:

Table 1 Comparison of HDI,GDI and IDG on BPS

	2010	2011	2012	2013	2014	2015	2016
PM	66.53	67.09	67.70	68.31	68.90	69.55	70.18
PG	65.30	66.38	66.77	67.20	67.80	68.52	69.60
DG	61.80	62.27	63.13	64.27	66.10	68.20	69.20

Inequality between HDI, GDI and IDG indicates that the successful development of human resources even though each year has increased, but not fully followed by the success of gender development. Efforts are being made to eliminate gender inequality is one of them is with the Republic of Indonesia issued Presidential Instruction No. 9 Year 2000 on Gender Mainstreaming in National Development. Gender Mainstreaming or often referred to as mainstreaming is a strategy that the Government sought to achieve gender equality by integrating the interests, aspirations and conditions of men and women in every stage of development management from planning, implementation, monitoring and evaluation. In PUG at least things are implied, the integration of the interests, aspirations and conditions of men and women in the development process, especially in the decision making process. Both the realization of gender equality means that men and women can equally enjoy the fruits of development.

Implementation of gender mainstreaming in development in Indonesia is a political move to minimize the issue of gender inequality, gender mainstreaming strategy relies on two approaches namely the First Government as an agent of change for development that can justice and gender equality. Secondly, the Government is involved into any PUG implementation process from planning, monitoring and evaluation.

In addition to the Presidential Instruction No. 9 of the implementing gender mainstreaming policies at both national and regional level, there are several laws issued by the government to support efforts in promoting Gender Equality. Legal products such as; Regulation of the Minister of Women's Empowerment and Child Protection No. 6 of 2009 on the Implementation of Data Gender and Children, Ministry of Home Affairs Number 67 Year 2011 concerning Amendment to the Regulation of the Minister of Home Affairs Number 15 Year 2008 on General Guidelines for the Implementation of Gender Mainstreaming in the Regions and the Joint Circular of Ministry state Development Planning, Ministry of Finance, Ministry of Interior and the Minister of Women's Empowerment and Child Protection No. 270/M.PPN/11/2012, Np.SE-33/MK.02/2012, No.050/ 4379 / A / SJ and No. In line with the Decree of the Minister of Interior No. 132 of 2003 on the General Guidelines for the Implementation of Gender Mainstreaming in Regional Development and Regulation Home Minister No. 67/2011 on General Guidelines for the Implementation of gender mainstreaming in the area, where in the Regulation instructed all units of Local Government under the coordination of the Agency Regional Development planning or the Agency for Gender Mainstreaming can integrate into the planning and budgeting. With the regulation is that each region in each line both at the provincial and district / municipal needs and are required to have a Local regulations or Programs on Gender Responsive area.

One success indicator of the quality of life there is the concept of the Gender Development Index (GDI) and Gender Empowerment measurment (GEM) are both quality of life based on gender equality. The phenomenon of persistence of the practice of gender injustice be obstacles to achieve equitable development and gender equality, which should both women and men have the same access to and participation in various areas of development

In Banten, inequality itself is visible through Average School graduating. male students - older men (8.90 years) when compared with women (7.82 years) Education attained the majority of women are also just the average only up to junior high school (66.6%), while men finish up high school (91.4%). Wages in the field of employment for men are also more than women by a ratio of 0.83%. in political affirmative policy of 30% quota for women are not able to be met because in the DPRD, female members only reached 18.8%. It is certainly a challenge for the members of these few women in parliament should be able to represent the interests of women in the legislative process. It is important to look at gender mainstreaming in accordance presidential instruction No. 9 of 2000 through portraits of women's participation in parliament.

This research is important to do because it is gender inequality still be found clearly. Strong patriarchy culture and religious values makes no other woman only has a function as a companion men without ever be seen as independent subjects who have authority over her body. In the political view, for example, the representation of women in the DPRD only 17 of the total 85 members, or only 10.8%. 17 represent the number of females in Banten province, which reached 5,981,508 inhabitants. Representative challenges brought these women are how to bring the interests of women in parliament to struggling their agenda.

The mainstreaming gender into the idea that no longer simply stop at the national policy level, but scaled back into policy development in the area. However, it is constrained by the level of political participation is still low for women in Banten. In Banten gubernatorial election in 2017, for example, from 3,767,852 female voters list, only 2.50963 million inhabitants are voting or about 66.6%. Challenges of women in Banten even greater because in addition to a patriarchy culture that must be dismantled its construction; the strong religious values also narrow the participation of women in the political sphere. Kiai and the winner have a very important position in the political agenda in Banten {3}. This research is interesting to study because the researchers attempting to capture gender mainstreaming conducted by a representative number of very little amid siege partiarki cultural and religious values were very strong.

Specifically, the study aims to look at some things start to explain the gender disparity in Banten. The initial data is then used to describe the political participation that are run by women legislators in the DPRD. Political participation data is used to elaborate further the political participation of women in implementing gender mainstreaming in the DPRD

2. Materials and Methods

Representation is making the voice, opinions and perspective present in the public policy process. As a modern form of democracy, representation is important to do to realize democracy Schumpeter's minimalist style. Western-style democracy is interpreted as a competitive mechanism for selecting leaders through institutional procedures in reaching a political decision by giving power to the individual to make decisions in order to gain the popular vote. Representative democracy is considered as the main starting point for democracy deemed not to be able to walk even when all the people govern, because according to Schumpeter will of the people is something that is very ambiguous. Therefore, the representation becomes important.

The representation is related with the guarantee of the ownership rights of every citizen as a representative has an obligation to bring the interests of the citizens they represent. The main element that must be present in the representation is a representative (who represent), *the represented (represented)*, interests are represented and socio-political context or setting. This element is the main element that no one of them then the representation will not be able to run well. Representation is divided into four forms.

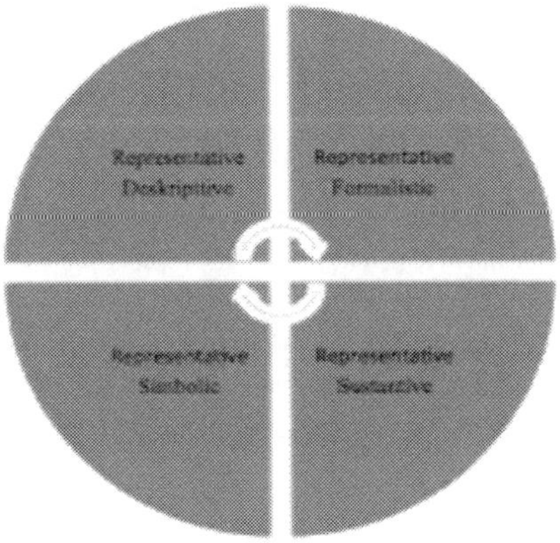

First, formalistic representation, understood in two dimensions, authority and accountability. Authority relating to any authorization granted by the constituents on the deputy, so that when a deputy to do something beyond his authority then they are deemed to no longer being run representation function. The second dimension is accountability describes the obligation of accountability of representatives to the constituency on the authority that has been given. This accountability involved by representatives does.

Second, descriptive representation, this representation requires representatives from the groups represented by (standing *for representative).* So it is regarded as the ideal representative is a representative that reflects the groups of people according to which it represents. Although it is sometimes representatives are not inherently do something for the benefit of the group they represent because they did not intend to act on behalf of the group. Third symbolic representation where the representative of a symbolic representing the group represented.

Fourthly, substantive representation. This form of representation is currently acting representative for (*act for)* in order to bring the interests of the represented to the realm of public policy. This representation refers to the quality of representation in which people are given the authority to represent the act in the person who is represented with full responsibility. This pattern moves parties represented, not just resemble or identified as in descriptive and symbolic representations.

This representation in the study is used to capture the gender mainstreaming in the policy making process in the legislative sphere. Gender mainstreaming appears because there is gender inequity. Injustice to women is reflected through the act of nominating women only because of their gender, such as marginalization, sub-coordination, stereotyping, violence and exploitation. Marginalization is a process of marginalization carried out by the community in order to hamper access to life. One form of marginalization is the closure of access to information and knowledge and limited space for democracy for women.

Stereotype is a labeling process in certain groups based on construction thinking based on life experience institutionalized for years. The effects of the labeling process are acts of subordination and discrimination. Labeling and stereotyping are the oppressive model in discrimination {4}. Subordination is the nomination of women only because of their gender who are considered not able to do certain jobs. This subordination will lead to discrimination, namely the limitation of opportunities and opportunities for women. Meanwhile women are also vulnerable to violence. Violence in women is perpetuated through cultural and structural factors. The cultural factor sees that the violence that befell women is due to the existence of values and norms that place men as parties who have full power over women so that legitimate men to act as arbitrarily as women. While structurally,

Violence is lasting because there are both formal and informal structures that perpetuate it. Gender mainstreaming is here to eliminate this imbalance. Because as a strategy to ensure gender equality, this principle offers a framework for understanding gender, namely a gender analysis framework in which the framework is based on awareness of possible differences in capacity, potential, aspirations, interests and needs of men and women at each stage of the development process.

Political Participation

Political scientists, formulating concepts and theories of political participation are behind the conditions of political culture that develop in society. Political culture as a pattern of individual attitudes and orientations towards politics in an internalized political system. Political culture provides regular subjective boundaries that underlie and give meaning to political actions.

Political culture in society consists of parochial culture, namely apathy towards government and feeling uninvolved, subject culture is the concern for politics but not balanced by participation in political processes, and participant culture in the form of individual active behavior and directly involved in the policy making process. Political culture as providing guidelines for effective political behavior control for individuals, as well as providing a systematic structure above. values and rational considerations that ensure the coherence of the performance of institutions and organizations. Basically political participation itself has the meaning as an activity of a person or group of people to participate actively in political life including choosing the leader of the country, and directly or indirectly influencing government policy, voting in elections, and others {2} .

Political participation activities can be explored from the following indicators:

1. Activities that can be observed, not just attitudes
2. It is the activity of individuals in the role of citizens, not a professional activity in the political field
3. Directed to influence government decision making both legal and illegal activities
4. Effective and failed activities categorized as political participation {3}

Initially political participation was limited to activities such as voting, political discussion, campaigns, forming and joining interest groups, and communicating with political officials. This is commonly referred to as conventional participation. In the late 20th century, the definition of political participation was expanded to include non-conventional participation activities including the submission of petitions, demonstrations, confrontations, mass protest movements, political violence on property, political violence in humans, guerrilla warfare, and revolution. Political participation is carried out as a symbolic reflection on the existence of the political system and on political events that have and will occur. Political participation provides an overview of the level of awareness of individual behavior in response to good socio-political symptoms caused by the application of government policies and consequences of the global situation.

The research method used in this research is descriptive qualitative method. This type of research will produce descriptive data about oral and written words and behaviors of the people in question. While the results of his research are descriptions of an internal phenomenon the context is different so the conclusions of each individual will differ depending on the point of view used. While the approach used is the phenomenology approach.

Especially hermeneutic phenomenology is the choice of methods that researchers use, because hermenutic phenomenology opens up space for writers to carry out interpretation research the problems studied. Hermeneutic phenomenology is used as a critical-reflective and interpretative analysis of the text.

Women in this study as a manifestation of political relations as a text to be interpreted, are not only interpreted as numbers, but further research is expected to reveal how women transcend power structure activities in the political reality of society so that they get equal opportunities in government politics.

The data collection technique used was in-depth interviews with key informants, namely members of the Banten Provincial DPRD, observations related to the representative ways women brought women's interests during the legislative process and recess period, and literature studies related to documents that were related to gender equality in Banten Province. Starting from policy documents that have been, are being or will be formulated and also libraries related to the concepts used.

3. Results and Discussions

The issues of Gender Inequality in Banten are still a problem in socio-political discourse in Indonesia. The discrimination and marginalization of women still often occurs in various fields: economic, social, cultural and political. Whereas when talking about gender equality, what should appear is a situation where men and women are in an equal position and have equal access in both the public and private spheres. Gender is not interpreted as a difference in terms of gender biologically, but is a difference in roles, functions, status and responsibilities in society.

Gender discrimination is a phenomenon, which is generally spread across the region, is no exception in Banten. Provincial formed since 2000, it still leaves the issue of gender inequality and marginalization of women. It can be seen from the Gender Development Index (GDI) in Banten Province. Based on data obtained from the Central Statistics Agency (BPS) Banten Province in 2016, figures Average Length of School (RLS) between men and women still show the gender inequality. As can be seen in the following figure:

Figure 4.1 On average Old School (RLS) Source: BPS Bantam, 2016

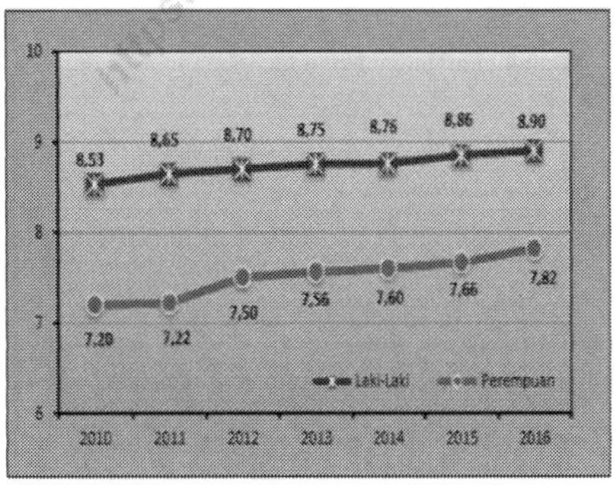

Based on the above data indicate that the Banten province still has serious problems regarding gender equality. In the field of education these data confirm that the gender gap to occur within a relatively long time without any significant change. Although these data describe the increasing number Average Length of School (RLS), among both men and perempun every year, but if we look in the eyes of gender equality, the data actually showed a gender imbalance, even this happens continuously every year. In 2016 showed the male RLS amounted to 8.90 while the female RLS by 7.82. That is, the gender inequality index in Banten in the field of education reached 1.08.

That is the state in this case the government is obliged to bring access to education for the widest possible with due regard to the principle of gender equality. Images in Banten gender equality in education confirm that the educational rights of citizens in Banten have not been fully met.

Gender disparity in education fields back emphasize that the existence of patrilineal culture work in the structure of society. Not only in the aspect of education, gender inequality in employment occurred in Banten. Gender equality requires the equal position and equal access of men and women in various fields, but in the field of employment, women are often discriminated against and positioned as subordinate to men. Companies often use the consideration of gender bias in recruiting workers. Gender stereotypes are constructed eg women workers should be attractive, willing to marry and give birth during the period of work.

In addition to these actions not only harm the spirit gender equality, but also human values (human rights) woman. Despite a ban on all forms of discrimination in employment, including gender discrimination is already contained in Law No. 13 of 2003 on Labor, but in practice it shows the opposite.

Gender-based discrimination in employment that occurred in Banten can be seen from the system of wages (salaries) are not equal between men and women. In many perusaahaan in Banten, the wages earned by women was lower than men without constituted by the reasons and considerations are obvious. Consider the following picture:

Figure 4.2 Average wages / salaries Net per month Full Banten, August 2010-2016

Source : BPS Banten, 2016

Periode	Upah (Rupiah)		Rasio Upah Perempuan Terhadap Laki-laki
	Laki-Laki	Perempuan	
(1)	(2)	(3)	(4)
Agustus 2010	1 789 034	1 370 610	0,77
Agustus 2011	1 922 067	1 465 094	0,76
Agustus 2012	2 068 546	1 631 239	0,79
Agustus 2013	2 537 784	2 060 932	0,81
Agustus 2014	2 533 041	2 104 649	0,83
Agustus 2015	2 675 104	2 264 734	0,85
Agustus 2016	3 755 778	3 111 622	0,83

In the picture above table shows a payroll system anatara unequal male and female. Unequal payroll system confirms that there is still a gender gap in the realm of employment, even this condition continues to occur each year as evidenced by the data in 2010 to 2016 women's wages are always in the second position under the male. Wage ratio of females to males increased every year.

This means that there is no significant change in the wage system, where women have always been a subject who experienced discrimination in the workplace. a married woman worker requires special treatment because they have to pass through pregnancy and childbirth as a natural natural (biological) that women can not be avoided. This erroneous stereotypes in the end while keeping the values of patriarchy. Unfortunately this patriarchy culture continues to be maintained because it can Discrimination was also evident when working women who are married and still lajing treated equally in terms of wages, while the male worker who is married and single reactors are treated differently.

Unfortunately this patriarchy culture continues to be maintained because it can Discrimination was also evident when working women who are married and still lajing treated equally in terms of wages, while the male worker who is married and single reactors are treated differently. In fact, a married woman worker requires special treatment because they have to pass through pregnancy and childbirth as a natural (biological) that women cannot be avoided. These erroneous stereotypes in the end while keeping the values of patriarchy. Unfortunately this patriarchy culture continues to be maintained because it can married women worked require special profitable *the status quo* because if women can feel gender equality in terms of remuneration, the position of men feel threatened because it can reduce their income because of budget politics run by the company does not put men as high wage earners.

Turning to another issue, the gender disparity in Banten also extends to politics. Most of the public views politics is the domain of men that women are considered inappropriate for practical politics. As a result, women's representation has not been able to work optimally. In the realm of parliament (DPRD Banten), for example, although it has been applied *affirmative policy* for women through the quota policy of 30% representation of women in parliament, but this policy has not been fulfilled. Of the 85 seats in DPRD Banten, only 18 women MPs (supposed to be 21, the minimum standard of 30%). Of course this has implications for the lack of pro-gender policies products produced DPRD Banten. Yet mainstreaming gender is important as a precondition for non-attendance equality of rights of women and men. In the realm of executive showed a slightly different pattern. The political leadership of women who served as head of the region in Banten province is quite a lot. From 8 districts / cities in the province of Banten, 4 (four regions) were led by a regent / mayor women. Consider the following table:

Table 2: Women Political Leaders in District /City in Banten Province

No	Kabupaten/Kota	Pemimpin Perempuan	Jabatan
1	Kabupaten Pandeglang	Irna Narulita	Bupati
2	Kabupaten Lebak	Iti Oktavia Jaya Baya	Bupati
3	Kabupaten Serang	Ratu Tatu Chasanah	Bupati
4	Kota Tangerang Selatan	Airin Rachmi Diany	Walikota

Source: Researcher, 2018

Based on the data presented in the table above explains that 50% of the number of districts / cities in Banten province is headed by a regent / mayor women. This means that the representation of women in political office in Banten is adequate. Nevertheless, the issue of gender inequality has not been a central issue in the regions they govern. This is because they do not focus fight for gender equality to then bring in a product that puts the political policy of gender mainstreaming. Four female political leaders in Banten not born crackled movement or ketertarikaan on women's issues. In the election dispute, they do not depart from the populist leadership from the grassroots level.

Four political leader who served as regent / mayor in the province of Banten in following successful contestation of the election to finally become a political official women can not be separated from the power source (*resources*) owned by the family. Political office achieved a leadership relay or inherited from family. Irna Nurilta a buapati Pandeglang is the wife of Dimyati is a former regent of Pandeglang two periods. Iti Ocativia Regent of Lebak Jaya Baya is the biological child of Jaya Baya who also served as a former regent of Lebak two periods. Similarly, Tatu Chasanah who served as regent Serang district is the younger brother of Atut Chosiyah, former governor of Banten, as well as Airin, a mayor of South Tangerang, still has a kinship with Atut Chosiyah. The linkage of women's political leadership,. That is the agenda of the struggle for gender equality is still far away.

Images gender gap in Banten as mentioned above cannot be separated from the social setting of Banten, which in turn restrict the movement of women to appear in public. Banten social settings are very close to the values of Islam. This is not only because the majority of the population Banten Islamic religion, but historically, Banten sultanate had experienced during the reign of the character of the religion (Islam).

Similarly, the government of Banten use Islamic slogans as regional identity, such as the Serang city Devoted, Tangerang City, "*berakhlakul Karimah*"(good personality) in Pandeglang, social setting Banten known have narrowed this Islamic movement of women. Islamic teachings that tend to promote the role and position of men in both the private and public sectors.

Consequently, women are increasingly marginalized. Condition this is often used by men to maintain their existence without the threat of domination of women. It can be seen from the exploitation of religion for political purposes. For example, create a social construction that women are prohibited from being a leader because it contradicts the teachings of Islam.

Women's Political Participation in the DPRD

Gender disparity in Banten requires policies that promote gender mainstreaming. Gender mainstreaming policy will be successful with the support of the local executive and legislative authorities as an institution that makes and passed legislation in the form of regulations. Local legislative role is very strategic because this representative body has the principal function of the legislation. Therefore, the initiative of the DPRD Banten in presenting gender-based policy is needed. Moreover, the involvement and political participation of women in the realm of parliament is expected to encourage the creation of a gender mainstreaming policy from various aspects.

Barriers to women's political participation in the realm of parliament can be seen from the very beginning the unfulfilled quota of 30% female representation. Of the 85 legislators Banten, which became the only female board members amounted to 18 people, if presented only some 21%. This would prevent women in Parliament impediment in promoting women's issues. The number of women MPs were 18 people in Banten DPRD if look at the distribution of each district / town area of their election the pleasure andadventuregendergapwillbefelt. Consider the following table:

Table 3: Number legislators regency / city in Banten

Kabupaten/Kota *Regency/City*	Jenis Kelamin/Sex		
	Laki-Laki *Male*	Perempuan *Female*	Jumlah *Total*
(1)	(2)	(3)	(4)
Kabupaten/Regency			
1. Pandeglang	9	1	10
2. Lebak	8	1	9
3. Tangerang	20	1	21
4. Serang	6	5	11
Kota/City			
1. Tangerang	10	3	13
2. Cilegon	2	1	3
3. Serang	2	4	6
4. Tangerang Selatan	10	2	12
Banten	67	18	85

Source: http://banten.bps.go.id (Accessed on 01.01.2018)

The portrait of how hard the women in the realm of parliament Banten. Based on the distribution of districts / cities, women who managed to occupy seats in parliament Banten no more than 5 people in fact most of the area only has one female board members (Pandeglang, Lebak regency, Tangerang, and Cilegon).

This reality reinforces the lack of participation of women in the political sphere. Number of board members of this minority will certainly hamper the struggle for gender equality agenda in the area of the constituency concerned. The number of board members to 18 people in Banten DPRD have implications for the unequal ratio between men and women in terms of charging pospos council leaders and fittings in BantenDPRD.

For example, part of the board fittings Budget Agency (Banggar) from 30 the number of members Banggar, only 5 women MPs (17%). This condition will impede the political policy of product produced. Politics budget emphasis on gender mainstreaming should be hampered because of the lack of voices bringing his colleagues fighting council on women's issues. During this time of political budget designed by the DPRD Banten they generally have not touched the specifics concerning women problems. Political policy anggraan built are still general. This was disclosed by Nuraeni, Deputy Chairman of Parliament Banten:

"Politics budget does not dissect each one specifically. Supposedly in the budget policy should lead to the problem of women's data. Program for the development of women, poverty alleviation programs and other women, it did not happen. So the policy of our budget is still general. Not to discuss specific women. (Nuraeni, Deputy Chairman of Parliament Banten)".

From 21 the number of members Baperda, only 5 women MPs. The sum of this minority will certainly provide barriers that mean for women MPs to prepare plans for gender-based local regulations. In arranging the program of local legislation, for example, they are difficult to insert gender equality issues in the meeting Baperda because so dominant male board members, while not all male board members have an interest in women's issues. The important is the turn out of 18 women legislators Banten, not all have a sensitivity to gender equality issues. Most of them departs from the general issues that are not specific to touch on women's issues are more substantive.

First, the political process in the realm of the Party. Before the female board members occupy seats in parliament Banten, they need to be involved in the organization of the party, or at least sign up for one of the parties. It is as set out in the regulations on the election that candidate board members must belong to political party recommendations. The party leaders must have an understanding that legislative candidates who would they recommend to follow the electoral contestation in the legislative election was a cadre or prospective qualified and if the leadership of the powerful. This is important because one day when the cadre or the recommended candidate by the party won seats in parliament Banten,

In addition, the patriarchy culture can actually be found in the body of the party, where the cadres or female politicians are often not given the opportunity equally with men in following electoral contestation. This can be seen from the predominance of the management structure in males than in females. Female politicians are not placed in strategic structure. The party leaders do not prioritize how to empower women cadres. The orientation of the party is still stuck at winning electoral strategy. This is as revealed by Encop Sofia (Member of Parliament Banten faction Gerindra) as follows:

> "I think this is not only happening in Gerindra not only, in general all of the parties did not think how to empower women. They (party leaders) only think how to win, and they are also not too think how to optimize the regeneration of an internal party. If the women of the early educated at the party certainly result will be good, but it is not. There are still a lot of homework (homework) does. So I think that gender equality propaganda must often be fought not over yet".

When a political party is not doing serious cadre against internal frightened female or female candidates on issues of gender equality, it becomes mustahail when female candidates managed to occupy DPRD Banten, they will have social sensitivity towards gender issues.

Second, the process of registration of candidates for the legislature. Problems and obstacles women legislators in memperjuangakan gender mainstreaming policies can be tracked from the process of registration of candidates for the legislature. From the observation of researchers and description of women MPs, in the process of registration of candidates the most obvious obstacle is at the stage of the struggle for the serial number of candidates through party approval. The sequence number of candidates has not only limited numbers and columns deratan names of candidates for the legislature. This sequence number has a political meaning that has a correlation to the political rights of women, especially women's political participation. The party leaders have enormous power in determining who the candidates are entitled to a number of first, second, third, and so on.

Defeating a potential cadres (chiefly women) in the internal party Such considerations confirms that there has been no commitment from the parties to seek qualified women candidates are considered to have strong leadership and has a specific issue to the struggle for gender pengarutamaan. Female candidates often simply placed in the bottom of the sequence number so impressed not only act as a complement. For if the party did not put forward women candidates in numerical order, the party threatened Banten disqualified by the Election Commission. This is as expressed by Ade Yuliasih:

> "The role of political parties in determining the qualified female candidates is very important. But the very fact that the role of political parties is lost in doing against women politicians. female politicians nominated as candidates cannot be separated from the party's concern that the electoral district will be crossed by the KPU. So actually just aborting the obligation, what is important is the name of the female candidates but not considering the quality of the female candidates themselves".

The third, The party of the political process at the stage of registration of candidates who do not consider the quality of female candidates first consideration of whether they have an interest in women's issues and gender equality, then the resulting candidates-candidates are women who act as a general in their political participation. This can be tracked in the campaign process, in which the voter segmentation targeting women candidates in general without touching specifically to the realm of gender issues and women's empowerment. This is due to the existence of the necessity to claim the candidates to mobilize as much as possible and therefore contributes to the achievement of the vote in the hope that voters can deliver them to DPRD Banten. Whereas when women candidates only focus on the empowerment of women with female-specific segmentation, it is considered will not be maximized in mobilizing voters. Yet this is a misperception, women candidates with a focus on women's empowerment issues there is no guarantee that the relevant candidates decreased voter. Comparison of the number of voters between men and women can be said to balance. In the 2014 elections and then for example the ratio of the number of male voters and women was 50.80% versus 49.20%. This means that women's voices can be said of potential voters to mobilize women voters so that it can deliver the gain seats in parliament. This mistaken perception in the end make no female candidates are focus on women's issues in the campaign.

Fourth, the political process in the realm of parliament. Explanation has previously been disclosed that the majority of Banten Parliament has no role or being involved in gender equality issues through tracking of the political process in the party, the nomination, until the campaign period, with all the obstacles, women candidates have not positioned themselves as fighters or activator (actor) on women's issues. The majority is still focused on general issues. Implications of late are women in parliament Banten will find barriers in realizing gender policy. This will affect the extent of political participation of women in Parliament in relation to promote gender equality. The political participation of women legislators in relation Banten realize gender mainstreaming can be tracked in two things: the political participation of women in the legislative process (policy formulation) and the participation of women in political organization for women politicians (political caucus of women).

Women's political participation in Banten Parliament in the legislative process has not fully towards gender mainstreaming. Female Assembly Members has not been able to play an active and responsive in making the discourse of gender equality as an important and strategic issues to be discussed and debated in discussion forums and meetings in parliament. It can be seen from the low initiations female legislators to encourage other board members to participate in creating a pro-gender policies (interview with Encop Sopiah, legislators Banten). as revealed by one informant as follows:

"So far I see friends who have become board in this province, I think it is not yet fully have *concern* on women's issues. Meskpun there were 18 female members of parliament, only if then everything then has lost interest to allocate time to fight for women's issues I think it still has not. Because this is related to partiality. "(Interview with Encop Sopiah, legislators Banten faction Gerindra).

Lack of participation Banten female legislators in gender mainstreaming can also be seen from the budget policies designed by members of the board. Gender-responsive budget policies are essential to promote the emergence of programs and activities that will be run by the government led to the creation of gender equality. That is, the program of gender equality is not only a discourse without any program or policy of the government, while the policies and programs need to be encouraged by the presence of gender-based budget politics.

Conditions worsened by the fact that the male board members do not have the sensitivity to issues of politics permpuan even consider gender-based budget is less important. This is as expressed by one of the members of Parliament Banten:

"Political of Banten parliamentary budget for granted, because it is considered too important yet still familiar wrote. Members of Parliament are also mostly fathers instead of women. And the key of budget politics in Parliament as the legislature, the executive if it's just running. Average budget planning it we are agreed as well, yes happens difficult also because his father bapak- more dominating. Politics budget does not dissect one by one specific (gender related). Kan should for example in the budget policy should lead to the problem of women's data, programs for development of women, women's poverty alleviation program, it did not happen, so policy is still the general budget. Not to discuss specific women ". (Interview with Nuraeni, Banten DPRD members of the Democratic Party).

Patriarchy Parliament culture in Banten became one of the other challenges that cannot be avoided. From several informant stated on certain cases, the male members of Parliament do not agree with their policy of gender-based politics with reason considered not so urgent. It can be seen from the minimal (if not non-existent) initiation of male legislators to membas women's issues in forums and meetings musyawah.

Meanwhile, Political participation of women in Parliament Banten in gender mainstreaming in addition to views of the political process that occurs in internal Parliament, can also be tracked on the extent of their involvement in the political organization of women MPs (Caucus Politicians Women's) influence on the creation of gender both in Parliament and in masyatakat , In this case, the political participation of women can be seen in the organizational wing of the party, KPP (Women's Political Caucus) and KKPPI (Indonesian Women's Political Caucus).

In general, political parties have a special wing of the party as the container moves women's political organization. For example in the United Development Party (PPP) mempiliki WPP (Wanita Persatuan Pembangunan), the Democratic Party has PDRI (Female Demokrtat Republic of Indonesia), Gerindra has PIRA (Female Indonesia Raya). Nevertheless, the existence of this wing of the party is not optimal in scoring political leaders and empowers women to society. One informant said that the presence of women's wing of the party is not effective, the women's movement is less optimal, and its activities have not been significant to women's issues. The presence of more women's wing of the party serves as a mobilization party before the election for voters (interview with Encop Sopiah, legislators Banten).

In reality, the purpose of the establishment of this KPP roles, responsibilities and functions (TOR), it still has not run optimally. Despite all the female legislators has been included in the KPP management, In addition the number of women MPs little in quantity is not supported by *political will* the equally important issue of gender equality, where the majority of women MPs not yet have a strong sense related issues so that very few women who do advocating, being vocal and critics of the importance of gender equality. It is, as revealed by one informant as follows:

> "LTO-operation in the realm of parliament. How build regulation that is friendly to women, how the budget pro oversees gender politics and so on. Again, I feel that the LTO is not optimal because the amount of her lacks much so that in the process *lobby-* is also weak. In addition, if then all women have lost interest legislators to allocate time to fight for women's issues, I think they have not, because this is related to partiality. Because not everyone considers that women's issues are important. If I was not it considers that it is important that women need the strength. "(Interview with Encop Sopiah, Member of Parliament Banten faction Gerindra)".

Not much different from the LTO, the political participation of women legislators in KKPPI (Political Women's Caucus of Indonesia) is deemed not significant in presenting gender mainstreaming agenda. If KPP a special container political participation work in parliament shutter, then KKPPI work in the realm of extra-parliamentary.

Political representation: Act for or Stand for?

Gender mainstreaming policies need to get a strong boost by the DPRD Banten as a representative body, which represents the interests of society in general. Engendering nt elevate the position of women in marginalized men, but strive to create equal rights and access in various aspects (politics, economics, social, cultural) between men and women. This is important because democracy without gender equality becomes meaningless.

But the political reality that happened just showed anomalies. There is no guarantee that legislators do function representations of women in parliament stands above polities in the interests of women. This is as disclosed in the foregoing discussion that not all women legislators have *concern* and interest in women's issues. In most regards the issue of gender equality is not too important. And if you observe in real susungguhnya reality happens in society, the problem of the gender gap is still found in Banten.

The lack of sensitivity to gender issues have implications to political participation Banten run by local legislators to exercise their functions of representation. Where issues regarding gender equality are not significant to be discussed in the legislative process in Parliament. It cause low motivation Banten DPRD members to conduct advocacy and criticism of pro-gender policies and the weakness of the process *bargaining* politics to legislators to persuade others (including the male board members) will bring the importance of gender mainstreaming. Not only in the realm of parliament, in organiasi political women ranging from women's wing of the party, KPP and KKPPI, political representation functions run by legislators Banten less a real contribution to the issue of women's empowerment.

Exposure explains that political Banten Parliament runs representation has just reached the level of symbolic representation. Symbolic representation as proposed by Pitkin (1967) is where the representative or representatives of a symbol or a symbol representing the group represented. In this case, members of Parliament Women acts as a representative of the people, especially women as equally as a woman., ie women legislators are supposed to represent women. Yet in reality, they do not perform the function of the representative in parliament to fight for the issues of women's empowerment. It can be seen from the lack of initiations of women legislators to control the budgets based on gender politics, weak lobbying to bring the agenda of policy formulation more pro gender in Banten. This confirms that the particulars representative function which is run by the local legislatures still bersfat deskripsif Banten (*Stand for)*.

In fact, when referring to the theory Pitkin (1967), form the most ideal political representation no substantive representation representative. Where members of parliament representing women as bringing the public interest in this case women into the realm of public policy through the functions carried out by the legislature (*regulating, budgeting, controlling)* so it is not just based on the reason for the similarity between the represented and equally represented, namely women. That is, members of parliament are women Banten not *act for* based on the fight for gender equality issues are not reflected because the presence of gender pengarutamaan policy. Despite existing ones that are considered pro-gender (Regulation keluaraga and Local Resilience Protection of Children and Women), but actually the regulations are not born of the initiation of the overall female legislators.

4. Conclusion

From all the view in this article, Gender mainstreaming in representative council in Banten Province show us that women still use to be the object of power. The formation of political agenda neither in the institution such as political party or the cultural institution stuck on the women as a central to gain the power, from the local election in Banten till the representative.

Our reflection was how to built the political agenda in Banten for gender mainstreaming. If we look backward, we must realize on differentiated policy as affirmative action (5). This means re-articulate of the gender agenda to bring equity and expanding public discourse for women agenda not only for the women in representative council but also for women in the whole context for inclusive movement (Young, 2000).

So in political agenda, gender mainstreaming must constructed in two stages. Substantive agenda where's inclusion for women on political party become substantive agenda and the second, cultural recognition for women become independent to decided and represent their selves to be a part of active citizen.

Reference.

[1] Arikunto, Suharsimi. 1998. Research Procedure A Practice Approach. Rineka Create. Jakarta

[2] Budihardjo, Miriam (1982). *Participation and Political Parties.* Jakarta: Gramedia Main library

[3] Hamid, Abdul (2010), *The Kiai in Banten: Shifting Roles in Changing Times* in Masaaki (eds) *Islam in Contention: Rethinking Islam and State in Indonesia.* Kyoto: Center for Southeast Asian Studies; Jakarta: Wahid Institute; Taipei: Center for Asia-Pacific Area Studies

[4] Hikmawan, M. D. (2017). Pluralisme Demokrasi Politik di Indonesia. *Journal of Governance*, *2*(2), 223–247. https://doi.org/http://dx.doi.org/10.31506/jog.v2i2.2678

[5] Hikmawan, M. D. (2017b). POLITIK PERBEDAAN: MINORITAS DALAM

IMPLEMENTASI KEBIJAKAN. *Journal of Indonesian Public Administration and Governance Studies (JIPAGS)*, *1*(1), 88–98.

[6] Huntington, SP & Nelson, J. (1994). *Political Participation in Developing Countries*. Jakarta: Rineka Reserved

[7] Mormel, Melanie & Schweisshem (2009). *Increased Women's Representation in Politics: Better Opportunities for Equality? The results of the legislative elections in Indonesia*. Jakarta: FES

[8] Ronald H, Chilcote (2006). *Search Paradigm Theory of Comparative Politics*. Jakarta: RajaGrafindo

[9] Suciptaningsih, Oktaviani Adhi (2010). *Women's Participation in Institutions Legislature in Kendal*. Journal Community (2) (2010)

[10] Tornquist, Olle (2013). *Indonesian Democratic stagnation - Introduction and Excecutive Briefing: Progress, Setbacks and options - options 2003 - 2007* in Democracy in Upper Sand. Yogyakarta: PCD Press and Demos

[11] The Central Bureau of Statistics (2017).

[12] *Banten province in Figures*. Banten: BPS Banten

[13] The Central Bureau of Statistics (2016). *Gender Development Index Banten Province*. Banten: BPS Banten

[14] Young, I. M. (2000). *Inclusion and Democracy. Oxford University Press*. Oxford University Press.

Problematic of Local Head Regulation As a Local Law Products in The Framework of National Legal Development

Abdul Aziz Nasihuddin[1], Tedi Sudrajat[2] and Sri Wahyu Handayani[3]

Faculty of Law, Jenderal Soedirman University

[1]ziz_lingk@yahoo.com (08122724640); [2]tedi.unsoed@gmail.com;
[3]ayu27fh@gmail.com

Abstract. In Indonesia, the main discusssion in the dynamics of local autonomy is the establishment of local law product after the discussion of general elections. How come? Article 18 paragraph (6) of the 1945 Constitution regulates that in the framework of implementing autonomy and co-administration duties, the local Government has the right to stipulate local regulations and other regulations. Practically, other regulations can be form as local head regulation and many kind of legal policy which made by head of local government. It can be imagined, how many local law products were established in one year in 34 provinces and 265 cities/regency. In this perspective, it is important to raises the object of Local Head Regulations as a miniature of government duties as a derivative of the Local Regulation. The problems which arrise is what is the position of local head regulation in the national legal development. This paper was conduct by doctrinal approach. The specifications which used is descriptive analysis by inventorying and identifying problems that related to harmonization and synchronization of local law products based on Law Number 12 Year 2011 concerning the Establishment of Regulations and Law Number 23 Year 2014 concerning Local Government. Analysing method is normative qualitative by using gramatical and systematic interpretation.In this study, it was found that local head regulation is a part of local law products in the form of legal policy which called as beleidregels. Based on that form, if there any substantive problems from the legal policy so it could not be tested in the legal procedure because the legal policy was made based by the freedom of government. Thats why the government should create an alternative solution to solve the substantive problems from the local head regulation as a legal policy.

Keywords: Local Head Regulations, Local Legal Products, National Legal Development

1. Background

On 11th October 2016, Joko Widodo, the president of Indonesia held a limited meeting of working cabinet which results in law revitalization and reformation program. It became government's strategic agenda to bring back public trust and creates justice as well as law certainty.[1] There were three things which were instructed within the reformation policy, First, regulations arrangement to produce a good quality regulation; Second, internal reformation in the prosecutorial institution, police, also Ministry of Law and Human Rights environment in order to generate a professional service and law enforcement as the result of legal revitalization; and third, development of legal culture.[2] Consequently, Farida and Arinanto said that it needs a thorough evaluation of various legislations, including Local Head Regulation.[3]

Sulityo, et.all stated that the implementation of local autonomy in Indonesia is aimed to encourage two points. The first one is to achieve equal people prosperity and the other one is to accelerate democracy in local level.[4] Besides, through wide autonomy area, in globalization strategic environment, region is expected to be able to increase competitiveness by paying attention to the principle of democracy, equity, justice, privilege, specialty, and potency of local diversity in Republic of Indonesia. Unfortunately, the big authority shift in region is not followed by region's readiness in form of human resources to manage the big authority. [5]

In this case, local regulation takes role as an instrument which is used to realize both of the vision. By the great amount of local law product due to the wide area of Indonesia, the study about local head regulation as local law product in developing effective national law is often ignored. The studies which have been examined so far are almost all about Local Regulation. Obviously local regulation receives more attention because it is a form of real product of democracy in local level. Local regulation is formed by two institutions which are directly chosen by community through Local Head Election.

Therefore, it is important to conduct a study on Local Head Regulation since in the local government structure, Local Head Regulation is the government regulation in Regency/City level below the Local Regulation. Thus, in fact, Local Head Regulation is the pillar in local regulation structure.

2. Research Problem

The problem is how to to implement local regulation or by the power of legislation, local leader stipulates Local Head Regulation. It means, local Head regulation only exists based on two matters, firstly, to carry out local regulation and secondly based on the legislation

Hence, it means that the problem is the position of Local Head Regulation as legislation as well as policy regulation.[6] Both of these positions certainly have implication on the different legal consequences as well.

3. Methodology

This paper used normative legal approach. It emphasizes the object on the legislation as positive law. The point of normative legal based on special legal science characteristic, which is placed on legal review or legal study about positive law, involving three layers of legal science. It consists of legal dogmatics study, legal theory, and law philosophy. In the level of legal dogmatics, the review is conducted on positive law identification especially on Local Head Regulation. Meanwhile, in the legal theory level, the theories which can be used to analyze the problem are reviewed. Furthermore, in the philosophy level, this paper is conducted to comprehend the perception of forming Local Head Regulation towards beneficial value, law certainty, and legal justice which develop among the society.

4. Result and Discussion

a. Theory about Legislation Hierarchy

According to legislation theory, legislation formation involves 2 (two) main points, they are: Firstly, material/substantial aspect, this aspect is related to the arrangement of legislation content that consists of legal principles and rules as well as guidance for concrete behavior in form of legal rules. As well as within the legal rules which will be defined in legislation; Secondly, formal/procedural aspect, this aspect is related to the activity of legislation formation (the efforts to comprehend method, process and technique of legislation) which takes place in particular country.[7]

Legislation holds an important role in democratic state of law in order to achieve people's prosperity. It can be seen that finally state's role is needed again after all this time people do not want their business to be interfered by the state. In the XIX century, state is requested to get involved in people's lives again which is more complex to create discipline and prosperity. [8]

In this case, law functions as the limitation of state's authority as well as people's freedom. Therefore, government and community's behavior which based on freedom rights will not disturb other people freedom and violates people's basic rights. In this condition, a good law is needed in which it does not just contain state's will but it also represents people's desire. The existence of norms differences and conflict between one legislation and other related legislations sometimes render difficulties for staffs in conducting their authority within government implementation. Consequently, to overcome the problem, a legal theory called "*Stufenbau Theorie*" is needed.

Stufenbau Theorie which is proposed by Hans Kelsen considers law process as it is depicted by norms hierarchy. Validity (mistake) from each norm (separated from the basic norm) depends on the higher norm. Hans Kelsen reveals that law arranges the formation of the law itself since a law norm determines how to form another law norm. One law norm is valid because it is made by another law norm determination and this another law norm becomes the validity of the first law norm. The relation between norms which arrange another norms formation is "super-ordination and sub-ordination". A norm which determines the formation of another norm is the higher norm while a norm which is made by another norm is the lower norm. Legislation stage is the order of level and degree of related regulations by remembering the authorized institution and the problem which is regulated within the regulation. Kelsen discusses the validity of legal norms by depicts it as validity chain which ends up to the state constitution. The first validity constitution is the last presupposition, the final postulate, in which the validity of all norms within law structure depends. A document which is the first constitutional form is the real constitution since a norm is binding only when it is presupposed as valid. This presupposition is called transcendental-logical pressupotion.[9]

Hans Kelsen theory is related to legal norm hierarchy which forms legal pyramid (stufenbau theory). One of the figures who develop this theory is Hans Nawiasky, a student of Hans Kelsen, who develops Nawiasky theory which is called as Theory *Von Stufenbau Der Rechtsordnung*. The structures of norm based on the theory are:

1) State fundamental norm (Staat Fundamental Norm);
2) State Basic Law (*Staat Grund Gesetz*);
3) Formal Regulation *(Formell Gezetz*);
4) Implementation rule and autonomous rule (*Verordnung En Autonome Satzung*).[10]

In line with matters above, in facing norm conflict the settlement using law principles may be applied. Purnadi Purbacaraka and Soerjono Soekanto try to introduce several principles in a regulation, they are:

1) Law should not be implemented retroactively;
2) Law which is made by the higher authority has the higher position;
3) Specific laws move aside general laws (*lex spesialis derogate lex generali*);
4) The latest law call of the previous law (*lex posterior derogate lex priori*);
5) Law cannot be contested;
6) Law is an instrument to maximally achieve spiritual and material well-being for community or individual through innovation and preservation (*welvaarstaats*).[11]

b. The Position of Local Head Regulation

One of the conseqeunces of the state of law is the development of positive law through legislation formation. In the hierarchy of legislation, Local Head Regulation may be interpreted through Article 8 Paragraph (1) and (2) Law Number 12 Year 2011 about the formation of legislation. It states that the legislations other than those mentioned in the hierarchy are still possible to exist as long as it is instructed by the higher regulation. This product of this legislation is included as delegated legislation/ secondary legislation. [12]

The first function of local head regulation is to specify the provision that still too general to be given the procedural guidance. By the role above, local head regulation is a subordinate regulation of Local Regulation. Hence, the content of local head Regulation should not against the regulation above it. In other words, Local head Regulation is one of delegated legislation products. As delegated legislation, the first area of this regulation is as legislation.

Furthermore, in the context of government implementation, Local Leaders run the function of public service in which to implement their authority, Governor / Mayor / Regent needs compelling legal instrument. The authority to form law product inherently attached to Government corresponding to the obligation to arrange and manage public interest. [13] According to Indroharto, three reasons which underlie this authority are; the broadness of administrative law environment so that it is impossible to spill all of it in the formal law; the very fast dynamics of public service and it is impossible to be followed by change of law, even public frequently demands for quick and responsive administrative action (emergency legislation); lastly, further regulation always related to the detailed and measurable technical evaluation. Giving this legislation authority for this administration often called as "regulation", since it has the element of "policy" and "to regulate". The term "legislation" is different since it tends to refer to "legislator/law maker" the representative institution/create the law.

As the law product which is published by Local Leader in order to run public service, Local Head Regulation is obviously placed as the Legislation as well as Policy Regulation. These two different positions of Local Head Regulation obviously have an implication on different law consequences.

Local Head Regulation is published over the higher regulation and as exit clause, in which the present study reveals that the differences of legislation ratio on this Local Head Regulation formation bring consequences on the different test. The position of Local Head Regulation as legislation is a bit non-debatable. The explanation of this matter has been detailed so that without further repetition, it can be directed to another variable in which the position of Local Head Regulation as Policy Regulation.

In general, policy regulation can be defined as general regulation which is published by government (in the Executive sense) related to the duty and responsibility which attached to the authority of the government. Thus, the legal basis of policy regulation is not legislation but the authority of the government.

By seeing the position of policy regulation which is "without a hook" on the higher regulation, there is a question appears on how policy regulation can be published. Oftenly, in government implementation practice, a gap between legality and reality happens. Meanwhile, public service dynamics demand for quick action and people need the solution immediately without waiting for the existence of regulation. In order to overcome this problem, inherent *freies Ermessen* is given with government authority as emergency exit. *Freies Ermessen* is used when; Firstly, emergency situation in which it is impossible to apply written provision; Secondly, the absence of regulation which arranges the matter; Thirdly, there is the regulation which arranges the matter but the norm causes bias.

Bagir Manan mentions the characteristics of policy are: policy regulation is not a legislation; principle of limitation and examination on legislation cannot be applied on policy regulation; Policy regulation cannot be tested through *wetmatigheid*, since there is no basis of legislation to make a decision on the policy regulation; Policy regulation is made based on *freies Ermessen* and the absence of related administrative authority to make legislation; Test of law on policy regulation is given to *doelmatigheid* so that the test is appropriate government general principles; In practice, it is given the format in various form and type of regulations, such as decision, instruction, circular letter, announcement etc. Even in form of regulation.

General rule of legislation formation is the function of the state. For a regulation which is applied "outside", its policy regulation formation is on the government hand in a narrow sense (Executive). The general/public authority of legislation material content is different from policy regulation material content.

Sanction within policy regulation can only list administrative sanction. Meanwhile, the legislation which its formation is through people representative approval may list a sanction which limits citizen's rights.

The scope of policy regulation material content is related to the authority to form decisions in the sense of *beschikkingen*. It is where the authority takes action in the field of private law and the authority to arrange plans which adhered on government institution.

The Function of Local Head Regulation as Local Regulation Implementation.

In order to run Local Government, Local Leader is given the authority to make local law product in form of regulation or even stipulation. Regulation law product is made in form of Local Regulation and Local Head Regulation which are used as the basic law of government implementation. [14]

As it is mentioned in Law Number 23 Year 2014, government gives authority for making local regulation in order to implement local autonomy and as co-administration duty or to further describe the provision of higher legislation.

According to Maria Farida Indrati, in legal norm group, there is implementation regulation (*verordung*) and autonomous regulation (*Autonome Satzung*). These regulations are placed below the Law. Implementation regulation comes from delegation authority and autonomous regulation comes from attribution authority. [15]

Therefore, it can be said that within legislation there are implementation regulation and autonomous regulation. The existence of the regulations is due to higher legislation and come from delegation and attribution authorities. Attribution authority in forming legislation (*attributie van wetgevingbevoegdheid*) is the authority to make legislation which is given by *Grondwet* (Constitution) or *wet* (Law) for a state institution/government. The authority attached continuously dan can be implemented on self-initiative everytime when it is needed, according to the limitation given. For example, the existence of attribution authority may be seen in Article 236 Law Number 23 Year 2014 about Local Governance. This Article gives attribution authority to Local Government to form Local Regulation in order to implement autonomy and co-administration duty. Besides, Article 238 also gives attribution authority to Local Government to form Local Regulation that contains imprisonment sanction at most 6 (six) months or fine at most Rp. 50.000.000,- (Fifty million rupiah).

What is meant by delegation authority in legislation formation (*delegatie van wetge-vingsbevoegdheid*) is delegation of authority to form legislation which is done by the higher legislation towards the lower legislation, whether the delegation is stated boldly or not. Meanwhile, attribution authority is different. The authority in delegation authority is not given but it is represented. Besides, this delegation authority is temporary as long as the delegation still exists.

Local Head Regulation Implementation in Banyumas Regency.

The authority of forming Local Regulation is a sign of region's independence in managing their own interest. Local Regulation is a strategic instrument to achieve decentralization purpose. In the context of local autonomy, the existence of Local Regulation is encouraging decentralization optimally. [16]

From the political empowerment point of view, decentralization may be seen from two sides, they are local government and central government. The purpose of decentralization from local government point of view is to realize political equality, local accountability and local responsiveness. Meanwhile, in central government point of view, the purpose of decentralization is to realize political education, provide training in political leadership and create political stability. [17]

To implement Local Regulation over legislation, local leader stipulates Local Head Regulation. The provision about formation principle and material content as well as local regulation formation as it is mentioned in Article 237, is applied mutatis mutandis over the formation principle and material content of Local Head Regulation.

Local Regulation and Local Head Regulation are prohibited to against the higher legislation, public interest and/or morality. What is meant to against public interest, are:

1) Disturbing harmony between citizens;
2) Disturbing access for public service;
3) Disturbing public safety and discipline;
4) Disturbing economical activity to increase people's prosperity; and/or
5) Discrimination over ethnic group, religion, belief, race, groups and gender. [18]

Based on the information from Head of Subdivision regulation and Head of Subdivision DH, Legal Division of Banyumas Regency, since 2010-2015 there are 406 Local Head Regulation, 80% (325 regulations) of them are made based on the delegation of the higher regulation and the rest (81 regulations) of them are not based on the higher regulation or it is published as exit clause. The delegation regulations do not cause any problem, but what about the exit clause? Theoretically, it can be understood because the dynamics of public service is very fast and it is impossible to be followed by the change of Local Regulation or any legislation above it. Even, public often demands for a quick and responsive administration action (emergency legislation); the other reason is that further regulation always related to detailed and measurable technical evaluation.

Conclusion

In this research, the result shows that Local Head Regulation is a part of local law product in form of policy regulation, so that it cannot be tested with *wetmatigheid*. Policy regulation is made based on *freies Ermessen* and the absence of related administrative authority to make legislation as long as it has a strong reason or emergency condition to make written provision; secondly, the absence of the regulation that arrange the matter; thirdly, there is a regulation which arranges the matter but its norm causes bias.

Acknowledgement

This article is intended to be obtained argumentation academically through the ideal concept local head regulation. Support from Ministry of Research and Higher Education is gratefully acknowledged.

References

[1] Thirzano, Yudie, (2017).Reformasi Hukum Bergulir Sesuai Nawacita, Tribun News 13 Oktober 2016, http://www.tribunnews.com/nasional/2016/10/13/reformasi-hukum-bergulir-sesuai-nawacita, accessed on 5th of October 2017.

[2] Kartika, Shanti Dwi, (2016). Pembentukan Kebijakan Reformasi Hukum, Majalah info singkat hukum, vol. VIII, No. 19/I/P3DI/October/2016

[3] Farida, Ike and Arinanto, Satya, (2017). Penataan Peraturan Pelaksana Putusan Mahkamah Konstitusi (Studi Kasus Putusan Nomor 27/PUU-XIII/2011) di dalam Penataan Ulang Jenis dan Hierarki Peraturan Perundang-Undangan Indonesia, Jakarta: Research Body of Indonesian People's Consultative Assembly

[4] Sulistyo, Yuri, et, all, (2014). Pengawasan Pemerintah Terhadap Produk Hukum Daerah (Perda) Melalui Mekanisme Pembatalan Perda Berdasarkan Undang-Undang Nomor 32 Tahun 2004 Tentang Pemerintahan Daerah, e-Journal Lentera Hukum, I (1): 1-12.

[5] Kurniawan, (2013) Pembentukan Produk Hukum dalam rangka Penyelenggaraan Kanun, Jurnal Ilmu Hukum XV (61) : 519-546.

[6] Syihabudin, (2003). Kajian terhadap Jenis dan Tata Urutan Peraturan Perundang-undangan Indonesia, Law Journal. 23 (10):46 – 71

[7] Faculty of Law Unpar, (1977). Keterampilan Perancangan Hukum. Bandung: PT Citra Aditya Bakti

[8] Hadjon, Philipus M., (1987). Perlindungan Hukum Bagi Rakyat, Surabaya: PT Bina Ilmu.

[9] Kelsen, Hans, (1945). General Theory of Law and State, New York:Russell & Russell.

[10] Attamimi, A.Hamid, (1990).Peranan Keputusan Presiden Republik Indonesia Dalam Penyelenggaraan Pemerintahan Negara; Suatu Studi Analisis Mengenai Keputusan Presiden yang Berfungsi Pengaturan Dalam Kurun Waktu Pelita I-Pelita IV. Jakarta: Legal Science Dissertation of Postgraduate Faculty of UI.

[11] Yuliandri (2011) Asas-Asas Pembentukan Peraturan Perundang-Undangan Yang Baik Gagasan Pembentukan Undang-Undang Yang Berkelanjutan, Cetakan ketiga, Jakarta: PT Raja Grafindo Persada.

[12] Jimly Asshiddiqie, (2005). Perihal Undang-Undang, Secretariat General of the Constitutional Court.
[13] Ridwan, (2009). Tiga Dimensi Hukum Administrasi dan Peradilan Administrasi, FH-UII, Press, Yogyakarta
[14] Aryani, Sylvia (2017). Eksistensi Peraturan Kepala Daerah Sebagai Peraturan Pelaksana Peraturan Daerah, Badamai Law Journal, 2 (1).
[15] Indrati, Maria Farida, (2007). Ilmu Perundang-Undangan I; Jenis, Fungsi dan Materi Muatan, Yogyakarta, Publisher Kanisius
[16] Rawasita, Reny, et.all (2009). Menilai Tanggung Jawab SosialPeraturan Daerah, Indonesian Center of Law and Policies Studies (PSHK), Jakarta
[17] Hidayat, Syarif, (2006). Desentralisasi untuk PembangunanDaerah, Jentera: Regional Regulation.
[18] Zein, Yahya Ahmad, (2017). Hierarki Peraturan Perundang-Undangan Indonesia (Studi Pasal 8 ayat (1) UU No. 12 Tahun 2011 tentang Pembentukan Peraturan Perundang-Undangan), Jakarta: Research Body of Indonesian People's Consultative Assembly.

Putting people in the proper position on policy process in rural area: A deliberation and empowerment process in village level of Indonesia

Sukarso; Swastha Dharma; Niken Paramarti Dasuki

Faculty of Social and Political
Sciences University of Jenderal
Soedirman
(sukarso_upb@yahoo.co.id) (swasthadharma@gmail.com) (nikenpd@yahoo.co.id)

Abstract. In public policy discourse, especially the classic one, the position of society as a policy target group is often, therefore, an object of policy even in some contemporary references. The position as an object thus tends to be relatively weak compared to other positions. A top-down policy process tends to put people in a weak bargaining position. On the other hand, empowerment policies often define target groups as groups that seem to be independent and sterile from other groups. From this assumption, this study tries to formulate an empowerment model that can reposition the "target group" of the policy from objects to subjects, including redefining "target groups" into compliancers, say it is, it refers to a multi group who has to comply to the policy with deferent ways. The research method uses a case study approach with data collection techniques are in-depth interviews and focus group discussions. The results of the study get an empowerment policy model that is more deliberative and comprehensive than a partial one. This model is different from the existing models so far, both from the formulator, output, delivery system, and of course the "target group" of the policy.

Keywords: deliberative; empowerment; policy process; reposition; target group

1. Introduction

The old position of people in the policy-making process is remaining to the recently anyway. They have been pressed to the edge of the power so they just merely an object than a subject of the policy process. From the models of public policy making so far, the position of ordinary citizens or citizens, implicitly or explicitly, turns out to be more an object, not a subject. Explicitly on the elite model, the authority of policy-making only exists in the elite group while the people become the target of the policy, as well as implicitly in the institutional and rational model. The policy process, especially in the policy-making one, has been elitist.

Issues of democratization and people participation have risen broadly. In the past two to three decades, the issue of more community involvement in the decision-making process or policy making has been stronger. As a result, more and more people are involved, directly or indirectly, in the policy-making process. The use of concepts such as public-hearing and transparency shows this direction. Decentralization era in Indonesia also strengthened especially in the last two decades. In Indonesia, the era of decentralization began in 1999 with the issuance of Law No. 22/1999 on regional

https://doi.org/10.1051/shsconf/20208601046

governance which was later corrected several times until now. In this era, the involvement of the community in the process (making) of policy was increasing. For example, all APBD products must go through a public hearing and must be published in the mass media. Finally, in 2014 Law No. 6/2014 was issued. about Villages. This law places the village as one of the spearheads of the government that is stronger in providing services to the community.

Being crucial position of the villages in Indonesia governmental, they should be compliance to the national government but they has to be accomodative to their people too.The position of the village in Indonesian governance is getting stronger, especially in budget allocations. The village government is given a budget allocation from the APBN directly in a relatively very large amount; with the aim is the welfare of the village community concerned. But, there is no deliberative and empowerment research of the policy process in the village level of Indonesia. So far the study of the involvement of the village community has tended to the policy process, especially participation in the policy process and not yet in the process of repositioning the village community in the policy process.

The research aim is to modeling policy process which is deliberating as well as empowering village people straightly and massively. In other words, the purpose of this study is to model a policy process that allows the community to become a policy subject, not an object of policy.

2. Methods

Research locations are two villages, Kotayasa and Banjarsari Kulon, in District of Sumbang, Banyumas Regency. The Kotayasa is representing a relatively big village while Banjarsari Kulon is representing the small one. The research focus is people perception and knowledge of budgeting formulation process in the village level.

Data sources of the research are informants which consist of the village government apparatus as well as the villages people, through routine neighborhood meeting while data collecting technique used both in-depth interview and focus group discussion (fgd). Data analysis technique is interactive from data collecting, condensation, display, and conclusion.

https://doi.org/10.1051/shsconf/20208601046

3. Results and Discussion

3.1 Brief description of the research location

Tabel 1. Brief Description of the Research Locations

No	Description	Research Locations	
		Kotayasa	**Banjarsari Kulon**
1.	Area size	505,99 Ha	212,279 Ha
2.	Population	9.157 persons	3.516 persons
	a. Male	4.651 persons	1.819 persons
	b. Female	4.506 persons	1.697 persons

Source: obtained from primary data

Pre-model of Community Participation in APBDes Formulation

From the results of the condensation and compilation of the data in advance, it appears that one of the most prominent in the process of formulating (and implementing) the APBDes is the low level of community participation, among others, marked by non-existent funds (zero rupiah) in the two villages of the research location. Likewise in the process of formulating activities and budgeting, community participation is very minimal, even almost none except for the real needs of physical development, such as village infrastructure. Beyond that, it seems that the village people as the subject of budgeting is still not visible. Therefore, this first year results are packaged in papers and presented at international conferences at the PublicAdministration Institute on August 21st and 22nd, 2017 in Jakarta. The title of the paper presented at the conference was Community Participation on In Local Government Budgeting Village of Indonesia: A Case Study in Banyumas Regency, Central Java.

From the results of the discussion at the conference, the pre-model of the Formulation of Village Budgeting can be formulated with an emphasis on Community Participation, or deliberative models. This model includes empowerment and community participation with an action research approach through intervention in the form of socializing aspects relevant to the formulation of the APBDes, with the following figure1.

Compilation and Condensation of Results of Interview and FGD

From the results of interviews and FGDs, there were important aspects of the potential for optimal community involvement in the APBDes formulation process, namely (1) legal aspects, or community understanding of the various rules in the APBDes formulation process; (2) administrative aspects, or people's understanding of administrative systems and procedures in the formulation of the Budgeting; and (3) socio-psychological aspects, or public awareness and understanding of the rights and obligations as citizens of the community dealing with the rights and obligations of the government towards their community. Each of these aspects was then translated into the main questions as instruments of the pre-test and post-test among the brief intervention socialization materials relevant to each of these aspects. The following questions are elaborated on each aspect and concise material of the intervention relating to that understanding.

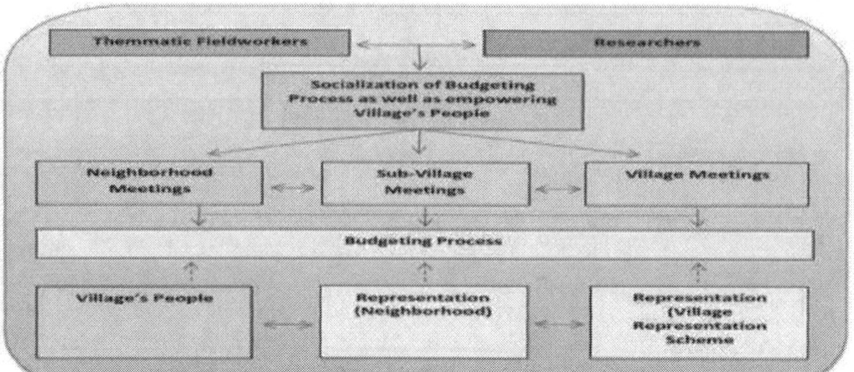

Figure 2. Pre-Model of Deliberative Budgeting Process

Figure 1. Pre-Model of Deliberative Budget Process

Results of the Pre-test and Post-test

The number of respondents selected by random cluster was 67 respondents, with details of 24 respondents were residents of Banjarsari Kulon Village and 43 respondents were residents of Kotayasa Village. The instruments are in the form of a scale with three answer choices 1 (low), 2 (medium), and 3 (high). The results of the pre-test and post-test can be examined in the following table 2.

From the table, it appears that the intervention in the form of socialization and increasing community understanding through FGD in these three aspects turned out to be very effective. Likewise, the potential for community involvement (item 3b; 3d; 3f; 3h; and 3j.) Turns out to be very high, in other words, through increasing community understanding of the APBDes formulation process can increase community involvement in the APBDes formulation process. Increasing community involvement in the process of formulating the APBDes is the first step to repositioning the village community from the APBDes object to be the subject of the APBDes.

Table 2. The Change of People's Perception and Knowledge Regarding Village Budgeting Process After "Intervention"

No.	Description	Pre-test	Post-test	Change
		f (x)	f (x)	%
1.	Legal Aspects			
a.	Do you know the rules about managing the Village Budgeting?	87 (1,29)	138 (2,05)	Naik 58%
b.	If you know, mention what are the rules?	77 (1,14)	146 (2,77)	89,6%
c.	Do you know the sanctions for violations or misuse in the management of APBDes?	92 (1,37)	163 (2,43)	77,1%
d.	If you know, mention these sanctions?	86 (1,28)	156 (2,32)	81,3%
e.	Do you know the difference in sanctions for criminal, civil and administrative?	94 (1,4)	158 (2,35)	68%
2.	Administrative Aspects			
a.	According to you, who is authorized and responsible for managing the Regional Budget?	88 (1,31)	186 (2,77)	111,3%
b.	Do you know about the APBDes formulation schedule?	80 (1,19)	146 (2,17)	82,5%
c.	Do you know about the types of funds available at the APBDes?	82 (1,22)	180 (2,68)	119,5%
d.	Do you know the allocation or allocation of these types of funds?	82 (1,22)	177 (2,64)	115,8%
e.	Do you know the amount of each type of fund that is in the APBDes?	79 (1,79)	156 (2,32)	97,4%
f.	Do you know the final balance becomes the starting balance for the following year?	80 (1,19)	140 (2,08)	75%
3.	Socio-psychological aspects			
a.	Do you understand the process of formulating the APBDes?	84 (1,25)	152 (2,26)	80,9%
b.	Do you want to participate in the APBDes formulation process next year?	142 (2,11)	189 (2,82)	33%
c.	Do you understand the implementation of the Regional Budget?	84 (1,25)	109 (1,62)	29,7%
d.	Do you want to participate in next year's APBDes?	130 (1,94)	188 (2,80)	44,6%
e.	Do you know the monitoring process in implementing the Regional Budget?	77 (1,14)	146 (2,17)	89,6%
f.	Do you want to participate in the process of monitoring the implementation of the Regional Budget next year?	143 (2,13)	192 (2,86)	34,2%
g.	Do you know the supervision in the implementation of the Regional Budget?	77 (1,14)	135 (2,01)	75,3%
h.	Do you want to participate in monitoring the implementation of the Regional Budget next year?	140 (2,08)	193 (2,88)	37,8%
i.	Do you know the results of the APBDes implementation?	93 (1,38)	156 (2,32)	67,7%
j.	Do you want to participate in measuring the results of the next year's APBDes?	81 (1,20)	136 (2,02)	67,9%

Note: n = 67; Scale = 1 to 3

Source: obtained from primary data

Discusion

Stakeholders in the formulation of APBDes.

The process of formulating the APBDes is actually directly or indirectly involving parties (1) Village Heads, (2) Secretaries and Village Devices, (3) Village Consultative Bodies (BPD), (4) sub-districts represented by Village Facilitators, and (5) Village communities in general. But it turns out that formally and real in the field, the parties involved are mainly Village Heads, Secretaries and Village Devices, and the BPD. The role of Village Facilitators turns out to be optional and the role of the community is directly very small or almost non-existent. The process of formulating the APBDes relatively only involves the Village Government and the BPD as well as (approval) from the District Government (via the sub-district).

3.5.1. Stakeholders Interaction.

The intense communication and discussion in the process of formulating the APBDes is mainly between the Village Government and the BPD and the District Government (through the District). The community does not directly communicate in the formulation process because it has been represented by the BPD. The following illustration explains the relationships between stakeholders in the process of formulating the APBDes.

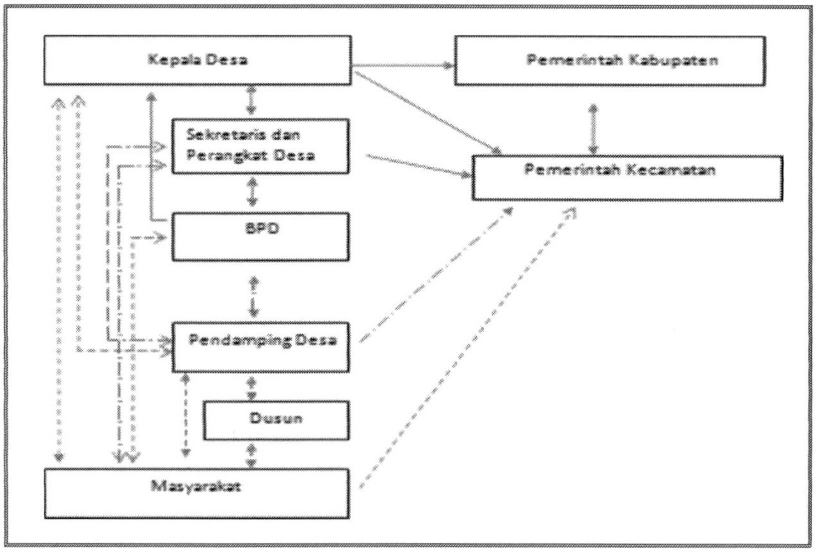

Source: obtained from primary data

Figure 2. The Stakeholders Interactions

Direct participation of village people: the subject of formulation.

From the results of the field study it appears that the role of the community directly in the process of formulating the APBDes is still very minimal and almost nonexistent, both formal and real access. The community tends to not know what and how the APBDes formulation process is, but actually the community needs and wants to know what and how it is. Therefore, there is a big possibility to increase community participation in the process of formulating the APBDes by giving opportunities to the community to be involved. The process of involving the community is more than just socialization, but there is a discourse on community empowerment in general on what and how to formulate the APBDes. The design in the pre-model is a scheme of how to increase community participation in the process of formulating the APBDes. The process of invasion and intervention, the final result that will be achieved is to place the village community as the subject of the formulation of the APBDes, not as an object as it has been.

Prospects of the People Position in the the Village Budgeting Process.
From the results of the pre-test and post-test, it appears that the intervention in the form of socialization and increasing community understanding in the APBDes process can increase community participation in the APBDes process. Therefore it can be strongly suspected that the position of the community can shift from the APBDes object to the subject of the Regional Budget with an increase in people's understanding. Efforts to improve people's understanding can be done by creating a community empowerment module covering legal, administrative and socio-psychological aspects. Increasing public understanding in legal aspects includes legal awareness and legal knowledge in the form of various regulations relevant to the formulation of the Regional Budget. Increasing public understanding in administrative aspects relates to community understanding in the APBDes formulation system and procedures. The increase in public awareness in the socio-psychological aspects is related to awareness of their rights and obligations in dealing with the rights and obligations of the government in the formulation of the APBDes.

4. Conclusion

From the results of the discussion, it appears that the need of the village community to be able to reposition them from the object of the formulation of the APBDes which is the subject of the formulation of the APBDes is very possible. From the results of a simple intervention in the second year, it was proven that increasing community understanding of the formulation of the APBDes could increase their participation in the formulation process and even the implementation of the Regional Budget. Therefore, in the third year, the final program in this study will be formulated in a more detailed and rigid manner and a retesting of the module to improve community understanding in the APBDes in the form of the Community Empowerment Module in the APBDes Process. This module formulation effort will be carried out in the third year, with an action research approach, which will include socialization and at the same time empowering the community about what and how to formulate the APBDes. These efforts, from the legal aspect, are efforts to increase legal knowledge and awareness, especially those related to the Regional Budget. In this case, the community becomes increasingly aware of what and how the rights and obligations of the community and government must be carried out in accordance with applicable regulations.

From the sociological aspect, these efforts are related to their awareness of the position that should be the subject of the formulation of the APBDes, not as an object. In this case, the community can increasingly realize what and how personal interests and the interests of society in general can walk together in balance. From the psychological aspect, these efforts are related to people's confidence and their existence in the process of formulating the APBDes. In this case, the community is increasingly able to actualize their interests in the process of budgeting their respective villages. Socialization and community empowerment in the APBDes formulation process became the central theme in the second year research activities. Thus, the third year research will contain inventions and interventions through the community empowerment module in the APBDes. To increase community involvement and continue to assist the process of involvement of the community concerned in the process of formulating their respective APBDes.

Acknowledgement

We wish to express our sincere thank to Ministery of Research, Technology and High Education of Indonesia for funding the research

We sincerely thank Mrs. Nani and Mr. Kokar, Head of Government of Kotayasa and Banjarsari Kulon Villages for facilitating the research where have been carried out.

References

[1] Albert-Roulhac, Catherine, 1998, *The Influence of EU Membership on Methods and Processes of Budgeting in Britain and France, 1970–1995, on Governance: An International Journal of Policy and Administration*, Vol. 11, No. 2, April 1998 (pp. 209–230). 1998.

[2] Dexter Whitfield, 2001, *Public Services or Corporate Welfare: Rethinking the Nation State in the Global Economy*, Pluto Press, London.

[3] Dinb, Hinh T., Abebe Adugna and Bernard MyersI, (2002), *The Impact of Cash Budgets on Poverty Reduction in Zambia: A Case Study of the Conflict between Well-Intentioned Macroeconomic Policy and Service Delivery to the Poor, on The World Bank Africa Region Poverty Reduction and Economic Management Division*, 1 October 2002

[4] Jacobs, Keith, 2001, *Devolved Budget Making within Local Authority Housing Departments: Staff Perceptions of Power and Control, on Local Government Studies*, Vol.27, No.2 (Summer 2001), pp.93–110.

[5] Mikesell, John, 2007, *Fiscal Administration in Local Government: An Overview*, on Anwar Shah (Ed.), 2007, *Local Budgeting (Public Sector Governance and Accountability Series),* The International Bank for Reconstruction and Development/The World Bank.

[6] Kelly, Joanne, 2000, *Budgeting and Program Review in Canada 1994–2000 on Australian Journal of Public Administration* Vol. 59(3):72–78, Sept-2000.

[7] Lewis, Carol W., *How to Read a Local Budget and Assess Government Performance*, on Anwar Shah (Ed.), 2007, *Local Budgeting (Public Sector Governance And Accountability Series),* The International Bank for Reconstruction and Development/The World Bank.

[8] Mullins, Danielr, 2007, *Local Budget Process*, on Anwar Shah (Ed.), 2007, *Local Budgeting (Public Sector Governance and Accountability Series)*, the International Bank for Reconstruction and Development/the World Bank.

[9] Miles, Matthew B., A. Michael Huberman and Johnny Saldana, 2014, *Qualitative Data Analysis: A Methods Sourcebook*, 3rd edt., Sage, Los Angeles.

Abbreviation:

APBDes: Anggaran Pendapatan dan Belanja Desa (The Village's budgeting)

Sport Development Index of Banyumas Regency

Ngadiman[1,*], I J Kusuma[1], Kusnandar[1], P Joko[1]
[1]Jenderal Soedirman University Purwokerto

*E-mail: ngadiman.unsoed@yahoo.com

Abstract. The purpose of this study is to evaluate the local government's success in its development, especially in the field of sports from the aspect of open space availability, community participation, sport resource, and community fitness level (Sport Development Index) that an evaluative research approach is conducted in this study. The research employs both stratified random sampling and cluster sampling technique. The population basic characters used include; the differences of regional progress level, gender between male and female, age ranging from children aged 7-12 years old, adolescents aged 13-17 years old and adults aged 18-40 years old, while the basic cluster used is district and sub-district or village. The research results show that the open space index is classified in low category (0.180), human resource index in medium category (0.551), public participation index in medium category (0.572), and physical fitness index in low category (0.488). Simultaneously, the Sport Development Index of Banyumas Regency in 2010 is classified in low category (0.487).

1. Introduction

Based on Law No. 3 year 2005 concerning on Indonesian National Sport System, it is mentioned that the pillar of sports is not only dealing with sport achievements, but also sport education and recreation. By simply measuring the sport success based on medals, we have belittled the meaning of sport education and recreation. Based on the national data taken from the Central Bureau of Statistics 2005, it shows that people who do sports for achievement purposes are only by 7.8%, while for health purposes by 65.2% and others by 27.0%. Thus, it is obviously inappropriate when the sport success is only measured from the sport achievements with the number of medals used as the indicator.

The sport index in European countries is developed based on seven indicators explaining the social significance of recreational sports, sport clubs, sport schools, population access for sports, socio-political and economic sport significance, as well as the importance of professional {17}. In Finland, the changes in success and development of sport policies cover three international sport phases: amateurism, totalization and professionalism (18). Meanwhile in Switzerland, the development of community sports has a relatively higher participation rate, yet the differences between regions in the original country are closely related to the historical, cultural and economic factors {7}.

The government policy on sport sectors as stipulated in Government Regulation No. 16 Year 2007 Article 92 has provided an operational explanation on the requirements of Sport Minimum Service Standards, including: open space for sports, human resources in sports, community participation, and community physical fitness level. In its development, those four dimensions become the bases for measuring the progress of sports development better known as Sport Development Index, hereinafter abbreviated as SDI (14}.

Nationally, the Indonesian SDI in 2005 is only at 0.220, which is classified, in low category (10). Meanwhile, the SDI in 2006, that figure increases up to 0.280, yet is still classified in low category (21). In the following years, some regions make similar studies. The results of a research conducted Jambi City Sport Development Index is in low category (0.318). Similarly, the Sport Development Index in Pontianak is also only at 0.302.

Banyumas regency is one of three major development areas for the achievement-based sports in Central Java, after Semarang and Surakarta. However, no studies have been conducted from the aspect of sport development index (SDI) as the indicator for a simultaneous sport development success.

2. Material and Methods

The focus of this research is to measure the sport development index in Banyumas Regency. The results of this research are expressed through index in the form of figures that this research is classified into a quantitative research type. As the purpose of this research is to evaluate the success of local government in the development, especially in the field of sports, an evaluative research approach is then conducted. Both stratified random sampling and cluster sampling technique are employed. The stratification is required to present the existing regional and community conditions in Banyumas regency that are greatly heterogeneous. Meanwhile, the cluster sampling is used considering to the research target area.

The basic population characters used: 1) differences in regional development levels: developed, developing, and underdeveloped. 2) gender differences of male and female, 3) age differences ranging from children aged 7-12 years old, adolescents aged 13-17 years old and adults aged 18-40 years old. The cluster area components used are district and sub-district/village. The population targets include 27 districts and 301 sub-districts/villages that 3 districts are randomly selected, consisting of North Purwokerto district representing the developed regional level category, Wangon district representing the developing regional level category, and Banyumas district representing the underdeveloped regional level category. Three sub-districts/villages are then also randomly selected from each selected district. Each sub-district consists of 30 people that the number of samples in one district is 90 people. Thus, the total research samples are 270 samples.

The research data sources consist of both primary and secondary data. The primary data consist of; community participation, open space availability, human resource in sports, and community physical fitness. Meanwhile, the secondary data cover number of population, regional area, and potential in sports. Thus, the data are collected with several sampling techniques: 1) Observation, used to obtain information about; regional area, number of population, potential in sports, human resource, and open space; 2) Questionnaire, used to obtain information about; community participation in sports; 3) Test, used to obtain information about; community physical fitness level, that Multi Stage Fitness (MFT) test is conducted; 4) Interview, used to obtain deeper information from the credible and related informants.

3. Result and Disscusion

a. Outdoor

Open space availability may be utilized by the communities for sport activities in those three districts is presented in the table below:

Table 1. Open Space Index Analytical Results

Aspect	Wangon District	North Purwokerto District	Banyumas District	SDI
Open Space	0.179	0.146	0.215	
Average				0.180

The results of data analysis from those three districts show that the open space index for sports in Banyumas Regency is 0.180. Based on SDI criteria, this figure is classified in low category.

b. Human Resources

The existing human resources, such as sport coaches, physical education teachers, and sport instructors in those three districts are shown in the table below:

Table 2. Human Resource Index Analytical Results

Aspect	Wangon District	North Purwokerto District	Banyumas District	SDI
Human Resource	0.865	0.514	0.274	
Average				0.546

The results of data analysis from those three districts show that the human resource index in Banyumas Regency is 0.546. Based on SDI criteria, this figure is classified in medium category.

c. Public participation

Public participation is the direct and active involvement of communities as the sport actors either formally or informally in those three districts as presented in the table below:

Table 3. Community Participation Index Analytical Results

Aspect	Wangon District	North Purwokerto District	Banyumas District	SDI
Public Participation	0.635	0.835	0.246	
Average				0.572

The results of data analysis from those three districts show that the community participation index in Banyumas Regency is 0.572. Based on SDI criteria, this figure is classified in medium category.

d. Physical fitness

Physical fitness in this case refers to the level of body's capacity to perform activities without significant fatigue experienced by the community in those three districts as presented in the table below:

Table 4. Community Physical Fitness Index Analytical Results

Aspect	Wangon District	North Purwokerto District	Banyumas District	SDI
Physical Fitness	0.609	0.473	0.469	
Average				0.517

The results of data analysis from those three sub-districts show that the physical fitness index in Banyumas Regency is 0.517. Based on SDI criteria, this figure is classified in medium category.

The results of data analysis obtained from each dimension in those three districts are then used as the basic calculation for SDI index in Banyumas Regency. Each dimension index may be seen in table 5 below.

Table 5. The Results of SDI Analysis in Banyumas Regency

No.	District	Open Space Index	Human Resource Index	Public Participation Index	Physical Fitness Index	SDI
1.	Wangon	0.179	0.865	0.635	0.609	0.572

2.	North Purwokerto	0.146	0.514	0.835	0.865	0.590
3.	Banyumas	0.215	0.274	0.246	0.469	0.301
Average		0.180	0.551	0.572	0.488	0.487

The results of data analysis from those three districts in Banyumas Regency show the average SDI index of 0.487 which is classified in low category.

The open space index of 0.180 indicates low category. In facts, this index is similar with that nationally in 2004 by 0.380, in 2005 by 0.110, and in 2006 by 0.266 (16). Open space refers to a place designated for sports activities by a number of communities in the form of buildings and or lands. Thus, open space is greatly required by the communities to perform physical activities. However, with the resulted low index, it can be concluded that the ratio between the availability of open space with the number of population above 7 years in Banyumas Regency is still lacking. This condition may be clearly explained that although there are still many open land areas in rural areas, yet not intended as places for sport activities. On the other hand, the open spaces in urban areas are greatly difficult to obtain for sport activities. Thus, it is not surprising that the parking lots, parks, and road bodies are used to perform sport activities, such as football by adolescents.

The human resource index of 0.551 is classified in moderate category. That index is higher than that nationally in 2004 by 0.115, 2005 by 0.047, and in 2006 by 0.099 {15}. The human resource figure is measured based on the ratio between the numbers of physical education trainers/instructors/teachers with the population over 7 years old in the related region. Thus, it can be concluded that in Banyumas Regency, the human resources are adequately provided in the form of physical education trainers/instructors/teachers expected to actively play their roles as the driving forces and facilitators for the communities and local government to develop sports in the related regions. This condition may be clearly explained as the annually increasing number of high school graduates go to college and choose to join the physical education or sport science study program.

The public participation index of 0.572 is classified in medium category. That index is higher than that nationally in 2004 by 0.345, in 2005 by 0.384, and in 2006 by 0.422 {15}. This condition shows that participation as a sport development dimension is already high. Thus, public awareness on healthy and active life throughout their life time is getting better. It means that the more individuals participating in sport activities in one area, the higher the public awareness in that area to have a healthy life and quality. Public higher participation reflects their interest and appreciation in sport activities. Thus, it can be concluded that participation is believed as a strong foundation as the prerequisite for sport buildings of not only the educational and recreational, but also the achievement-based sports.

The fitness index of 0.488 is classified in low category. The index is similar to that nationally in 2004 by 0.540, in 2005 by 0.352, and in 2006 by 0.335 {15}. This figure shows that the level of body's capacity to perform activities without experiencing significant fatigue is still low. This condition may explain that science and technology may impact the patterns of community activities. The results of survey conducted by the Health Department in 2001 report that 56% (27% male and 29% female) of people aged 25 years and over suffer from hypertension, 3% with heart disease, 1.2% with diabetes, and 5.9% with obesity (overweight).

The Sport Development Index of Banyumas Regency in 2010 is 0.487 that it is classified in low category based on SDI criteria: 0 – 0.499: low; 0.500 – 0.799: medium; and 0.800 - 1: high. Thus, Banyumas Regency which is at the point of 0.487 is in SDI index ranging from zero to one. An index point addition of 0.513 is still required to achieve the condition considered high or perfect for the sport development. This figure also provides guidance for the policy makers in the field of sports to make various improvements, especially on open space availability aspect. Some efforts may be performed by increasing the availability of open areas provided for the communities to perform sports activities. With these efforts, the Sport Development Index of Banyumas Regency may be continuously increased from time to come.

4. Conclusion

Banyumas Regency open space index is classified in low category (0,180), human resource index in medium category (0,551), public participation index in medium category (0,572), and physical fitness index in low category (0,488). Simultaneously, the Sport Development Index of Banyumas Regency in 2010 is classified in low category (0.487).

References

[1]. BPS [Central Bureau of Statistics]. 2005. *Statistik Indonesia* 2005 [Indonesian Statistics of 2005]. Jakarta. Badan Pusat Statistik [Central Bureau of Statistics]

[2]. Brown, D.R., 1990. *Exercise, Fitness and Mental Health*. Champaign, IL. Human Kinetics

[3]. Departemen Kesehatan [Health Department]. 2002. *Profil Kesehatan Indonesia [Indonesian Health Profile]*. 2001. Jakarta. Depkes RI [Health Department of the Republic of Indonesia].

[4]. Departemen Kesehatan [Health Department]. 2004. *Rencana Strategis Pembangunan Kesehatan 2004 [Health Development Strategic Plan of 2004]*. Jakarta. Depkes RI [Health Department of the Republic of Indonesia].

[5]. Dirjen Olahraga-Depdiknas [Directorate General of Sports- National Department of Education], (2004). *Sport Development Index Nasional Tahun 2004 [National Sport Development Index of 2004]*. Jakarta. Proyek PPKO-Dirjen Olahraga[PPKO Project-Directorate General of Sports]. Depdiknas [National Department of Education].

[6]. Gammon, M.D.,John, E.M., & Britton,J.A., 1998. Recreational and Occupational Physical Activities and Risk of Breast Cancer. *Journal of The National Cancer Institut*. 90 (2). 100-117

[7]. Hanspeter Stamm &Markus Lamprecht. 2011. Swiss sports participation in an international perspective. *European Journal for Sport and Society* Volume 8.

[8]. International Olympic Comittee, 2000, *Olympic Charter*, Lousanne, Switzerland, IOC

[9]. International Olympic Comittee, 2002, *OlympicReview*, Edisi XXVII-38 April-Mey. Official Publication of The Olympic Movement

[10]. Kemenegpora [State Ministry of Youth and Sports]. 2005. *Laporan SDI Nasional Tahun 2004 [National SDI Report of 2004]*. Jakarta. Kementerian Negara Pemuda dan Olahraga [State Ministry of Youth and Sports]

[11]. KONI [Indonesian National Sport Commitee], 2010. *Laporan Porprof 2009 [Porprof Report of 2009]*. KONI Banyumas [Indonesian National Sport Commitee of Banyumas].

[12]. Lawson, H.A., 2003. Empowering People and Advancing Community Development. *Paper*. Presented in International Conference on Sport and Sustainable Development Yogyakarta.

[13]. Maksum, 2002. *Pola Partisipasi Masyarakat Dalam Berolahraga [Public Perticipation Patterns on Sports]*, Jakarta, Ditjen Dikti [Directorate General of Higher Education]

[14]. Mutohir,T.C. dan Lutan, R. 2001. *Olahraga dan Transformasi Nilai [Sports and Transformation of Figures]*. Jakarta, Ditjen Dikti [Directorate General of Higher Education]

[15]. Mutohir, TC. 2003. Sport Development Index. *Majalah* Forum Olahraga. Edisi IX [Sport Forum Magazine Edition IX]. Jakarta. Ditjen Olahraga [Directorate General of Sports].

[16]. Mutohir, TC., dan Maksum, A, 2007. Sport Development Index: Konsep, Metodologi dan Aplikasi [Sport Development Index: concept, Methodology, and Application]. Jakarta. PT Indeks

[17]. Otmar Weiss,Gilbert Norden,Michael Nader & Florian Arnusch. 2016. European Sport Index: the social significance of sport in 28 European countries. *European Journal for Sport and Society* Volume 13.

[18]. Pasi Koski &Jari Lämsä. 2015. Finland as a small sports nation: socio-historical perspectives on the development of national sport policy. *International Journal of Sport Policy and Politics* Volume 7.

[19]. UNESCO, 1979. *Learning to Live Together in Peace and Harmony, Velues Education for Peace Human Rights, Democracy, and Sutainable Development for The Asia-Pacific Region*. Bandung. UNESCO-APNIEVE Conference.

[20]. Undang-Undang No. 22 tahun 1999 tentang Otonomi Daerah [Law No.22 Year 1999 concerning on Regional Autonomy]. Jakarta. Depdagri [Department of Home Affairs]

[21]. Undang-Undang Nomor 3 Tahun 2005 tentang Sistem Kelolahragaan Nasional [Law No. 3 Year 2005 concerning on National Sport System]. Jakarta. Kemenegpora [State Ministry of Youth and Sports]

Traditional Transport Tools as Supporters of the Regional Economy through Tourism and Culture Activities

E R Gultom[1]

[1]**Law Faculty, Trisakti Jakarta university**

Abstract. Some regions in Indonesia such as West Sumatra, Yogyakarta, Jakarta and other regions, often found traditional Bendi transportation which is used to transport people or goods from one place to the agreed destination. Apart from functioning as a public transportation tool, Bendi also has an important role in tourism, which is used around to transport tourists both from within and outside the country. The activity carried out by Bendi can increase the income of the local people who use Bendi as a regional tourism promotion tool. How is the existence of a Traditional Transport Tool in the midst of modern transportation in Indonesia?; How is this traditional Bendi transport regulated in Indonesia's national regulations? and What is the support of traditional Bendi transportation for regional economic development where its position is as a means of transportation? Is the subject matter in writing this paper. The research method used is normative, using secondary data and primary data as supporters, through interviews with some observers of traditional transport equipment. Then it will be analyzed descriptively. The conclusion obtained is that in a certain area, there are provisions for regional regulations that regulate the existence of these Bendi traditional transportation equipment, such as in the Yogyakarta area. Traditional Bendi transportation support as a supporter of the regional economy through tourism activities is very large, such as those found in the Yogyakarta area, where Bendi is always there to surround tourist areas that deliver domestic and foreign tourists in visiting tourist areas, as well as in West Sumatra , Padang, Bendi still exists even though it is rare, but it is still used as a tourism and cultural transport tool.

1. Introduction

Transportation in essence is to move people or goods from the place of origin to the destination safely, and the transfer is needed to achieve and increase benefits and efficiency {4} . Indonesia knows many transportation or transportation tools, whether it's short or far away. The types of transportation known are land, sea and air. There are transportation equipment or modern transportation and traditional transportation that have their own characteristics .

The diversity and richness of Indonesian culture is invaluable, from customs, folk songs, food, tourist attractions, and not including transportation in Indonesia. All of them each have their own characteristics and are truly authentic from Indonesia. Traditional Indonesian transportation equipment has a very valuable value and has an important position because it is part of Indonesia's cultural heritage.

At present there are so many modern modes of transportation popping up, even online and this makes the traditional transportation modes more backward. Just mention the traditional transportation modes such as rickshaws, bicycle ojek, delman, gig or bendi which in this paper are referred to as one of the designations of a trandisical conveyor with three or four wheels, not using engines but horses as a substitute for the subject. The distinctive feature of this traditional conveyance is that it has four wheels to be stable and balanced when walking, with a relatively large shape, and is sometimes pulled by two horses or just one horse, controlled by a driver behind the horse.

The names of this traditional transport equipment vary according to the area where the transport is located, in the Sukabumi area called Nayon, in Yogyakarta called Dokar, in Solo or Klaten, Malioboro, Ngasem Market, and Kotagede called Andong. In West Sumatra, precisely Minangkabau and North Sulawesi are called Bendi.

This traditional conveyance has the potential that can be used optimally, but currently it cannot be implemented, even though the uniqueness and originality of the transportation of the dam can be a symbol or symbol of an area that can boost the economy of the area where the transportation is available, namely increasing the number of foreign or domestic tourists to come to tourist cities in Indonesia. Indirectly the existence of a bendi as a traditional conveyance is a form of uniqueness from the area where the transportation equipment is located. This traditional transportation tool is used for transporting short distances or in rural areas that are regional or in rural areas.

In some regions, it is precisely in the tourist area and permitted to operate, but must follow the existing rules, such as paying attention to the cleanliness of the environment in place. The Regional Government recognizes the existence of traditional transportation equipment with the issuance of regional regulations by the Local Government.

These things are the rationale for discussing the potential of traditional bendi transportation equipment to increase the income of the surrounding community from the area where it operates in certain areas that have been regulated by the local government through its regional regulations.

A. Problems

Based on the background above, the formulation of the problem from this paper is:

1. What is the existence of this traditional transport vehicle in the midst of the flow of modern transportation in Indonesia?
2. How is this traditional transportation regulated in the regulations in Indonesia?
3. How is this traditional transportation support for regional economic development where its position is as a means of transportation?

B. Purpose of Writing

Based on the formulation of the problem above, the purpose of this paper is to describe the following:

1. Regarding the existence of traditional transportation equipment that uses animal horses in Indonesia;
2. Regarding Indonesia's national rules that underlie the operation of traditional transportation;
3. About the potential of this traditional transportation to the economic development of the region where its position is as a means of transportation.

C. Benefits of Writing

The writing of this paper is useful to illustrate the potentials that can be extracted from the existence of traditional bendi transportation, among others for the community, namely increasing the income of the community around the area where the dam is operating and benefiting from increasing community activities that automatically increase local income through regional tourism potential. seeded and can attract foreign and domestic tourists who automatically increase the potential of the area's natural tourism.

2. Results and Discussion

1. About traditional transportation equipment that uses animal horses

Transportation plays an important role in social life and society and is very helpful in providing facilities and facilities, namely (benefits of transportation-www.pinterdw.blogspot.co.id):

a. Services for individuals and groups;
b. Exchange and delivery of information;
c. Personal and social travel;
d. Shorten travel time between home and work place;
e. Support the expansion of the city or the spread of the population into smaller groups.

To carry out all the activities of interest above, transportation or transportation tools are needed. Transport is a means to help individuals or groups reach out to various destinations, or send goods from the place of origin to the place. Every transportation activity that is categorized as both personal and public has the purpose of carrying out good and proper transportation services for the community, and especially important is to get to the destination safely.

Transportation has an important role in people's lives. Law No. 22 of 2009 concerning Road Traffic and Transportation, that "road traffic and transportation has a strategic role in supporting national development and integration as part of efforts to advance public welfare as mandated by the 1945 Constitution of the Republic of Indonesia".

Bowersox states, "Transportation is the transfer of goods or people from one place to another, where the product is moved to the required destination" (Bowersox-1981). Salim stated, "transportation is a means for humans to move something either people or objects from one place to another, with or without using tools". The tools used can be in the form of human, animal, nature or other objects using machines or not machined.

The history of mobile transportation starts from carrying goods or passengers using animal power facilities, to make it easier to move goods, using animal power such as cattle, buffaloes and horses. But what has survived until now is the horse. Used horsepower to facilitate the movement of people and goods from one place to the destination. The traditional conveyance is a bendi, besides the name applies another name based on the known transport area.

For example, in the Yogyakarta area, the area uses transportation with the help of horsepower as a public transportation tool. The name known is andong, is one of the transportation transports in Yogyakarta. Andong is a transport or transportation vehicle that uses or utilizes animal power, namely horses. At present, andong can be enjoyed from all walks of life regardless of social status, as in the past, only officials and courtiers can use it. Andong can be found along Malioboro, Bringharjo market, the north square of Yogyakarta city and Parangtritis tourism to surround the beautiful beach. Andong in Yogyakarta can be divided into 2 (two) types, namely horse cart for tourism and non-tourism {15}

2. Traditional Transport Equipment that uses Horse Animal Power drawn by a Coach, About its existence in various regions, Symbol of Specificity and Uniqueness in a Region

Several researchers, including, have carried out several studies on traditional transportation equipment: Yuni Astuti {5}, entitled "The Existence of Becak Transportation in the Development of Transportation in Yogyakarta". The study concluded that pedicabs in the Yogyakarta area are still used by the community as an alternative means of transportation even though there are many modern transportation. Rismawati {2}, who conducted a study on "The Existence of Delman (Keretek) Transportation Equipment on the Life of the Coachman

Delman in Majalaya District, Bandung Regency". This study revealed the social and economic conditions of the coachman.

The passage of time proves, even though many modern transports have sprung up even online, the existence of traditional dam transportation is not lost by time, even strengthened by the existence of regions, which further strengthens it as a symbol of the distinctiveness of the region. This is evident, the existence of traditional transportation equipment that uses horsepower and is controlled by a coachman.

Even though at present the traditional transportation equipment is marginalized due to modern transportation equipment, there are many benefits to using traditional transportation equipment, including:

a. Environmentally friendly and does not cause air pollution because it does not use fuel;
b. More economical in terms of manufacturing costs and in operation;
c. Its existence as a means of cultural preservation that has distinctiveness and uniqueness

The name of the Traditional Transport that uses Horses as its attracting force varies from region to region, namely:

a. Andong is the name in Yogyakarta, Solo and the surrounding areas such as Klaten, Karanganyar, Sukoharjo, Boyolali and Salatiga. Regional officials and Dutch people can only use ancient times. But now, as time goes on, its existence is actually used for the benefit of the local community, which requires a dam service to deliver to certain areas that pass through alleys or narrow roads that are not too far away, or to deliver mothers and children. children in the neighborhood. Andong continues to exist, although there are many modern transportation equipment everywhere. It still survives because there are still many people preserving its existence because it is one of the Javanese cultural heritages.
b. In the Minangkabau area, no less important is the function and potential of this bendi. The existence of bendi has a very important role in the history of transportation in the Minangkabau in the past. Even though it was once an important transportation tool in Minangkabau, even in a Minang song that accompanies the paying dance, there is the lyrics "Babendi-bendi ka tanang river, stop by mangiak bungo lambayuang". The lyrics suggest that the bendi used to be a traditional conveyance that was very popular with Minangkabau people.
c. In Jakarta the name Bendi is known as Delman. Its existence is still recognized by the community with frequent use in the afternoon or when school children are enjoying school holidays. At around four to five in the afternoon, there was a clattering sound accompanied by the run of the horse's legs which were controlled by a driver sitting in the front, carrying passengers consisting of children and even mothers, to surround the village, say in the Warakas area in North Jakarta. Even at Monas, Delman or Bendi will appear on certain days, especially when the city of Jakarta welcomes its birthday.
d. In West Sumatra, wealthy merchants, regional officials such as demang, assistant demang and so on, only used horse-powered transportation controlled by a coachman. However, at this time with the passing of time, the Bendi has turned functions into public vehicles that are used by the community in short-distance environments, such as delivering mothers to the Market and children going to school.

e. In the area of West Nusa Tenggara it is known as Cidomo, which stands for Cikar-Dokar-Mobil ". Its use is more emphasized on short-distance environmental transportation or in rural areas that are regional in nature between villages.

f. In North Sulawesi, Manado traditional transportation is known as Bendi. Its presence attracted the attention of many people, and was used to get around in nearby areas for sightseeing around villages in certain areas.

g. Dokar is the term in West Java. Dokar is one of the developments of the Delman or Bendi transportation equipment. In the past this vehicle was called Dog-Car in the colonial era. The mention of the animal is wrong, because what's interesting is the horse. But along with the time the replacement of the designation became Dokar, whose existence spread almost all over Java Island with some form modifications, but with the same pattern of course.

3. Traditional Transport Tools that use animal horses and their localized wisdom

Traditional transport equipment that uses trains with the help of animal power as a puller and is controlled by a coachman has high potential. This is because its uniqueness and distinctiveness can take the hearts of its users to use its services to deliver to the desired areas of the environment.

In Minangkabau, these transportation equipment are very popular with the people even though many modern means of transportation are growing and developing. The local City Government because of its uniqueness and originality preserves its existence. It is imperative that this can be a potential in the city of Minangkabau and other areas in West Sumatra.

In the city of Manado, the official of the Governor of North Sulawesi, Soni Sumarsono, was very enthusiastic about the existence of these traditional transportation equipment. Even lifting the traditional transportation equipment called Bendi as the local wisdom of the area, as one of the means of transportation of the Minahasa community, which is now starting to fade.

4. Traditional Transport Tools and Tourism Potential

Although the existence of these traditional transportation equipment continues to be pushed in with the presence of modern transportation equipment, its existence is still maintained in several important cities, for example in West Sumatra, including among them in the city of Payakumbuh. Sometimes as a traditional conveyance, Bendi in West Sumatra has *neofungsionalisasi*, namely sometimes as a means of human transportation, and sometimes also as a support for tourism developed by the city government {6}. Tourism potential in West Sumatra is very high, let alone enlivened with the uniqueness and distinctiveness of Bendi as a traditional conveyance that has high added value for tourism development.

Many tourist areas in West Sumatra where bendi can operate, say in the Bukit Tinggi area, tourist areas include Banto Market, Pasar Bawah, Pasar Atas, Jam Gadang, and others. For the stop or base of the bendi has been provided in this tourist city. So the Bendi is not only a transportation for the local community, it is also used for tourism transportation and

https://doi.org/10.1051/shsconf/20208601048

the Bukittinggi city government is very supportive for preserving and cultivating the potential of the dam as a means of transportation.

In Yogyakarta, many efforts have been made to increase the potential of bendi to support regional tourism and to be developed in conjunction with efforts to organize the local area's historic area. At present Yogyakarta is conserving traditional bendi transportation empowerment, because it is a traditional transportation that is feasible and environmentally friendly. So bendi is an emission free vehicle and must be conserved (16).

5. Traditional transportation equipment and community income

The existence of Traditional Transport Equipment pulled by Horse animals, greatly helps the economic wheel of the people in the area where the conveyance operates. Like in the area of West Sumatra, with the existence of bendi the community was greatly helped because of the great interest of the tourists to use this traditional transportation service that will transport where these tourists want to visit places of interest. The local city government, to facilitate and this have also provided the place of dismissal in order to support tourist charm in West Sumatra.

Environmental hygiene is also noted, how the horse that pulls this bend when removing waste does not pollute the streets and reduce environmental pollution. The City Government also provides solutions that make it easier for the Coachmen, namely people who drive a bendi to help maintain the cleanliness of the environment by keeping the stools from Horses whose energy is used to pull the dam not scattered to pollute the environment that is passed.

Likewise in Yogyakarta tourist areas, such as in the Malioboro area, Parangtritis Beach has been provided by the local city government for a stop from Andong, so that when Andong services are needed by tourists both outside and within the country it is not difficult, because the special stops are .

In the tourist area of Lombok, West Nusa Tenggara, Cidomo operates in predetermined places, both for tourism and places to meet the needs of the people who need Cidomo's services.

In terms of tariffs, the city government in each region does not participate in determining the amount. This transport Tariff is determined based on an agreement between the driver, namely the Coach and Passenger or Consumer, while the government only conducts supervision.

6. Regional Regulations and Existence of Traditional Transportation

The city government in the area of each of these transports has established the existence of this traditional transportation, and is a local wisdom that must be preserved as a national cultural heritage, among others:

a. Governor of the Special Region of Yogyakarta, Special Region of Yogyakarta Regulation Number 5 of 2016 concerning Moda of Traditional Pedicab and Andong Transportation.
b. Article 1 number 5 states: "Andong is a Traditional Transportation Mode with wheels of 2 (two) or 4 (four) wheels drawn by horses.

In Article 2 it is regulated about Traditional Transportation, namely: "guaranteeing the safety and smoothness of traffic on the road; guarantee the sustainability of the preservation of Traditional Transportation; regulate the arrangement and implementation of Traditional Transportation; improve the welfare of Operators and / or Drivers, "and other Articles which emphasize the importance of the existence and operation of traditional transportation.

c. Regional Regulation of Padang City Number 11 Year 2005 concerning Public Order and Peace of Society.
   Article 1 number 13 states: "Bendi is a traditional vehicle that uses a horse as a driving force".
d. East Lombok Regency Government, East Lombok Regency Regional Regulation Number 7 of 2004 concerning Retribution for Non-Motorized Vehicles.
   Article 1 number 6 states: "Non-motorized vehicles are vehicles driven by animal power;
   Article 1 number 7 states: "Cidomo Dokar is a non-motorized vehicle drawn by animals and used to transport people".
e. Regional Regulation of Payakumbuh City Number 6 of 1998
f. Regional Regulation of Payakumbuh City Number 8 of 2006

3. Conclusion

Traditional transportation using horse power in Indonesia varies in its name according to the area where the transport is located, is highly recognized as a symbol or symbol of regional cultural heritage that must be preserved because it has a unique and unique as a means of transportation used to transport people from one place to the destination safely.

Indonesia's national rules that underlie the operation of traditional transportation are made to legally reinforce the legitimacy of this traditional transportation equipment to operate mainly in potentially tourism areas, which aim to develop regional tourism potential in Indonesia. Each region that permits the operation of traditional transport issues its regional regulations to support it, because its uniqueness can support the tourism potential of each region in Indonesia to attract tourists from both inside and outside the country to use traditional transportation services while enjoying the desired tourist areas in Indonesia. . Thus, it can support regional income through the tourism sector.

This traditional transportation has the potential to lift the economy of the region or region where its position is as a means of transportation, and to help the surrounding community to have income from the operation of these transport equipment.

Reference

[1]. Bowersox, C. 1981. Introduction to Transportation. New York: Macmillan Publising Co, Inc.
[2]. Rismawati, Lina. 2011. "Eksistensi Alat Transportasi Delman (Keretek) Terhadap Kehidupan Kusir Delman di Kecamatan Majalaya Kabupaten Bandung". Skripsi FPIPS
[3]. Salim, Abbas, R.1973. Manajemen Transportasi. Jakarta: PT.Grafindo Persada
[4]. Soekardono.2986. Hukum Dagang Indonesia, Cetakan ke-III. Jakarta:Rajawali
[5]. Astuti, Yuni, 2010, "Eksistensi Angkutan Becak dalam Perkembangan Transportasi di Yogyakarta". Jurnal Ampta
[6]. Sofyan, Fikrul Hanif dan Arsa Dedi. 2017. "Kuta Bendi di Kota Payakumbuh, Sumatera Barat Hingga Akhir Abad ke-20".Jurnal Mozaik Humaniora
[7]. Nugroho, Arinto. 2017. "Aturan Penggunaan Becak Tradisional di Kota Surabaya". Jurnal Unes

[8]. Undang-Undang Nomor 22 Tahun 2009 Tantang Lalu Lintas dan Angkutan Jalan
[9]. Gubernur Daerah Istimewa Yogyakarta, Peraturan Daerah Istimewa Yogyakarta Nomor 5 Tahun 2016 Tentang Moda Transportasi Tradisional Becak dan Andong.
[10]. Peraturan Daerah Kota Padang Nomor 11 Tahun 2005 Tentang Ketertiban Umum dan Ketentraman Masyarakat.
[11]. Pemerintah Kabupaten Lombok Timur, Peraturan Daerah Kabupaten Lombok Timur Nomor 7 Tahun 2004 Tentang Retribusi Kendaraan Tidak Bermotor.
[12]. Peraturan Daerah Kota Payakumbuh Nomor 6 Tahun 1998
[13]. Peraturan Daerah Kota Payakumbuh Nomor 8 Tahun 2006
[14]. Http://www. Pinterdw.blogspot.co.id/manfaat-transportasi/ diakses tanggal, 5 November 2018
[15]. Http://yogyakarta.panduwisata.id/tranportasi/andong-kereta-kuda-yang-merakyat/diakses tanggal, 5 November 2018
[16]. Harian Bernas, Hendra Kurniawan, Senin, 15 Februari 2016

SHS Web of Conferences 86, 01049 (2020) https://doi.org/10.1051/shsconf/20208601049
ICORE 2019

Developing Cooperative Law Culture at Bmt (Baitul Maal Wa Tamwil) to Remove The Rente System in The Traditional Market

N Triana[1], Naqiyah[1]
[1]IAIN Purwokerto

Email: triananita@ymail.com, Naqiyah.mukhtar@gmail.com

Abstract. This study aims to uncover and build *Baitul Maal wa Tamwil* (BMT) with the cooperative legal culture to eliminate rent systems in traditional markets. The research questions are: how is the existing condition of BMT legal culture and how to build new construction of cooperative legal culture on BMT. This research uses qualitative type of research, a traditional non-doctrinal legal research, with a socio legal approach. The results showed that the business of borrowing money with a very high interest or the practice of loan sharks until now is still rampant in a number of traditional markets, the practice of loan sharks in the form of *Plecit Bank* or *Ucek-Ucek Bank* remains in demand by the public. The existence of *Baitul Maal wa Tamwil* (BMT) apparently has not been able to play a optimal role in traditional markets. BMT is under the legal umbrella of cooperatives in practice apply to the Bank. such as borrowing money with a high margin, there is still a guarantee, a long process and consider the customer as a banking customer, not as a member, so there is no cultural closeness. For this reason, it is necessary to establish a legal culture in BMT that has a cultural spirit and cooperative spirit, namely mutual cooperation and togetherness, from members to members. all the benefits of the cooperative achieved by the group, can always be felt directly by all members.

https://doi.org/10.1051/shsconf/20208601049

1. Introduction

The business of lending money at very high interest rates or the practice of loan sharks is still prevalent in a number of traditional markets. Many traders are in debt because it is difficult to repay loans to loan sharks. The existence of loan sharks in traditional markets is inseparable from the paradigm that developed among traders. The process of applying for loans in banks is complicated. For example, the process of applying for a loan at a bank must be accompanied by appropriate collateral.(https://ekonomi.kompas.com/read/2017/05/06/143000626/mengapa.praktik.rentenir.masih.membelenggu.pedagang.pasar).

Difficulties in accessing market traders and / or doing business in the market to sources of capital are often the reason why many people are trapped in loan sharks who provide various facilities, such as being able to borrow without interest, visiting market communities with a friendly cultural approach, but behind that it brings difficulties borrowers money because of the high interest costs that must be returned.

The research result from Tini Hayatur Rohmah {13} entitled Plecit Bank Practices in Traditional Markets in Patikraja Subdistrict, Banyumas, showed the size of the practice of loan sharks in the form of *Plecit Bank* or *Ucek-Ucek Bank* in traditional markets. *Plecit Bank* is still in demand by the community, because it is easy for the loan process, and the market community is close enough to Plecit Bank officers who come to greet them every day. The existence of financial institutions around the market such as BMT *Baitul Maal wa Tamwil* has not been able to play an optimal role in this market, eventhough the existence of Shari'ah Financial Institutions, such as BMT is needed to be able to provide solutions to problems faced in eliminating the entrapment of market communities with loan sharks, while reducing level of poverty among rural communities.

Baitul Maal wa Tamwil (BMT) or also called "Sharia Cooperative", is a sharia financial institution that functions to collect and channel funds to its members and usually operates on a micro scale. BMT consists of two terms, namely "*baitul maal*" and "*baitul tamwil*". Baitul maal is a term for organizations which play an important role in collecting and channeling non-profit funds, such as *zakat, infaq and shadaqoh*. Baitul tamwil is a term for organizations that collect and channel commercial funds. thus BMT has a dual role, namely social functions and commercial functions {16}. Based on this definition, BMT can be maximized as a financial institution that can eliminate rent and loan sharks.

Based on this background, it is very important to conduct research to build the BMT *Baitul Maal wa Tamwil* as a strategic financial institution to eradicate rent systems in traditional markets.

2. Materials and Methods

https://doi.org/10.1051/shsconf/20208601049

This research is a non-doctrinal legal research, with a socio-legal approach {8} see also {15}. Through a sociolegal approach, the legal object is positioned in a broad social context, by not placing it as material isolated from culture (systems of thought, knowledge systems) and power relations between law makers, law enforcement, parties and the wider community. Understanding that law is a set of norms that are separated from social unity will only deny the relevance of law as the norm from the social basis of the birth of the law and the place where the law works. Through this approach, the assessment is carried out by describing the substance of legal norms and social reality, and the interrelationships between the two objects of study.

3. Results and Discussion

1. Loan Sharks on Traditional Markets

The economic process of the community is largely sustained in a buying and selling process and this occurs in traditional markets. The traditional market is a market that is managed simply with a traditional physical form that implements a direct transaction bargaining system where the main function is to serve the needs of the community both in villages, sub-districts and others {11}. Prices in traditional markets have uncertain properties, therefore bargaining can be done. Buyers in traditional markets (usually mothers) have behaviors that are happy to transact with communication / dialogue in terms of pricing, looking for the quality of goods, ordering desired items, and the development of other prices.

Some traditional market consumers are middle to lower class people who have very price-sensitive characteristics. When the low price factor which was previously the superiority of the traditional market was able to be torn down by the modern market, there was relatively no reason for consumers from the lower middle class not to go shopping in the modern market and leave the traditional market.

Goods sold in traditional markets are generally local goods and are viewed in terms of quality and quantity, goods sold in traditional markets through less stringent sorting. The distribution chain in traditional markets consists of producers, distributors, sub-distributors, retailers, consumers {1}. Constraints faced in traditional markets include payment systems to distributors or sub distributors in cash, sellers cannot carry out promotions or provide commodity discounts. They can only reduce the price of goods that are less attractive to consumers. In addition, it can experience difficulties in fulfilling the continuity of goods, weak in mastering technology and management, thus weakening competitiveness (12).

Traditional market traders face almost the same problem, namely the inability to develop because of lack of support both materially and skillfully. Small traders are constrained in the velocity of money and eventually owed. The most familiar financial institution in the traditional market is a daily bank with a loan system with a very high interest rate. With this system, many small traders are trapped by loan sharks.

Loan sharks are people who lend money to customers in order to earn profits through interest withdrawals. People who make a living by lending money; handyman usury; money release; loan sharks. Renten activities are a form of activity that provides loans to people in need in the form of interest payments determined by the renters {2}. The target of loan sharks is small traders in the markets and poor people in the villages. These loan sharks make use of the innocence of these poor people to reap huge profits. They offer convenience in obtaining any amount of money without any guarantee and anytime-anywhere.

This convenience is the reason for the majority of residents to accept money lenders and borrow money from them, even though they have already known the amount of interest they have to pay. The amount of interest is charged, resulting in more debt. The original debt was only Rp. 1,000,000.00 in one month to Rp. 1,200,000.00. This is known as the "*rolasan*" system. If it is due, it can not pay, then the debt and interest will be repaid, so from Rp. 1,200,000.00 to Rp. 1,400,000.00. So when we reached Rp 10,000,000.00, then within one month he has to return Rp 12,000,000.00.

The loan system provided by loan sharks is the Daily Bank / Package Loans, Payday Payments Weekly / Monthly Interest, and Rent Pawn. Based on the results of the study {10} the practice of borrowing money from the community does not have to borrow money from loan sharks, meaning the debtor or the borrower himself comes to borrow from the loan sharks and agrees to the interest set by the loan sharks. This illustrates that between the two agreements have occurred and have been equally willing about the interest set, the flowering of money (rent) that is practiced has multiplied because the creditor must pay his debt every day along with interest set, during the specified time and if it is too late to pay the next day you have to pay two days and so on, this causes the debtor to object and it is difficult to pay. Moreover, if the debtor is not good at using or borrowing it is not used for productive needs, it will have a negative impact and result in the family's economy being poor because they have to share income for family needs and to pay installments to loan sharks.

2. The Challenge of BMT Against Moneylenders

BMT is a micro-economic actor and can be stated as a new economic actor in the national economy, but seeing its vision and mission, BMT has a very strategic role, by contributing to the movement of the small economy that is very real, namely by participating in empowerment poor people and MSMEs.

The characteristics of BMT are as follows:

- Business-oriented, looking for mutual profits, increasing economic utilization the most for members and the community and environment.
- Not a social institution, but it is useful to make effective use of social funds for the welfare of many people and can organize educational activities can empower its members in order to support the economy.
- Grow from below based on the participation of the surrounding community. Owned by small communities from the BMT environment itself, not belonging to individuals or people from outside the community. On this basis BMT cannot be incorporated into the company.
- BMT is committed to meeting allcomponent of society under the layers through forums for recitation, da'wah, education, and socio-economic activities that have implications to productive activities in the economic field. BMT management and operations are carried out according to a professional approach with Islam methods.

The operating principle is based on the principle of profit sharing, buying and selling (ijarah), and deposit (wadiah). Therefore, even though it is similar to an Islamic bank, it can even be said to be the forerunner of an Islamic bank, BMT has its own market share, namely small communities that are not affordable banking services and small business actors who experience "psychological" obstacles when dealing with banks. Baitul Maal Wat Tamwil has several functions, including the following:

- Grouper and channeling funds, by way of saving money in BMT, the money can be enhanced utility, which raised the surplus units (those who have excess funds) and unit deficit (the party that lack of funds).
- The creator and provider of liquidity, can create legal payment instruments capable of providing the ability to fulfill the obligations of an institution / individual.
- Source of income, BMT can create jobs and provide income to its employees.
- Information sharing, giving information to the public about the risks of the advantages and opportunities that exist in the institution.
- As an Islamic microfinance institution that can provide funding for small, micro, medium and also excess cooperatives, it does not require burdensome guarantees for these MSMEs.

In carrying out its business, various contracts that exist in BMT are almost the same as contracts that exist in banks financing Islamic people's banks. The contracts are: in the BMT operational system, the owners of capital invest their money in the BMT not with the motive of getting interest, but in order to get profit-sharing profits. The product of collecting Islamic financial institutions is:

- Wadiah demand deposits, which are savings products that can be withdrawn at any time. Customer funds are deposited in BMT and may be managed. Every time the customer has the right to take it and has the right to get a bonus from the benefits of the use of demand deposits by BMT. The amount of the bonus is not set at the beginning but is truly the policy of the BMT. Even so the nominal is sought in such a way as to be always competitive.

- Mudarabah Savings, funds deposited by customers will be managed by BMT, to obtain profits. Profits will be given to customers based on customer agreement. Customers act as shahibul mal and Islamic financial institutions act as mudarib.
- Mudarabah deposits, BMT is free to carry out various businesses that are not contrary to Islam and develop it. BMT is free to manage funds (mudarabah mutaqah). BMT functions as a mudarib and customers are also Shahibul Maal. There are also customer funds that are deposited for certain businesses. Customers limit the use of funds for certain types and places. This type is called mudharabah muqayyadah.

In fact, in the field, some BMT managers still use conventional methods of marketing their products. They collect mudharabah savings and deposits by promising customers a fixed profit every month, whether requested or not, for a certain period of time, for example three months, six months, one year and so on. The amount of this fixed profit is usually aligned with the amount of interest in the conventional bank market at that time.

The reason they put forward was because such a method was considered to be simpler and more effective when most of our society did not yet understand the financial techniques developed by BMT. Another fact also shows that the average BMT manager finds it difficult to offer mudharabah fund collection products to the public. They almost always fail to provide satisfactory answers when prospective customers ask questions about the amount of profit sharing (profit) that will be received each month if the funds are deposited in the BMT.

This has a huge impact, for the next BMT activities, namely for financing activities. This advantage must be given to customers, which is ultimately borne by the Borrowers of funds (Debtors) at the BMT. So that the margin that must be paid by the Borrower (Debtor) to the BMT, is quite large, it can even be greater than the conventional bank interest, even if compared to the loan shark, it is still smaller.

Facts on the ground show, some BMTs apply provisions, recipients of financing facilities are required to return the following principal for the results gradually every month. Even if there is a delay in installment payments, one month, two months and so on, he will be subject to sanctions for payment of fines in the amount of the installment for the month. This is the reason why customers face great difficulties in developing their business. Ironically, although in aqad it is called the portion of profit sharing that can be obtained by both parties if the customer's business is profitable, in the practice the BMT asks for the profit in a fixed amount of rupiah figures. Another fact that has been collected in the

https://doi.org/10.1051/shsconf/20208601049

field is related to the practice of binding guarantees and the provisions regarding the existence of a debt statement for customers for the payment of mudharabah received.

Margin problems or profit sharing like this, are still considered not attractive enough by traders in traditional markets, BMT has not become an alternative choice in financing, so there are still many small traders more interested in daily banks. Because assuming the same between BMT and daily bank (loan sharks) there is still interest in financing.

Another challenge that must be faced by BMTs is the problem of inflexibility (not easy) in carrying out the distribution of funds (financing) BMT. The facts show that BMTs still behave like banks that have a lot of rules so that sometimes they cannot touch small entrepreneurs who are hampered by the rules in banking institutions (such as banks) even though they actually deserve assistance. Problems that generally block small entrepreneurs in asking for assistance in banking institutions (such as banks), namely the issue of credit guarantee eligibility, ownership of business licenses and various other administrative requirements must be met.

Administratively, the following are ways to borrow money from BMT, including:

1. Register for membership, the BMT will only provide loan funds to members and prospective members.
2. Fill in the Loan Form, is the first step as a loan application requirement.
3. Submit a photocopy of your own KTP and a photocopy of your husband and wife (if you are married).
4. Submit photocopies of family cards (KK), electricity bill accounts, salary slips and collateral.

The terms of this BMT when compared to moneylenders are more complicated, especially if this condition has to be done at the BMT office, then the billing is done in the market by uniformed BMT people, then traders are reluctant to do so. because there is a culture of shame found out in debt. They prefer to go to moneylenders, borrow like to neighbors, through chatting for a while and behind the public, nobody knows. This is what causes small traders in traditional markets to depend on moneylenders because, traders are very spoiled with a loan shark system. The reasons are:

1. Moneylenders are people they already know, because loan sharks get along long enough in the market. Greet each other between moneylenders and traders and understand the difficulties of traders
2. Moneylenders come to the merchants kindly (persuade) to provide capital loans.
3. Can be done at any time and under any circumstances even though in urgent circumstances money can be disbursed, even at night.
4. The loan process does not require a long time.
5. The requirement to apply for a loan is not so difficult, without going through complicated forms. different from the bank the conditions must have SIUP, Guarantee and others

6. To pay installments for moneylenders to the sword, it means that the trader does not have to waste his time paying installments by visiting loan sharks.
7. Without providing collateral in advance (collateral) because of mutual trust
8. Low transaction costs sometimes even without transaction fees
9. Funds obtained are not limited to economic business activities, but the funds are also for urgent activities such as medical expenses, education, and the like.
10. Can be done by all groups, not limited to members only.
11. The borrowing party does not need a KTP, possesses securities or other collateral

For this reason, traders prefer moneylenders as a place to borrow compared to BMT. In fact, moneylenders until now are one of the drivers of ribawi practice in the market. The dependence of traders on the need for money for capital or consumption becomes a profitable business for moneylenders. Even though the system carried out by loan sharks greatly facilitated traders, traders also felt burdened by the existence of loan sharks. The burden that must be borne by the merchant against loan sharks is: 1. Interest that is too large compared to borrowing a bank, cooperative, BMT, etc., 2. The nominal that is forged is relatively small, ie the maximum loan ranges from Rp. 500,000 - Rp.1,000,000, - 3. Social sanctions for traders if they cannot afford to pay.

The reason for loan sharks choosing traders as the target of financing is:

1. The possibility of bad credit is small because if the trader is unable to pay social sanctions, the merchant will accept it. For example, if a loan shark collects installments in front of a customer, the trader will immediately pay because he is embarrassed if he does not pay the installments.
2. With a small loan nominal, the trader prefers to pay off the debt rather than close the business, meaning that the trader will not run away from debt because he considers the business more than the debt
3. The velocity of money will be more due to small loans lent with a relatively short period of time.

Based on the facts in the community, it appears that there has been a shift in legal culture from BMT, as a non-bank financial institution with a cooperative legal umbrella, but in the field it still applies as a bank that is capital-based. Legal culture as part of the legal system is important for the operation of law in society. Legal culture which is the human attitude of the legal actors concerned in carrying out the law in society, because as well as any arrangement of legal structures to carry out the stipulated legal rules and as good as any quality of legal substances made without legal culture supported by the people involved in the system and the community, law enforcement will not run effectively {4}

BMT with Hybrid Rules of Cooperatives and Islamic Law

Lawrence M. Friedman (5) suggests that the effectiveness and success of law enforcement or the running of law depends on three elements of the legal system, namely the legal structure (structure of law), legal substance (substance of the law) and legal culture (legal culture) . The legal structure of law enforcement officials or perpetrators of legal actors in an institution, legal substance including the legislation and legal culture is the living law that is adopted in a society.

Law as a tool to change society or social engineering is nothing but merely the ideas that the law wants to realize (9). To ensure the achievement of legal functions as a better engineering society, not only is the availability of law required in terms of rules or regulations, but also a guarantee for the realization of the rule of law into legal practice, or in other words, guarantee of law enforcement (good law enforcement). So the operation of the law is not only a mere legislative function, but also an activity of the executor. The implementation of the law will be very effective if the values contained in the rule of law are the manifestation of the intrinsic values that exist in the community concerned which is usually a description of the state ideology.

Philosophically BMT is the embodiment of the Cooperative which is a form of economic activity, mandate of the 1945 constitution. The 1945 Constitution clearly has provided guidance on developing an economic system based on Article 33. At least there are 3 verses in the article that must be operationalized in the economic system. The three verses are: (Paragraph 1) The economy is structured as a joint effort based on the principle of kinship; (Paragraph 2) Production branches that are important to the state and which control the livelihood of many people are controlled by the state; and (Paragraph 3) Earth and water and natural resources contained therein are controlled by the state and are used for the greatest prosperity of the people. The three verses provide 2 important messages, namely (i) family principle is the foundation of economic activity (not economic competition); and (ii) the state has an active role in economic activities because it is given economic control rights in fields of production that are important to the state and which control the livelihoods of many people, as well as natural resources.

The extent of economic democracy based on the constitution contains at least three main things (3). First, the participation of all community members in the national production process. With this way of thinking there should not be members of the community who do not work, all must work and carry out production activities so as to produce economic value. Socially, people who work will have dignity and self-esteem so that they become independent individuals. Second, economic democracy is also understood as community participation in contributing to the enjoyment of national production. Even though they have worked, they will not automatically enjoy the production, either directly or indirectly, if the wages given do not take into account the value of "just and civilized humanity". Third, the activities of establishing production and sharing of national production must take place under the leadership and ownership of community members. In other words, all members of the community must also have a national production factor.

Based on this, cooperatives are considered as business buildings that are in accordance with the constitution, because in cooperatives they contain business practices that are in accordance with the soul and constitutional value. First, cooperatives are a collection of ideas / ideas about a movement or economic business organization and contain the principles of economic struggle. This is certainly different from other business entities, such as PT or CV which are individual based and profit oriented. Second, the principles of the economic movement and economic struggle can be seen, among others, from the character of cooperatives which are a collection of people (not capital), equality of votes, and ideals of shared prosperity. Third, the nature of the economy is actually interaction between humans, not capital relations. The implication is that the bargaining position is not determined by the amount of capital, but togetherness relations are framed in mutual welfare. The cooperative clearly shows the accentuation of human relations.

The location of the principle of kinship when associated with cooperative business practices is that the Spirit of Article 33 paragraph 1 actually does not permit the practice of economic competition / competition, but an economic activity that encourages the emergence of "economic cooperation" (co-operation). Economic cooperation in the smallest unit is the joining of people in a business building, just as a cooperative practices. With business practices like this, the dichotomy of relations between workers and owners has never emerged which in economic practice are now often involved in disputes. Next, this collaboration in the meso and macro context operationally brings together the following axis: relationships between businesses (cooperatives) in small / medium and large scale (in production, distribution, etc.), and cooperatives with BUMN / BUMN and the private sector . All relations are framed in matters that have been divided according to Article 33 of the 1945 Constitution.

Finally, the spirit of the cooperative is nothing but togetherness. In short, togetherness is reflected in these three things, namely ownership, determination, and responsibility {14}. Co-operative ownership is none other than each member is a business owner (cooperative), which is not divided in terms of majority or minority shareholders. Meanwhile, co-operative determination is the manifestation that each decision is taken jointly and each member has the same voting rights. Voting rights are determined based on people, not the number of shares. Finally, co-operative responsibility is a consequence of the ownership process and the joint decision. All the benefits and risks of the cooperative business movement are borne together. All of these manifestations of effort are the characteristics of cooperatives and are not found in other business entities.

Philosophical The idea of establishing BMT as a sharia financial institution was born from a sharia economic law system based on three fundamental principles of Islamic teachings, namely Tawhid (Unity of God), Khilafah (Representative) and Is (Justice). The three fundamental principles are applied to sharia financial institutions in the form of the principle of ridha'iyyah (willing to be willing), the principle of benefit, the principle of justice and the principle of mutual benefit. In addition, the principles of muamalah such as family, mutual cooperation, taking advantage and avoiding harm and caring for the weak economic groups became the main basis for the interests of establishing BMT in Indonesia.

Baitul Maal Wat Tamwil (BMT) consists of two terms, namely baitul maal and baitul tamwil. Baitul Maal is more directed at efforts to collect and channel funds that are non-profit, such as: zakat, infaq, and sedeqah. While baitul tamwil as a business of collecting and distributing commercial funds. These efforts become an inseparable part of BMT as an institution supporting economic activities of small communities based on Islam. This institution was established with the intention of facilitating the lower classes of society that is not accessible to the services of Islamic banks or Islamic BPRs.

While juridically the Forms of BMT Legal Entities are diverse, there are BMTs that are legally cooperative, AD / ART refers to Law No. 25 of 1992 concerning Cooperatives. Furthermore, BMTs with legal entities are foundations, AD / ART refers to Law No.28 of 2004 concerning Foundations and finally BMTs that are not incorporated can be categorized as associations or associations.

At present the majority of BMTs that have a cooperative legal umbrella, refer to Law No. 25 of 1992 concerning Cooperatives and PP No. 9 of 1995 concerning the Implementation of Savings and Loan Business Activities by the Cooperative, BMT conducts savings and loan practices, which are implemented in various forms of savings and deposits for storage, and various forms of loan financing products. As Koper BMT has characteristics such as financial institutions that do collect funds and channel them, even if only limited to and for members of cooperatives, prospective members, or other cooperative members. What is meant by prospective members here is to have registered with the cooperative, but have not paid off the deposit of principal savings and other obligations. If the obligations have not been fulfilled within 3 months, it must be removed from the cooperative membership. Savings and Loans Cooperative The regulation is held so as not to clash with the Banking Law and show the Cooperative's existence.

Sociologically, the establishment of BMT in Indonesia is based more on the demands and support of Muslims for the existence of sharia-based financial institutions. As is known, Muslims constitute the majority of the population of Indonesia, and there are still many Muslims who still live in the middle to lower classes, but there is no sharia-based financial institution that can reach middle to lower class Muslims, the majority of whom are in villages. The formation of BMT in the centers of life of rural communities is the answer to the needs of these Muslims.

Based on the philosophical, juridical and sociological basis, the ideal BMT is BMT that has the characteristics in accordance with the cooperative. BMT is based on Pancasila and the 1945 Constitution and is based on Islamic sharia principles, faith, integrated cohesion, cooperative kinship, togetherness, independence and professionalism. BMTs must cling to sharia principles. Faith becomes the foundation of the belief to want to grow and develop (6). Integration implies the hope of achieving success in the world and in the hereafter there is integration between the maal side and the social and business tamwil. Kinship and togetherness means that efforts to achieve success are achieved together. Independence means that BMT cannot live only by relying on a helping hand of the government, but must develop from increasing participation of members and the community, for this reason the management pattern must be professional.

In practice, BMT can remove the rent system that is run by daily banks, bank plecit, banks and others by the following way: For the initial stage in managing Baitul Tamwil is to enter the market as a circulation place for financial circulation, goods and services. The main competition for BMTs is moneylenders who have the ease of borrowing and disbursing processes. The process in loan sharks management is very easy, it only takes less than 1 day, sometimes with only a matter of minutes, the money is disbursed. The weakness of the Baitul Tamwil institution is at the speed of disbursement which sometimes takes 1 to 3 days. Therefore, to implement the channeling approach of the Baitul Tamwil strategy, the loan process only lasts one day, the submission of today and tomorrow is disbursed with the guarantee of the original KTP of the borrower.

Ball pick-up service is a service that is one of the main concerns for BMTs so that most of the BMT-BMTs in Indonesia have office locations that are close to markets. In addition to the strategic location, BMT marketing also went directly to the field to receive prospective depositors and customers who made financing. Do not hesitate in learning to loan sharks in terms of relationships and their closeness with traders in the market. Family-friendly approach, greeting kindly in the daily life of the market, will facilitate BMT to attract prospective members.

Keep in mind that BMT is not a Bank. BMT must be born from the community for mutual needs, so BMT members are members of a group of people who feel the same fate, know each other, and feel they have BMT as an institution of unity in the economic and social fields. BMT staff and employees act actively and dynamically, have a positive outlook, and are productive in attracting and managing public funds. BMT offices are opened at certain times and are supported by a number of staff and employees to provide services to customers. Others go directly to the field, go to the market, go to homes to find customers, withdraw, and channel funds to customers, deposit funds into BMT cash, monitor, and supervise. This way will be able to slowly remove moneylenders from market life.

Sharia rules must be maintained in BMT financial institutions, as difficult as any conditions faced, BMTs still maintain an economic system that is in accordance with sharia. One of them is the contract system (agreement) that exists in financial institutions that apply sharia patterns to the rules of financing namely mudharabah contract (profit sharing agreement) and murabahah agreement (sale and purchase agreement). contract (agreement) to provide mutual benefits for both parties as partner transactions or cooperation. Agreements (agreements) that have been agreed upon are made in advance so as not to tyrannize both the BMT and debtor customers in the future.

Mudharabah as an aqad that is hard to be implemented by BMT but actually is aqad that can be used and used as a mainstay of BMT in providing capital loans to market traders, to fight (eliminate) loan sharks, This contract is a business partnership agreement between two parties where the first (Shahibul Maal) provides all capital, while the other parties become managers and business profits are divided according to the agreement set forth in the contract. In the mudharabah contract relationship is not between the capital provider, but between the provider of funds (Shahibul Maal) and the entrepreneur (mudharib). From the above understanding it can be seen that BMTs bear all capital while the customer only has expertise capital (but does not have funds). Business profits are divided according to agreement while the overall loss is borne by the capital owner (BMT) as long as it is not due to the manager's negligence. Applications in BMT for mudharabah in terms of financing are: 1. Working Capital Financing, such as trade working capital and services. Special investment (mudharabah muqayyadah), where special funding sources with special distribution with terms set by Shahibul Mall.

Proportional margin from BMT will be able to attract market people in the product of financing BMT. In determining margin rules, BMTs must be able to compete with loan sharks. As a comparison, the cost of loans to moneylenders is very high. One-time transaction Rp. 1,000,000, then the loan interest is Rp. 200,000, - for 100 days, some even only 50 or 75 days. This is of course very burdensome for the economy of traders and the community. In installments, loan sharks directly deduct payments in front of 100,000 to 200,000 per transaction. Borrowers only get Rp. 900,000 to 800,000, and must last up to 9 to 10 times taken every week.

BMTs must be able to manage finances so that the proportion is proportional and smaller than moneylenders. The advanced BMT started by way, for loans under 500,000, - Baitul Maal did not take the margin of this as an initial step to help traders to be released from moneylenders. The return system uses a daily system, 2 weeks or once a week this depends on the characteristics of the business community. At this stage the Baitul Tamwil provided allowance for financing for 14 days since the financing was disbursed. While the amount of money given is not cut at all where the loan is Rp. 1,000,000 - immediately given without deductions. The borrower only has the obligation to open a savings account of Rp. 10,000, and also an administration fee of Rp. 10,000, - to purchase matrei and print sale / financing contracts. This method provides great benefits for the market community to be able to develop the business. 80% of financing customers have savings along with installment payments. Through this facility there is an increase in business scale, increased savings and also infa rate.

BMT can serve all levels and all people who need it, even if for example the community members are non-Muslims. Although BMT is a financial institution that implements sharia rules in its management, BMT still adheres to the principles of tolerance, togetherness of economic prosperity, where transactions are not only specific to serving Muslims. The principle of service to all people is in line with Rahmatan lil 'Alamin's mission which is the main principle in carrying out sharia.

BMT becomes an educational tool to practice bermuamalah in Islamic teachings. BMT together with mosque administrators, mushalla, through pamphlets, leaflets distributed every Friday and also through regular community meetings, conducting direct education to both customers financing the Baitul Tamwil and prospective BMT customers in the market and residents around the market regarding the system for results, various kinds of aqad in BMT and also buying and selling. The latest approach adopted in financing the Baitul Tamwil is the participation of religious leaders and community leaders. This is in the form of an announcement agreement for customers who break promises and cancel payments.

BMT can expand to other business sectors not only in financial services because BMT is a non-bank financial institution, and has two orientations, roles and functions at the same time As mentioned in the definition of BMT in a literal sense, BMT has two orientations, roles and functions at once namely in addition to productive economic profit, BMT also has orientation, role and function in the social (baitul mall). The funds in the BMT come from zakat, infaq and alms funds, in the sense that the funds come from the people and are also channeled to the benefit of the people. From the character and role of BMT that really goes according to the mandate of the constitution and Islamic law, it is optimistic that BMT can eliminate moneylenders, and become heroes of the people's economy. BMT helps people from the downturn and despair of business capital needed by small and medium market traders.

4.Conclusion

The business practice of lending money with a very high interest is still rampant in traditional markets. The practice of moneylenders in the form of daily banks, Plecit Bank, Ucek-Ucek Bank, are still in demand by the community because moneylenders have a legal culture that is close to market traders. They also know each other closely and its procedure of financing is very easy. The existence of Baitul Maal wa Tamwil BMT has not been able to play an optimal role in traditional markets. BMTs that are within the cooperative legal umbrella in practice apply as a bank, such as lending money with a high enough margin, there is still a guarantee, long process and it consider customers as banking customers, not as members, so that there is no cultural closeness. For this reason, it is necessary to rebuild the legal culture in BMT based on the Pancasila philosophical values, the juridical basis of cooperative law, and the Islamic legal basis in making the BMT an ideal legal entity that has a cultural spirit and cooperative spirit, namely mutual cooperation and togetherness, from member to member, all cooperative benefits achieved by the group, can always be felt directly by all members. BMT with this mutual cooperation, togetherness and Islamic business will be able to erase the rent system in traditional markets.

Reference

[1] Aliyah, I., Setioko, B., Pradoto, W. (2016). *The Roles of Traditional Markets as the Main Component of Javanese Culture Urban Space (Case Study: The City of Surakarta, Indonesia). IAFOR Journal of Sustainability, Energy & the Environment*, 3(1). https://doi.org/10.22492/ijsee.3.1.06

[2] Arief Moh Zainol Sutrisni, *Praktek Rentenir Penghambat Terwujudnya Sistem Hukum Perbankan Syariah di Kabupaten Sumenep.* Article · September 2013 *with* 76 Reads DOI: 10.24929/feb.v3i2.103

[3] Baswir,Revisond.. *Koperasi Indonesia* Edisi Kedua".Yogyakarta: BPFE.2013

[4] Chambliss J William and Robert B. Seidman , *Law, Power and Order*, Philipine, Addison-Wesley Publishing Company, 1971.

[5] Friedman.M.Lawrence. *The Legal System: A Social Science Perspective.* New York, Russel Sage Foundation, 1975.

[6] Hudiyanto, *Berislam Dengan Koperasi*".Yogyakarta:EKPI IPIEF UMY.2014

[7] Imaniyati Neni Sri . *BAITUL MAAL WAT TAMWIL (BMT) Sebagai Pelaku Ekonomi Dalam Perspektif Hukum Ekonomi Nasional.* Ringkasan Disertasi Program Doktor Ilmu Hukum Universitas Diponegoro Semarang .2009.

[8] Irianto Sulistyowati , *Memperkenalkan Studi Sosiolegal dan Implikasi Metodologisnya, dalam Metode Penelitian Hukum, Konstelasi dan Refleksi.* Editor Sulistyowati Irianto dan Sidharta, Jurnal JHMP-FHUI, Yayasan Obor Indonesia, Jakarta, 2009..

[9] Nonet Philippe and Philip Selznick *Law and society in transition: toward responsive law.* Front Cover. Harper & Row, Apr 1, 1978 - Law –

[10] Parlina Yeyen, *Praktik Pinjaman Rentenir Dan Perkembangan Usaha Pedagang di Pasar Prapatan Panjalin Majalengka* . Jurnal Inklusif, Jurnal Pengkajian Penelitian Ekonomi dan Hukum Islam, Vol.2. No.2. 2017

[11] Rohmah Tini Hayatur, *Praktik Bank Plecit di Pasar Tradisional Kecamatan Patikraja Kabupaten Banyumas dalam Perspektif Hukum Islam.* Thesis IAIN Purwokerto, 2017.

[12] Sinaga, Pariaman. 2008. Menuju Pasar yang Berorientasi Pada Perilaku Konsumen.

[13] Siswanto, Rudi. *Traditional Market Development Study Based Local Wisdom in Indonesia*. Sumatra Journal of Disaster, Geography and Geography Education, [S.l.], v. 1, n. 2, p. 211-217, dec. 2017. ISSN 2580-4030. Available at: <http://sjdgge.ppj.unp.ac.id/index.php/Sjdgge/article/view/72> doi: https://doi.org/10.24036/sjdgge.v1i2.72.

[14] Swasono,Sri-Edi, Kembali ke Pasal 33 UUD 1945 Menolak Neoliberalisme, Jakarta: Yayasan Hatta, 2010.

[15] Wignjosoebroto Soetandyo, *Penelitian Hukum dan Hakikatnya sebagai Penelitian Ilmiah, , Ragam-Ragam Penelitian Hukum,* Jurnal JHMP-FHUI , 2009

[16] Yaya Rizal , Aji Erlangga Martawireja, dan Ahim Abdurahim. *Akuntansi Perbankan Syariah : Teori dan Praktik Kontemporer*.Jakarta:Salemba Empat. 2009

Initial Implementation of Productive Migrant Village (PMV) Programs in Banyumas Central Java Indonesia

Muslihudin [1,*], **T R Wulan**[1], **T Sugiarto**[1], **S Wardhianna**[1], **S Wijayanti**[2].

[1] *Sociology Department, Faculty of Social and Political Sciences, Jenderal Soedirman University*, Purwokerto, Indonesia.

[2] International Relation Department, Faculty of Social and Political Sciences, Jenderal Soedirman University, Purwokerto, Indonesia.

*Corresponding author: muslihudin1963@yahoo.com

Abstract. Since 2017 Banyumas Regency has received a program from the Ministry of Manpower and Cooperatives to become a pilot program for productive migrant village programs. The village that was used as a pilot was Cihonje village, Gumelar district and Losari village, Rawalo district. The program has been determined that is to realize the four productive migrant village pillars which include; 1). Providing migration information and services, 2). Developing productive business, 3). Facilitate the establishment of community care and growth (community parenting), 4). Facilitating the establishment and development of cooperatives/financial institutions. The four programs have been strived to run quickly by the two pilot villages, with various tools being formed from personnel, software, and hardware. The implementation of productive migrant village program in Banyumas Regency is still not implemented on the four pillars. Productive business programs are the most desirable programs by both villages in implementing productive migrant village. On the contrary, cooperative formation programs are programs that have not been carried out. The information-giving program also includes programs that are not going well because there is resistance to the interests of private employment agencies that have been sending Indonesia migrant worker (IMW) so far. Community parenting programs are also classified as programs that are still not optimal because they do not involve the community as caregivers of IMW's children. The positive expectations from the Cihonje village community and Losari village are also the initial capital for the success of the productive migrant village program. However, the success of the productive migrant village program remains with the government because this program is a top-down planning program. For that sustainability must be done if the program does not want to be just a memorable program as other projects over the years.

1. Introduction

Indonesia as a country with a large population has employment problems is a common thing. The high workforce demands many jobs. When domestic conditions are unable to meet these demands, the alternative is to seek employment abroad or become an Indonesian migrant worker (IMW). It has been a long walk, that domestic employment coupled with wage disparities that much different from abroad. Based on data from the National Agency for the Placement and Protection of Indonesian Migrant Workers (BNP2TKI), in 201 5 there were 6, 5 million Indonesian migrant workers (IMW) working in 142 countries around the world. The high number of IMWs working abroad was not offset by adequate protection by the state for IMWs. In 2013, based on data from the BNP2TKI Crisis Centre, there were at least 13 thousand cases affecting IMW in various countries, 300 of which were threatened with death sentences (9).

The number of cases that have been fall IMW abroad is due to the lack of knowledge, experience, and information possessed by workers (Latifah et al, 2016). IMW usually departed is still not fully prepared about various matters related to regulations, culture, values , and customs in the destination country. In other words, people who want to go to work abroad have not received accurate information to work abroad in accordance with the legislation (5).

In connection with the income earned from working abroad, IMW generally has the aim of being able to improve the standard of living later when it is finished work. Even many of them are aiming for money for business capital. But in reality, the purpose of the destination remains because a lot of money from his work is spent only on the consumption needs of the family left behind. In other words, many IMWs have not been able to utilize the work they have obtained for business ventures that are productive, but more consumptive.

Another problem that is common in the IMW family is related to the children they left behind. We know that most IMW are women. Their children are abandoned by their mothers for years, so IMW's children automatically lose guidance and care from their mother or father. In other words, PM I's children do not get guidance and education the good one because his parents aren't complete. On this basis, the government feels the need to create a coordinative program integrated to answer all problems in above that is with the PMV program.

The intent and purpose of the productive migrant village itself specifically (5). juggling a number of non-procedural migrant workers; 2) realize a productive society and a family of prosperous Indonesian migrant workers in the village of origin; 3) provide labour information and migrant services at the village level; 4) empower family of TKI and ex TKI and their families; and 5) encourage the active role of village government in the villages originating from TKI and all stakeholders.

PMV is in line with Law No. 6 of 2014 concerning Villages which provides space for villages to increase the broadest possible village initiatives to protect and prosper the villagers.Program Target The number of villages is 400 villages by 2019 with details: In 2017 there were 120 villages, covering 100 villages in 50 regencies/cities of Indonesian migrant workers, and 20 villages in 10 regencies/cities of NTT Province. In 2018 there were 130 villages. In 2019 there were 150 villages.

2. Materials and Methods

This research was conducted since February 2017 in Banyumas Regency, which is one of the districts that received Productive Migrant Village Programs from the Ministry of Manpower and Cooperatives. The research site is Cihonje Village, Gumelar District and Losari Village, Rawalo District. The location selection was based on the consideration that the PMV Program had been implemented in the two villages. In accordance with the provisions stipulated by the Ministry of Manpower and Cooperatives, these two villages are the representation of villages in Banyumas Regency. the two villages are a PMV program pilot project in Banyumas Regency. Thus the results of this study can be used as a reference for adoption in other villages that are implementing the PMV program.

Data collection methods are in-depth interviews, observation and focus group discussions (FGD). Data analysis was carried out both in the field and after data was collected. Data that has been collected, then processed to be systematic. Processes start from writing interviews, observing, editing, classifying, reducing, presenting data, and summarizing data (Miles, 2014).

3. Results and Discussion

3.1. PMV as a National Program

PMV is the central government program, especially the ministry of labour and cooperatives. Thus the productive migrant village program is a top-down planning program. As the top-down program (1) will be conducted effectively as planners, implementers and assessors also the government itself. Thus it does not need too much community participation except only as a program object. In addition, the community is also not required to contribute to its programs because the costs are borne and budgeted by government funds. Goals and targets mainly quantitatively would be achieved with to fast as planned. Muslihudin et al (6) also said that the top-down implementation is marked by a very big role of central government as a "commander", so that it needs technical guideline in the village society.

On the other hand, a top-down program has consequences that the community is not active so that the community in the region does not have creativity, lack initiative and tend to come from the road even ignorant. According to Marini (4), there are three approaches to development planning, namely a technocratic approach that is top down, a combination of top-down and bottom-up approaches, and a community-based/participatory approach. A technocratic or top-down approach is more dominant in various ministries including in the forestry ministry.

In the top-down planning system, on the side the community cannot monitor, also tends to be reluctant to criticize, does not know the process of running a program, the program is difficult to be fully achieved because the community does not know from the start, the community also lacks participation because their votes are not taken into account in the process of running a process. Rosyadi (8) said that the fundamental weakness of development in Indonesia was due to bad governance. The dominance of the role of the bureaucracy feels so strong in determining the process of procuring a decision until the implementation of various development programs. The role of the community is more at the level of implementation. As a result, the level of community dependence on the bureaucracy is increasing.

As a national program, the program content is the same for all the PMV pilot program villages. PMV program consists of 4 (four) main integrated activities implementation, support and expected on going.

3.1.1. Providing Information and Migrant Services.

Through the Employment Information and Migration Services Centre, villagers who want to work at home and abroad get services on labour market information, work guidance/counselling (i.e. information through training), information on how to become procedural migrant workers, and document making services for prospective labour migrants as requirements for the creation of a passport (such as KTP, KK, certificate, and other documents) carried out at the village hall through an active role from the village government. The centre also works to help resolve the problems of migrant workers by providing a reference to the competent authorities (i.e. police, labour offices, BP3TKI, or NGOs).

3.1.2. Develop Productive effort.

Helping Indonesian labour migrants and their families to have the skills and willingness to grow productive businesses through training, mentoring, and assistance in productive business facilities to marketing.

3.1.3. Facilitating Community as Parenting (Community Parenting).

Assist the community in the formation of a community whose task is to provide guidance to the families of Indonesian labour migrants in terms of educating, nurturing and guiding children properly and correctly. Through this activity, the TKI children are cared for together by the community in a teaching and learning centre called "learning house" productive migrant village. Parents and couples who live at home are given training on how to raise, care for, educate, and guide children properly and correctly so that they can continue to go to school and develop their creativity.

3.1.4. The government established and Developing Cooperative Institute/Finance.

Establish and develop cooperatives/financial institutions that aim to strengthen people's productive efforts for the long term and sustainable.

3.2. Implementation in Banyumas Regency

3.2.1. Cihonje village Gumelar district

Base on prominent PMV program implementation in Cihonje village, Gumelar District are:

Raising Rabbit. Rabbit cultivation as a form of developing productive businesses is carried out by several IMW families. The activity is to utilize the resources in the hilly area so that raising livestock is a conducive business. The aspect of distribution or sale of livestock products is very easy because the demand for rabbit meat is very high from the cities of Purwokerto and Baturraden. Even the high demand for rabbit meat cannot be fulfilled by rabbit farmers in Cihonje village. The problem that is often faced by rabbit breeders is related to the pests or health of rabbits which are sometimes in pain even to death. The knowledge of raising rabbits is required to know in more detail so that it can overcome if there are problems with rabbits.

Snack Craft. Preparation of snacks made from cassava is also a mainstay of the productive effort of the IMW. The results of the training were called "Cantirku". Actually the food craft is quite appropriate for the PMVs of Cihonje village because cassava is indeed produced in the village. The constraint is the raw material to make my certificate not from the results of the process itself, but the results of the purchase of a private company, IMW only fry and then packed. Even though there was already assistance for the equipment to process from cassava to semi-finished material which was a help from the government. The equipment is quite a lot, until now only stored in the PMV study room. Another obstacle is the marketing of handicrafts, some of which are former IMW mothers. Because marketing while still in the village itself, has not been able to penetrate into other areas. In this case, people who have an entrepreneurial spirit are needed.

Community Parenting, the care of IMW's children is carried out in the form of providing courses in subjects that have been difficult for elementary and junior high school students, such as courses in English, mathematics and science. It was also only limited to being close to PMV learning houses. Care in the form of advocacy for problematic IMW children in schools has also been assisted by a PMV program assistant. However, parenting by the community itself and not by assistants has not emerged as expected by the PMV program. Even people still have the view that IMW children are indeed worthy of trouble.

3.2.2. Losari village Rawalo district

Some prominent PMV program implementation in Losari Village, Rawalo District are:

- Producing Chips Tempe.

The craft done by the former female IMWs is tempe chips with the brand "Citra Sari". Tempe chips are not really new in the Banyumas area because tempe chips have become a home industry. Marketing the craft is still very limited and there is no potential to be great because the packaging is still modest.

- Cultivation of Dragon and Papaya Fruit.

Planting dragon fruit marsh that is intended as marsh grow productive enterprises Losari village seems not obtain significant result. Even the planting of fruit yet to obtain results, the land transfer has occurred. It seems that the planting of dragon fruit is rather forced so it has not shown any meaningful results.

- Information Services.

The provision of information about the process to become an active IMW was carried out by the PMV facilitator of the village of Losari. Although there are obstacles in conveying this information because recruitment agencies / agents of Indonesian Private Employment Agencies (IPEA) who were once known as IESP (Indonesian Employment Service Providers) do not want the program. The information service program makes IMW candidates become aware of clear procedures so they tend not to want to go through private channel institutions.

- Community Parenting.

In terms of the care of IMW children, the Losari village PMV is also the same as the Cihonje village which is the provision of English and mathematics courses at the productive migrant village (PMV) learning house. This means that the care of children by the community has not yet been realized as in other villages.

The implementation of several programs in two PMV villages is still very poor to say successful. This is understandable because it has only been running for one and a half years. However, if it is read from the implementation of the two villages, the productive business program is the program most in demand by former IMWs. The program that seems not to be in demand or is running is the formation of cooperatives. Community parenting programs also seem to be far from expectations. However, this kind of program needs to be considered about program optimization and sustainability. Because often these top down programs need serious and sustainable efforts so that the program does not end in vain.

Leksono et al (3) who reviewed the Mandiri Energy Village Program from the Ministry of Energy and Mineral Resources also produced a recommendation that the success of the program must be completed by the mentoring program. Supported by subsidies from the government to actually succeed in accordance with was expected. Because it is not complete, the assistance and facilities will only stall as in the energy independent village program.

4. Conclusion

Productive migrant village program is a program of nation all applicable in all Indo n e waste. PMV is applied to villages that contribute the largest Indonesian migrant workers. Targets are planned until 2019 are 400 villages that will receive PMV program. Because this program is a top down planning program, it is uniform in all villages that get the program. The effectiveness can be achieved in accordance with the benchmarks of the government. Community participation in the PMV program is also estimated to be less optimal.

Implementation in Banyumas regency, especially in Cihonje village, Gumelar district and Losari village, Rawalo district observations in the same outline, is that productive business programs are the most popular programs to run. On the contrary, making cooperatives is a program that has not yet appeared. The information-giving program for prospective IMWs is a program that deals with private employment agencies that have been mastering the departure of workers abroad. Community parenting programs are

https://doi.org/10.1051/shsconf/20208601050

also programs that seem difficult to materialize well.

Acknowledgment

Our thanks to the Institute of Research and Public Services (IRPS) Jenderal Soedirman University, the funders of this research. We would also like to thank the office of the Ministry of Manpower in Banyumas Regency for cooperating in strengthening the PMV program. Thank you also to the Village Head of Cihonje and Losari who gave permission for the research team. Also for companion PMV Cihonje Village (Yuli and Indra) and companion PMV Desa Losari (Renate and Arif).

References

[1] Kusumanegara S., Muslihudin, Ahdiati T. 2015. Carrying Forest Village Community Institution of Program Energy Independent Village, LPPM, Jenderal Soedirman University, Purwokerto.

[2] Latifah SW., Waluya AJ., Retna ER. 2016. Empowerment of Ex-Tki Through Local Potential-Based Business Development Journal of Business, Management & Banking Vol. 2 No. 1.2016: 67-80

[3] Leksono B., Muslihudin, Kusumanegara S. 2014. Development of Nyamplung Social and Economic Studies in Cilacap and Purworejo, Swara Politika Journal, Faculty of General Soedirman University, Purwokerto, edition vol. 13 pp. 69-79.

[4] Marini A., 2013. Study of the Effectiveness of Forestry Development Planning in the Perspective of Strategic Planning, Scientific Journal of Public Administration, Volume 14, Number 1, June 2013, p.398-411

[5] Ministry of Manpower, 2017, PMV: Productive Rural Migrants and Protect Indonesian Migrant Workers Building from Villages, Head of Research and Development Ministry of Manpower Jakarta.Miles BM., Huberman AD,, Saldana J. 2014. Qualitative Data Analysis: A Sourcebook method. Third Edition Copyright SAGE Publications. Inc. Washington.

[6] Muslihudin M., Kusumanegara S., Ahdiati T. 2015. The Program Implementation of Energy Self Sufficient Village based on *Calophyllum Inophyllum* in Purworejo, Central Java, Indonesia. Proceeding of International Conference on Green Technology ISSN:2355–3456

[7] Republic of Indonesia, 2014. Law Number 6 of 2014. concerning Villages.

[8] Rosyadi, S. 2010. The New Paradigm of Development Management, First Edition, Gava Media Publisher, Yogyakarta.

[9] Wulan T.R., Muslihudin M., Sugiarto T., Wardhianna S., Wijayanti S. 2018. Gender Relations in Implementation of "PMV" (Productive Migrant Village). International Conference on Gender and Development, Makassar, Indonesia.

Kyai and Power in Banten, Indonesia

M. Dian Hikmawan[1], M. Rizky Godjali[2], Ika Arinia Indriyany[3]
[123]Government Department, Universitas Sultan Ageng Tirtayasa
dian.hikmawan@untirta.ac.id, rizky.godjali@untirta.ac.id, ikaarinia@fisip-untirta.ac.id

Abstract. In Indonesian socio-cultural society, Kyai is an important indicator to perceive the diversity of the socio-cultural. Kyai is not only a leader in religious rituals but also kyai also be used as symbols in political roles. This research tries to elaborate on the role of Kyai in political structuration in Banten,. This research tries to analyze cultural aspects to elaborate on the socio-political function and role of the Kyai, it becomes a very interesting matter for further review. Therefore, these studies involve the role of Kyai in political Power in Banten. by using a phenomenological approach in understanding the realities that occur in Socio-cultural. This research was conducted to understand, first, understanding and interpretation this by using the structure of society in Banten to understand the role of Kyai in socio-cultural, how the agent perceives the political agenda. Secondly, morality or the right direction, it is means how the influences of morality to the political decisions on the socio-cultural agenda. Third, power relation, how to perceive regulation on the socio-cultural and its influence on government policies in Banten
Keywords: *Kyai, Power, Socio-Cultural, Social Change*

Introduction

Kyai with all existence has become a symbol of the social agenda politics in Banten, even shifting between Champion and the Ulama (religious scholars) become an interesting study, at least can be traced from some literature from research (Hamid, 2010) about the process of marginalization of the political role of the clergy on the one hand and the domination of warlords on the other hand, both the functions, responsibilities, gait, also in mindset. Basically gait Kyai is not only seen from the excitement of transforming religious values in society, but also on sociopolitical persistence in the struggle alone. See figure Kyai, then that is in our minds is: First, as founder boarding schools, that schools can be said to be aspects that are almost always in the development of society, especially concerning the influence of Kyai.

Kyai schools are those whose daily activities teach religious sciences in schools, both in formal educational institutions and non-formal schools and other programs. Second, as a teacher of the congregation, as the leaders of prayer groups. Third, as defenders of the ummah, as a preacher, and others. In the context of Kyai as a preacher, Kyai becomes a figure of inspiration to his community, so that in the totality of a preacher must be professional. As a social construction, the concept of Kyai certainly different for each community, including in Banten. Kyai concept for society Banten become a symbol of the continuity of the existing culture. the influence of a traditional agrarian culture that characterizes the lives of the vast majority of Banten gives meaning and value appreciation of Kyai higher.

Historically, Kyai is part clerical position figure quite encouraged within the context of the various considerations concerning policies, especially policies that involve the community, especially when dealing with the needs and mobilization movements. Then Kyai assessed and placed as a determinant of progress and whether or not a civilization. Therefore, progress is always synonymous with change, then arises whether Kyai first accusation to be changed or society. Kyai is required to perform a variety of breakthroughs in conveying values and symbols. Traditional lifestyle agrarian society Banten contribute greatly to the preservation of local wisdom values compared to other regions were more modernist Banten and metropolis (Muslim, Kolopaking, Dharmawan, & Soetarto, 2016). For that understanding of the community, Kyai In Banten had a much deeper meaning than just a group of people who are considered to have any depth of knowledge of Islamic teachings. Kyai had much to do movement even struggle, one through its political role by carrying various Islamic parties. Kyai figure is quite encouraged within the context of the various considerations concerning policies, especially policies that involve the community, especially when dealing with the needs and mobilization movements.

Then Kyai assessed and placed as a determinant of progress and whether or not a civilization. Therefore, progress is always synonymous with change, then comes the accusation whether Kyain first to be changed or society (Yasin, 2016). The allegation was then placed in a variety of interests when Kyai was always placed as the guardian of a tradition that should be stagnant and not dynamic so that conflicts of social change must be addressed in the name of religion. Or vice versa, Kyai which should be in the vanguard of the front to adjust to various upheavals changes due to the demands of the times. The position makes Kyai are in important strategic (Abdussalam, 2015).

Kyai, Power and Political Impact

Roles are participation in contributing to or contributing to the activities or developments in political and social life. (Linton, 1956) describes the role of the dynamic aspects of status. A role is a behavior that is expected of a person occupying the role becomes an active verb and contributive. The role of being a set of behaviors that are expected by others of one's corresponding position in the system. Thus the role is very strongly influenced by the social situation both inside and outside and is stable. A role is a form of behavior that is expected in certain social situations. According (Soekanto, 2007) the role of the dynamic aspect of the position. Between the role and the position can not be separated for each other to have a mutual dependence. Someone said to perform a role when he was exercising its rights and its obligations. In the roles there are two (2) kinds of expectations are the expectations from the public on the role holder or the obligation of stakeholders and expectations that are owned by the holder of the role of the community or on people related to him.

Everyone has a variety of roles derived from the patterns of social life. It was at once means that the role of determining what he had done for the community and what opportunities are given by the people to him. The importance of the role of a community is because he set the behavior of someone who is at the same time in certain limits can be a resource in predicting the actions of others (Arinia, Hikmawan, & Ma'asan Mayrudin, 2019). The person concerned will be able to adjust to the group of their behavior. The role more points to the function, adjustment, and as a process. According to Levinson's role may include three things:

1. The role includes norms connected person in society. The role in this sense is a series of rules that guide a person in public life

2. The role is a concept of what can be done by individuals in the community as an organization
3. The role can also be regarded as individual behavior that is important to the social structure of society According (Soekanto, 2007) in Indonesia there is a tendency to emphasize the position rather than the role.

The term 'Kyai' has much in common with the ulama, although sometimes there are significant differences. The word 'ulama' is the plural of 'alim meaning 'people who have knowledge'. In the original context, ulama means scientists in a social, religious, or cultural sense. In the Quran, the cleric is not the predicate of monopoly for interpreters, hadith, fiqh, Sufism, and so on which is called al-'ulum al-diniyyah religious sciences (Noeh & Othman, 2012)"

This is thought to be due to a strong tendency to emphasize the value of materialism than spiritualism. Value materialism in most cases measured on the attributes or characteristics of certain outward like titles, honors, luxury homes, vehicles, uniforms, and so forth. Kyai the social status of a person which in itself is attached to a variety of roles. One of the roles held by Kyai is your role as a leader in the midst of the cultural community. Because outside the formal structures of government, the Kyai can be categorized as an informal leader (Jannah, 2015).

As a center of propaganda, Kyai quite a lot of contributions in shaping the life of society, it is very influential in government policy since the Dutch colonial era to the present. Likewise with modern Indonesian civilization, that spearheaded the development of the Malay archipelago civilization contained in Kyai existence itself. Kyai be the main actors of all social conception of religion as well as an institution, so Kyai had a significant influence on the development of the social, political, and economic, especially for local communities (Noeh & Othman, 2012). It can be said that the Kyai and their influence in it become a pattern-forming system in community life, but it also can be a determinant of culture Kyai is currently not only dominant in the religious social development of society, but also involved in politics and bureaucracy. Confidence in the Kyai compared to the other leaders, according to (Noeh & Othman, 2012) is seen that the orientation of the rural population is Kyai and not on bureaucratic leadership. That's because Kyai has its own charisma that is hardly owned by other leadership.

So, Max Weber thought that the patrimonial structure which is based on the charisma of the ruler is the main factor of the cultural system of Islam, so we need a charismatic and sacred ruler to be a leader or ruler. While the authorities are patterned to be a smart guard of honor and the trust of his people. not the bureaucratic leadership. That's because Kyai has its own charisma that is hardly owned by other leadership. So, Max Weber thought that the patrimonial structure which is based on the charisma of the ruler is the main factor of the cultural system of Islam, so we need a charismatic and sacred ruler to be a leader or ruler. While the authorities are patterned to be the smart guard of honor and the trust of his people. not the bureaucratic leadership. That's because Kyai has its own charisma that is hardly owned by other leadership. So, Max Weber thought that the patrimonial structure which is based on the charisma of the ruler is the main factor of the cultural system of Islam, so we need a charismatic and sacred ruler to be a leader or ruler. While the authorities are patterned to be the smart guard of honor and the trust of his people.

Method

The method used in this research is a descriptive qualitative method. This kind of research will generate descriptive data on oral and written words and behavior of people. The average results of his research are the description of a phenomenon in different contexts so that the conclusions of each - each individual will vary depending on the viewpoint used. While the approach used is the phenomenological approach. in particular hermeneutic phenomenology into a selection method that the researchers use, because phenomenology hermeneutic open space for authors in conducting research on the problem studied interpretation. Hermeneutic phenomenology is used as a critical analysis-reflective and interpretive of the text. Women in this study as a form of reality politics into the text to be interpreted, not only interpreted as a number, but further research is expected to reveal how the women go beyond the activities of the structure of power in the political realities of society so that they get the same opportunities in politics.

Hermeneutic phenomenology has the virtue of a method, whereby this method can be done in a comprehensive approach to the concepts and facts emerge. The interpretation offered an opportunity for researchers to explore the facts and concepts. At the end of this hermeneutic phenomenological method delivers the researchers to rational thinking and argumentative in sorting out facts and concepts that manifested itself not as an object but as a subject under study so as to provide a wealth of knowledge for researchers in uncovering the true meaning. Eidetic reduction in hermeneutic phenomenology allows research to reveal the meaning of which is not only visible on the surface of the political realities (Hikmawan, 2020). Thus the hermeneutic phenomenology as a method considered appropriate in the present study in order to get the essence (eidos) which in fact became a research question in this study (Hikmawan, 2014, 2017).

The data collected in this study comes from the first, in-depth interviews conducted (depth) on the informant who linked understand this context Kyai as pesantren leaders became a subject in the interview this time. Second, the observation by observing the phenomenon in order to obtain the data from the first source. The observations were made related to the structure of Kyai influence in the public sphere and Banten district administration. Third, the analysis of literature do is a study related to previous research that has been studied as well as from studies on particular offerings culturally in Banten. To maintain the validity of the data, researchers used the data triangulation technique. Triangulation is checking power back to the correctness of the data and their interpretation (Arikunto, 1998). The literature states that there are four triangulation techniques that can be used in research. First triangulation with the source, comparing and checking the degree of confidence in information obtained through time and different tools. Second, the triangulation method that can be applied with a view of checking the degree of confidence discovery research some power collection techniques or checking the degree of confidence multiple data sources with the same method. Third, triangulation with the investigator which utilize other researcher or analyst for the purposes of checking back in the degree of confidence of the data. Fourth,

Results and Discussions

Some public entities in Indonesia have a certain social construction involving Kyai figure. As the region has much in common, including placing Kyai as an intellectual in the field of Islamic religious sciences. Kyai a role model should be respected (Bahar, 2014), not only by his disciples (students) but also by society. Research (Horikoshi, 1987) mentions that Kyai a charismatic leader in the field of religion, he is fluent, and has the ability to careful in reading the minds of their followers. The other characteristics of the Kyai are forthright, bold, and forthright in attitude, and even as an expert he is far superior to the scholars in applying the principles of ijtihad (recognize the teachings of Islam logically). What are the findings of (Horikoshi, 1987) aligned with (Dhofier's, 1982) which saw Kyai an honorary degree for something that is considered sacred. Kyai title also honors for the elderly, as well as an expert for an Islam / have schools and teaching classics. Kyai notch as stated above can be found in some regions that have a tradition of boarding school education is still thick as the region "Horse hooves" in East Java, Madura, West Nusa Tenggara, a half of Central Java, and West Java. Differences between the position Kyai-region lies in the degree of influence and how to perform other functions in addition to religious leaders in the political and social dynamics of the community. Some studies were able to write the Kyai's role as a political broker,

Kyai in hierarchy of environmental education Boarding School. The hierarchical structure of the highest levels of the Guardians, Abuya, Kyai, Ustadz, and Pupils. Educational hierarchy is based on the level of understanding of religious knowledge, piety, and the ability to manage the patron-client relationship in stages. Muslim Research et al (2015) provide information that the capability of running the dichotomy of worldly and religious affairs (hereafter) to determine a person entered the hierarchical levels in question. This means that a guardian as hierarchy shoots will give a very small portion of the world affairs than affairs of religion in everyday activities. Likewise happens to give students more time for a larger part of world affairs than affairs of the hereafter.

The hierarchy structure of education at the boarding school environment in Banten region is dominated by the traditional Salafi pesantren Pesantren. The structure of the hierarchy is open to students until Kyai position. Each individual can be students, Ustadz, until Kyai gradually by running the activities and rituals that have been determined. To be able to rise to the next level, one should also run the patron-client relationship. Much like the cells of social construction. An Ustadz students can be when they are in senior positions (based on the ability to perceive religious knowledge and have practice time learning much longer than other students). Ustadz has students as disciples. Ustadz as teachers notch technical groups of informal learning in the boarding school. An Ustadz still must learn to deepen the religious knowledge of Kyai. Ustadz will rise if the level is Kyai existing students who become Ustadz.

Ustadz transformation into Kyai has to touch it to someone who is in a position senior Ustadz based on the ability of religious understanding and experience of time. A pattern like this will continue until the highest level. A mayor has pupils that Abuya. An Abuya has pupils that group Kyai. Kyai has pupils that Ustadz groups, as well as Ustadz who have few students. However, this kind of hierarchical structure is closed to the position Abuya and the Mayor. Not everyone had the opportunity to become Abuya and the Mayor. Only those who have lineage owner and founder of a boarding school can be Abuya. Even the mayor barely designation was given to those who are still alive. Only the great scholars who have died who received the title of mayor of the local community.

Kyai in Banten has a basic position as a religious leader who has students some Ustadz. Kyai Someone should build a career from the ground as a student regardless he has the lineage of a large Kyai or not. Over the development time, socio-political dynamics in Banten has put Kyai not just a religious studies teacher and caregiver Ustadz and Pupils at boarding school. Kyai has another role and influence in society. (Tihami, 1992) on the Kyai and Jawara in Banten give the results of a study that the knowledge of religion, the power of magic, and leadership pattern has formed a considerable influence on the Kyai and Jawara. Kyai influence on society has already begun in the struggle for the independence movement. Kyai able to appear as a scholar and "umaro" in the communities. History records several major Kyai like KH.Tubagus Ahmad Chatib able to become the first resident in the territory of the Residency Banten.

Other names that Brig KH.Sjam'un as the leader of freedom fighters, KH. Tb.Abdoel Halim as Regent Pandeglang, KH.Abdul Fatah Hasan as a member BPUPKI. Not only the leaders at the highest levels of government in the area, yet the underlying positions as the district officer, district, and head of the police force, mostly filled by the Kyai (Suharto, 2001). In the study (Suharto, 2001), the appearance of the ulama (represented by the figure of Kyai) in Banten local politics is a business in which they have been destroyed by the fighting after the position of Dutch colonialism. People want the appearance of the scholars as to the commander of the believers so that the vacuum of government in the early days of independence is the best opportunity for the scholars to show again as "umaro". The desirability group of scholars as the leader of the regional administration in Banten is a result of its ability to defend the interests of the people since the era of colonialism. Kyai became the leader of a small army unit that contains the students who are ready to fight against government abuses of colonialism. Boarding school environment not only as a means of informal education but became the center of propaganda, scholars consolidation arena champions and builders institutions caring for the oppressed people through moral and material assistance food. Kyai as pesantren leaders certainly has a positive effect on the existence of pesantren in the middle of residential areas (Fadhilah, 2011).

Colonial Government Kyai well aware of the strategic role in society. So they made a maneuver to separate Kyai role in public life. Colonial Government began to introduce a modern system of government bureaucracy. The influence and authority of the bureaucracy in larger units to be able to control people's lives as a policy field crops, application of the law of the colonial state in relation to civil society, taxation, trade controls, and more. Colonial Government sends leaders of the central civil service to fill positions including leader of government bureaucracy. Norms and values it tries to be separated from the horizontal relationships in the community. At this point, the degradation of Kyai role in public life in Banten by (Suharto, 2001) due to the efforts of the colonial government made gradually, in addition to several events that show their repressive policies against Kyai outspoken resistance. Kyai but still has its own place in public view. Its main activities as a religious studies teacher and leader of the Pesantren keep running. Pesantren still run social functions.

The golden age of Kyai performs its role as "umaro" in Banten in the early era of independence halt when the national political spectrum is in control of the leadership of President Suharto. Pancasila as the sole principle of policy implementation, simplifying the political groups, the hegemony of the power of the military-bureaucratic-Golkar, and the state policies that tend Islamophobia had an impact on the changing role of Kyai in Banten, not least Banten. Kyai was shaped by the political system into a state agency for the success of government programs. Even for a program that remains controversial to those who believe as the Family Planning Program contraceptive use, or prohibition of religious symbols in public spaces, Kyai still has to tell the good of the community. Kyai becomes a tool of legitimacy on government decisions aimed at strengthening the power of the regime. Kyai lined up into knots pollster for Golkar in every electoral victory of the New Order era. How the government to rake Kyai form of repression or cooperative.

 https://doi.org/10.1051/shsconf/20208601051

The government uses the instruments funding for Islamic Schools to control and control of Kyai influence. The government will carry out arrests and disappearances without the judicial process against Kyai who do not follow the instructions of the ruling regime. Stereotypes as a lackey of the PKI or the Islamic State of Indonesia movement establishing a flavor in the state's repressive measures against Kyai in Banten. Kyai lined up into knots pollster for Golkar in every electoral victory of the New Order era. How the government to rake Kyai form of repression or cooperative. The government uses the instruments funding for Islamic Schools to control and control of Kyai influence.

The government will carry out arrests and disappearances without the judicial process against Kyai who do not follow the instructions of the ruling regime. Stereotypes as a lackey of the PKI or the Islamic State of Indonesia movement establishing a flavor in the state's repressive measures against Kyai in Banten. Kyai lined up into knots pollster for Golkar in every electoral victory of the New Order era. How the government to rake Kyai form of repression or cooperative. The government uses the instruments funding for Islamic Schools to control and control of Kyai influence.

The government will carry out arrests and disappearances without the judicial process against Kyai who do not follow the instructions of the ruling regime. Stereotypes as a lackey of the PKI or the Islamic State of Indonesia movement establishing a flavor in the state's repressive measures against Kyai in Banten. The government uses the instruments funding for Islamic Schools to control and control of Kyai influence. The government will carry out arrests and disappearances without the judicial process against Kyai who do not follow the instructions of the ruling regime. Stereotypes as a lackey of the PKI or the Islamic State of Indonesia movement establishing a flavor in the state's repressive measures against Kyai in Banten.

The government uses the instruments funding for Islamic Schools to control and control of Kyai influence. The government will carry out arrests and disappearances without the judicial process against Kyai who do not follow the instructions of the ruling regime. Stereotypes as a lackey of the PKI or the Islamic State of Indonesia movement establishing a flavor in the state's repressive measures against Kyai in Banten. After the collapse of the New Order, the role of Kyai in Banten experienced a shift. The reform era provides open space for Kyai to perform a strategic role in society without losing their autonomy. Unfortunately Kyai, not the only actor who has that opportunity. There are actors politicians, businessmen, and a group of champions who are at the same time compete to seize influence in society. (Hamid, 2010) on local politics in Banten show is to shift the role of cultural broker Kyai into a political broker.

On the other hand, the winner getting a stronger political role that is largely facilitated by the dynamics of the political system favorable position in the midst of a democratic transition and economic crisis situation of citizens. Kyai group stuck in the competition and preserve the power of local political relations with the group of Champions. Kyai does not have enough resources to build influence in society autonomously in order of procedural democracy that focuses on the popularity and ownership of capital. Meanwhile, Kyai is still faced with the norm for "istiqomah" maintain morale and propaganda movement-based theology. So the choice to be an actor in the political constructs impractical or directly involved in the decision making the process so dilemma. Kyai group uncertainty in the transformation of the political system in his capacity as a subordinate cause appears logical interest groups in the electoral event or mass support purposes. Nevertheless, the influence of Kyai not cracked swallowed the changing times and the social dynamics of politics.

Kyai as a Social Agent

The existence of Kyai in Banten is very vital. Kyai is a solicitor de problems for various problems that afflict the people. The problem-solving approach always refers to the religious provisions contained in Al-Corán and As-Sunnah. Various problems ranging from private affairs in a small scope such as family to national scale public affairs, Kyai always provide solution considerations. Local society calls it with the term looking for "Barokah" or "Nyareat" to Kyai every time it collides with world affairs that cannot be resolved through normal procedures. The influence of Kyai in Banten in such a way is not created through a simple process. The position of Kyai is not just a religious leader or a pesantren leader and then automatically has a big influence in the community. An academic explanation is describing the role of Kyai in social relations and power relations that are authentic traits in the area of Banten.

As stated earlier, the Pandeglang community is in a traditional agrarian order. It is not just a matter of agriculture as a foundation for citizen economy, but it is related to social interactions, perspectives, and political perceptions of the community towards something. Traditional agrarian societies have different patterns of social relations with communities in modern industrial areas or trade-in services. In traditional agrarian societies, community ties within the village are still very close. This bond will be seen from empathy and a feeling of affinity between citizens. One person who has difficulty will be assisted by many residents to provide help. The value of tolerance, mutual cooperation, and helping become capital in building togetherness in the life of the village community.

Rural communal life will focus on many similarities compared to differences. Similarities in aspects of livelihood, economic activity, education, religious life, and matters in private areas such as property owners are always sought to be carried out together. A person will become isolated when he has a difference in certain aspects. The next characteristic of the life of traditional agrarian societies is the bond to their culture and region.

A culture that is nurtured by the local community will be maintained as part of the norm and identity. Community culture will involve regional aspects. Places to stand, move, get the benefits of nature and rituals of worship are one of the instruments of carrying out a recognized culture. So that traditional agrarian societies will not leave the village where they live too far to cross cities, unless they have been influenced by the value of modernity. A traditional agrarian society was so religious. Maintenance culture will go hand in hand with the implementation of religious activities. Islamic religion that developed in Banten society, has an impact on social life. Understanding of the value of religion in synergy with the custodian of culture and patterns of communal rural community life. Social activities In Banten hard everyday citizens released from a religious value. Thus, the figure of Kyai will have a special place in the public eye. Kyai advice and action not merely admonished religious leaders, but the basis for the preservation of the culture and characteristics of a communal society. Based on the demands of religion, Kyai would advise people to be a deep relationship with God (vertical) and worship the relationships between people (horizontal).

Based on the literature references on social structuration, (Bagguley, 2003; Clark, 1994; Giddens, 1987, 1990) in describing structuration not be equated with a bridle (constraint) but always curb (constraining) and delivered (enabling). Kyai group in its cultural influence deliberate action to accomplish their goals, at the same time act Kyai also have unintended consequences (unintended consequences) on the establishment of structures that have an impact on subsequent community action. Kyai action and discourse to the students and the local community are actually part of doing greatness.

Kyai was aware that the people should be directed to the appropriate kindness commands and avoid a ban on religion. Kyai with routine patterned giving lectures and examples of good deeds based on religious norms. What does Kyai not only to complete the tasks and obligations of connective teachings of the Prophet? However, there is the impact of unintended consequences (unintended consequences). This impact is formed from Kyai policy determination to consistently be on the road charity the good and prohibiting unjust society directs its activities. What was delivered Kyai will affect people's actions in the future.

Kyai greeting along with his actions is not a restraint (constraint). Kyai never imposes an encouragement to students or the public. But what do Kyai can be analyzed as one who always curb (constraining) and delivered (enabling) as what is revealed by (Giddens, 1990). Kyai use the arguments of religion as an instrument to make the students and the community are in a state must follow what it conveys. The structure described as a continuous flow behavior such as cognition, support or even break during the sense they awarded him (Giddens, 2011).

Conclusion

Kyai relationship patterns around the hierarchy kept waking repeatedly and maintained. Kyai as an agent that reproduces activity to express himself as an actor of social and political change in Banten. Kyai behavior would be understood as a command to perform an action that is considered to bring the truth and ideal conditions. Such relationships include various aspects of life in the social, political, and economic in Banten.

Reference.

Abdussalam, I. (2015). Reproduksi Kekuasaan Kyai Dan Blater Di Kabupaten Bangkalan. *DIMENSI : Journal of Sociology*, *8*(1).

Arinia, I., Hikmawan, I. M. D., & Ma'asan Mayrudin, Y. (2019). *The Movement of Women with Disabilities on Indonesia's Legislative Election 2019.*

Bagguley, P. (2003). Reflexivity contra structuration. *Canadian Journal of Sociology*, *28*(2), 133–152. https://doi.org/10.2307/3341456

Bahar, M. S. (2014). KIYAI AND BEJINGAN: LOCAL POLITICAL ELITE AND THE HEGEMONIC CULTURE. *Journal of Islamic Civilization in Southeast Asia*, *3*(1), 61–86.

Clark, J. (1994). Anthony Giddens: Consensus and Controversy. *Social Forces*, *73*(1), 324. https://doi.org/10.2307/2579935

Dhofier, Zamaksar. 1982. Islamic Boarding School Tradition: Study of the Kyai Life View. LP3ES. p. 55. Jakarta.

Fadhilah, A. (2011). Struktur dan Pola Kepemimpinan Kyai dalam Pesantren di Jawa. *Hunafa: Jurnal Studia Islamika*, *8*(1), 101–120.

Giddens, A. (1987). *The Nation-State and Violence: Volume Two of a Contemporary Critique of Historical Materialism. Social Forces* (Vol. 66). Polity Press. https://doi.org/10.2307/2578926

Giddens, A. (1990). *Structuration Theory: Anthony Giddens and the Constitution of Social Life.* (I. J. Cohen, Ed.), *Social Forces* (Vol. 69). Macmilan. https://doi.org/10.2307/2579624

Hamid, Abdul (2010), The Kiai in Banten: Shifting Roles in Changing Times in Masaaki (eds) Islam in Contention: Rethinking Islam and State in Indonesia. Kyoto: Center for Southeast Asian Studies; Jakarta: Wahid Institute; Taipei: Center for Asia-Pacific Area Studies

Hamid, A. 2011. Pergeseran Peran Kyai dalam Politik di Banten Era Orde Baru *dan Reformasi.* Al Qalam. Vol. 28, No. 2, pp. 339-364.

Hikmawan, M. D. (2014). *Politik Perbedaan : Demokrasi dalam Paradoks.* Universitas Gdjah Mada.

Hikmawan, M. D. (2017). Politik Perbedaan: Minnoritas dalam Implementasi Kebijakan. *Journal of Indonesian Public Administration and Governance Studies (JIPAGS)*, *1*(1), 88–98.

Hikmawan, M. D. (2020). *Consensual Democracy: A Challenge for Differentiated Citizenship. International Journal of Innovation, Creativity and Change. www.ijicc.net* (Vol. 11). Retrieved from www.ijicc.net

Horikoshi., Hiroko., Kyai dan Perubahan Sosial, Jakarta: P3M, 1987.

Jannah, H. (2015). Kyai, Perubahan Sosial Dan Dinamika Politik Kekuasaan. *FIKRAH: Jurnal Ilmu Aqidah Dan Studi Keagamaan*, *3*(1), 157–176.

Muslim, A., Kolopaking, L. M., Dharmawan, A. H., & Soetarto, E. (2016). Dinamika Peran Sosial Politik Ulama dan Jawara di Pandeglang Banten. *Mimbar*, *31*(2), 461–474.

Noeh, M. F., & Othman, M. R. (2012). The Role Of Kyai In The 2004 Presidential Election : A Case Study In East Java, *27*(2), 96–107.

Suharto. 2001. Banten in Revolution, 1451-1949: Process Integration *In the Unitary State of the Republic of Indonesia* (Disertation).Depok: UI.

Tihami, M.A. 1992. Kiyai dan Jawara in Banten: Study of Religion, Magic and Leadership in Pasanggrahan village (Thesis). Depok: UI.

Yasin. (2016). Kiai Lokal Antara Penghambat Dan Penyelamat Dalam Pelaksanaan Undang-Undang Perkawinan No. 1 Tahun 1974. *Jurnal Pemikiran Hukum Dan Hukum Islam*, *7*(1), 57–73.

AUTHOR INDEX

AUTHOR INDEX